PROCEEDINGS OF THE FOURTH SYMPOSIUM ON THE GEOLOGY OF THE BAHAMAS

Editor

John E. Mylroie

Production Editor

Donald T. Gerace

Sponsored by the Bahamian Field Station

June 17 - 22, 1988

ISBN 0-935909-31-1

Printed by Don Heuer in the U.S.A.

PREFACE

The Fourth Symposium on the Geology of the Bahamas was held on San Salvador Island, Bahamas June 17-22, 1988. A total of 59 participants listened to 36 presentations on the geology of the Bahamas, including reports on paleontology, diagenesis, depositional environments, stratigraphy, geochronology, and groundwater processes. This proceedings volume contains 34 of the presentations made at the symposium. The symposium series began in 1982, when 9 papers were published in the first proceedings volume. In 1984, the second symposium generated 18 papers, and in 1986 the third symposium produced 22 papers. The papers from the four symposiums represent a balance of work done on San Salvador Island at the Bahamian Field Station, and work done elsewhere in the Bahamian Archipelago. The proceedings volumes provide a single source review on the geology of the Bahamas.

The purpose of the symposium series is to provide a forum where workers in Bahamian geology can meet in an informal atmosphere and share results, ideas and critiques. It has been particularly useful for the many geologists working at the Bahamian Field Station on San Salvador, generating a synthesis and overview of activity. The Bahamian Field Station draws workers from over 80 academic institutions in the United States, few of whom have field seasons on the island that overlap sufficiently to allow sharing of ideas and results while on the island. Of the 34 papers in this volume, 18 represent the combined efforts of graduate students and their advisors. The growth of graduate research, often supported by the Bahamian Field Station, is a measure of the commitment geologists have made to the outstanding carbonate geology of San Salvador Island.

The production of a symposium proceedings is a lengthy and involved task. The Executive Director of the Bahamian Field Station, Dr. Donald T. Gerace, has devoted endless hours and unlimited energy to the organization and execution of the Fourth Symposium on the Geology of the Bahamas. With the help of his wife, Kathy, they have worked very hard to publish this proceedings volume. The efforts of Don and kathy Gerace in operating the Bahamian Field Station, and conducting activities such as the symposium series, demand our strongest recognition. My fellow colleagues on the Geology Committee of the Field Station, Bob Adams, Roger Bain, Mark Boardman, Jim Carew, Al Curran, Jim Teeter, and Brian White made the operation of the Fourth Symposium an easier and more enjoyable task than otherwise possible.

The Fourth Symposium on the Geology of the Bahamas produced presentations that represent both the most modern and sophisticated of laboratory analyses, and the classic direct and uncomplicated observations made in the field. Work by teams of senior academics is balanced by data generated in student projects. The result is this proceedings volume, in which each of the papers represents an advance in our understanding of Bahamian geology.

John E. Mylroie
Mississippi State University

REMARKS

With each geology symposium held at the Bahamian Field Station, I become more proud of not only the increased quantity of research papers presented, but find the quality continually improving. This, the results of the Fourth Symposium on the Geology of the Bahamas reflects both the larger number of papers delivered, and an even broader range and specificity of topics covered. I am particularly impressed with the number of excellent papers by graduate students, and realize these are the results of dedication by both the students and their professors. And I admit a deep personal pride when I realize that the Bahamian Field Station has played such an important part in the lives of these young scholars. It is thus my sincere pleasure to have these proceedings published and distributed throughout the scientific world.

I would like to extend my sincere appreciation to Dr. John Mylroie, the Chairman of the Fourth Symposium and editor of this volume. His dedication to these tasks were outstanding, and working with him is always a welcomed pleasure. He devoted untold hours to the editing of these papers, corresponding with the authors, and finalizing the papers for this volume. He and I were assisted by a most competent Scientific Committee: Drs. Robert Adams, Roger Bain, Mark Boardman, James Carew, Allen Curran, James Teeter, and Brian White.

I would also like to express my gratitude to the many people who assisted with bringing this book to press, especially my wife, Kathy, and our dear friend, Betty Ann Hughes, who were both instrumental in the word processing of this volume. Francine Hughes spent many hours proof reading and producing the figures for the final version. And finally, Don Heuer must be thanked for his hours of work in publishing this Proceedings volume.

Donald T. Gerace
Executive Director
Bahamian Field Station

TABLE OF CONTENTS

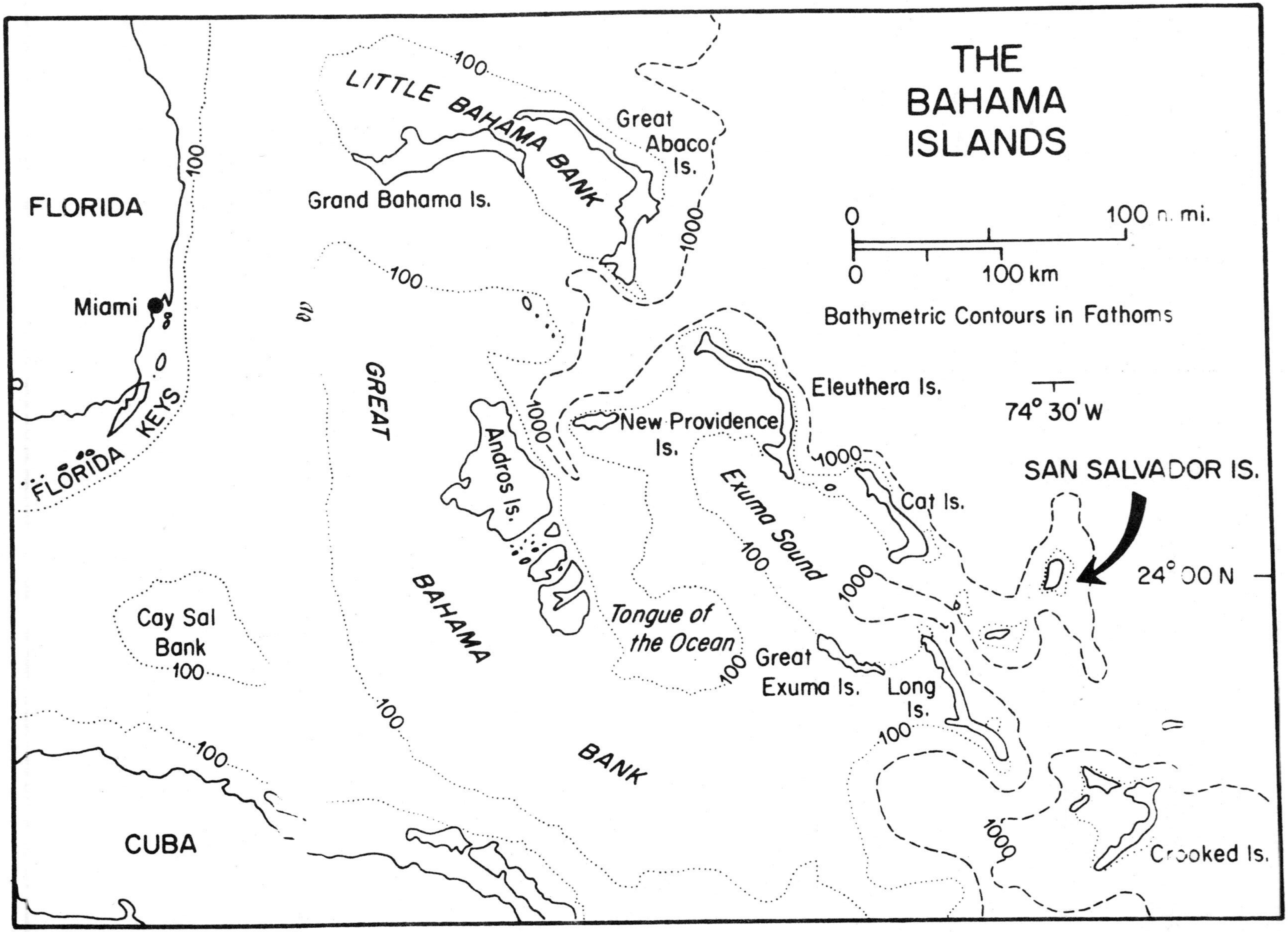

INDEX MAP TO LOCATION OF SAN SALVADOR, BAHAMAS

BIOTURBATION OF BEACH SANDS BY BETTLES AND ANNELID WORMS SAN SALVADOR ISLAND, BAHAMAS

Robert W. Adams
Department of the Earth Sciences
State University of New York College
Brockport, New York 14420

ABSTRACT

Beetles, tentatively identified as rove beetles, are present as surface burrowers on three San Salvador Island beaches. Raised "tunnels" 1 mm in diameter, up to 10 cm in length, are common on the high tide swash zone surface. The beetles tunnel and burrow through upper fine- to lower medium-grained, well-rounded beach sand. The burrows are ephemeral features, being destroyed during a subsequent high tide swash event. If the burrows were preserved in the geologic record they would indicate the position of a beach swash zone.

Annelid worms up to 10 cm in length are present in the supratidal and intertidal beach sand of French Bay and East Beach. Sand particles within the worms indicate ingestion and excretion processes while they burrow through the beach material. The vertical and horizontal distribution of the worms suggests probable significant bioturbation and obscuring of stratification within the upper beach zone. Such zones may be interpreted as a shallow marine bioturbated facies rather that a beach facies when observed in the sedimentary record.

INTRODUCTION

Many tracks, trails, burrows and other organic markings have been observed in the shallow marine, beach and dune environments of San Salvador. Curran and White (1988) have documented the characteristic trace fossils in the shallow subtidal, upper beach and dunal zones of the island. Additional organic activity within the supratidal and foreshore zone (intertidal) by burrowing beetles and worms can be added to the Curran and White model for San Salvador and elsewhere.

TUNNELING AND BURROWING BEETLES

Field Observations

Burrowed tunnels are present on the middle to upper foreshore surface of the beach immediately east and west of the Government Dock, Grahams Harbor (Figs.1 and 2). The tunnels are approximately 1 mm in inside diameter and 2 mm in outside diameter (Fig. 3). They range from a few to over 10 cm in length, commonly are oriented normal to the water line and are often sinuous and curved. They appear at about the time of low tide, extending from the middle to upper part of the foreshore.

The tunnels are formed by an insect which lands on the sand, burrows into and through it, and then breaks through to the surface and flies off. The insects are 1.7 to 2.0 mm long and 0.6 to 0.7 mm wide and resemble beetles (Fig. 4). Normally they are found singly along the tunnels (not at the end); rarely two insects may be in the same tunnel.

The above observations were made over several years during the winter months of December and January. In June 1988 additional aspects of these insects were noted. During a period of mixed tides, when the two high tides are of significantly different height, the insects have an opportunity to occupy the intertidal zone between the two high tides for a longer period of time (that portion of the beach not reached by the swash of the lower high tide). The extent of burrowing is greater with large areas of the beach surface completely tunneled including vertical burrows to a depth of 1 cm. Associated with this evidence of intense insect activity are slightly narrower tunnels occupied by insect larvae.

On San Salvador the insects have been found at Grahams Harbor Government Dock, the pocket beach on Cut Cay and the small beach west of the building at the entrance to Pigeon Creek. The sand in which they tunnel is bioclastic

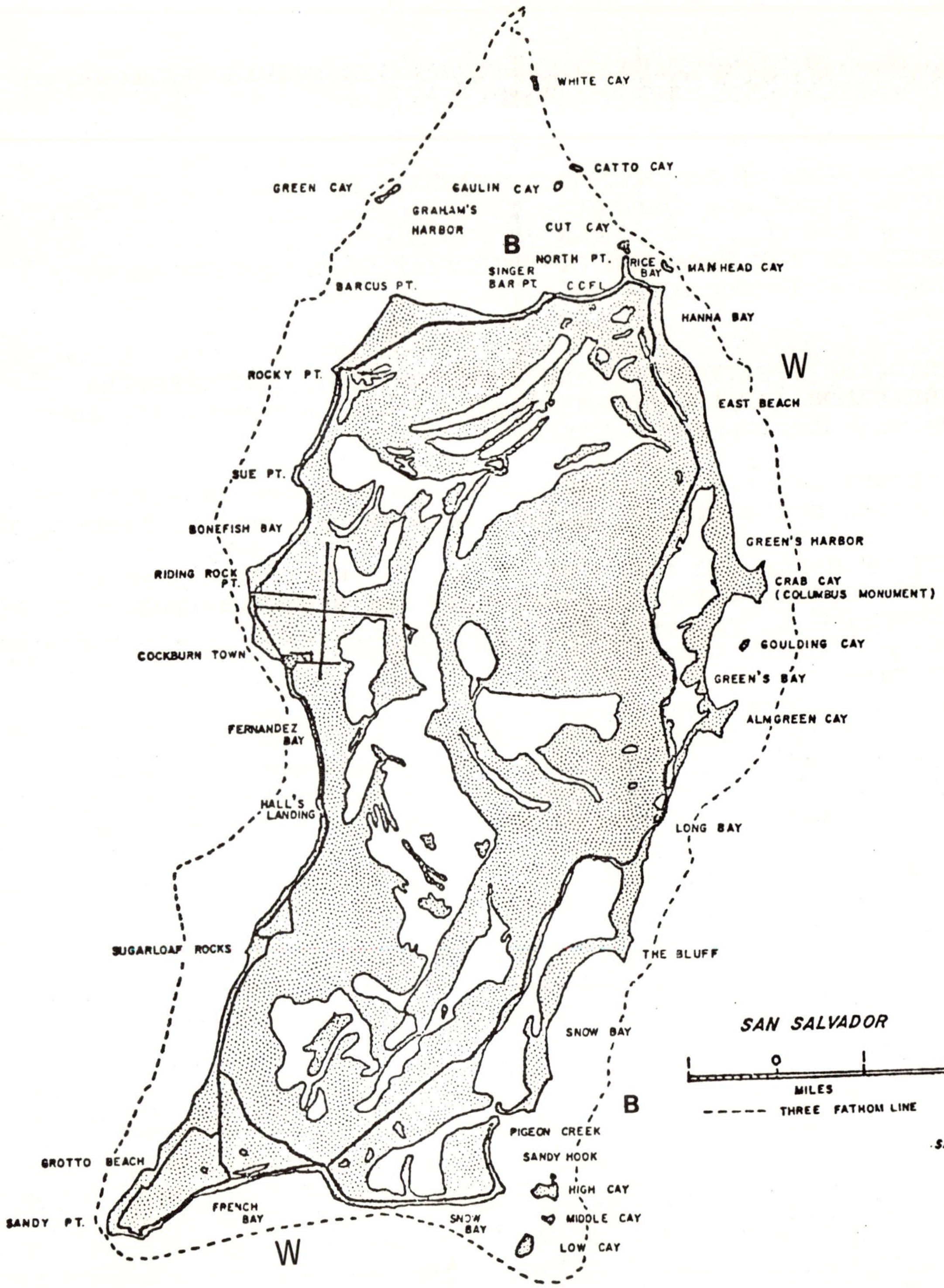

Fig. 1. Index map of San Salvador. W: Annelid locality, B: Beetle locality.

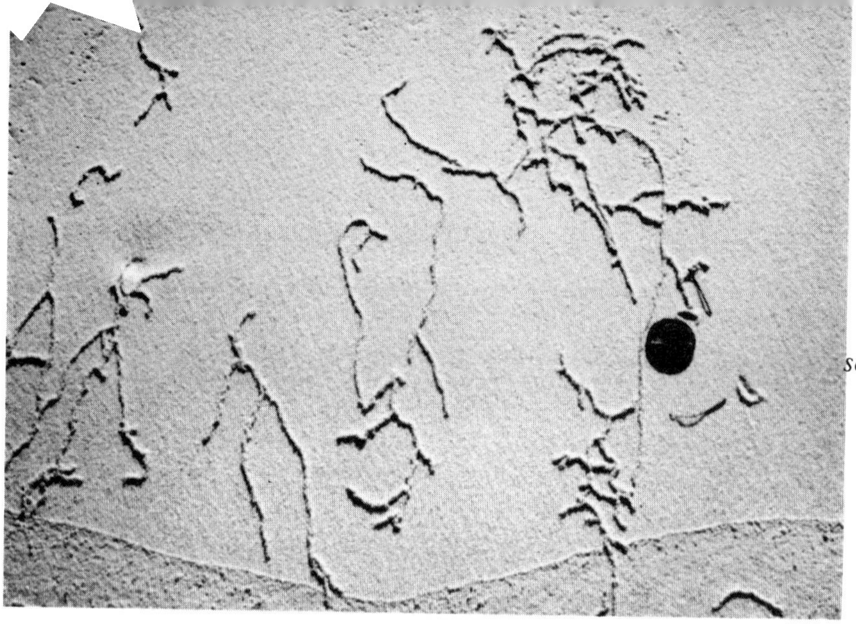

Fig. 2. Insect burrowed tunnels. U.S. penny scale.

Fig. 3. Exposed insect tunnels, mm scale.

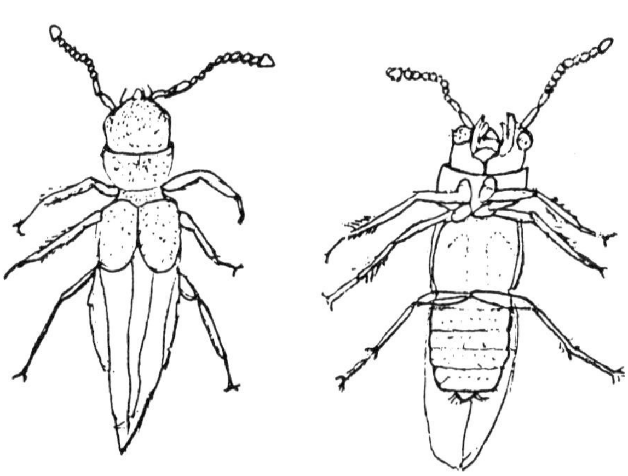

Fig. 4. Sketch of burrowing insect. 2 mm length.

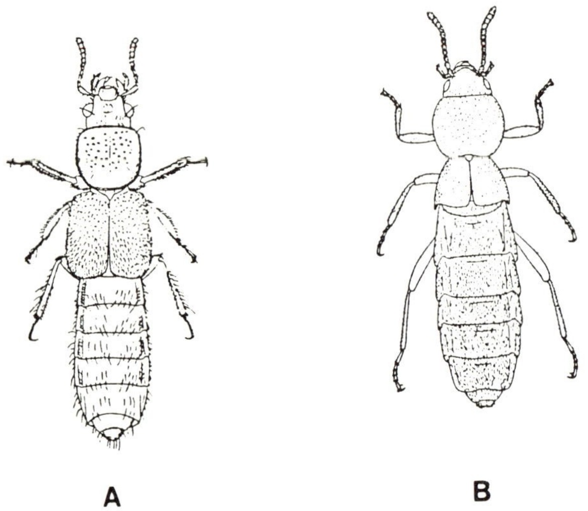

Fig. 5. Beetles: A. Bledius *sp., B.* Pontamalota *sp. (from White, et al., 1984)*

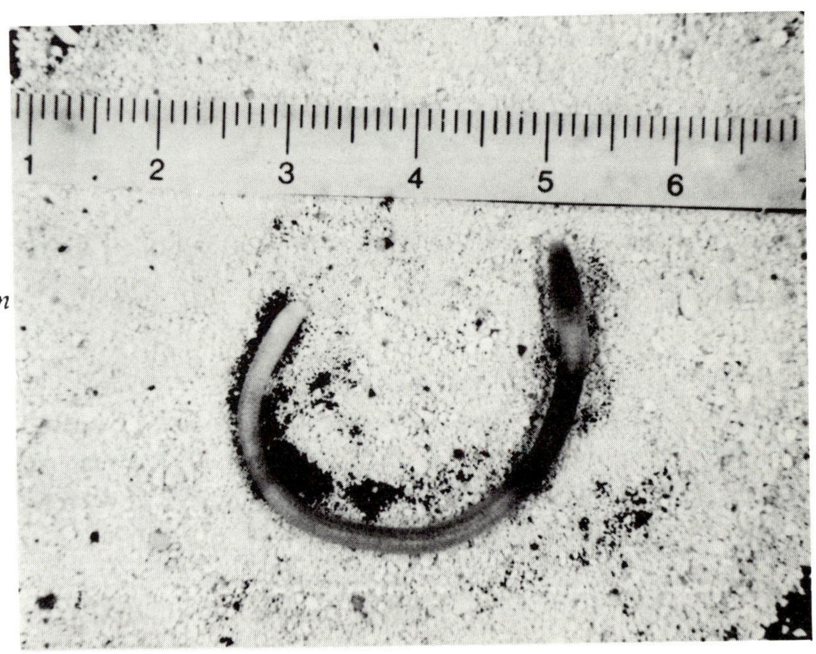

Fig. 6. French Bay beach earthworm (dead).

Fig. 7. Live earthworm with internal sand grains. French Bay.

lower medium (250-350 microns, 2 to 1.5 phi) to upper fine (177-250 microns, 2-2.5 phi) and is rounded to subrounded. Active burrowing takes place after the sand dewaters to an appropriate degree during a falling tide.

Interpretation

Examination of the insects by an entomologist suggests that the insect is an intertidal rove beetle of the order Coleoptera, family Staphylinidae, genus *Bledius*. (A. Bratt, personal communication, July 26, 1988). This family of beetles contains 3000 species in North America (Arnett, 1985) many of which are found in the intertidal marine environment. Two Staphylinidae species are shown in figure 5.

Intertidal rove beetles range in size from 1 to 20 cm (Arnett, 1985) and are predators on insect larvae, snails, insects and crustaceans (Moore and Legner, 1976).

Significance

The beetle structures have been observed only on the finer-grained beaches of San Salvador. The control on their presence could be grain size, however, sand moisture, food availability and reproductive cycle are additional considerations for their distribution. Such features formed in the intertidal zone are subject to destruction during a subsequent high tide event. The presence of "tunneling" larvae that may have hatched from buried eggs and vertical burrowing suggest that bioturbation of beach laminations in the upper centimeter of the beach has taken place. Rare preservation of the tunnels, vertical burrows or an extensive bioturbated thin bed would indicate the position of the middle to upper foreshore of the beach.

ANNELIDS

Field Observations

An annelid is present in the intertidal foreshore beach sand immediately to the east of the Government Pier in French Bay and in the sand of the high tide flat at East Beach (Fig 1.). In the middle and lower foreshore the worms are found in a vertical position with their heads 3 to 9 cm below the surface. In the upper foreshore the worms are about 3 cm below the surface in a horizontal position. At East beach they are found to a depth of 15 cm in apparently homogenous sand. They are up to 10 cm long (alive, 5.5 cm dead) and 2 mm in diameter (Fig. 6). Close examination shows a "trail" of sand grains within and extending the length of each worm (Fig. 7). The grain size of the beach sand ranges from 1 mm to .25 mm (0 to 2 phi) in diameter with the mean .5 to .35 mm (1.5 to 1 phi). The worm resembles the common earthworm (Class Oligochaeta) of which there are over 3000 species. It has been identified as an earthworm (K. Fauchald, personal communication, July 18, 1988).

Significance

The bioturbation by worms of the beach sand can be significant. In one study of such worms (Dales, 1967) it was reported that small worms of this type eat about 84 grams of sand per year, replacing what is in them every 15 minutes. The San Salvador worms are probably larger (the content of one would be significant) and they must "turn over" a large amount of sand annually. Homogenous or bioturbated zones found in the upper beach facies of the Pleistocene and Holocene rocks of the area may be caused by such organisms.

CONCLUSION

The beaches of San Salvador contain significant evidence of burrowing by insects and worms. Documentation of what is present will provide new data which may be useful for interpreting ancient marine sequences. The Curran and White trace fossil model of the upper beach rocks and eolianites on San Salvador can be modified to include such new features.

The author has found a polychaete in the East Beach area and Curran has reported a worm at Bonefish Bay (H. A. Curran, personal communication, July 1988). Careful examination of the beaches of San Salvador and other Bahamian islands should extend the known distribution of these faunal elements and their record in the sediments and rocks of the region.

ACKNOWLEDGEMENTS

Field work for this research was conducted during the residence of the Brockport Department of the Earth Sciences Oceanography Practicum courses at The College Center of the Finger Lakes Field Station. Dr. Donald T. Gerace provided logistical support and encouragement during these periods. Dr. H. A. Curran suggested additional locations for annelids. Dr. Alan Bratt, Biology Department, Calvin College, Michigan and Dr. Kristian Fauchald, Smithsonian Institution, assisted in insect and worm identification respectively.

REFERENCES CITED

Arnett, R.H., Jr., 1985, American Insects: New York, Van Nostrand Reinhold, p. 301.

Curran, H.A. and White, B., 1987, Trace fossils in carbonate upper beach rocks and eolianites: recognition of the backshore to dune transition, in Curran, H.A., ed., Proceedings of the third symposium on the geology of the Bahamas: Fort Lauderdale, Florida, CCFL Bahamian Field Station, p. 243-254.

Dales, R.P., 1967, Annelids: London, United Kingdom, Hutchinson University Library, p. 47.

Moore, I. and Legner, E.F., 1976, Intertidal rove beetles (Coleoptera: staphylinidae), in Cheng, L., ed., Marine Insects: Oxford, United Kingdom, North-Holland, p. 521-551.

White, D.S., Brigham, W.U., and Doyen, J.T., 1984, Aquatic Coleoptera, in Merritt, R.W. and Cummins, K.W., eds., An introduction to the aquatic insects of North America, second ed.: Dubuque, Iowa, Kendall Hunt, p. 361-437.

THE DEPOSITIONAL EVOLUTION OF SNOW BAY, SAN SALVADOR

C. Brannon Andersen* and Mark R. Boardman
Geology Department
Miami University
Oxford, OH 45056

ABSTRACT

Snow Bay, located on the southeast windward margin of San Salvador, is a 1x2km inner shelf lagoon rimmed by a barrier reef and cays. The lateral and vertical Holocene facies mosaic of Snow Bay records the banktop response of a high-energy lagoon to the Holocene sea-level rise. Cluster analysis of surface and subsurface sediments reveals four distinct marine facies - an abraded grain "grainstone", and abraded grain "grapestone", a foraminiferal "packstone", and a *Halimeda*-rich "packstone/wackestone". Although the facies are distinct, enough overlap exists to consider the sedimentary gradients gradational. The most important variables for determining sedimentary gradients are mud, mud fraction mineralogy, sorting, *Halimeda*, aggregates, and abraded grains.

The four facies are recognized in sediment cores and are combined to form a composite high-energy shallowing-upwards sequence. The sequence is composed of a basal transgressive abraded grain "grapestone" containing intraclasts overlain by a fining-upwards (shallowing-upwards) subtidal *Halimeda*-rich "packstone/wackestone" and foraminiferal "packstone". The regressive intertidal/subtidal abraded grain "grainstone" of the ebb-tidal delta abruptly overlies the *Halimeda*-rich "packstone/wackestone" of the *Thalassia* meadow.

The high-energy sequence in Snow Bay records the banktop response to a sea-level rise. Radiocarbon dating of peat deposited in poorly drained topographic lows and sediment within cores link the banktop response to sea-level rise. Rapid transgression and deepening of the lagoon produced the abraded grain "grapestone" facies. The mobile sand sheet of Snow Bay is a relict of this transgression. With continued sea-level rise, stabilization of the sandy lagoon substrate by benthic flora produced the shallowing-upwards *Halimeda*-rich "packstone/wackestone". During the relative stillstand of the last 1500 years, the ebb-tidal delta abraded "grainstone" prograded over the *Thalassia* meadow forming a regressional cap of redeposited sand.

INTRODUCTION

Changes of depositional style as a result of sea-level rise are an important feature of carbonate platform growth. Sediments characteristically build up to sea-level during transgressions because sedimentation rates are generally much greater than the rate of subsidence and/or sea-level rise (Schlager, 1981). The resulting sediment accumulation often forms a shallowing-upwards sequence (James, 1984) which is a special type of punctuated aggradational cycle produced by a base-level fluctuation (Goodwin and Anderson, 1985).

Shallowing-upwards sequences are classified as low-energy and high-energy sequences (James, 1984), and are distinguished by intertidal environment. Low-energy lagoonal sequences are capped by tidal flats, and often the shallow subtidal sediments contain stromatolites (James, 1984). High-energy sequences are capped by beaches, which often lack distinctive characteristics and are difficult to recognize in the rock record (Inden and Moore, 1983; James, 1984).

Facies models of carbonate platforms have concentrated on low-energy facies relationships. As a result, more emphasis is placed on models of low-energy banktop responses to sea-level rise (tidal flats and muddy lagoons) than high-energy banktop responses (beaches and sandy lagoons; Enos, 1983; Wilson and Jordan, 1983; James, 1984; Read, 1985). This is not surprising for two reasons. First, our understanding of cyclic carbonate facies models comes from half a dozen modern examples (James, 1984), of which the most herald-

*Present Address: Heroy Geology Laboratory, Department of Geology, Syracuse University, Syracuse, NY 13244-1070

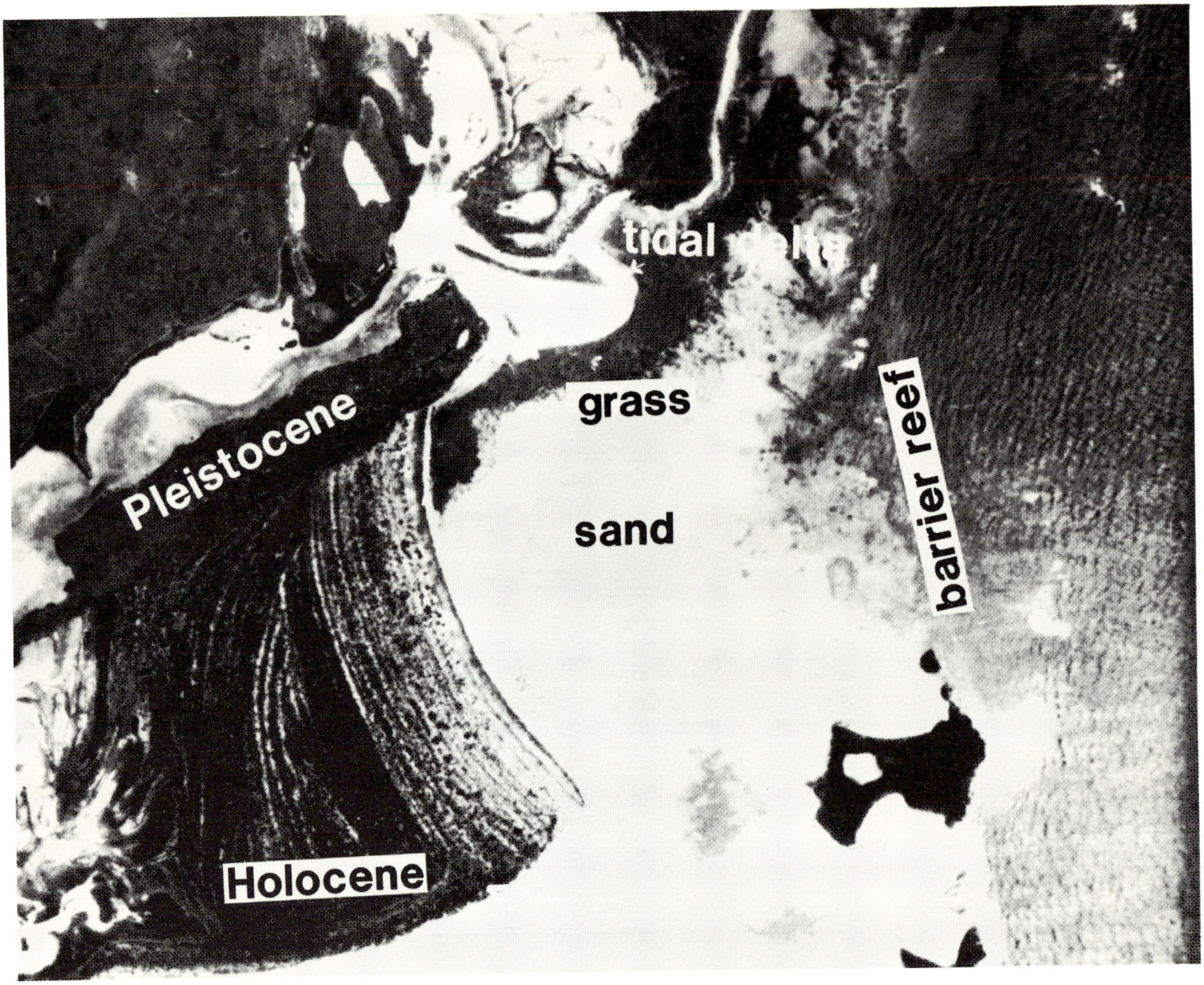

Fig. 1. Air photo of Snow Bay, San Salvador showing the major subtidal depositional environments. The lagoon is rimmed by a barrier reef and cays to the east of the photo. Note the prograding beach ridges of Sandy Hook on the north.

ed are tidal flat studies (e.g., Shinn, Lloyd and Ginsburg, 1969; Logan, Davies, Read, and Cebulski, 1970; Purser, 1973; Hardie, 1977). Second, high-energy facies associations appear to be rare in the geologic record, probably as a result of poor recognition criteria (Inden and Moore, 1983).

Few modern analogues of the high-energy shallowing-upward model exist, and geologists encounter difficulty when trying to interpret ancient carbonate packstone and grainstone facies associated with a high-energy environment (Inden and Moore, 1983; McKee and Ward, 1983). Since no single criterion is diagnostic for an environment as diverse as lagoons, lateral and vertical sequence transitions are the most reliable guide (Enos, 1983).

Lagoons are the carbonate "factories" of platforms (Neumann and Land, 1975; James, 1984), and the sediments form a vertical sequence which records the flooding history of the platform (Boardman, 1976; Boardman, Evans, and Andersen, 1986; Andersen and Boardman, 1987; Rasmussen and Neumann, 1988). A variety of factors control the depositional evolution of modern lagoons including benthic flora, water depth, and antecedent topography.

Few studies have concentrated on lateral and vertical sedimentary gradients in modern windward high-energy lagoons (e.g., the Florida reef tract and northeast of Abaco Island; Ginsburg, 1956; Swinchatt, 1965; Locker, 1981). In contrast, other studies have primarily concentrated on the relationship of benthic communities and their sediment record (Taylor and Lewis, 1970; Miller, 1988). Only a few studies have examined the origin and time frame for the development of

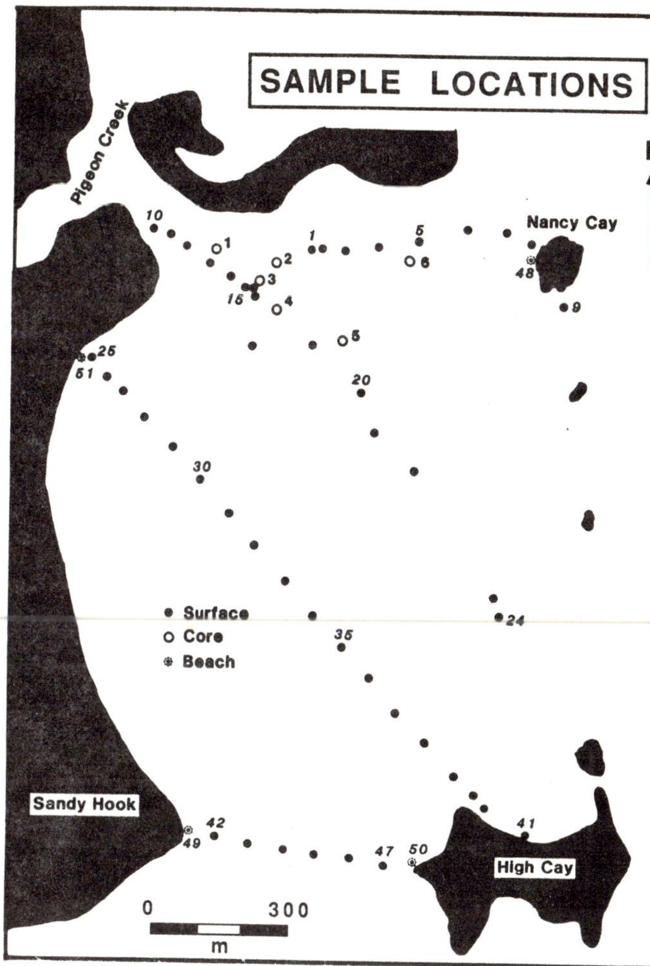

Fig. 2. Snow Bay sample locations. Filled circles represent surface samples. Open circles represent cores. Circled stars represent beach samples.

high-energy environments (Enos, 1977; Hine, 1977; Locker, 1981; Harris, 1984).

The objective of this study is to examine the lateral (Andersen and Boardman, 1987) and vertical sedimentary gradients in order to understand the relationships and controls of high-energy lagoonal sedimentary characteristics. The relationship between the response to sea-level rise and the time frame for the development of the high-energy shallowing-upward sequence is of special interest.

METHODS

Field Work

Snow Bay is a 1x2km high-energy inner shelf lagoon located on the southeastern corner of San Salvador, Bahamas. The lagoon is rimmed by a barrier reef and small cays (Fig. 1). Water depth and a general description of flora, fauna, and sedimentary structures were recorded for two of the northern transects (sites 1 to 8, and 15 to 24). In addition, four beach samples were collected for comparison to the subtidal samples.

Six sediment core sites were chosen based on sediment thickness and environment (Fig. 2). One core is from the center of the tidal delta, two cores are from the tidal delta edge, and three cores are from the *Thalassia* meadow. The core sites form a transect which crosses the thickest accumulation of sediment. All cores were brought back to the lab and described megascopically. Four cores were sampled at 10cm intervals for an additional 67 samples.

Sample Analysis

Textual analysis. All sediment samples were prepared for textural analysis by soaking in a 25% Clorox solution followed by three rinses of pH 9 water. Textural data were collected using standard wet sieving at one phi intervals and pipette techniques for <63μm,<16μm, and <4μm.The method of moments was used to determine sorting (Folk, 1966; Lewis, 1984).

Mineralogical analysis. The aragonite:calcite ratio of the mud fraction (<63μm) was calculated from x-ray diffraction data using the peak area method (Chave, 1954). Position of the 104 peak of calcite determined whether the calcite composition was of the high- or low-magnesium variety (Goldsmith, Graf and Joensuu, 1955).

Compositional analysis. Composition was determined by identifying approximately 300 grains per sample of the coarse sand fraction 1-2mm). The coarse sand fraction was chosen to maximize the ease of identification (minimize the amount of grains of unidentifiable origin). Also, *Halimeda* is concentrated in the coarse sand fraction, so it could be used as an environmental indicator (Basan, 1973; Mallik, 1976).

Petrographic analysis. Composition of intraclasts and eolian rocks were determined using standard petrographic techniques. Cement types of the intraclasts, eolian rocks, and aggregates were confirmed using SEM in combination with EDX analysis.

Statistical analysis. Relationships between surface and core samples were determined by cluster analysis. The program CLAP (written by Sepkoski and Sharry, 1976; modified by Arnold Miller, 1987) provided Q-mode and R-mode analysis of the data using the UPGMA method of clustering. The Q-mode and R-mode analyses were combined to form a two-way cluster analysis (Miller, 1988). The samples were compared using textural, compositional, and mineralogical data (a total of 10 variables per sample).

The cosine theta similarity coefficient used in this analysis is described elsewhere in the literature (Imbrie and Purdy, 1962; Purdy, 1963; Imbrie and Van Andel, 1964). Since the variables were of different units (e.g., phi and percentage) the data were standardized (Imbrie and Van Andel, 1964; Harbaugh and Merriam, 1968). This technique in combination with descriptive statistics of variables (mean, standard deviation) is used to define sedimentary gradients. The student t-test is used to confirm that the clusters are statistically different. Sedimentary gradients are considered gradational when variable means are statistically different, but the standard deviations overlap.

Radiocarbon dating. Selected surface, core, and rock samples were chosen for radiometric carbon-14 analysis. For all sediment samples, the sand fraction analyzed is 250-1000μm. In addition, the mud fraction (<63μm) was dated for two samples to provide a sand-mud couplet.

Bulk rock samples were prepared for radiocarbon dating using a stainless steel drillbit. The rocks were studied using standard petrographic techniques, and all contained less than 10% cement. The minor amount of cement is not considered to significantly affect the rock ages since carbonate eolianites lithify soon after deposition (McKee and Ward, 1983; Carew and Mylroie, 1985; Boardman and others, 1987).

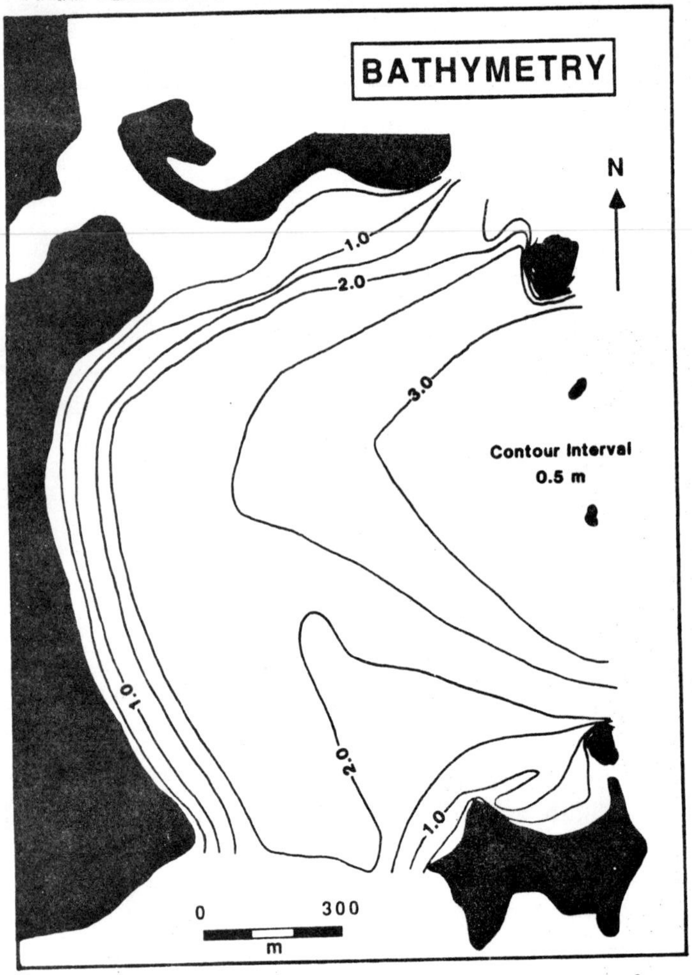

Fig. 3. Snow Bay is slightly shallower to the north and south, but a pronounced sill which would restrict circulation with the open ocean does not exist.

RESULTS

Environments of Deposition

The bathymetry of the lagoon slopes seaward to a maximum depth of 3.3m (Fig. 3). The northern and southern ends of the lagoon are slightly shallower than the center of the lagoon, but a pronounced sill was not found. Snow Bay has excellent exchange with the open Atlantic and maintains normal salinity. The coast is dominated by beaches and fits the criteria for a microtidal, wave-dominated coast (Davis and Hayes, 1984).

Transects across the lagoon distinguish three major subtidal environments (Fig. 1). An ebb-tidal delta located in the northwestern corner of the lagoon forms a 2m thick lobe of sand prograding into the *Thalassia* meadow. A 200-400m wide *Thalassia* meadow rims the tidal delta, and forms a 2m thick sediment package deposited in a topographic low. The rest of the lagoon is dominated by a mobile sand sheet (equivalent to a platform interior sand blanket; Illing, 1957; Ball, 1967) less than 50cm thick. Patch reefs dot the mobile sand sheet between the cays. A fourth major subtidal environment, the barrier reef, is considered a separate sedimentary system beyond the scope of this study.

Two-way cluster analysis of surface and core samples distinguishes four sediment facies based on composition, texture, and mud fraction mineralogy (Fig. 4). Facies names are based on variable means (Table 1). T-test statistics indicate that the means of the most abundant variables for each facies are distinct, even though overlap occurs between the sediment facies. The four sediment facies are: 1. abraded grain "grainstone", 2. abraded grain "grapestone", 3. foraminiferal "grainstone/packstone", and 4. *Halimeda*-rich "pack

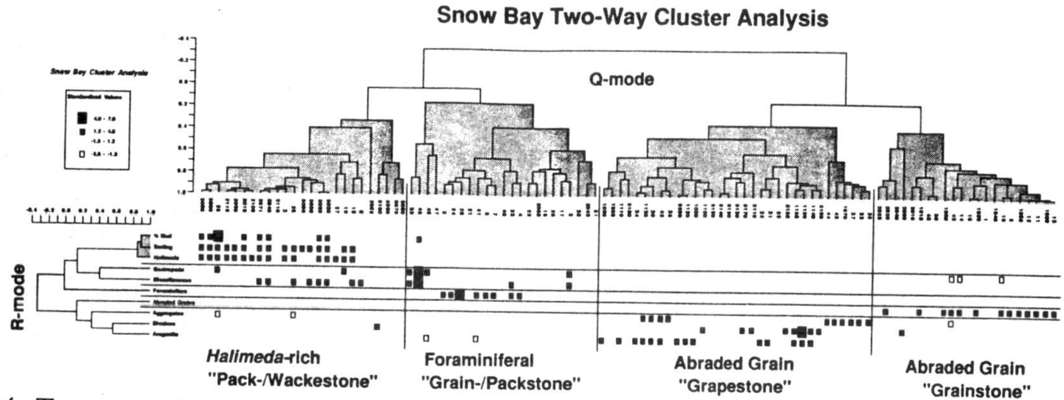

Fig. 4. Two-way cluster analysis of Snow Bay surface and core samples using ten variables. Filled squares depict variables well above the standardized mean and open squares depict variables well below the standardized mean. No square indicates variables near the mean.

Snow Bay Cluster Analysis
Variable Means of Cluster Facies

	Cluster A N=22	Cluster B N=33	Cluster C N=23	Cluster D N=25
% Mud	**0.89 ± 0. 89**	**3.03 ± 2.02**	**6.40 ± 6.96**	**23.01 ± 12.87**
Sorting (ø)	0.92 ± 0.20	1.34 ± 0.28	1.56 ± 0.43	2.71 ± 0.49
Halimeda	6.50 ± 4.35	8.43 ± 4.11	19.83 ± 9.64	**48.11 ± 16.74**
Bivalves	5.07 ± 3.72	12.03 ± 8.50	5.28 ± 3.33	6.10 ± 4.07
Gastropods	5.28 ± 2.40	6.55 ± 1.91	8.69 ± 5.20	5.75 ± 2.30
Foraminifera	12.04 ± 5.76	9.45 ± 3.34	**28.43 ± 14.76**	12.65 ± 3.76
Aggregates	13.31 ± 5.05	**28.08 ± 8.76**	12.49 ± 6.75	11.08 ± 6.00
Abraded Grains	**53.51 ± 11.69**	**30.83 ± 9.78**	17.22 ± 11.32	9.39 ± 6.10
Miscellaneous	3.98 ± 2.59	4.08 ± 1.68	7.39 ± 3.81	6.80 ± 3.55
Aragonite	55.72 ± 6.76	66.05 ± 8.13	47.70 ± 5.95	50.83 ± 3.03
Water Depth (cm)	49.33±44.85	199.58±32.92	192.30±87.41	161.57±74.90
	Abraded Grain "Grainstone"	Abraded Grain "Grapestone"	Foraminiferal "Grain-/Packstone"	Halimeda-rich "Pack-/Wackestone"

Table 1. Variable means of two-way cluster analysis sediment facies. Bold face indicates variables used for facies name designation.

stone/wackestone". Of these facies, all four are important surficially, but the foraminiferal "grainstone/packstone" is only a minor component of the vertical sequence.

Vertical Sedimentary Gradients

Megascopic descriptions of the six sediment cores from the ebb-tidal delta and *Thalassia* meadow show that the vertical sequence is composed of "grainstone", "packstone", and "wackestone" sediments (Fig. 5).

The analysis of cores 1, 2, and 5 show that the four sedimentary facies which form the lateral sedimentary gradients are also present in the vertical sedimentary sequence (Fig. 6). The vertical sedimentary sequences from all cores are combined to form a composite sequence.

Ebb-tidal delta cores. The sediments underlying the ebb-tidal delta are comprised of a lower abraded grain "grapestone" facies which is overlain by the abraded grain "grainstone" facies (core 1). There is a distinct upward increase in abraded grains and sorting; however the mud content is consistently low (Fig. 7).

The abraded grain "grapestone" of the ebb-tidal delta contains three intervals of intraclasts (Fig. 5). The intraclasts, which range in size from 0.5-3.0cm, are composed of moderate to poorly sorted, well-rounded micritized grains and cortoids cemented by isopachous rims of aragonite needles.

At the ebb-tidal delta edge, 60cm of abraded grain "grainstone" overlies sediments of the *Thalassia* meadow (core 2). A thin foraminiferal "grainstone" forms a transition between the two sediment types. The abrupt transition between sediments of the *Thalassia* meadow and the ebb-tidal delta is recognized by abrupt changes in texture and composition (Fig. 8).

Thalassia Meadow Cores. The sediments underlying the *Thalassia* meadow are composed of a basal abraded grain "grapestone" facies which is overlain by the *Halimeda*-rich "packstone/wackestone" facies (core 5). The upper part of the

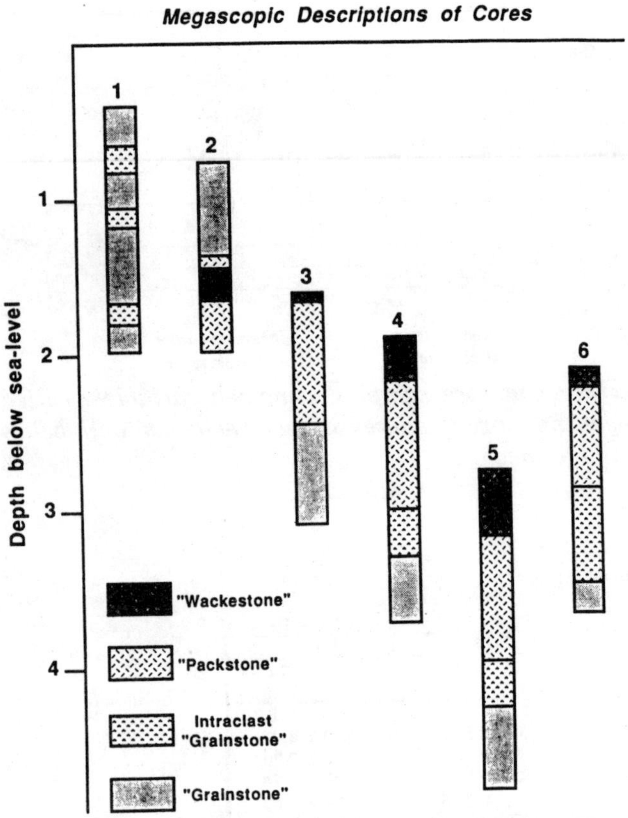

Fig. 5. Megascopic core descriptions. Textural terminology from Dunham (1962). Intraclastic "grainstone" contains intraclasts ranging in size from 1-10cm.

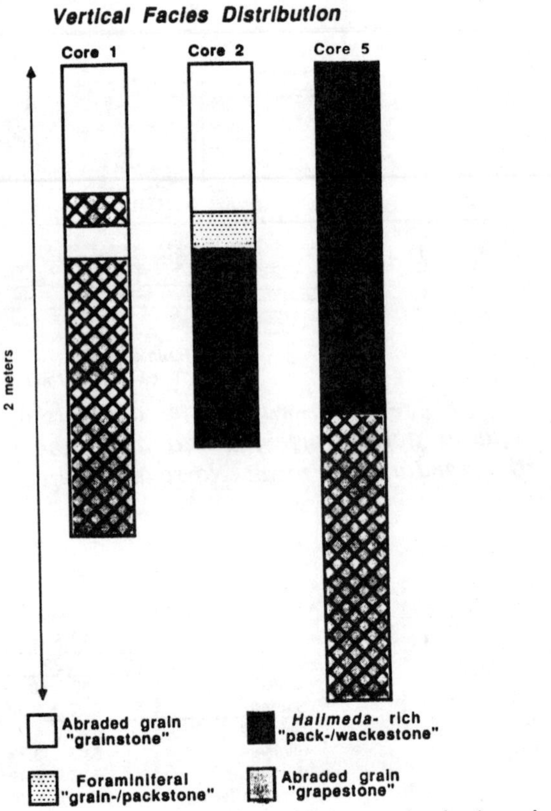

Fig. 6. Vertical distribution of facies in cores 1, 2, and 5. Core 1 is from the center of the ebb-tidal delta, core 2 is from the seaward edge of the ebb-tidal delta, and core 5 is from the Thalassia *meadow.*

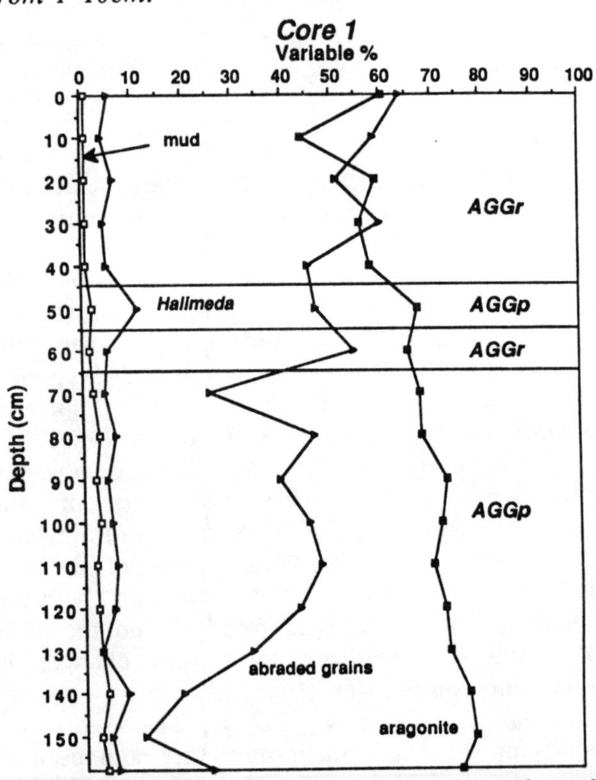

Fig. 7. The low proportion of mud and Halimeda *in core 1 suggest a lack of substrate modification by benthic flora. AGGr = abraded grain "grainstone" and AGGp = abraded grain "grapestone".*

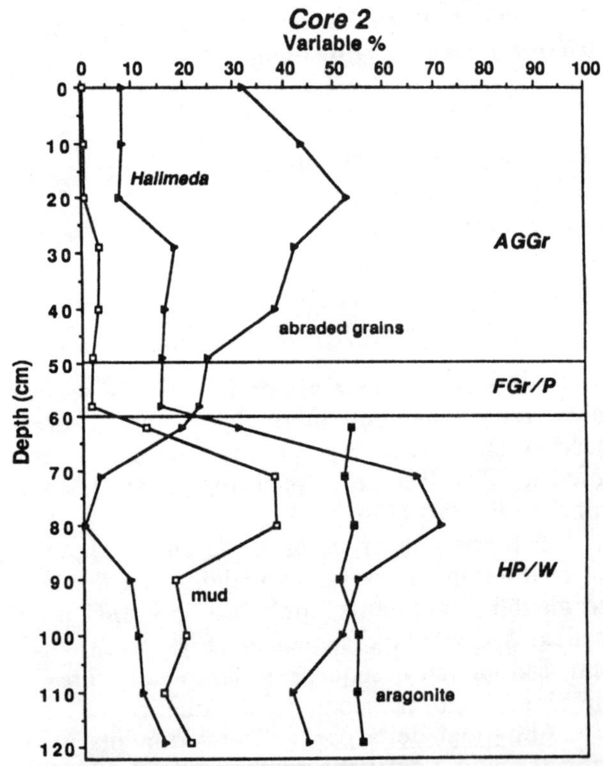

Fig. 8. The abrupt change in texture and composition at 60cm in core 2 reflects progradation of the tidal delta over the Thalassia *meadow. AGGr = abraded grain "grainstone", FGr/P = foraminiferal "grainstone/packstone", and HP/W =* Halimeda*-rich "packstone/wackestone".*

abraded grain "grapestone" from the *Thalassia* sequence also contains intraclasts of the same type as the tidal delta (see previous section).

The *Halimeda*-rich "packstone/wackestone" fines upwards and becomes more poorly sorted (Fig. 9). Compositionally *Halimeda* increases upwards while aggregates and abraded grains decrease. Also, the mud fraction mineralogy changes upwards from aragonite-dominated to approximately equal amounts of aragonite and high-magnesium calcite.

Radiometric Dating

Radiometric dating using carbon-14 was performed on selected surface, core, and rock samples and a peat sample (Table 2).

Surficially, carbon-14 dating of the abraded grain "grapestone" yields an age of 3090 y.b.p. for this sediment. A single peat sample was recovered *in situ* from the Pigeon Creek tidal channel where tidal currents have eroded the overlying sediments. Carbon-14 dating yields and age of 6280 y.b.p. for this peat.

Carbon-14 dates for cores were obtained for the abraded grain "grainstone" of the ebb-tidal delta sequence and the abraded grain "grainstone" and *Halimeda*-rich "packstone/wackestone" of the *Thalassia* sequence. The ebb-tidal abraded grain "grainstone" overlying the *Thalassia* sequence yields ages of 3350 y.b.p. 30cm below the surface and 3470 y.b.p. 50cm below the surface.

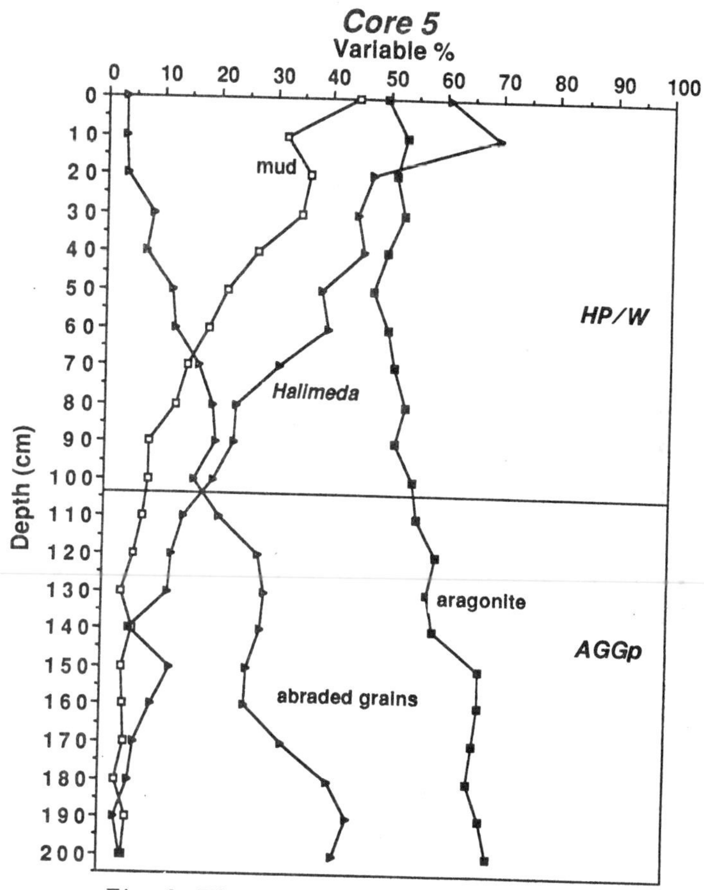

Fig. 9. The textural and compositional trends in core 5 reflect the effect of substrate modification by benthic flora. HP/W = Halimeda-*rich "packstone/wackestone" and AGGp = abraded grain "grapestone".*

SNOW BAY RADIOCARBON DATES

Core Samples

Core	Sample Type	Sample Depth	Water Depth	Total Depth	Age
2	sand	30 cm	70 cm	100 cm	3350±60
2	sand	50 cm	70 cm	120 cm	3470±90
2	mud	66 cm	70 cm	136 cm	490±70
2	sand	70 cm	70 cm	140 cm	3510±320
2	sand	120 cm	70 cm	190 cm	3630±120
5	mud	34 cm	250 cm	284 cm	1410±470
5	sand	34 cm	250 cm	284 cm	2540±140
5	mud	63 cm	250 cm	313 cm	3690±240
5	sand	63 cm	250 cm	313 cm	4280±240
5	sand	196 cm	250 cm	446 cm	3850±190

Surface Samples

Sample Type	Sample Depth	Water Depth	Total Depth	Age
peat	surface	500 cm	500 cm	6280±740
sand	surface	190 cm	190 cm	3090±80

Rock Samples

Sample Type	Sample Location	Age
bulk rock	western most sand ridge of Sandy Hook	3650±60
bulk rock	SE corner, Sandy Hook	1990±70
bulk rock	eolianite, Nancy Cay	4220±80
bulk rock	eolianite, Snow Bay Ridge	3700±70

Table 2. Radiocarbon dates from the Snow Bay area of San Salvador.

A carbon-14 date from the abraded grain "grapestone" near the base of core 5 yields an age of 3850 y.b.p., which is nearly the same age as the surficial abraded grain "grapestone".

Radiocarbon dating of sand-mud couplets from the *Halimeda*-rich "packstone/wackestone" show large discrepancies in sand and mud ages. Mud dates range from 600 to 3140 years younger than the sand dates. The discrepancy between the ages increases upwards with the largest difference being at the tope of the *Thalassia* sequence in core 2 where ebb-tidal delta sands directly overlie *Halimeda*-rich "packstone/wackestone" (Table 2).

DISCUSSION

Examination of the vertical sedimentary gradients reveals the influence of benthic flora, water depth, and antecedent topography on facies relationships in high-energy carbonate environments. The lateral facies distribution (Andersen and Boardman, 1987) and the vertical sedimentary sequence provides a record of banktop responses to the Holocene sea-level rise.

Vertical Facies Distribution

The depositional sequence in Snow Bay includes a pulse of transgressive sand (abraded grain "grapestone") overlain by shallowing-upward regressive deposits. The regressive deposits include subtidal muddy sands (*Halimeda*-rich "packstone/wackestone") and upper regressive sands (abraded grain "grainstone"). The foraminiferal "grainstone/packstone" is a minor constituent of the vertical sequence.

Basal transgressive sand. The basal sediments from cores in the *Thalassia* meadow and ebb-tidal delta are comprised of the abraded grain "grapestone" facies which is up to one meter thick (cores 1 and 5, Fig. 6). The abraded grain "grapestone" facies represents a subtidal sand and/or beach deposited during the initial strandline migration.

The upper 50cm of abraded grain "grapestone" contains 1-10cm intraclasts in cores 1, 4, 5, and 6. The grains comprising the intraclasts are not aggregates, but are similar to the abraded grain "grainstone" of the beach and ebb-tidal delta. The sand matrix surrounding the intraclasts contains a proportion of aggregates similar to the lower 50cm of the basal abraded grain "grapestone" (20-25%).

Typically, transgressive beach deposits are not preserved; rather, erosion of a beach deposit leaves behind a winnowed surface capped by a thin lag of coarse intraclasts of fossil debris (Adams, 1983; Inden and Moore, 1983). The composition of the intraclasts suggests they represent beachrock lag from the erosion of the beach during strandline migration. Although the intraclasts fit the criteria for being beachrock fragments (Stoddart and Cann, 1965; Scoffin, 1970b; Alexandersson, 1972; Beier, 1985), these same criteria can be applied to non-beach, submarine cementation (Shinn, 1969). Cores from Joulter's ooid shoal show dramatic increases in aggregates where submarine crusts occur (Harris, 1979). The increased abundance of intraclasts from the Snow Bay cores are not associated with an increased abundance of aggregates. This distinction further suggest beachrock origin rather than a submarine crust. Therefore, we interpret the sediment to represent a reworked beach (abraded grain "grainstone") and beachrock deposit. As sea-level rose, this sediment became subtidal and was cemented by aragonite forming an abraded grain "grapestone".

Radiometric dates from the surficial abraded grain "grapestone" and from the basal abraded grain "grapestone" (bottom of core 5) yield ages of 3090 y.b.p. and 3850 y.b.p. respectively. The closeness of the two ages suggests two possibilities. First, that the abraded grain "grapestone" is a transgressive lag deposit, a "relict" foreshore deposit from 3000 to 4000 y.b.p. Alternatively, the mobile sand sheet is receiving significant quantities of "old" sand from the present day erosion of dunes (Boardman et al, 1987), which is being cemented subtidally. Nearby dunes such as Nancy Cay (C-14 age=4220 y.b.p.) and Snow Bay Ridge (C-14 age=3700 y.b.p.) show appreciable erosion.

Subtidal muddy sands. Cores from the *Thalassia* meadow contain a fining-upwards *Halimeda*-rich "packstone/wackestone" facies which represents gradual substrate modification by benthic flora. Substrate modification by benthic flora results in changes in texture, composition, and mud fraction mineralogy (cores 2 and 5, Fig. 6; Andersen and Boardman, 1987). An interesting feature of the vertical sequence is the general absence of the transitional environment (foraminiferal "grainstone/packstone") between the basal transgressive sand and the subtidal muddy sands.

The source of sand in *Thalassia* meadows is thought to be produced and deposited *in situ* (Scoffin and Tudhope, 1984; Boscence, 1986). In the *Thalassia* meadow of Snow Bay the coarse sand fraction is angular which indicates that transport and mechanical abrasion of the coarse sand is minimal. While minimal transport is recognized for the coarse sand fraction, significant transport is recognized for the medium sand fraction. Analysis of the medium sand fraction (0.5 to 1.0mm) of core 2 shows an average of 29% more abraded grains per sample than in the coarse sand fraction.

Radiometric dating of medium sand and the mud fraction from the same sample (a sand-mud couplet) shows that there is a difference in ages. The discrepancy between the medium sand and mud ages increases upwards, with the sand always being older than the mud (Boardman and others, in prep.). Therefore, a significant portion of the medium sand fraction is "old sand which has been redeposited, i.e., there is an additional (allocthonous) source of sand (Boardman and others, 1987); Boardman and others, in prep.). This may indicate that only the coarser sand fractions contain abundant sand which has been produced *in situ*. The significantly higher proportion of abraded grains in the medium sand fraction further supports this hypothesis. The abundance of "old" sand is also evidence of the bankward transport of sand characteristic of windward-oriented high-energy lagoons (Halley and others, 1983).

Upper regressive sand. Only cores from the ebb-tidal delta contain an abraded grain "grainstone" facies (cores 1 and 2, Fig. 6). In the center of the ebb-tidal delta, this facies overlies the abraded grain "grapestone" (core 1, Fig. 6); whereas at the seaward edge of the ebb-tidal delta, the abraded grain "grainstone" abruptly overlies the *Halimeda*-rich "packstone/wackestone" with a thin foraminiferal "grainstone" between the two facies (core 2, Fig. 6).

The upward transition to the abraded grain "grainstone" represents the formation or progradation of the ebb tidal delta. In the Florida Reef Tract, ebb-tidal deltas only form where shoals at the shelf margin damp large oceanic waves (Enos, 1977). Although subtidal shoals do not exist in Snow Bay, the growth of the barrier reef to sea-level should have the same dampening effect on wave energy.

The C-14 ages of the abraded grain "grainstone" (3350 and 3470 y.b.p.) suggest that the ebb-tidal delta is old (>3000 y.b.p.). However, a radiometric date of mud within the *Halimeda*-rich "packstone/wackestone" sediment just below the ebb-tidal delta sediments is 490 y.b.p. and suggests progradation of the delta within the last 500 years. The medium sand fraction of the same sample from the *Halimeda*-rich "packstone/wackestone" yields an age of 3510 y.b.p. which is similar to the ebb-tidal delta sediments. The similarity of the medium sand ages suggests that the ebb tidal delta sands are comprised primarily of redeposited "old" sand (Boardman and others, in prep.).

Composite high-energy shallowing-upwards sequence. The shallowing-upward sequence is an important concept in carbonate geology, but few modern analogues for the high-energy shallowing-upward sequence exist. The composite vertical sequence from Snow Bay is a high-energy shallowing-upward sequence (Fig. 10) and provides a modern analogue to ancient high-energy deposits.

Goodwin and Anderson (1985) propose that in all sedimentary environments, sediments accumulate episodically as thin (1-5m) shallowing-upward cycles separated by sharply defined non-depositional surfaces. These shallowing-upward cycles are called punctuated aggradational cycles (PAC). Each individual PAC is characterized by gradational internal facies boundaries similar to those recognized in cores from Snow Bay. In the special case of shallow marine carbonate sequences, PAC's form as a direct response to sea-level fluctuations and have exposure surface boundaries, such as the Pleistocene exposure underlying the Snow Bay sediments.

James' (1984) model of a high-energy shallowing-upward carbonate sequence is a special case of the PAC hypothesis. James proposes four distinctive sedimentary units: a thin basal transgressive grainstone, a subtidal fossiliferous mudstone, intertidal grainstones of the lower and upper foreshore, and a supratidal exposure surface. The composite high-energy vertical sedimentary sequence observed in Snow Bay is remarkably similar to this model (Fig. 10). The 2-3 meter thick high-energy sequence from Snow Bay is characterized by:

1. An intertidal basal transgressive sand (~90cm thick); the abraded grain "grapestone" facies which contains intraclasts,

2. subtidal muddy sands (~100cm); the fining-upwards *Halimeda*-rich "packstone/wackestone" facies,

3. subtidal/intertidal regressive tidal delta sands (~50cm); the abraded grain "grainstone" facies.

The transitional environment (foraminiferal "grainstone/packstone") is a minor component of the

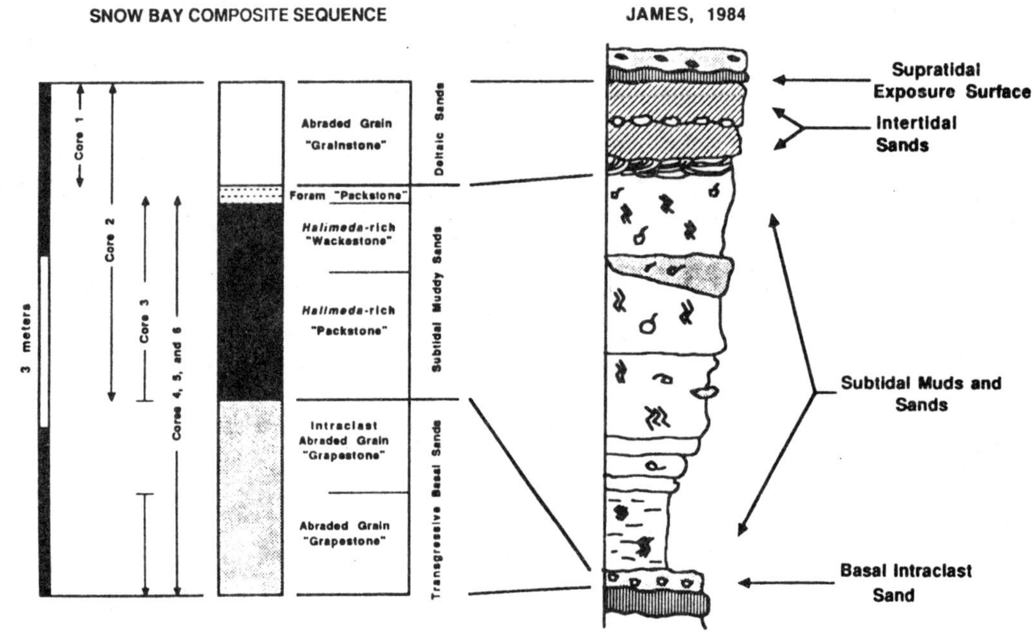

Fig. 10. Comparison of the composite high-energy shallowing-upwards sequence for Snow Bay to the model proposed by James (1984).

vertical sequence.

The two primary differences between James' model and the cores from Snow Bay are the substitution of tidal delta sands for beach sands and the incomplete nature (in each core) of the Snow Bay shallowing-upward sequence.

The composite Snow Bay high-energy shallowing-upward sequence supports the PAC hypothesis and provides a modern example for James' (1984) model. While a sedimentary sequence containing *Thalassia* can be properly applied only to late Cretaceous and younger rocks (Brasier, 1975), the principles of the high-energy shallowing-upward sequence as a punctuated aggradational cycle should be applicable to Paleozoic and early to mid-Mesozoic rocks. Algae, bryozoans, and crinoids have acted as sediment traps and baffles in the past (Harris and Martin, 1979; McKinney, McKinney, and Listokin, 1987; Carbonate Seminar, 1987); so similar high-energy sequences should have formed in topographic lows of ocean-facing environments in the past.

The Depositional Evolution of Snow Bay

The previously described lateral and vertical sedimentary gradients of Snow Bay record the banktop response to a sea-level rise. The depositional package contains both eolian and marine facies which accumulated within the last 6300 years (Andersen and Boardman, 1988).

A peat deposit in a topographic low five meters below sea level records banktop flooding 6300 years ago (Fig. 11). The peat formed in a low, poorly drained area and overlies a hard, well-cemented surface which we presume is the Pleistocene contact.

The rest of the Pleistocene surface in Snow Bay is covered by a poorly sorted transgressive sand which includes intraclasts of probable beachrock origin. This transgressive sand, similar to the basal unit in the high-energy shallowing-upward sequence described in James (1984), records strandline movement across the bank as sea-level rose. A large amount of sand was produced during the initial transgression, and redeposition of this sand is an important feature of the evolution of Snow Bay and other high-energy lagoons (Trumbel and Neumann, 1986; Boardman and others, 1987; Boardman and others, in prep.).

Temporally associated with this transgressive sand is eolian dune formation (Carew and Mylroie, 1985; Boardman and others, 1987; Andersen and Boardman, 1988). Dunes up to twenty meters high form the cays and cliffs along the eastern margin of San Salvador. Well preserved eolian cross bedding which dips below sea level and C-14 ages from 5400 to 3700 y.b.p. indicate the dunes formed rapidly during initial banktop flooding. Erosion of these dunes provided additional sand to the strandline (Fig. 11).

Stabilization of the sandy lagoon floor by

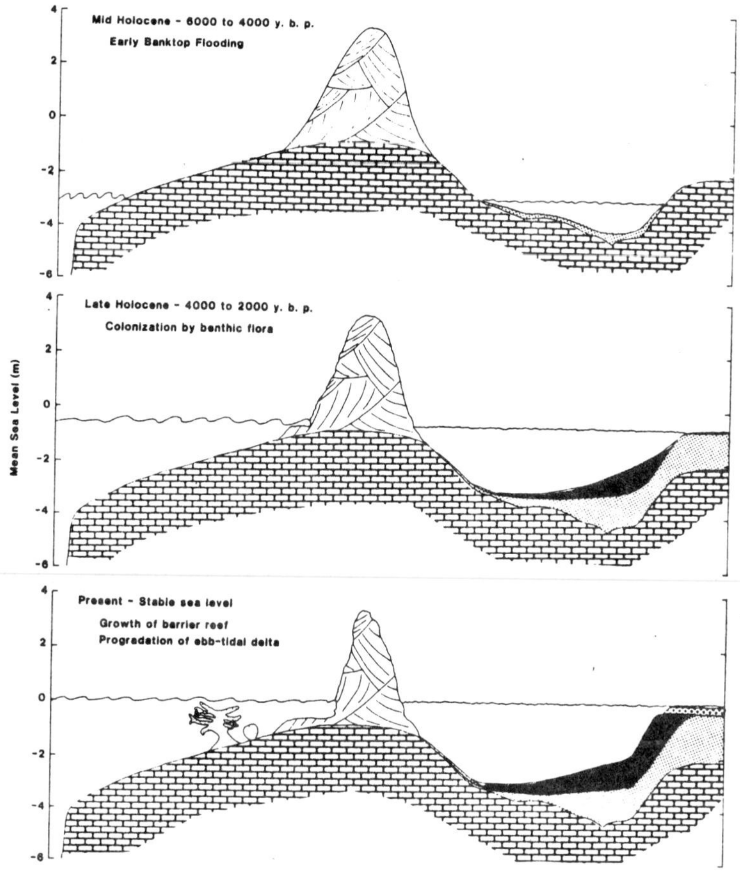

Fig. 11. The depositional evolution of Snow Bay. Mid Holocene - initial banktop flooding and temporal dune formation on Pleistocene bedrock (limestone pattern). Late Holocene-strandline migration forms the abraded grain "grapestone" (light stipple). As sea-level rose, this sediment was colonized by benthic flora forming the Halimeda-*rich "packstone/wackestone" (black). In the last 1500 years, growth of a barrier reef allowed coastal beach ridges and the ebb-tidal delta to prograde into Snow Bay forming the abraded grain "grainstone" (heavy stipple).*

benthic algae, most notably *Halimeda*, coincided with the continued rise of sea-level. *Thalassia* was able to colonize areas with sufficient sediment thickness (at least 7cm, Scoffin, 1970a) and sufficient sediment stability (Moore, 1963). Substrate modification by benthic flora caused gradational changes in the sediment characteristics.

Regressive deposition of sand along the strandline began when sea-level reached near present level (Fig. 11). The style of deposition changed from dune formation to prograding beach ridges (Boardman and others, 1987; in prep.). Air photos and geologic surveys clearly indicate that Sandy Hook is a recently prograded beach ridge system.

In their stratigraphy of San Salvador, Carew and Mylroie (1985) place Sandy Hook in the youngest member (Hanna Bay Member) of the Holocene Rice Bay Formation. They proposed that the Hanna Bay Member was deposited in equilibrium with present sea level sometime after 5000 years ago. Sea-level in the Bahamas has been within 70cm of present for at least 1400 years (Winter, 1987; Donn and Boardman, 1988; Boardman, Neumann, and Rasmussen, 1988); yet whole rock C-14 ages of the oldest and youngest sand ridges of Sandy Hook are 3650 and 1990 y.b.p. respectively (Table 2). Therefore, either 1. the sand ridges were formed when sea level was between 1.2 and 0.5m below present (i.e. Carew and Mylroie, 1985, are wrong), or 2. the sand comprising the ridges is older (>1400 y.b.p.) than the progradation of Sandy Hook. We favor the second alternative for two reasons. First, the very friable nature of the Sandy Hook sand ridges suggests a shorter time for cementation than the better cemented rocks of the nearby North Point Member (High Cay, Nancy Cay, Snow Bay Ridge) which range in age from 3700-4200 y.b.p. (Table 2). Second, C-14 dates of sand-mud couplets suggest a source of "old" sand which could easily contribute to the formation of Sandy Hook and "contaminate" the C-14 ages of the Sandy Hook sand ridges.

The ebb-tidal delta, formed at the mouth of the Pigeon Creek tidal channel, is a prominent feature of Snow Bay. The ebb-tidal delta may have formed any time since the area was flooded (-2m at 4500 y.b.p.). Our data however, suggest that the tidal delta formed recently from "old" sand. The form of the ebb-tidal delta requires that sand dispersal by ebb currents dominate over dispersal by wave energy. The growth of reefs to sea-level may have reduced wave energy enough to allow the dominance of ebb-tidal currents, in a manner similar to that described by Enos (1977). Today the line of barrier reefs and cays substantially reduces wave energy entering the lagoon. The high, ocean-facing cliffs of the North Point Member cays suggest they are absorbing wave energy in addition to supplying sand to the strand line.

Time Frame for Sequence Development

Ancient shallowing-upward sequences developed in 10,000 to 250,000 years (Goodwin and Anderson, 1985; Hardie, 1985; Heckel, 1985; Grotzinger, 1986). However, this time frame includes time of subaerial exposure, so the actual time for sediment accumulation is unknown. Studies of modern carbonates indicate that sediment accumulation occurs rapidly. Six to eight meters of oolitic sediment formed in 1500 years on Lily Bank (Hine, 1977), and the four meter thick ooid shoal of Joulter's Cay formed in less than 5000 years (Harris, 1984). The sediment package of Snow Bay, a modern example of a high-energy shallowing-upward sequence, formed in less than 6300 years.

CONCLUSIONS

The composite vertical sequence of Snow Bay is remarkably similar to the high-energy shallowing-upward sequence proposed by James (1984). The sedimentary sequence includes a transgressive basal sand of possible lower foreshore/beach origin, fining-upwards subtidal muddy sands representing substrate modification by benthic flora, and regressive intertidal/subtidal ebb-tidal deltaic sands. The sequence contains an abundance of "old" sand transported bankward into the *Thalassia* meadow, ebb-tidal delta, and prograding beach ridges.

The lateral and vertical facies mosaic of southeastern San Salvador records the banktop response to the Holocene sea-level rise. The complete depositional package accumulated in the last 6300 years and contains both eolian and marine facies. The summary of events which formed the Snow Bay sedimentary sequence are:

1. The initial banktop flooding 6300 y.b.p. resulting in peat formation in topographic lows five meters below present sea level.
2. The deposition of a transgressive abraded grain "grapestone" as the strandline migrated across the banktop in response to continued sea-level rise. Eolianite deposition is closely associated with early strandline migration.
3. Colonization and stabilization of the sandy lagoon floor first by benthic algae, then by seagrasses. These benthic flora trap and bind increasingly muddier sediments and form the *Halimeda*-rich "packstone/wackestone".
4. Recent (since 1400 y.b.p.) progradation of Sandy Hook and formation of the ebb-tidal delta. Ebb-tidal formation is probably due to growth of the barrier reef to sea level.
5. Progradation of the ebb-tidal delta over the *Thalassia* meadow.

ACKNOWLEDGEMENTS

We would like to thank all the field assistants who have participated in this research-Pam Alley, Paul Bergstrand, Norm Colby, Dave Daugherty, Carol Gerber, Wayde Hartwick, Rick McCartney, Ron Pelle, Bill Schmidhorst, and the participants of the 1986 Modern Carbonate Deposition Workshop.

We are indebted to the College Center of the Finger Lakes' Bahamian Field Station and its staff on San Salvador Island, Bahamas for logistical and financial support for this research.

Equipment, radiocarbon dating, and travel for this study were funded by a Grant-in-Aid of Research from Sigma Xi to CBA, National Science Foundation Grants OCE-83-15203 and OCE-85-40732 to MRB, and funds provided by Miami University to the Modern Carbonate Deposition Workshop. The Miami University Geology Department provided funds for travel and radiocarbon dating.

REFERENCES CITED

Adams, R.W., 1983, General guide to the geological features of San Salvador, in Gerace, D.T., ed.: Field Guide to the Geology of San Salvador, CCFL Bahamian Field Station, p. 1-66.

Alexandersson, T., 1972, Intragranular growth of marine aragonite and Mg-calcite: evidence of precipitation from supersaturated seawater: Journal of Sedimentary Petrology, v. 42, p. 441-460.

Andersen, C.B. and Boardman, M.R., 1987, Lateral and vertical sedimentary gradients in a high-energy carbonate environment, Snow Bay, San Salvador, Bahamas: Abstracts of Society of Economic Paleontologists and Mineralogists Annual Midyear Meeting, v. 4, p. 4.

Andersen, C.B. and Boardman, M.R., 1988, The depositional evolution of southeastern San Salvador, Bahamas: A high-energy carbonate environment: Abstracts of Society of Economic Paleontologists and Mineralogists Annual Midyear Meeting, v. 5.

Ball, M.M., 1967, Carbonate sand bodies of Florida and the Bahamas. Journal of Sedimentary Petrology, V. 37, p. 556-591.

Basan, P.B., 1973, Aspects of sedimentation and development of a carbonate bank in the Barracuda Keys, South Florida: Journal of Sedimentary Petrology, v. 43, p. 42-53.

Bathurst, R.G.C., 1975, Carbonate sediments and their diagenesis. Developments in Sedimentology 12, 2nd. edt.: Elsevier, New York, 658 p.

Beier, J.A., 1985, Diagenesis of Quaternary Bahamian beachrock: petrographic and isotopic evidence: Journal of Sedimentary Petrology, v. 55, p. 755-761.

Boardman, M.R., 1976, Lime mud deposition in a tropical island lagoon, Bight of Abaco, Bahamas: Unpublished M.S. Thesis, University of North Carolina, Chapel Hill, N.C., 121 p.

Boardman, M.R., Evans, L.A., and Andersen, C.B., 1986, Sedimentary gradients of Holocene Bahamian lagoons: control by sills and sea-level: Abstracts of Society of Economic Paleontologists and Mineralogists Annual Midyear Meeting, v. 3, p. 11.

Boardman, M.R., Carew, J.L., and Mylroie, J.E., 1987, Holocene deposition of transgressive sand on San Salvador, Bahamas: Geological Society of America Abstracts with Program, v. 19, p. 593.

Boardman, M.R., Carew, J.L., Mylroie, J.E., and Andersen, C.B., in prep., Holocene deposition of transgressive sand on San Salvador, Bahamas.

Boardman, M.R., Neumann, A.C., and Rasmussen, K.A., 1988, Holocene sea level in the Bahamas: The 4th Symposium on the Geology of the Bahamas Abstracts and Program, CCFL Bahamian Field Station, p. 4.

Bosence, D.W.J., 1986, Controls on lateral gradients in carbonate grains and mounds, Florida Bay: Abstracts of Society of Economic Paleontologists and Mineralogists Annual Midyear Meeting, v. 3, p. 13.

Carbonate Seminar, 1987, Ramp Creek and Harrodsburg Limestones: A shoaling-upward sequence with storm-produced features in southern Indiana, U.S.A.: Sedimentary Geology, v. 52, p. 207-226.

Carew, J.L. and Mylroie, J.E., 1985, The Pleistocene and Holocene stratigraphy of San Salvador Island, Bahamas, with reference to marine and terrestrial lithofacies at French Bay, in Curran, H.A., ed., Pleistocene and Holocene carbonate environments on San Salvador Island, Bahamas - Guidebook for Geological Soc. Amer. Orlando annual meeting field trip: Ft. Lauderdale, FL, CCFL Bahamian Field Station, p. 11-61.

Chave, K.E., 1954, Aspects of the biochemistry of magnesium 1. Calcareous marine organisms: Journal of Geology, v. 62, p. 266-283.

Davis, R.A., Jr. and Hayes, M.O., 1984, What is a wave-dominated coast? Marine Geology, v. 60, p. 313-329.

Donn, T.F., 1986, Erosion of a rocky carbonate coastline: Andros Island, Bahamas: Unpublished M.S. Thesis, Miami University, Oxford, OH.

Donn, T.F. and Boardman, M.R., 1988, Erosion of a rocky carbonate coastline: Andros Island, Bahamas: Journal of Coastal Research, v. 4, p. 381-394.

Enos, P., 1983, Shelf, in Scholle, P.A., Bebout, D.G., and Moore, C.H., Carbonate Depositional Environments: American Association of Petroleum Geologists Memoir 33, p. 267-296.

Goldsmith, J.R., Graf, D.L., and Joensuu, O.I., 1955, The occurance of magnesian calcites in nature: Geochimica et Cosmochimica Acta, v. 7, p. 212-230.

Goodwin, P.W. and Anderson, E.J., 1985, Punctuated aggradational cycles: a general hypothesis of episodic stratigraphic accumulation: Journal of Geology, v. 93, p. 525-533.

Grotzinger, J.P., 1986, Cyclicity and paleoenvironmental dynamics, Rocknest platform, northwest Canada: Geological Society of America Bulletin, v. 97, p. 1208-1231.

Halley, R.B., Harris, P.M., and Hine, A.C., 1983, Bank margin, in Scholle, P.A., Bebout, D.G., and Moore, C.H., eds. Carbonate Depositional Environments: American Association of Petroleum Geologists Memoir 33, p. 463-506.

Harbaugh, J.W. and Merriam, D.F., 1968, Computer Applications in Stratigraphic Analysis: Wiley, New York, 282p.

Hardie, L.A., ed., 1977, Sedimentation on the modern carbonate tidal flats of Northwest Andros Island, Bahamas: The John Hopkins Press, Baltimore, MD, 202p.

Hardie, L.A., 1985, Carbonate tidal flat cycles: episodic "growth bands" of carbonate platforms: Abstracts of the Princeton Symposium on Cycles, Princeton, NJ, p. 12.

Harris, F.W. and Martin, W.D., 1979, Benthic community development in limestone beds of the Waynesville (Upper Dillsboro) Formation (Cincinnatian Series, Upper Ordovician) of southeastern Indiana: Journal of Sedimentary Petrology, v. 49, p. 1295-1306.

Harris, P.M., 1984, Cores from a modern carbonate sand body; the Joulter's ooid shoal, Great Bahama Bank, in Harris, P.M., ed., Carbonate sands: Society of Economic Paleontologists and Mineralogists Core Workshop no. 5, p. 429-464.

Heckel, P.H., 1985, Middle and Upper Pennsylvanian cyclothems of midcontinent North America, a record of glacial-eustatic periodicity: Abstracts of the Princeton Symposium on Cycles, Princeton, NJ, p. 16.

Hine, A.C., 1977, Lily Bank, Bahamas; case history of an active oolite sand shoal: Journal of Sedimentary Petrology, v. 47, p. 1554-1581.

Hine, A.C., Wilbur, R.J., and Neumann, A.C., 1981, Carbonate sand bodies along contrasting shallow bank margins facing open seaways in northern Bahamas: American Association of Petroleum Geologists Bulletin, v. 65, p. 261-290.

Illing, L.V., 1954, Bahamian calcareous sands: American Association of Petroleum Geologists Bulletin, v. 38, p. 1-95.

Imbrie, J. and Purdy, E.G., 1962, Classification of modern Bahamian carbonate sediments, in Ham, W.E., ed., Classification of Carbonate Rocks: American Association of Petroleum Geologists Memoir 1, p. 253-272.

Imbrie, J. and Van Andel, Tj.H., 1964, Vector analysis of heavy-mineral data: Geological Society of America Bulletin, v. 75, p. 1131-1156.

Inden, R.F., and Moore, C.H., 1983, Beach environment, in Scholle, P.A., Bebout, D.G., and Moore, C.H., eds., Carbonate Depositional Environments: American Association of Petroleum Geologists Memoir 33, p. 211-267.

James, N.P., 1984, Shallowing-upward sequences in carbonates, in Walker, R.G., ed., Facies Models, p. 213-228. Geoscience Canada, Reprint Series 1.

Locker, S.D., 1981, Origin and depositional history of a semienclosed windward lagoon off Great Abaco Island, Bahamas: Unpublished M.S. Thesis, University of North Carolina, Chapel Hill, NC, 181 p.

Logan, B.W., Davies, G.F., Read, J.F., and Cebulski, D., 1970, Carbonate sedimentation and environments, Shark Bay, Western Australia: American Association of Petroleum Geologists Memoir 13, 223 pp.

Lowenstam, H.A., 1955, Aragonite needles secreted by algae and some sedimentary implications: Journal of Sedimentary Petrology, v. 25, p. 270-272.

Mallik, T.K., 1976, Grain-size variation in the Kauaratti Lagoon sediments, Lak Shadweep, Arabian Sea: Marine Geology, v. 20, p. 57-75.

Maxwell, W.G.H., Jell, J.S., and McKellar, R.G., 1964, Differentiation of carbonate sediments in the Heron Island Reef: Journal of Sedimentary Petrology, v. 34, p. 294-308.

McKee, E.D. and Ward, W.C., 1983, Eolian, in Scholle, P.A., Bebout, D.G., and Moore, C.H., eds., Carbonate Depositional Environments: American Association of Petroleum Geologists Memoir 33, p. 131-170.

McKinney, F.K., McKinney, M.J., and Listokin, M.R.A., 1987, Erect bryozoans are more than baffling: Enhanced sedimentation rate by a living unilaminate branched bryozoan and possible implications for bryozoan mudmounds: Palaios, v. 2, p. 41-47.

Miller, A.I., 1988, Spatial resolution in subfossil molluscan remains: Implications for paleobiological analyses: Paleobiology, v. 14, p. 91-103.

Moore, D.R., 1963, Distribution of the seagrass, *Thalassia*, in the United States: Bulletin of Marine Science in the Gulf and Caribbean, v. 13, p. 329-342.

Neumann, A.C. and Land, L.S., 1975, Lime mud deposition and calcareous algae in the Bight of Abaco, Bahamas: a budget: Journal of Sedimentary Petrology, v. 45, p. 763-786.

Purdy, E.G., 1963, Recent calcium carbonate facies of the Great Bahama Bank: Journal of Geology, V. 71, p. 334-355, 472-497.

Purser, B. H., ed., 1973, The Persian Gulf: Springer-Verlag, New York, 471 pp.

Rasmussen, K.A. and Neumann, A.C., 1988, Holocene overprints of Pleistocene paleokarst: Bight of Abaco, Bahamas, in James, N.P. and Choquette, P.W., eds., Paleokarst: Spring-Verlag, New York, 132-148.

Read, J.F., 1984, Carbonate platform facies models: American Association of Petroleum Geologists Bulletin, v. 69, p. 1-21.

Reddering, J.S.V., 1987, Subtidal occurances of ladderback ripples: their significance in paleo-environmental reconstruction: Sedimentology, v. 34, p. 253-257.

Schlager, W., 1981, The paradox of drowned reefs and carbonate platforms. Bull. Geol. Soc. Amer., 92, 197-211.

Scoffin, T.P., 1970a, The trapping and binding of subtidal carbonate sediments by marine vegetation in Bimini Lagoon, Bahamas: Journal of Sedimentary Petrology, v. 40, p. 249-273.

Scoffin, T.P., 1970b. A conglomeratic beach rock in Bimini, Bahamas: Journal of Sedimentary Petrology, v. 40, p. 756-759.

Scoffin, T.P., and Tudhope, A.W., 1985, Sedimentary environments of the central region of the Great Barrier Reef of Australia: Coral Reefs, v. 4, p. 81-93.

Shinn, E.A., 1969, Submarine cementation in the Persian Gulf: Sedimentology, v. 12, p. 109-144.

Shinn, E.A., Lloyd, R.M., and Ginsburg, R.N., 1969, Anatomy of a modern carbonate tidal flat, Andros Island, Bahamas: Journal of Sedimentary Petrology, v. 39, p. 1201-1228.

Stoddard, D.R. and Cann, J.R., 1965, Nature and origin of beach rock: Journal of Sedimentary Petrology, v. 35, p. 243-247.

Swinchatt, J.P., 1965, Significance of constitutent composition, texture, and skeletal breakdown in some recent carbonate marine sediments: Journal of Sedimentary Petrology, v. 35, p. 71-90.

Taylor, J.D. and Lewis, M.S., 1970, The flora, fauna, and sediments of the marine grass beds of Mahe Seychelles: Journal of Natural History, v. 4, p. 199-220.

Trumbel, W. and Neumann, A.C., 1986, The depositional history of an open windward marginal lagoon, northern Great Bahama Bank: Society of Economic Paleontologists and Mineralogists Annual Midyear Meeting Abstracts, v. 3, p. 110-111.

Wilson, J.L. and Jordan, C., 1983, Middle shelf, in Scholle, P.A., Bebout, D.G., and Moore, C.H., eds., Carbonate Depositional Environments:

American Association of Petroleum Geologists Memoir 33, p. 267-296.

Winter, J., 1987, Speculations on prehistoric coastal topography of the Bahamas, in Curran, H.A., ed., Proceedings of the Third Symposium on the Geology of the Bahamas: Fort Lauderdale, FL, CCFL Bahamian Field Station, p. 205-214.

MODERN CARBONATE SEDIMENT PRODUCTION AND ITS RELATION TO BOTTOM VARIABILITY GRAHAMS HARBOR, SAN SALVADOR, BAHAMAS

Michael E. Armstrong and Arnold I. Miller
Department of Geology (ML 13)
University of Cincinnati
Cincinnati, Ohio 45221

ABSTRACT

In this study, estimates are provided of carbonate production by epibionts on *Thalassia* and five genera of calcareous green algae (*Penicillus, Halimeda, Rhipocephalus, Udotea, Acetabularia*), for a nearshore portion of Graham's Harbor, San Salvador, with emphasis on spatial variability in production associated with bottom variability in carbonate producers.

Field work was conducted during January, 1988 utilizing 11 randomly-spaced north-south transects. Each transect consisted of four stations at 10 meter intervals. At each of the 44 stations, bottom vegetation was censused by counting seagrass blades (*Thalassia, Syringodium, Halodule*) and thalli of macroaglae (*Penicillus, Halimeda, Udotea, Rhipocephalus, Acetabularia*) within a 10 x 20 cm quadrat haphazardly tossed on the substrate. *Thalassia* blades were collected from each station for subsequent laboratory analyses of epibiotic carbonate by Chittick gasometric analysis.

Preliminary results estimate the annual carbonate sediment production in the study area by calcareous green algae and epibionts on *Thalassia* to be 136 $g/m^2/y$. Principle contributors to carbonate sediment accumulation include *Halimeda* (58 $g/m^2/y$), *Penicillus* (38 $g/m^2/y$), and epibionts on *Thalassia* (28 $g/m^2/y$). Bottom cover within Graham's Harbor is patchy on a small scale, and the bottom variability results in a highly variable rate of carbonate production.

INTRODUCTION

Purpose

Previous studies of carbonate sediment production associated with seagrass ecosystems have generally been limited in both their treatment of spatial variability in bottom composition and the diversity of carbonate producers evaluated. The principal contribution of this study is to provide estimates of carbonate production by epibionts on *Thalassia*, and five genera of calcareous green algae (*Penicillus, Halimeda, Rhipocephalus, Udotea, Acetabularia*), for a nearshore portion of Graham's Harbor, San Salvador. Emphasis will be given to the relationship between carbonate production rates and the variable nature (i.e., spatial patchiness) of the bottom.

Background

Substantial research has been conducted on the productivity of modern carbonate banks. Lowenstem (1955) was among the first to propose an allochemical origin for lime muds, mainly through the post-mortem disintegration of lightly calcified green and red algae. The first rigorous attempt to quantify the contribution of algal sources to Holocene carbonates was provided by Stockman *et. al.* (1967), who concluded that production by the green alga *Penicillus* was by itself more than adequate to account for the amount of Holocene lime mud observed in south Florida.

Neumann and Land's (1975) study in the Bight of Abaco introduced the concept of a modern "carbonate budget". They found that all of the Holocene lime mud presently within the basin could be derived from four common algal genera (*Penicillus, Halimeda, Rhipocephalus*, and *Udotea*), and, further, that there is a significant overproduction of sediment by these sources relative to the amount of material presently contained in the Bight of Abaco Basin.

Algal studies stimulated research on the relative contribution of other calcified plants and animals to Holocene sediments. Land (1970) undertook a study in Discovery Bay, Jamaica, in which he quantified the sediments produced from the post-mortem decay of *Thalassia* blades, and the subsequent release of epibionts (mainly red algae and serpulid worms) to produce lime mud. Patriquin (1972) established a similar estimate for a *Thalassia* meadow in Barbados. Nelson and Ginsburg (1986) quantified lime mud production by *Thalassia* epibionts in Florida Bay (same study area as Stockman *et. al.*, 1967). Each of these studies indicated that production by epibionts on *Thalassia* may equal or exceed that of calcareous green algae as a source of fine carbonate.

Numerous other investigations have attested to the significance of calcareous green algae as producers of carbonate sediment. Bach (1979) studied the carbonate sediment production of four calcareous green algae genera, (*Halimeda, Penicillus, Rhipocephalus*, and *Udotea*), in Card Sound, south Florida. Wefer (1980) estimated production by the green algae *Halimeda* and *Penicillus*, as well as, the brown alga *Padina*, in Harrington Sound, Bermuda. Bosence *et. al.* (1985) studied a modern carbonate mud mound in the Florida Keys, and established sediment production estimates for the green algae *Halimeda, Penicillus*, and *Acetabularia*, as well as for epibionts on *Thalassia*, molluscs, and two genera of coral.

Despite the magnitude of this earlier work, small-scale variations in benthic environments, and potentially associated variations in carbonate production rates, have often been ignored. Thus, production rates calculated for a single location, or for widely spaced localities, are often extrapolated to perhaps unrealistically large areas of the seafloor. In contrast, this study establishes carbonate production rates for both green algal and epibiotic carbonate sources for a relatively confined study area, while considering the small-scale variable nature (i.e., spatial patchiness) of the bottom.

STUDY AREA

Graham's Harbor is a windward, high energy lagoon, located at the northeastern end of San Salvador (Fig. 1). This shallow basin is bounded by San Salvador to the south, and North Point to the east. A barrier reef protects the lagoon along its northern margin, and it is open to deeper waters to the west. The study area incorporates a nearshore *Thalassia*-dominated environment that covers approximately 2 x 10^5 m^2. Maximum water depth within the study area is 3 m; Colby and Boardman (1988) have determined

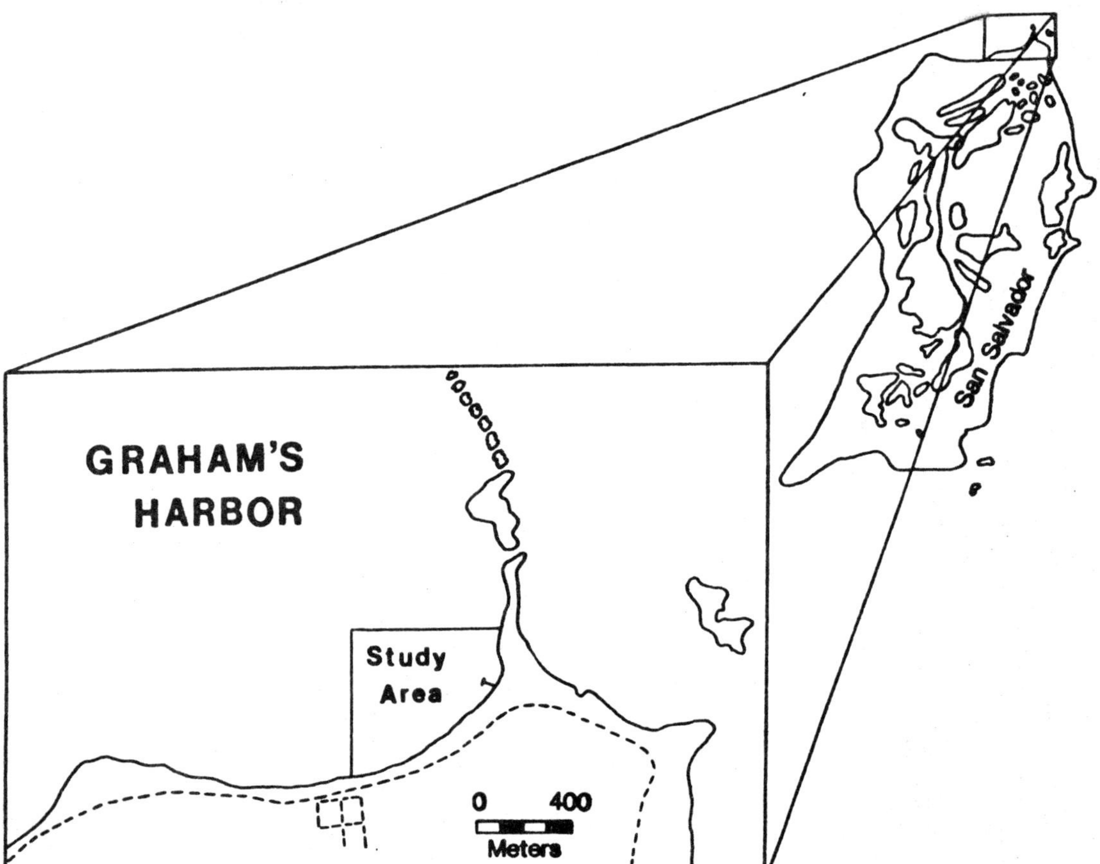

Fig. 1. Map of Graham's Harbor, located at the northeastern end of San Salvador; note the limits of the study area within Graham's Harbor.

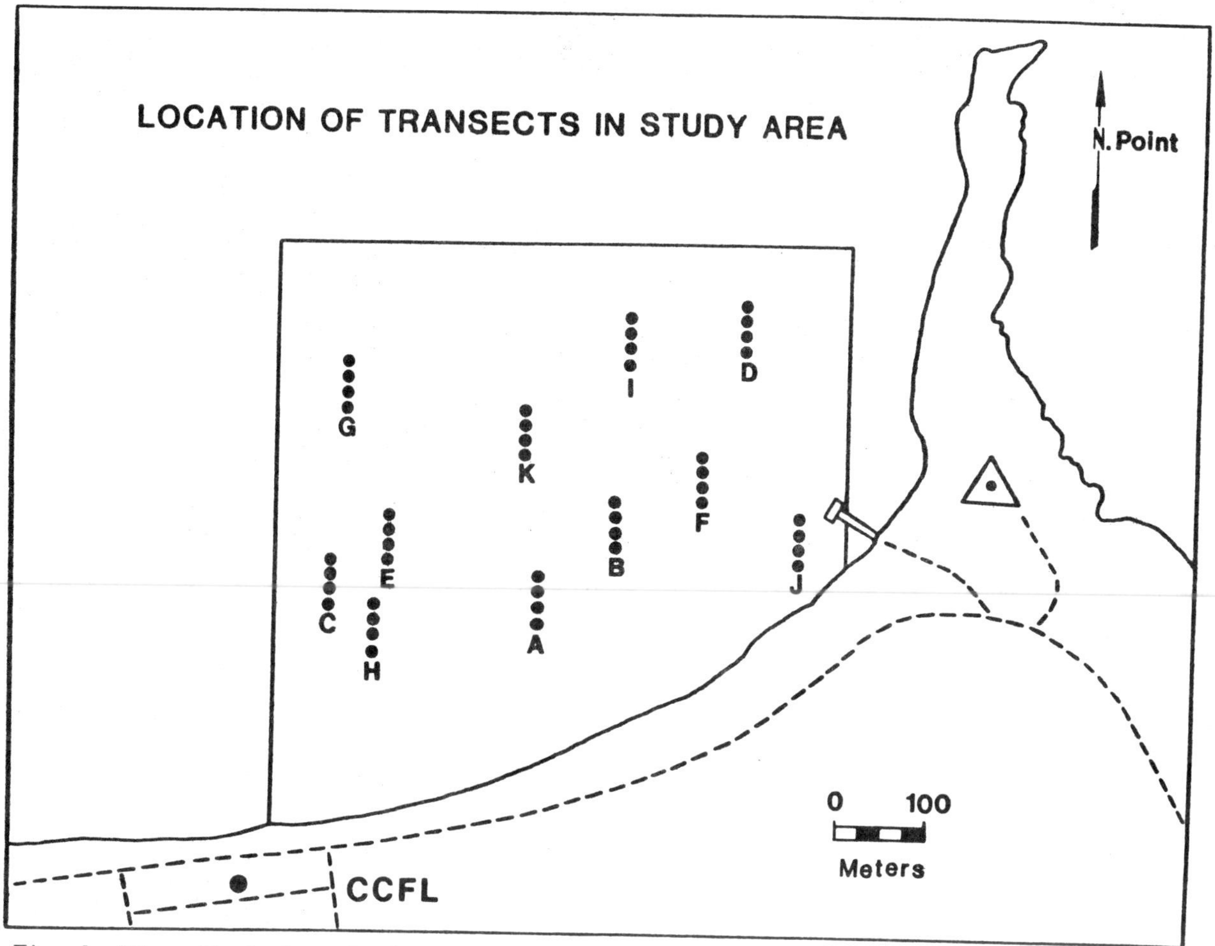

Fig. 2. Map displaying the location of transects within the study area. Transects are oriented north-south, and spaced to sample the entire study area.

the maximum water depth within the remainder of Graham's Harbor to be about 6 meters.

Colby and Boardman (1988) have reported on the texture, mineralogy, and grain types of Holocene sediments within Graham's Harbor. Sediment thickness reportedly varies between 1.5 and 4 m, and studies of sediment cores have indicated a coarsening upward sequence for these deposits.

Peat samples recovered from immediately above the Pleistocene bedrock have been radiocarbon dated and indicate initial flooding of Graham's Harbor approximately 5000 y.b.p. Colby and Boardman (1988) have used sediment thickness, along with peat dates, to estimate a Holocene rate of sediment accumulation of between 30 and 80 cm/1000 years.

METHODS

Field Methods

Field work was conducted during January, 1988. A second field season was carried out during June, 1988 to assess the impact of seasonality on epibiont and macroalgal sediment production; results of this summer field work are not included in this report.

Within the study area, censusing of macroflora was accomplished utilizing 11 transects oriented north-south; transects were spaced so that the entire study area was covered (See Fig. 2). Each transect consisted of four stations at 10 meter intervals. At each station, a 10 x 20 cm. quadrat was haphazardly tossed on the substrate, and bottom vegetation was censused by counting grass blades (*Thalassia, Syringodium, Halodule*) and thalli of macroalgae (*Penicillus, Halimeda, Udotea, Rhipocephalus*, and *Acetabularia*) within this quadrat. This procedure was repeated eight times at each station to determine station-level variations in abundance.

Approximately 60 *Thalassia* blades were collected from each station by clipping the blades at the sediment-water interface. Thereafter,

Thalassia blades were dried and bagged for subsequent laboratory analysis of epibiotic carbonate.

Laboratory Analysis

Laboratory analysis of epibiotic carbonate on *Thalassia* blades was accomplished by CO_2 evulsion in a Chittick gasometric apparatus (Dremanis, 1962). This apparatus measures the volume of CO_2 evolved by reacting grass blades with 6M HCl. Using weighed quantities of standard $CaCO_3$, the apparatus was calibrated based on the volume of $CaCO_3$ evolved. A correlation coefficient (r) between $CaCO_3$ reacted, and CO_2 evolved, was calculated as .99, attesting to the accuracy of the method.

Production Calculations

Fig. 3 illustrates the series of steps involved in the calculation of annual carbonate sediment production from census data. The eight rectangles in the upper left corner of the diagram represent the eight values obtained by repetitive censusing at each station. Census values for each repetition are averaged, and a mean for the standing crop of each organism/station is obtained (Step 1 in Fig. 3). Annual carbonate production rates are generally reported as g $CaCO_3/m^2/y$, and the area within a 10 x 20 cm. quadrat is .02 m^2. Thus, the mean abundance of each organism at each station must be multiplied by 50 to determine the number of organisms/m^2 (Step 2 in Fig. 3).

Next, these values are multiplied by the weight (in grams) of $CaCO_3$ produced by postmortem, disintegration of green algae (Step 3a in Fig. 3), or release of epibionts from *Thalassia* blades (Step 3b in Fig. 3). Values for calcification per alga were obtained from the literature (Table 1, depicted by a periodical in Fig. 3), and for carbonate associated with epibionts by Chittick gasometric analysis (depicted by the schematic Chittick apparatus in Fig. 3).

Finally, this value must be multiplied by the estimated turnover rate for the organism being considered. The turnover rate for a particular

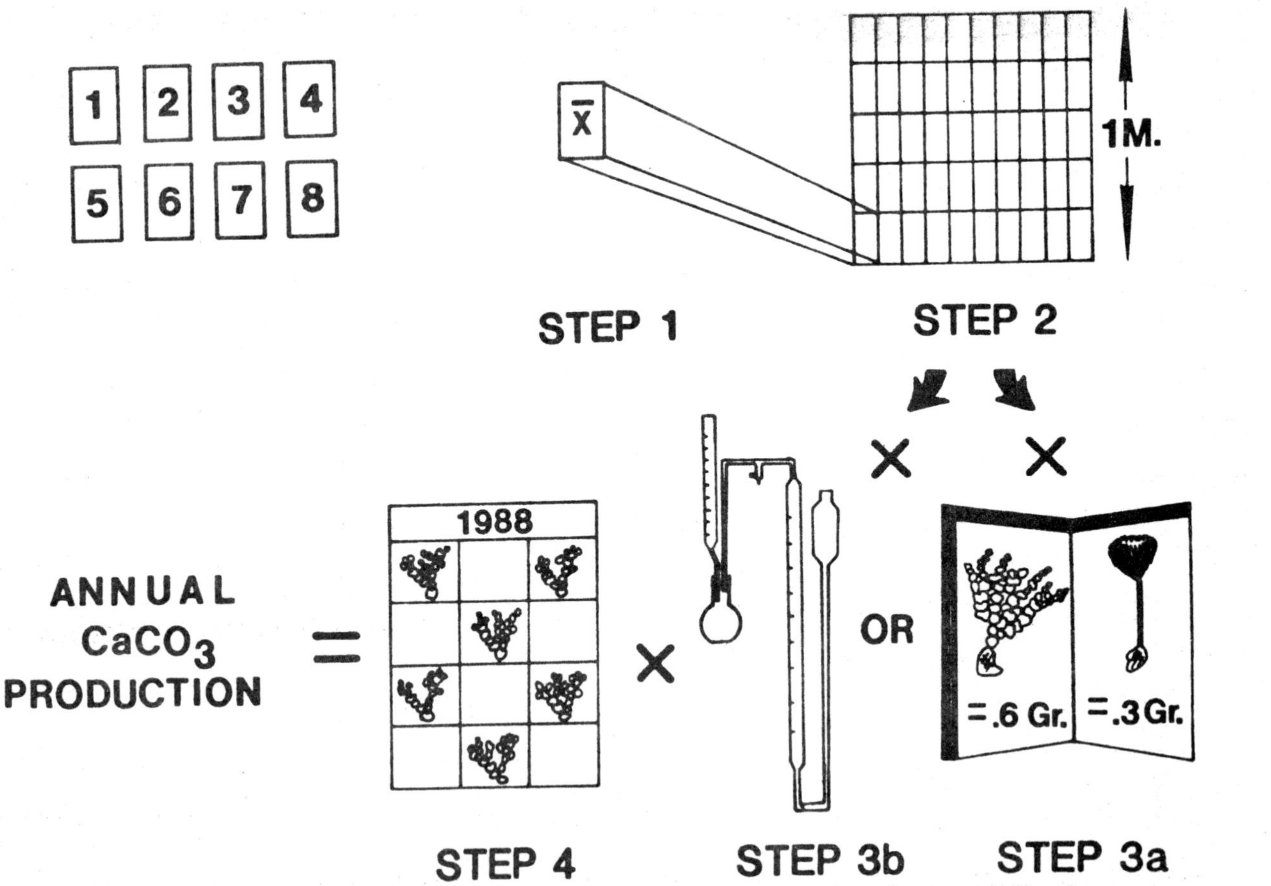

Fig. 3. Summary of steps involved in the calculation of annual carbonate production rates from census data.

Table 1: Production values for various carbonate producing organisms.

	NELSON (1986)	STOCKMAN (1967)	NEUMANN (1975)	BACH (1979)	WEFER (1980)	BOSENCE (1985)	MARSZALEK (1975)	THIS STUDY (1988)
Thalassia								
Turn Over	6 crops	---	---	---	---	---	---	6 crops
Penicillus								
Turn Over	---	6-9 crops	6-12 crops	68 days	4 crops	5 crops	---	6 crops
Calcification	---	.17-.52 g	.274 g	---	---	---	---	.274 g
Halimeda								
Turn Over	---	---	6-12 crops	---	7 crops	2 crops	---	6 crops
Calcification	---	---	.628 g	---	---	---	---	.628 g
Rhiphocephalus								
Turn Over	---	> P. cap.	6-12 crops	50 days	---	---	---	6 crops
Calcification	---	---	.413 g	---	---	---	---	.413 g
Udotea								
Turn Over	---	---	6-12 crops	87 days	---	---	---	6 crops
Calcification	---	---	.213 g	---	---	---	---	.213
Acetabularia								
Turn Over	---	---	---	---	---	---	3-5 crops	4 crops
Calcification	---	---	---	---	---	---	74 %	74%

organism is based on the number of times its standing crop is thought to replicate itself each year (Table 1; step 4 in Fig. 3). Thus, $CaCO_3$ production, for each organism at each station, is determined as follows:

$$\text{Annual } CaCO_3 = \bar{X}*50*(\text{gr.}CaCO_3/\text{organism})*\text{turnover rate}$$

RESULTS AND DISCUSSION

Standing Crops

The average abundance of seagrasses and calcareous green algae (standing crop) are summarized in Table 2. Upon casual observation of the substrate, the wider-bladed variety of seagrass, *Thalassia*, may appear to be more prevalent than *Syringodium* or *Halodule*. Census work reveals that *Syringodium* is actually the most abundant variety of seagrass, with an average abundance of 810 blades/m^2. *Thalassia* is substantially less abundant, with an average density of 516 blades/m^2. *Halodule* averaged 170 blades/m^2.

Penicillus was determined to be the most abundant genus of green algae with an average estimated abundance of 23 plants/m^2. Less common genera include *Halimeda* at 15 plants/m^2, *Acetabularia* at 6 plants/m^2, *Rhipocephalus*, and *Udotea* (both averaging 3 plants/m^2).

Table 2: Standing crops of Seagrasses and calcareous green algae, January 1988

SEAGRASSES	RANGE/M^2	AVERAGE/M^2
Syringodium	0-1587	810 +/- 124
Thalassia	0-1156	516 +/- 78
Halodule	0-806	170 +/- 53
MACROALGAE		
Penicillus	0-43	23 +/- 5
Halimeda	0-94	15 +/- 7
Acetabularia	0-43	6 +/- 4
Rhipocephalus	0-31	3 +/- 2
Udotea	0-13	3 +/- 2

Bottom Variability

Fig. 4 is a map of the study area that illustrates the relationship in floral composition of sample localities, as determined by Q-mode cluster analysis. Q-mode cluster analysis is a multivariate statistical technique, that hierarchically groups samples together on the basis of similarities in composition within a given data set.

Initially, (as seen in Fig. 4) all floral variables were considered, and the 44 stations were grouped into four clusters: *Halodule*-dominated, *Syringodium*-dominated, *Thalassia*-dominated, and Intermediate. Fig. 4 suggests some degree of non-randomness in the distribution of bottom types within the study area, in that a few transects are dominated by the same bottom type at three or four stations. However, other transects are more highly variable.

In Fig. 4 the substantially greater abundances of seagrass blades relative to macroalgae,

CLUSTER ANALYSIS MAP: ALL FLORAL DATA

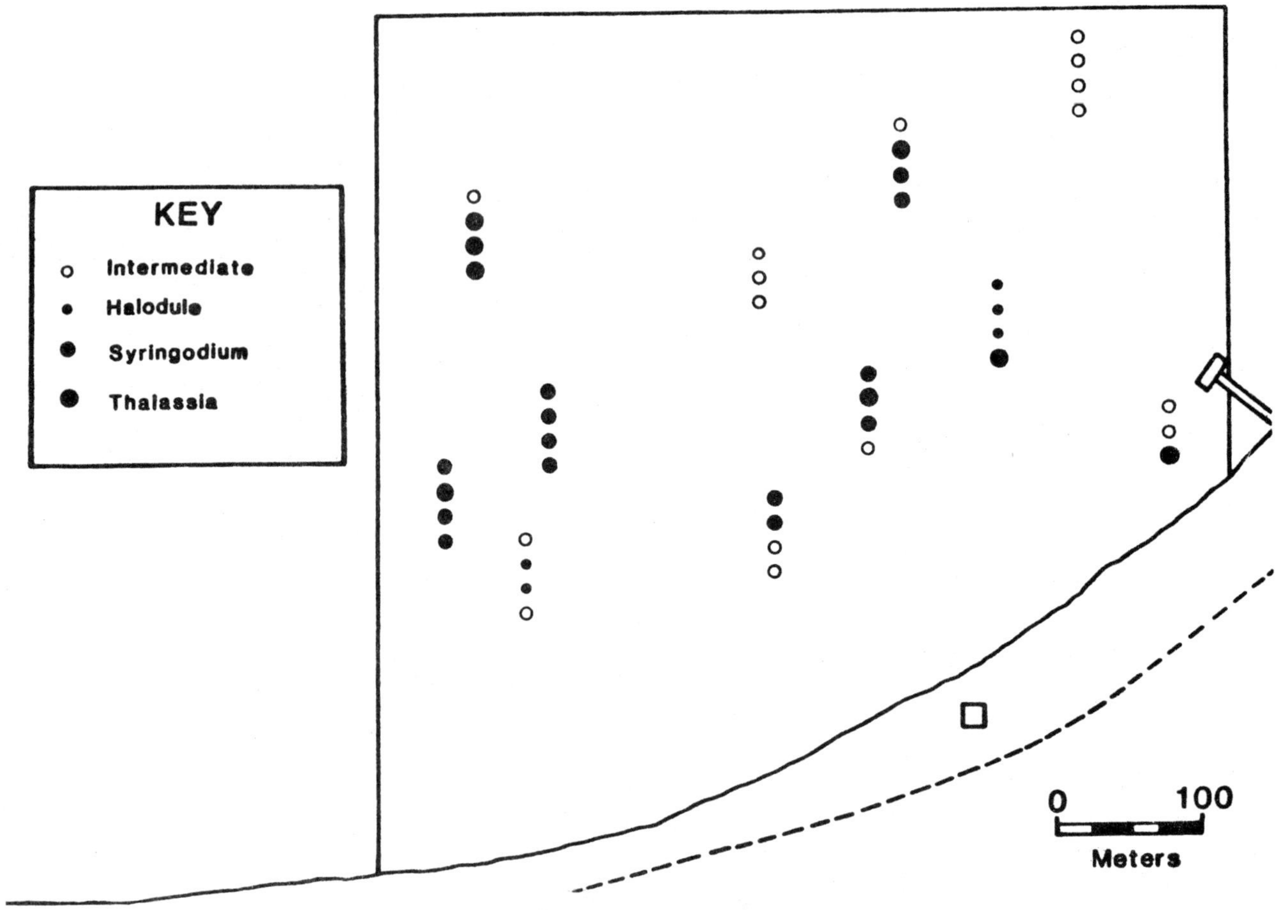

Fig 4. Distribution of census stations determined by cluster analysis to have similar faunal compositions. All floral variables are considered; blank stations denote areas of bare sand.

effectively overshadowed recognition of bottom variability in macroagal abundances. Clusters (bottom types) were determined primarily by similarities in abundances of seagrass genera. Thus, a second cluster analysis was conducted using macroalgal data only. Fig. 5 is a map showing the distribution of stations determined by Q-mode cluster analysis to have similar macroalgal bottom types. As with seagrasses, clustering of macroalgal bottom types suggests a pattern that is not entirely random. Certain macroalgal bottom types persist on the scale of 2 or 3 stations, but more generally, the distribution shows little or no pattern.

Production Rates

Epibionts on *Thalassia*

The annual production associated with epibionts on *Thalassia* has been calculated using data for standing crop (from census work), epibionts per blade of *Thalassia* (from Chittick analysis), and turnover rates (from literature), to be 28 g $CaCO_3/m^2/y$, with a range at individual stations of 0 to 68 g $CaCO_3/m^2/y$. Production varies substantially even on the scale of individual transects (40 meters), and there do not appear to be any onshore-offshore or east-west gradients in epibiont production.

Table 3 compares epibiont production in Graham's Harbor to that reported for other modern carbonate banks. The epibiont production estimated for January 1988 in Graham's Harbor is approximately five times less than that reported by Land (1970) in Discovery Bay, Jamaica, approximately three times less than that reported by Nelson and Ginsburg (1986) (for areas of similar *Thalassia* density) in Florida Bay, and two orders of magnitude less than that reported by Patriquin (1971) for Barbados.

When samples collected in June are analyzed, production values may be more in line with

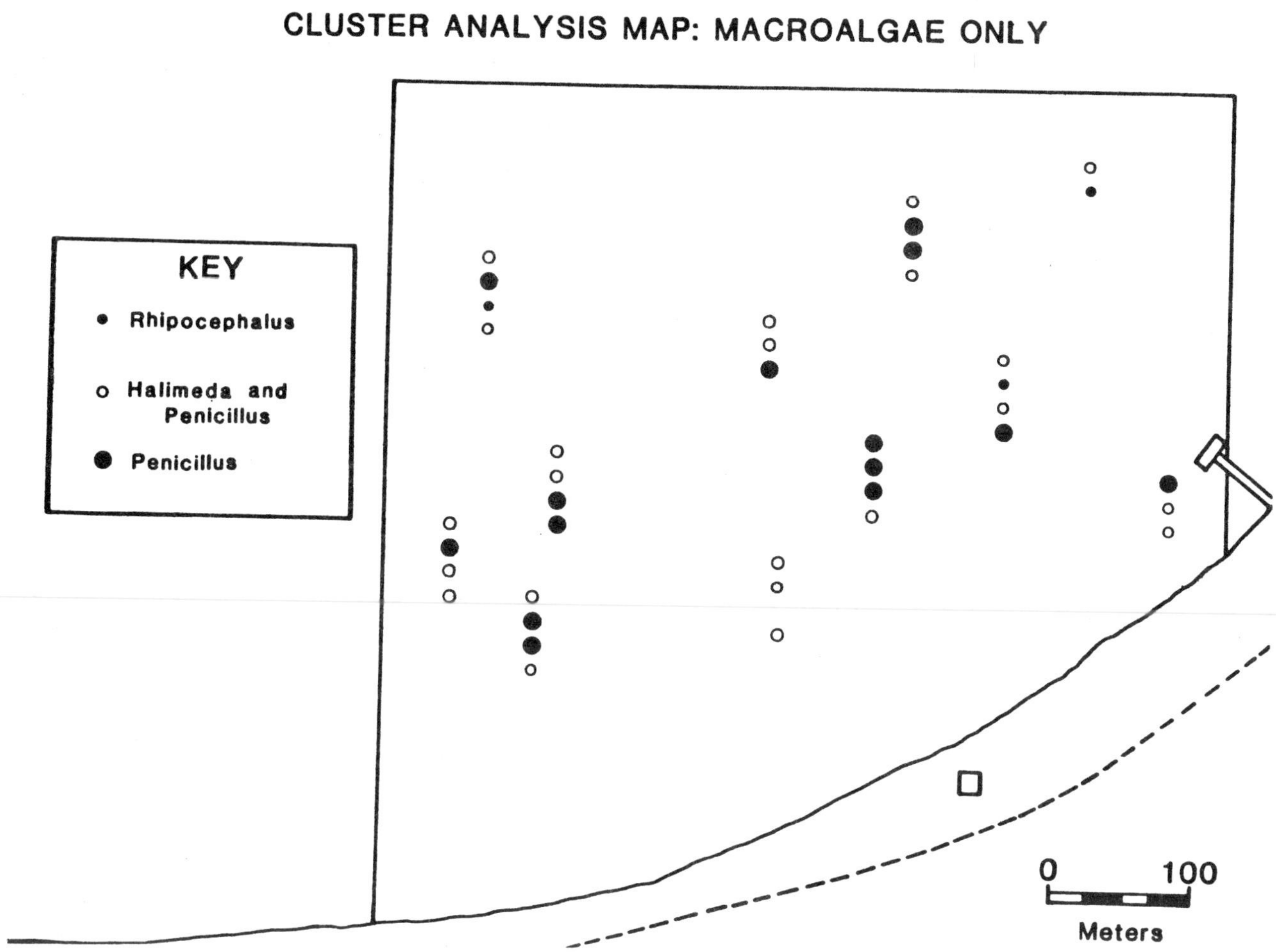

Fig. 5. Distribution of census stations determined by cluster analysis to have similar faunal compositions. Only green algal genera considered; blank stations denote areas with no algal plants.

others reported in the literature. However, preliminary results suggest that production will increase only slightly above that estimated with just the January data. The disparities between epibiont production rates reported here and those of previous authors may be the consequences of differences in *Thalassia* (eg. Land, 1970), lagoonal circulation and energy (eg. Nelson and Ginsburg, 1986) and methodology (Patriquin, 1971).

Green Algae

The annual production associated with various genera of green algae has been calculated using data for abundance (from census work), data for weight of carbonate/plant (from the literature), and turnover rates (from the literature). Although the abundance of *Halimeda* was found to be less than that of *Penicillus*, carbonate production associated with *Halimeda* was greater (58

Table 3: Comparison of reported carbonate production rates associated with epibionts on Thalassia.

AUTHOR	LOCATION	$g/CaCO_3/m^2/y$
Patriquin (1972)	Barbados	2800
Land (1970)	Jamaica	180
Nelson & Ginsburg (1986)	Florida Bay (Entire Field Area)	118
	Florida Bay (0-1000 Blades/m^2)	73
** This Study (1988)	Graham's Harbor	28 +/- 6

g $CaCO_3/m^2/y$). This discrepancy results from the greater mass of skeletal carbonate associated with *Halimeda*, compared to *Penicillus*.

Production of sediment associated with *Penicillus* averaged 38 $g/m^2/y$ across the study area. Other green algal genera contributing to annual sediment production included *Rhipocephalus* (6 g $CaCO_3/m^2/y$), *Udotea* (3 g $CaCO_3/m^2/y$), and *Acetabularia* (1 g $CaCO_3/m^2/y$). At most stations these last three genera were absent or rarely present, but occasionally *Rhipocephalus* and *Udotea* were significant contributors, with production rates as high as 60 g $CaC0_3/m^2/y$, and 26 g $aC0_3/m^2/y$, respectively. Marszalek (1975) reported that periodic blooms of *Acetabularia* contribute significantly to local sediment accumulation; no *Acetabularia* blooms were observed during field work, and thus, production by this genus was minimal.

Table 4 compares the carbonate production associated with calcareous green algae to that reported by previous authors. Production associated with *Halimeda, Penicillus, Rhipocephalus*, and *Udotea* is comparable or sightly larger than that reported by Neumann and Land (1975) in the Bight of Abaco, and within the range of values reported for the Florida Keys and other areas.

Total Carbonate Production

The total carbonate sediment production in Graham's Harbor is estimated to be 136 g $CaCO_3/m^2/y$, (detailed in Table 5). It is recognized that this value may be subject to seasonal fluctuations (Bach, 1979; Wefer, 1980). Such variability will ultimately be accounted for using data collected during the summer of 1988.

Fig. 6 is a map showing the variability of total carbonate production rates within the study area. This figure conveys the high degree of spatial variability in carbonate production which can exist even between such closely-spaced census localities. Thus, for *Thalassia*-dominated lagoonal environments, as well as other localities and depositional environments exhibiting a high degree of spatial variability in carbonate producers, it is suggested that census data for single stations not be extrapolated to extensive areas of that environment.

How does total carbonate production in Graham's Harbor compare to that of other localities and environments of modern carbonate deposition? Direct comparison is difficult; Neumann and Land did not consider epibiont production in the Bight of Abaco, and even with the combination of production estimates from Stockman et. al.

Table 4: Comparison of reported carbonate production rates associated with calcareous green algae.

GENUS/AUTHOR	LOCATION	g $CaCO_3/m^2/y$
Halimeda		
Bosence (1985)	Florida Keys	105
** This Study (1988)	Graham's Harbor	58 +/- 25
Neumann & Land (1975)	Bight of Abaco	24-48
Bach (1979)	South Florida	2-11
Penicillus		
Bosence (1985)	Florida Keys	135
** This Study (1988)	Graham's Harbor	38 +/- 8
Neumann & Land (1975)	Bight of Abaco	22-44
Wefer (1980)	Bermuda	30
Stockman & Ginsburg (1967)	Inner Reef Tract	25
	Florida Bay	3
Bach (1979)	South Florida	1-3
Rhipocephalus		
** This Study (1988)	Graham's Harbor	8 +/- 6
Neumann & Land (1975)	Bight of Abaco	2-3
Bach (1979)	South Florida	1-2
Udotea		
Bach (1979)	South Florida	.1-6
Neumann and Land (1975)	Bight of Abaco	2-4
** This Study (1988)	Graham's Harbor	3 +/- 2
Acetabularia		
Marszalek (1975)	Florida Keys	720
** This Study (1988)	Graham's Harbor	0.5 +/- .3

Table 5: Total carbonate sediment production, Graham's Harbor, January, 1988

SOURCE	$g\ CaCO_3/m^2/y$
Halimeda	58
Penicillus	38
Epibionts on Thalassia	28
Rhipocephalus	8
Udotea	3
Acetabularia	1
Total	136

(1967), and Nelson and Ginsburg (1986), data are still lacking for *Halimeda* and other major green algal sources in Florida Bay. Clearly, more study is needed to see how total carbonate sediment rates vary with locality, space, and time.

Bosence (1985) has produced one of the most comprehensive studies of carbonate sediment production to date. He estimated sediment production for all major producing organisms on Tavernier mound (a modern carbonate mound in the Florida Keys) to be 500 $g/m^2/y$. While considerably higher than the value calculated in the present study, such a high rate might be expected because carbonate mounds occupy the end spectrum of reefs, an environment known to be populated by exceedingly high carbonate producers (Stern, *et. al.*, 1977).

CONCLUSIONS

Annual carbonate production by calcareous green algae and epibionts associated with *Thalassia* is estimated at 136 $g/m^2/y$ for a 3.6×10^5 m^2 area of Graham's Harbor. Principle contributors to carbonate sediment accumulation include *Halimeda* (58 $g/m^2/y$), *Penicillus* (38 $g/m^2/y$), and epibionts on *Thalassia* (28 $g/m^2/y$).

Comparison with previous studies suggests

TOTAL CARBONATE SEDIMENT PRODUCTION

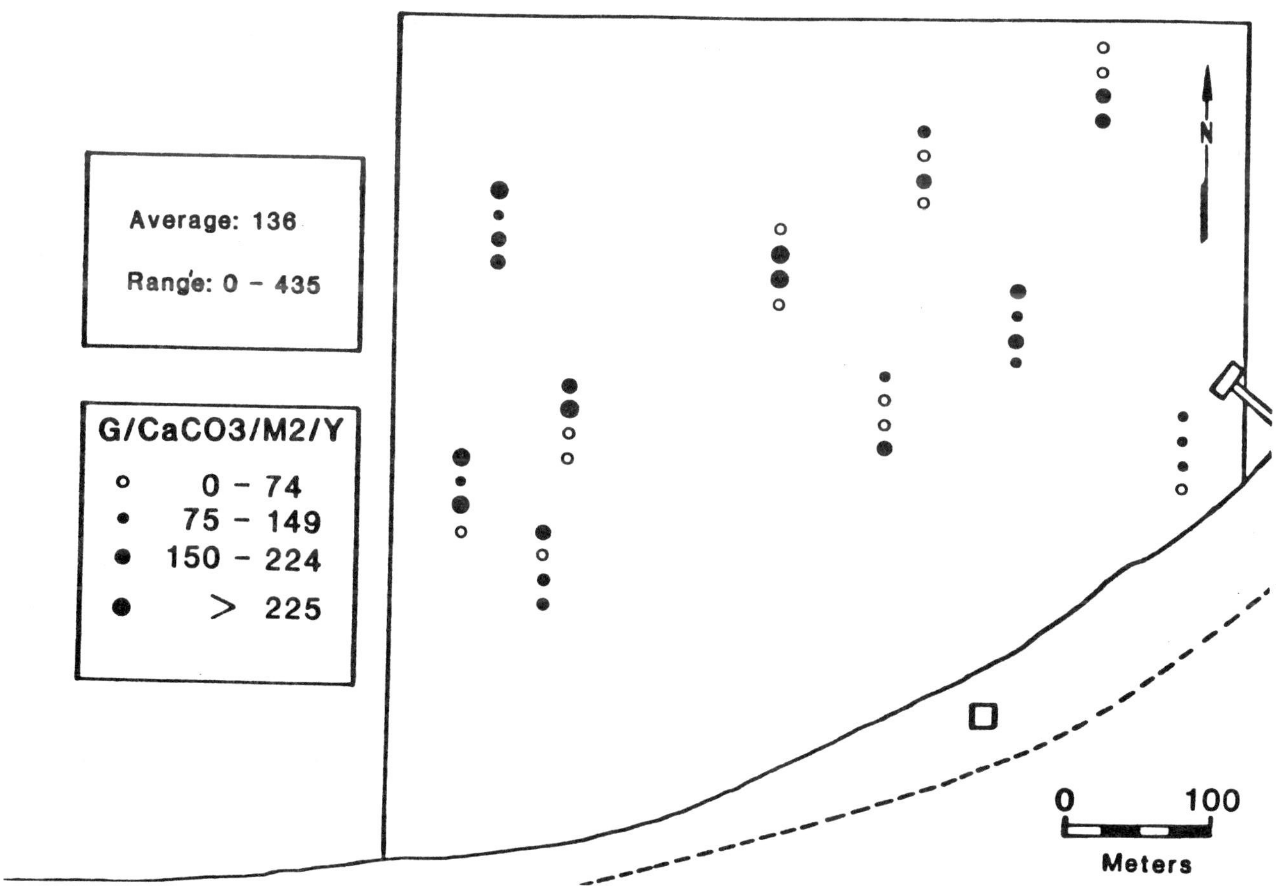

Fig. 6. Map of the study area which illustrates the variability of total carbonate production within the study area. Transect A is not represented because values for epibionts on Thalassia *were not obtained.*

that carbonate production associated with epibionts on *Thalassia* is approximately three times less than that reported for similar *Thalassia*-covered bottoms in Florida Bay, and is substantially less than those values reported for Jamaica, and Barbados. Carbonate production associated with various macroalgae is comparable to sightly higher than that reported in the Bight of Abaco, and within the range of values reported for other areas.

Substantial insight into the production variability within a modern lagoonal environment can be determined by closely spaced sampling. Bottom cover within Graham's Harbor is patchy on a small scale, and the bottom variability results in a highly variable rate of carbonate production. This suggests that production rates determined at individual sampling and census sites should not be extrapolated far beyond the bounds of such stations.

ACKNOWLEDGEMENTS

We are indebted to Don Gerace for providing the research facilities and accommodations which made this study possible, including the resources of his equipment room, the exploitation of his "John-boat", and the patience of he and his wife Kathy. We thank Mark Boardman for his continued interest in this study, and for his consultation on many matters. Our colleague Sue Gaffney provided data on the average dry weight of *Acetabularia*. We also thank all those divers who "volunteered" their time in the sun. This study was partially supported by the Kenneth E. Caster Fund, Department of Geology, University of Cincinnati.

REFERENCES CITED

Bach, S.D., 1979, Standing crop, growth and production of calcareous Siphonales (Chlorophyta) in a South Florida lagoon: Bulletin of Marine Science, v. 29, p. 191-201.

Bosence, D.W.J., Royston, J.R., and Quine, M.L., 1985, Sedimentology and budget of a Recent Carbonate Mound, Florida Keys: Sedimentology, v. 33, p. 317-343.

Colby, N.D., and Boardman, M.R., 1988, Holocene deposition in a high-energy carbonate lagoon: Graham's Harbor, San Salvador, Bahamas: Abstracts and Programs for the Fourth Symposium on the Geology of the Bahamas.

Dremanis, A, 1962, Quantitative gasometric determination of calcite and dolomite by using chittick apparatus: Journal of Sedimentary Petrology, v. 32, p. 520-529.

Land, L.S., 1970, Carbonate mud: production by epibiont growth on *Thalassia testudinum*: Journal of Sedimentary. Petrology. v. 40, p. 1361- 1363.

Lowenstam, H.A., 1955, Aragonite needles secreted by algae and some sedimentary implications: Journal of Sedimentary Petrology, v. 25, p. 270-272.

Marszalek, D.S., 1975, Calcisphere ultrastructure and skeletal aragonite from the alga *Acetabularia antillarium*: Journal of Sedimentary Petrology, v. 45, p. 266-271.

Nelson, J.E., and Ginsburg, R.N., 1986, Calcium carbonate production by epibionts on *Thalassia* in Florida Bay: Journal of Sedimentary Petrology, v. 56, p. 622-628.

Neumann and Land, L.S., 1975, Lime mud deposition and calcareous green algae in the Bright of Abaco, Bahamas: a budget: Journal of Sedimentary Petrology, v. 45, p. 763-786.

Patriquin, D.G., 1972, Carbonate mud production by epibionts on *Thalassia*: an estimate based on leaf growth rate data: Journal of Sedimentary Petrology, v. 42, p. 344-366.

Stern, C.W., Scoffin, T.P., and Martindal W., 1977, Calcium carbonate budget of a fringing reef on the west coast of Barbados: Bulletin of Marine Science, v. 27. p. 479-510.

Stockman, K.W., Ginsburg, R.N., and Shinn E.A., 1967, The production of lime mud by algae in South Florida: Journal of Sedimentary Petrology, v. 37, p. 633-648.

Wefer, G., 1980, Carbonate production by algae *Halimeda, Penicillus*, and *Padina*: Nature, v. 285, p. 323-324.

EXPOSED BEACHROCK: ITS INFLUENCE ON BEACH PROCESSES AND CRITERIA FOR RECOGNITION

Roger J. Bain
Department of Geology
University of Akron
Akron, OH 44325

ABSTRACT

Beachrock forms where sediment is cemented within the beach deposit beneath a thin cover of sediment. In this buried state, sand no longer experiences transportation. Carbonate-rich water from adjacent land migrates through and cements these sands. Short-term erosional conditions along a coast exhume this lithified sand and result in the presence of exposed beachrock.

Beachrock modifies the beach environment by providing a rocky-bottom near shore where previously sandy sediment existed. This change in substrate causes a change in the shorezone population from one that preferred a sandy bottom to one which prefers a firm substrate. As a result encrusting and boring organisms invade and leave their skeletons in the accumulating beach sediment.

As a result of the lithified buttress which beachrock provides, the beach is somewhat more protected from erosion, however water depths increase more rapidly offshore because unlithified subtidal sands are eroded. Wave energy therefore decreases more drastically over the beachrock resulting in an accumulation on the beach of less sorted, coarser bioclastic material which is derived from the rocky-bottom population. Clasts of beachrock likewise become incorporated in beach sediment.

Criteria for recognizing the presence of exposed beachrock in a beach deposit include: 1) occurrence of rocky bottom fauna and flora, 2) low profile growth forms of larger fauna such as coral, 3) micritized bioclastic texture, 4) recognition of marine cement, 5) borings into beachrock, 6) increase in grain size, 7) decrease in sorting, 8) inclusion of chaotic blocks of beachrock, and 9) possible darkening of beachrock surface. These features must occur within a beach-subtidal sequence of deposits otherwise they might represent merely a marine hardground.

INTRODUCTION

Beachrock was first described more than 80 years ago (Branner, 1904) and since then innumerable papers have described beachrock in various settings from around the world. While beach sediment can be cemented rapidly under a wide range of conditions, beachrock most commonly occurs along carbonate coasts in tropical and subtropical climates. Most papers published in the past 50 years have suggested mechanisms for rapid cementation of beach sediment. Bathurst (1975) reviews proposed mechanisms for beachrock formation. Scoffin (1987) provides a brief but more up-to-date review of mechanisms but overemphasizes the importance of unusual or less common occurrences of beachrock and also fails to recognize that intertidal beachrock is commonly associated with supratidal-cemented sediment which Kuenen (1933) referred to as "cay sandstone". It is indeed common to find beachrock cemented by aragonite and/or Mg-calcite as Scoffin indicates but it is also common in tropical-subtropical carbonate beach deposits to find associated coastal deposits cemented by sparry calcite. It is this author's opinion that the mechanisms described by Moore & Hanor (1974) and Hanor (1978) adequately explain such cementation. Hanor (1978) and Moore & Hanor (1974) suggest that meteoric water which resides in landward carbonate deposits dissolves aragonitic carbonates. Upon migration seaward, possibly due to recharge of groundwater, this carbonate-rich water precipitates $CaCO_3$ within shallow-buried beach and lower dune sediment, either as a result of mixing water (Moore & Hanor, 1974) or degassing of CO_2 (Hanor, 1978). The purpose of this paper is not to explain mechanisms of cementation but rather to analyze the importance of beachrock in terms of beach processes, beach deposits and the interpretation of beach deposits in the geologic record.

In order to appreciate the role of beachrock one must first recognize that beachrock does not form in the setting in which it is generally observed. The exposed nature of beachrock is secondary to its original creation. Beachrock forms when beach sediment, generally medium- to coarse-grained carbonate sand, is cemented in the shallow subsurface (<1 meter). Cementation occurs beneath a thin cover of sediment where it is not actively being transported or reworked. The zone of cementation extends from subtidal to lower dune. In a matter of a few years inactive beach sediment can be cemented sufficiently such that upon exhumation it resists erosion by normal beach energy. Until such time it is exhumed, it does not represent beachrock, only rapidly cemented beach sediment. Its impact on beach processes and beach deposition occurs only upon exposure.

Removal of the thin cover of sediment thus exposing weakly cemented beachrock initiates modification of the beach rock. If cementation is insufficient to withstand erosive forces of the beach environment there will be little effect and no record of this early cementation. If, on the other hand, cementation is sufficient to hold sand grains in place, exposure can lead to numerous changes in the beachrock, the associated beach environment and beach deposit.

Beachrock exposed to sea water undergoes several modifications which readily distinguish it from beach sediment which was cemented but never experienced later exposure. Exposed beachrock is hardened by micritization (Bathurst, 1975). Micritization is the result of biochemical activity of organisms, fungi and algae in particular, and possible physio-chemical precipitation of additional micritic cement. The hard substrate of beachrock provides a surface which is inhabited quickly by encrusting and boring organisms. Probably as a result of endolithic algae and fungi the exposed beachrock is darkened to a depth of several millimeters. While beachrock may gain additional material during exposure as a result of encrustation and additional cementation, erosion dominates in the total modification picture and, in general, beachrock volume decreases. Bioeroders bore, scrape and bite to reduce the volume of rock but, to even a greater degree, physical forces erode and transport beachrock.

Exposure of beachrock also results in modification of the beach environment in terms of substrate, biota, sediment grain size and composition, and shoreline progradation rates. Prior to the exposure of beachrock the beach/shore environment consists of a sandy substrate. With removal of the thin sediment cover and exposure of the semi-lithified beachrock, a hard substrate is present inviting an invasion by a new population. The sandy bottom likely would be dominated by burrowing organisms whereas the hard substrate is populated by encrusting coral, barnacles, algae, fungi, pelecypods, gastropods and worms. Boring organisms such as barnacles, pelecypods, worms, sponges, algae and fungi as well as grazing gastropods, echinoids and chiton also inhabit the rocky bottom. The presence of these organisms in turn modifies the composition of sediment produced locally since their skeletons will be either encrusted onto the beachrock or incorporated into associated sediment. As a result of the incorporation of large skeletal fragments and large clasts of beachrock into beach sediment, grain size increases and the degree of sorting decreases.

On a larger scale shoreline progradation rates may also be altered in favor of more rapid shoreline accretion. Where beachrock is exposed by temporary erosional conditions, the beach will be somewhat protected and shoreline erosion will be less than on a beach that is not protected by exposed beachrock. Therefore when depositional conditions return the net effect will be a more rapid seaward advance of beach sediment.

Failure to recognize beachrock in a sedimentary sequence could lead to the misinterpretation of conditions and timing of deposition as suggested by Strasser and Davaud (1986). A rocky substrate provided by beachrock and associated change in biota could be misinterpreted as a major change in water depth and energy. The textural and stratigraphic features could be mistaken for a record of sea level change, a diastem or perhaps an unconformity.

BEACHROCK ON SAN SALVADOR

General

Exposures of beachrock are common along the coasts of San Salvador. Exposures along the shore of Grahams Harbour near the CCFL Field Station and near the Government Dock on French Bay on the southwest shore were studied in detail. Both areas offer not only beachrock as described by other workers (ie. cemented intertidal sediment) but also the "cay sandstone" of Kuenen (1933) where sediment of the upper beach and lower dune is cemented.

At French Bay, depending on existing conditions, beachrock may be exposed from subtidal to lower dune or it may be nearly entirely buried. Conditions vary tremendously such that sediment cover changes daily. Generally cemented supratidal beach sediment is covered by less then .5 meters of sand and weakly cemented dune sediment is exposed at several locations 200 meters west of the dock (Fig. 1). Subtidal and intertidal beachrock is exposed for a distance of several hundred meters west of the dock (Fig. 2). The intertidal beachrock of French Bay is a textbook example. It consists of three sets of laminated carbonate sandstone beds which lie perfectly parallel to the present beach face, dip seaward at the same angle and consist of sediment nearly identical to modern beach sand. Fenestral porosity indicative of intertidal deposition is common (Fig. 3). While laminae are present in beachrock from subtidal to supratidal (Fig. 4), the lithified dune deposits generally lack laminae perhaps as a result of root action or bioturbation (Fig. 1).

At Grahams Harbour directly north of CCFL beachrock is widely exposed and lithified sediment is continuous from subtidal to supratidal dune deposits. It appears that several generations of lithified deposits exist along the coast, however it is believed that conditions under which lithification occurred did not differ from those presently observed.

Cement

Samples of lithified beach and dune sediment were studied in thin section in order to examine the nature of cement. Samples were collected from subtidal to dune at both locations. At French Bay sand grains in the subtidal to lower supratidal beach are cemented by isopachous micritic Mg-calcite. Upper beach and dune sand is weakly cemented by isopachous and meniscus calcite microspar. Grahams Harbour beachrock is cemented by isopachous acicular aragonite. Subtidal samples are weakly cemented by discontinuous less well-developed cement whereas intertidal (yellow zone) and supratidal (black zone) samples show a more extensive growth of isopachous aragonite cement. Upper beach and dune samples are cemented by both isopachous and meniscus sparry calcite. Grahams Harbour dune samples are older and much more indurated than French Bay samples. Sand of French Bay dunes is still being reworked.

In reviewing mechanisms for cementation, Hanor (1978) provides an adequate model which will produce the characteristics of these deposits. Because cements are forming above high tide as well as below and because cements are not only aragonite and/or Mg-calcite but also calcite, evaporation of sea water is not a satisfactory mechanism by itself. At both French Bay and Grahams Harbour broad areas landward of the beach provide porous reservoirs for groundwater. As groundwater migrates into the subsurface and seaward, CO_2-content increases and temperature decreases resulting in the dissolution of less stable carbonate materials. When these carbonate-enriched waters reach the coast, temperature increases, CO_2-content decreases (Hanor, 1978) and possible mixing of fresh water and seawater (Moore & Hanor, 1974) results in the precipitation of calcium carbonate which cements sand grains of the basal dune to subtidal sediment. Other mechanisms might work also but mechanism of beachrock formation is not the topic of this paper. Instead let us look at the influence of beachrock on beach processes and beach deposits.

Exposed Beachrock

The exposed nature of beachrock does not represent the environment in which it forms but instead results in modification of beachrock which was previously cemented beneath beach sediment. This modification includes a slight increase in volume of rigid beachrock through encrustation by organisms, a more significant reduction in beachrock volume as a result of erosion and bioerosion, and increased cementation resulting from biochemical and physio-chemical precipitation and alteration.

Where beachrock becomes exposed because surficial beach sand is eroded, encrusting organisms quickly occupy the newly exposed firm substrate. Vermetid gastropods (Fig. 5) occur most commonly encrusting beachrock in areas below low tide. As a result of this encrustation, beachrock increases in volume as encrusting masses of aragonitic skeletons attain thicknesses of 10 cm or more. In addition to the skeletal mass, sediment which fills between and within the gastropod skeletons also increases beachrock volume.

Red algae likewise quickly inhabit the new rocky substrate adding to rock volume and rigidity. Serpulid worms cement their calcareous tubes to the exposed underside of beachrock slabs. Barnacles locally can be quite abundant. Coral encrust subtidal exposures (Fig. 6) and spread in low-profile growth forms taking advantage of the high energy and firm substrate. Pelecypods attach to the beachrock especially wedging into depressions and crevices.

While encrustation is most common on the upper surface of beachrock, lower exposed surfaces are repleat with borings (Fig. 7). These organisms by their action remove rock material and their occurrence leads to further physical erosion. Boring pelecypods, barnacles, algae, fungi, worms and sponges weaken beachrock which results in more rapid erosion. Grazing organisms, in particular echinoids, chiton, gastropods and parrot fish, feed on encrusting organisms and reduce beachrock exposures.

Beachrock exposed as a result of erosion of beach sand is itself subject to erosion within the wave zone. Physical erosion reduces beachrock volume by dislodging weakly cemented grains as well as large slabs. A characteristic of beachrock is closely spaced joints which enable storm waves to erode and transport beachrock. As a result clasts of beachrock become incorporated in beach sediment. Large slabs of beachrock act as buttress on a sandy beach, thereby reducing erosion of sand.

Beachrock is diagenetically altered by endolithic fungi and algae. Clastic and/or oolitic textures become micritized adjacent to borings and, in intensely infested areas, total micritization has led to obliteration of original texture and increased cementation of the rock. Furthermore, as a result of endolithic algae and fungi beachrock surfaces are darkened.

The exposed beachrock with its population of encrusting, boring and feeding organisms changes the nature of sediment accumulating on the adjacent beach. Prior to beachrock exposure the shoreline probably was accumulating sandy bioclastic sediment derived from perhaps a barrier reef, patch reefs and a sandy carbonate shelf. Bioclastic grains were reworked before being deposited on the beach as fairly well-sorted sand. Sorting and grain size were the result of beach energy and distance of beach from the source. Generally few coarse fragments reached the beach from reefs. Bioclastic grains were dominated compositionally by fragments of burrowing pelecypods, worm tubes, and foraminifera.

With the introduction of a rocky substrate adjacent to the beach, the nature of beach sediment changes noticeably. Most striking is the introduction of large clasts of beachrock. More subtle but also easily recognizable is the compositional change of bioclasts. Included in the beach sand are remnants of the rocky bottom fauna and flora including echinoid spines, red algal fragments (*Goniolithon* primarily), patellid gastropods, *Halimeda*, coral and clusters of vermetid gastropods. Due to the proximity of the beachrock and its associated organisms, lithic and biologic clasts are not reworked to the same degree as on a sandy beach. Also because beachrock prevents beach erosion, water depths offshore from the beachrock increase slightly more rapidly than in a sandy shelf-beach environment. As a result, waves washing over the beachrock and onto the protected beach experience a more rapid energy reduction. These circumstances produce a deposit which is coarser and less sorted than on a sandy beach. Since the exposed beachrock represents lithified sediment which accumulated on a sandy beach, sediment deposited during the period of beachrock exposure will be coarser and less sorted than the sediment within the beachrock. Sand-sized particles will likely be nearly identical in both beach and beachrock however composition will differ.

On a larger scale the presence of beachrock in a depositional setting results in more rapid beach progradation. Figure 8 illustrates two coasts, one where sediment remains uncemented (A) and a second where rapid cementation lithifies beach sand beneath the surface (B). Under conditions dominated by deposition both coasts prograde equally (Fig. 8-1). If both coasts experience temporary erosional conditions (Fig. 8-2) erosion will be greater along the coast where no beachrock is exposed (Fig. 8-2A). The exposed rampart of beachrock (Fig. 8-2B) protects, to some degree, sand of the adjacent beach and dune. When both coasts revert to depositionally dominated conditions the coast protected by beachrock (Fig. 8-3B) begins progradation at a more seaward position. After repeated erosional-depositional events shoreline progradation should be noticeably more advanced in areas of beachrock formation.

RECOGNITION OF HOLOCENE BEACHROCK

In a subtidal-beach-dune sequence of sediments or sedimentary rocks where beachrock formed due to early cementation, exhumation and later burial by prograding coastal deposits, evidence of its presence may not be striking or noted at first glance. Beachrock consists of beach sediment and is therefore compositionally, texturally, and stratigraphically very similar to the host rock in which it occurs. Beachrock dips

Fig. 1. Exposure of cemented dune sand located approximately 200 m west of Government Dock, French Bay, San Salvador Island.

Fig. 2. View of exposed beachrock west of Government Dock, French Bay, San Salvador Island.

Fig. 3. Fresh surface of beachrock at French Bay displaying intertidal fenestral porosity in lower half of exposure and borings of pelecypods and barnacles in upper half.

Fig. 4. Freshly exposed supratidal beachrock which was uncovered by recent storm. Note preservation of laminae, French Bay. Scale = 15m.

Fig. 5. Block of exposed beachrock which was eroded from subtidal location and deposited on beach at French Bay. Of particular interest is the dark encrustation of vermetid gastropods which increases beachrock volume.

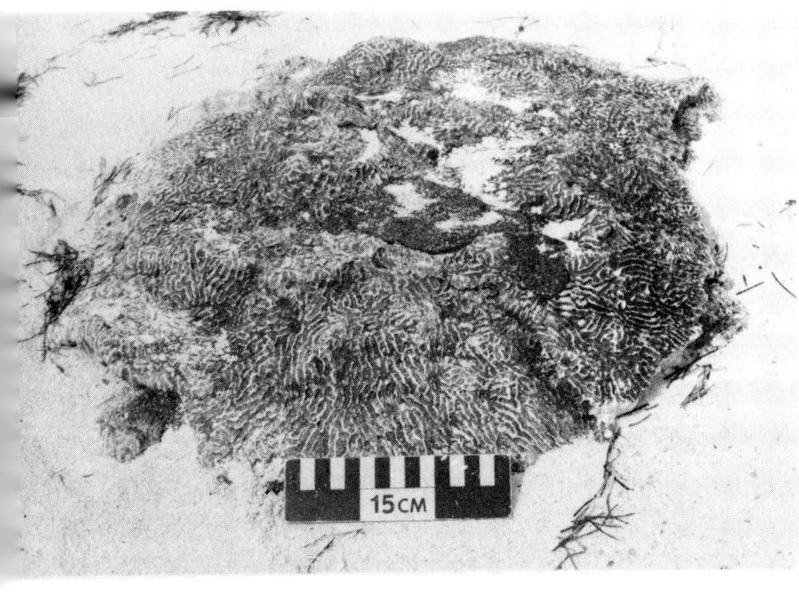

Fig. 6. Large low profile Diploria *which was eroded from shallow subtidal environment in French Bay where it encrusted exposed beachrock. Upward growth is restricted by water depth at low tide.*

Fig. 7. Closely spaced borings of pelecypods and barnacles exposed in broken sample of French Bay beachrock. Borings generally are concentrated on the underside of beachrock beds.

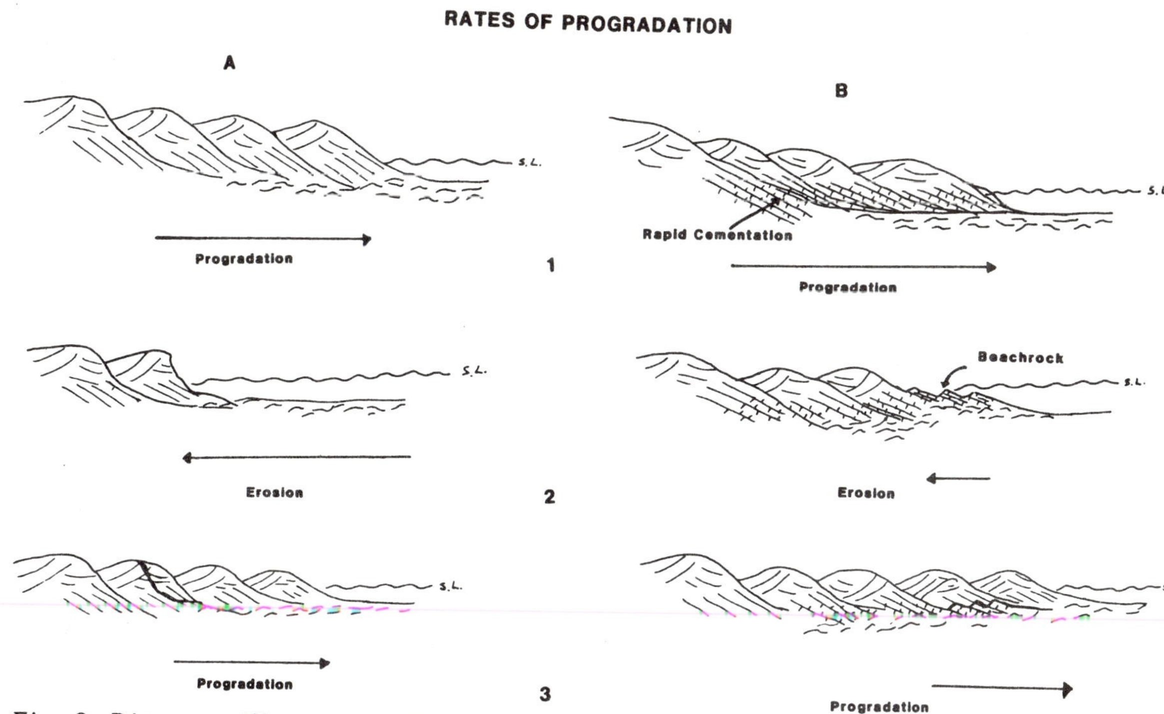

Fig. 8. Diagrams illustrating relative rates of progradation of two coasts, one which is unprotected by beachrock (A) and one which is protected (B). At stage 1 both coasts have prograded equally and early cementation has lithified sand within beach B. In stage 2 temporary erosional conditions prevail; beach A is eroded whereas beach B experiences erosion to a lesser extent as a result of protection by the beachrock rampart. When depositional conditions return to both coasts, B begins progradation at a more seaward position resulting in more rapid progradation rates through time.

parallel to beach strata and short term exposure at sea level does not create high relief erosion.

Upon closer inspection of exposures and petrographic examination the following features indicate the presence of beachrock (Fig. 9). Locally laminae of dislodged beachrock will occur in angular discordance with laminae of undisturbed beach sediment. Sediment is more coarse-grained and less sorted. Most striking is the localized occurrence of rocky bottom fauna and flora, in particular, flat growth forms of coral, vermetid gastropod encrustations, abundant *Goniolithon* fragments and colonies, patellid gastropods and *Halimeda* plates. Within the bioclastic host sediment or rock, beachrock may be more tightly cemented and commonly is micritized. This cemented, micritic rock displays borings and encrustations. Petrographic examination may reveal the presence of marine cement if diagenesis has not altered the original cement.

Most of these characteristics would also be present in any shallow-water carbonate hardground. Recognition as beachrock, therefore, depends on determination of its occurrence within a beach sequence.

Holocene Example of Exposed Beachrock

At quarry A (index map) on San Salvador Island, excavations for road material have exposed deposits which have been interpreted as representing eastwardly prograding dune-beach-subtidal marine environments (Fig. 10). Dune sediments consist of well-sorted, fine-grained oolitic sand which displays sweeping crossbeds typical of eolian deposits. In basal dune deposits, fine-grained bioclastic material increases in abundance and ooid percentages decrease downward into what is interpreted as beach sediment. Beach deposits contain a low percentage of ooids and are primarily composed of medium-grained bioclastic sand. Alternating fine-grained and medium-grained, planar laminae dip consistently to the east. A horizontal, 10 cm zone of fenestral porosity occurs throughout the north lobe of Quarry A and is interpreted as having formed by trapped air in intertidal sands during falling tides.

Beneath these beach sands occurs a poorly

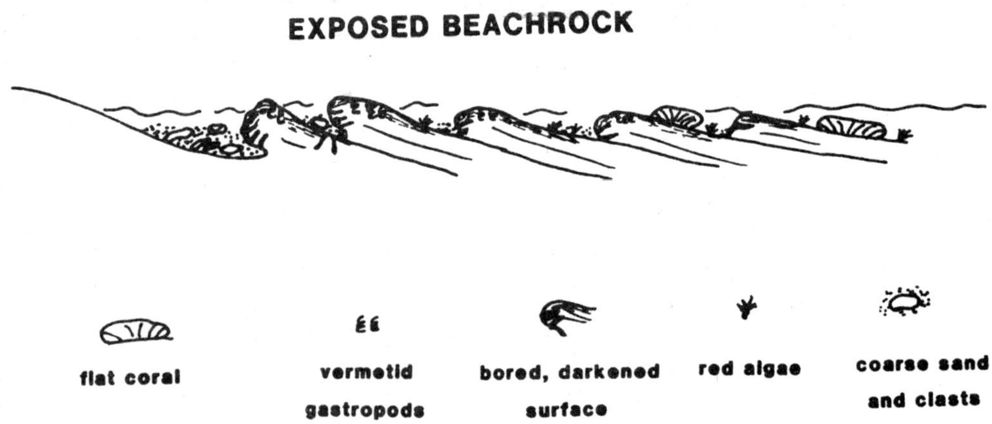

Fig. 9. Diagram illustrating features of beachrock exposed to shallow marine, nearshore conditions.

sorted, medium- to coarse-grained bioclastic sand and gravel which varies considerably both compositionally and texturally and which lacks bedding. This coarser facies is interpreted as subtidal and offers a diverse accumulation of marine fossils. Ooids are almost totally lacking in subtidal deposits and occur only in beach-dune sediment as reported elsewhere (Florentino & Bain, 1984; Lloyd, Perkins, and Kerr, 1987). Of particular interest within lower beach and subtidal facies is the presence of large (>40 cm) *Diploria* which in most instances are relatively thin (<15 cm)(Fig. 11). The flat, low-profile corals are somewhat irregular, especially their lower surfaces, and occur, rather curiously, in medium--grained, bioclastic sand. In several associations, sand beneath coral is considerably more lithified than that overlying the coral.

At several locations, particularly in the south lobe of the quarry, are masses of vermetid gastropods which occur at the same approximate horizon as the coral (Fig. 12). Carbonate sand containing vermetid gastropods generally is much more lithified and at several locations the sands are intensely micritized. Gastropod encrustations are commonly 20 cm thick. In addition to gastropods and corals, micritic rock in Quarry A also contains boring pelecypods, perhaps *Lithophaga* (Fig. 13) and organisms which prefer rocky substrate such as *Fissurella* and *Brachiodontes*.

Within the lower beach-subtidal facies, excavation of loose carbonate sand has exposed and dislodged not only the micritized rock but also well-lithified slabs of bioclastic limestone. These slabs display borings, both filled and open, and encrustations of serpulid worms on the lower

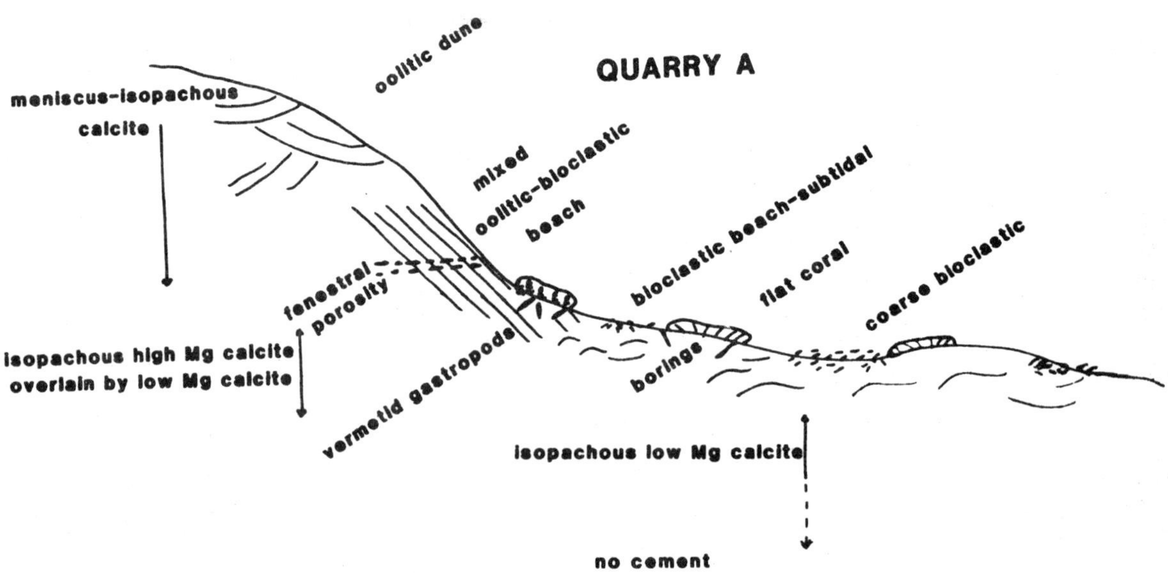

Fig. 10. Diagram illustrating lithologic and biologic features of Holocene sediment exposed in Quarry A, San Salvador Island. (Refer to index map)

Fig. 11. One of many Holocene Diploria *exposed in Quarry A, San Salvador Island. Coral measures approximately 30cm.*

Fig. 12. Vermetid gastropod encrustation of Holocene beachrock in Quarry A, San Salvador Island. Scale shows cm.

Fig. 13. Skeleton of boring pelecypod in micritized Holocene beachrock exposed in Quarry A, San Salvador Island. Pelecypod measures 12cm.

Fig. 14. View of underside of well-lithified slab of Holocene beachrock from Quarry A, San Salvador Island. Note the encrustations of serpulid worms. Bar scale = 3cm.

Fig. 15. Photograph of slab of well-lithified Holocene beachrock overlain by coarse-grained, poorly sorted beach sediment which includes several clasts of beachrock, Quarry A, San Salvador Island. Scale shows cm.

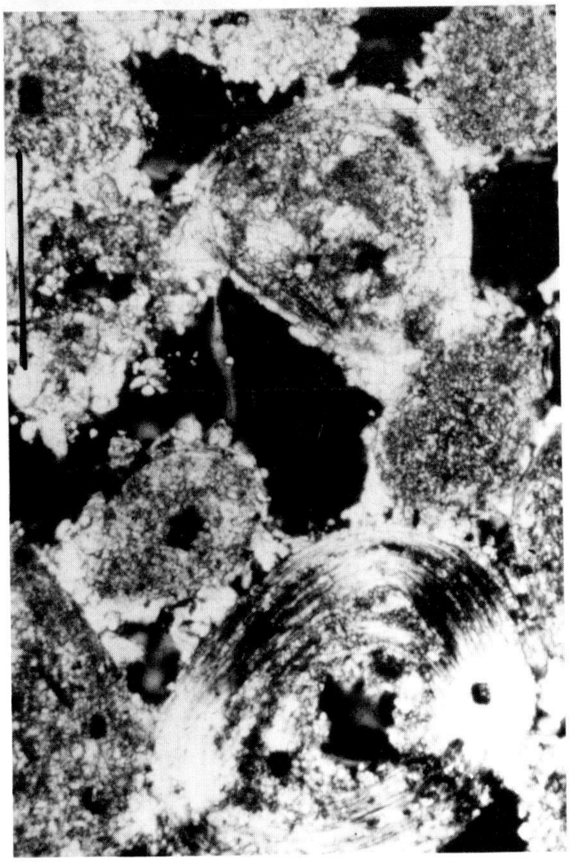

Fig. 16. Photomicrograph of Holocene ooids from dune facies, Quarry A, San Salvador Island. Note cementation of ooids by sparry calcite. Bar scale = .5mm.

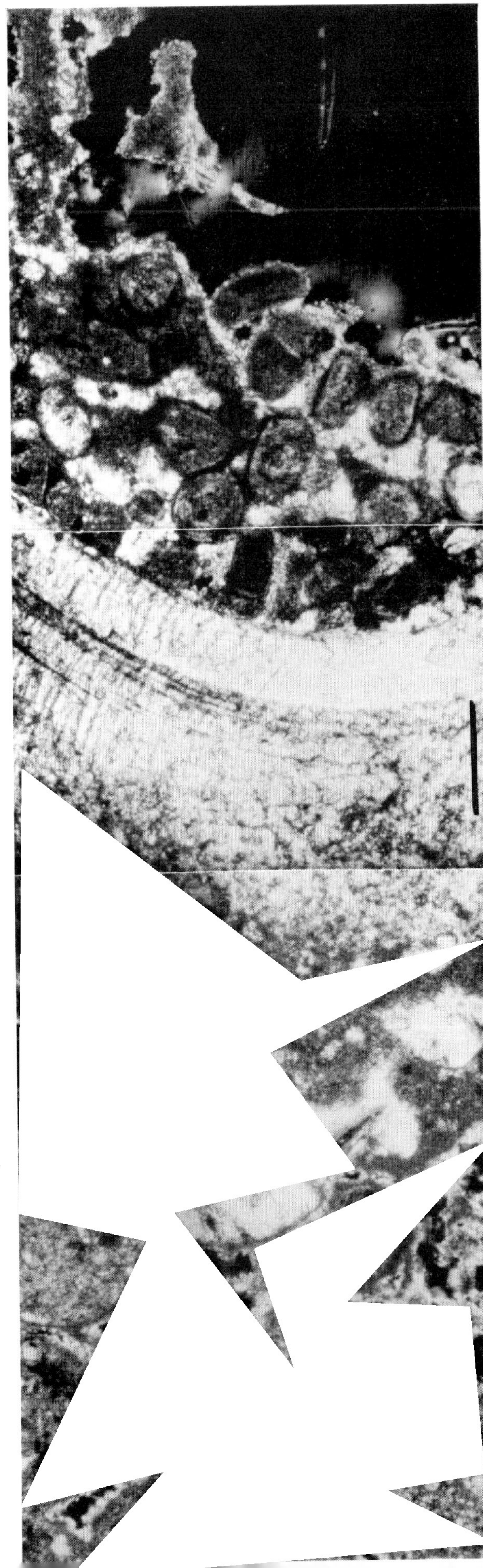

Fig. 17. Composite photomicrograph of Holocene vermetid encrustation on beachrock, Quarry A. Thin section displays body cavity and skeletal wall of vermetid gastropod and the beachrock on which the gastropod encrusted. Note ooids of beach sand within gastropod are cemented by sparry calcite whereas bioclastic grains of beachrock are cemented initially by micritic, isopachous Mg-calcite which is overlain by sparry calcite cement. Bar scale = 1mm.

surfaces (Fig. 14). Clasts of similarly lithified material, ranging in size up to 20 cm, occur throughout the lower beach-subtidal facies in the north lobe of Quarry A (Fig. 15). Associated sediment is dominated by fragments of echinoid spines, *Goniolithon*, *Porites* and fragments of *Halimeda*.

Petrographic and microprobe analyses of cements present within sediment of Quarry A indicate a variation in cement types including meniscus, isopachous, low-Mg calcite, and isopachous, micritic, high-Mg calcite. Distribution of these cements is illustrated in Figure 10. All carbonate deposits in Quarry A are at least partially cemented by sparry low-Mg calcite (Fig. 16). High-Mg calcite is restricted to lower beach--subtidal sediments and where it occurs it pre--dates low-Mg calcite. This relationship is especially evident in Figure 17 which traverses a thin section of beachrock through the vermetid gastropod encrustation. Oolitic beach sediment which infilled gastropods is cemented by isopachous calcite whereas bioclastic beachrock, to which vermetid gastropods attached, is cemented first by micritic Mg-calcite and later by calcite. Similar cement relationships were noted associated with coral occurrences. Also whereas low-Mg calcite cement occurs both as meniscus and isopachous cement, high-Mg calcite occurs only as isopachous cement. At a depth of several meters into subtidal sediments, cement is absent.

The occurrence of low profile *Diploria*, serpulid worms and vermetid gastropods within what appears to be nearshore bioclastic sand suggests that these sands were lithified at the time of coral, worm and gastropod encrustation. Presence of other rocky-bottom organisms, namely *Fissurella*, echinoids, and *Brachiodontes* likewise indicates a hard substrate existed. Boring organisms within micritic limestone proves sediment was lithified. Cement variation and cement stratigraphy strongly suggest marine cementation. Low-Mg calcite cement was precipitated from vadose and phreatic fresh water and represents late- stage and present-day cementation. Isopachous, micritic high-Mg calcite cement was precipitated from seawater and resulted in the formation of beachrock. Clasts of early cemented, exhumed and eroded beachrock are scattered through lower beach-subtidal sediment. These clasts, as well as the exposed beachrock, provided the rocky substrate for organisms of Quarry A.

CONCLUSIONS

Beachrock results from the exhumation of early-cemented beach deposits and provides not only a rocky-bottom for fauna and flora but also alters the beach environment as a result of its presence. Beaches are less subject to erosion because of the lithified rampart of beachrock and rates of progradation are accelerated. Beach and subtidal sediment associated with beachrock differs from that of a sandy beach which lacks beachrock. Composition of beach sand is modified by the influx of nearby rocky-bottom organisms and texture of beach deposits becomes coarser and less sorted as a result of a change in wave energy.

These alterations of beach fauna, flora, sediment texture and composition are preserved in the sedimentary record and can be used to recognize ancient beachrock. Quarry A on San Salvador Island provides an excellent example of a Holocene beach where beachrock was exposed and beach sediment was thereby modified.

REFERENCES CITED

Branner, J.C., 1904, The stone reefs of Brazil, their geological and geographical relations, with a chapter on coral reefs.: Bulletin Museum of Comparative Zoology, Harvard College, v. 44, p. 1-285.

Bathurst, R.G.C., 1975, Carbonate sediments and their diagenesis (2nd ed): New York, Elsevier, 658 p.

Florentino, E. and Bain, R., 1984, Environment of deposition of the Granny Lake Oolite, San Salvador, Bahamas: p. 187-196, in Teeter, J.W., (ed.) Proceedings of the Second Symposium on the Geology of the Bahamas; CCFL Bahamian Field Station, San Salvador, Bahamas.

Hanor, J.S., 1978, Precipitation of beachrock cements: Mixing of marine and meteoric waters vs. CO_2-degassing: Journal of Sedimentary Petrology, v. 48, p. 489-501.

Kuenen, P.H., 1933, The Snellius-Expedition, 5. Geological Results, 2: Geology of Coral Reefs, Kemick, Utrecht, 125 p.

Lloyd, R.M., Perkins, R.C., and Kerr, S.D., 1987,

Beach and shoreface ooid deposition on shallow interior banks, Turks and Caicos Islands, British West Indies: Journal of Sedimentary Petrology, v. 57, no. 6, p. 976-982.

Moore, C.H. and Hanor, J.S., 1974, Boiler Bay beach rock, St. Croix: West Indies Lab Special Publication 5, p. 71-75.

Scoffin, T.P., 1987, An Introduction to Carbonate Sediments and Rocks: Blackie, London/Chapman & Hall, New York, 274 p.

Strasser, A. and Davaud, E., 1986, Formation of Holocene limestone sequences by progradation, cementation, and erosion: Two examples from the Bahamas: Journal of Sedimentary Petrology, v. 56, no. 3, p. 422-428.

HOLOCENE SEA LEVEL IN THE BAHAMAS

Mark R. Boardman
Geology Department
Miami University
Oxford, OH 45056

A. Conrad Neumann and Kenneth A. Rasmussen
Curriculum in Marine Science
University of North Carolina
Chapel Hill, NC 27514

ABSTRACT

A sea-level curve covering the late Holocene (10,000 years to present) for the Bahamas emerges, based on C-14 dating of basal peat samples from cores recovered from lagoons and tidal flats and on measurements of the rates of coastal erosion. This sea-level curve shows that Bahamian platforms were flooded earlier than predicted, using the Bermuda sea-level curve, and somewhat later than predicted, using the sea-level submergence curve for South Florida.

Using this new sea-level curve, the temporal framework of late Holocene depositional processes can be constrained. The record of lagoon filling is better integrated with the Holocene sea-level rise, and the signal of banktop emergence/submergence in the periplatform record is supported by this new curve. An episode of early Holocene formation of eolian dunes concurs with the concept of a Holocene "high-energy interval." These data enhance our understanding of late Quaternary fluctuations of sea level.

INTRODUCTION

Sea level exerts a profound influence on sedimentary systems for several reasons. Basically, it determines the upper limit of marine processes (erosion and deposition) and the lower limit of subaerial processes (erosion and deposition). Sea level and the rate of sea-level rise control the modern sedimentary sequences of lagoons (Boardman and Neumann, 1976; Bosence et al., 1985; Rasmussen and Neumann, 1987; Andersen and Boardman, 1989; Colby and Boardman, 1989), sand shoals (Hine, 1977; Harris, 1979), reefs (Adey and Burke, 1976; Macintyre and Glynn, 1976; Adey, 1978), tidal flats (Shinn et al., 1969; Shinn, 1986), and eolian dunes (Boardman et al., 1987; Vacher and Hearty, in press). Ancient sequences are also primarily controlled by sea level (Grottzinger, 1986, Pratt and James, 1986, Strasser, 1988; Read, 1988). Sea level is perhaps the most significant single parameter in evaluating the accumulation of carbonates (Schlager, 1981; Schlager and Ginsburg, 1981; Kendall and Schlager, 1981).

There is overall agreement that during the Pleistocene, the volume of water in the ocean has fluctuated in response to glacial advances and retreats, and this has caused worldwide eustatic fluctuations in sea level on the order of 100 to 150 meters (Matthews, 1984). However, this general agreement is moderated by the understanding that a single eustatic sea-level curve is not the panacea for studies of shallow-water processes (Wanless, 1982). Local effects control the vertical stability of the crust and thus control relative sea level. Subsidence (or uplift) can result from loading (or unloading) of water, sediment, and ice (Bloom, 1967; Matthews, 1984) and crustal rebound (Walcott, 1970). Bulges (geoid changes) may propagate through the earth (Walcott, 1970; Mörner, 1980) and cause local changes in relative sea level which are unrelated to any local surficial process.

Evaluation of Holocene carbonate depositional processes in the Bahamas are increasingly asking more detailed questions, and more detailed comparisons are being made. These studies are increasingly more dependent on accurate and specific information on the relative rise of sea level. For example, a large portion of the tidal flats on northwest Andros Island may have been eroded during the past 1000-2000 years because of a small and unknown rise of sea level (Shinn, 1986). The onset of marine conditions in silled lagoons has been shown to be controlled by sea level (Boardman, 1976; Boardman et al., 1986a; Rasmussen and Neumann, 1987; Colby and Boardman, 1989). The exact timing and extent of Holocene flooding of entire platforms is a major point of contention in evaluating the signal of periplatform sedimentation (Droxler et al., 1983; Droxler, 1985, 1986; Boardman and Neumann, 1984, 1985, 1986; Boardman et al., 1986b). The development and evolution of Holocene eolian dunes suggests there was an interval during which sea level had flooded the platforms but was rising too fast to allow dune progradation (Boardman et al., 1987). Studies such as these require an accurate Holocene sea-level curve for the Bahamas.

There are many excellent sea-level curves that have been used for studies in the Bahamas. The sea-level curve from Bermuda (Neumann, 1972) is perhaps the most carefully constructed and complete Holocene record of sea-level rise. It is comprised only of dates of peat-on-bedrock and contains 31 sample points from a small area. For this reason, it has been used in several studies in the Bahamas and other areas (Adey and Burke, 1976; Macintyre and Glynn, 1976; Hine, 1977; Boardman and Neumann, 1984; Droxler et al., 1983; Rasmussen and Neumann, 1987). The curve for south Florida (Scholl et al., 1969) is from an area more similar to the Bahamas and is closer to the Bahamas than Bermuda. Because of this, several researchers have used the south Florida curve for Bahamian studies (Harris, 1979). The south Florida curve, however, is based on C-14 dating of several types of sea-level indicators other than samples of peat-on-bedrock and suffers because of the inherent unreliability of using shells as sea-level indicators. Other curves from nearby areas are likewise candidates for use in Bahamian studies (e.g., Fairbridge, 1974; Lighty et al., 1982; Digerfeldt and Hendry, 1987). However, each of these curves must be used with a clear understanding that sea-level fluctuations in the Bahamas are likely to be different, perhaps by a lot. The fact is we don't have a sea-level curve for the Bahamas, and because minor changes in sea level may be critical to processes of deposition, a Bahamian sea-level curve is needed.

METHODS

Sediment cores from shallow-water sediments sometimes contain peat deposits lying directly on top of the Pleistocene bedrock surface. This peat is the last surviving portion of terrestrial or mixed terrestrial/marine material and is usually preserved in shallow depressions ("pocket peat" of Wanless, 1974) or in well-protected, quiet-water environments. Peat is an intertidal deposit which can be dated with C-14 methods. Thus, it is an excellent record of sea level. The C-14 age of ten samples of peat from cores from lagoons on San Salvador, Andros Island, and Bight of Abaco, Little Bahama Bank in the Bahamas have been determined. Only peat directly in contact with the bedrock surface is included in order to limit problems caused by compaction (Bloom, 1964; Kaye and Barghoorn, 1964; Van de Plassche, 1980). To confirm that the peat directly overlies the bedrock, either pieces of rock were recovered or the nose of the aluminum core tubing was bent. In addition, marine shells which directly overly the peat have been dated, where possible, to confirm the timing of initial marine sedimentation. A computer-generated, third-order, best-fit line is drawn through the data points.

RESULTS

The results of C-14 analyses are included in the curve of Bahamian sea level (Figure 1). Evidence of sea-level rise during the last 2000 years is particularly difficult to determine with accuracy because sea-level rise has been slow, and peat has a higher likelihood of being eroded by waves and currents. The sea-level data from studies of erosion of limestone coastlines in the Bahamas (Donn and Boardman, 1988) is used to confirm that sea level has been within 50cm of present for the past 1400 years.

DISCUSSION

The Bahamian sea-level curve is different than other curves of late Holocene sea level (Figure 2). It is higher than the Bermuda curve for all but the last 2000 years which means that depositional events recognized in the Bahamas occur earlier than would be expected or predicted (by up to 1000 years) if the Bermuda curve were used. It is lower than the curve of early Holocene sea level from south Florida which means that Bahamian depositional events occur later than would be predicted, using the south Florida curve.

Lagoon Evolution

The evolution of lagoonal sequences is strongly controlled by sea level and the antecedent topography of the newly-flooded platform. A record of the Holocene rise of sea level is recorded in changing mineralogy and constituent

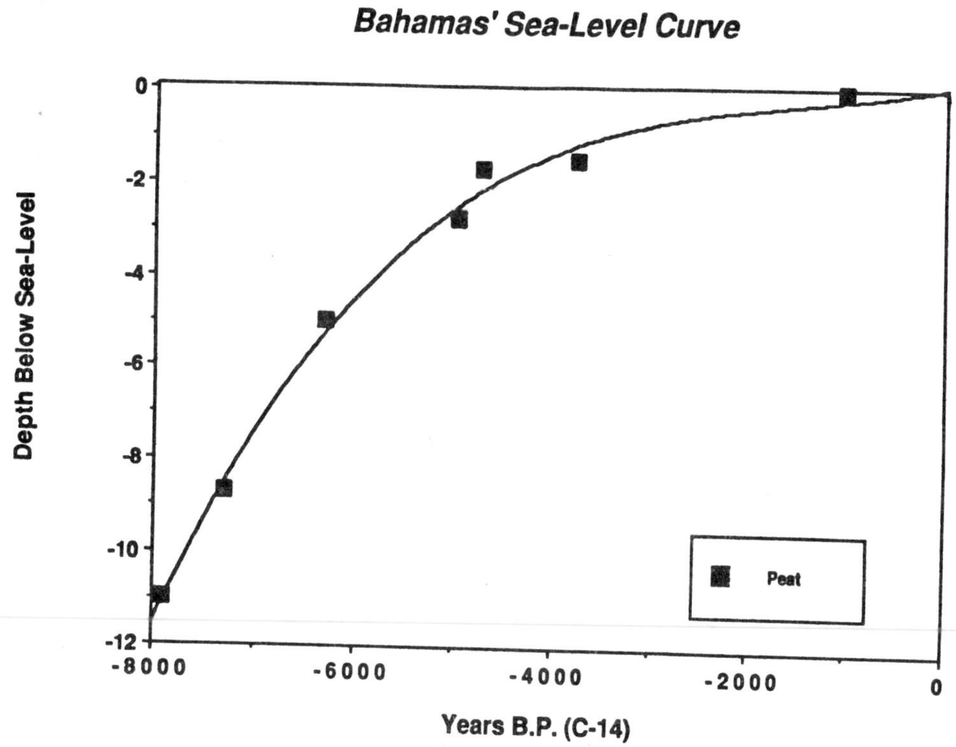

Fig. 1: The Bahamas' sea-level curve is derived from C-14 dates of peat samples recovered from the bedrock interface in sediment cores from San Salvador, Andros Island, and Bight of Abaco.

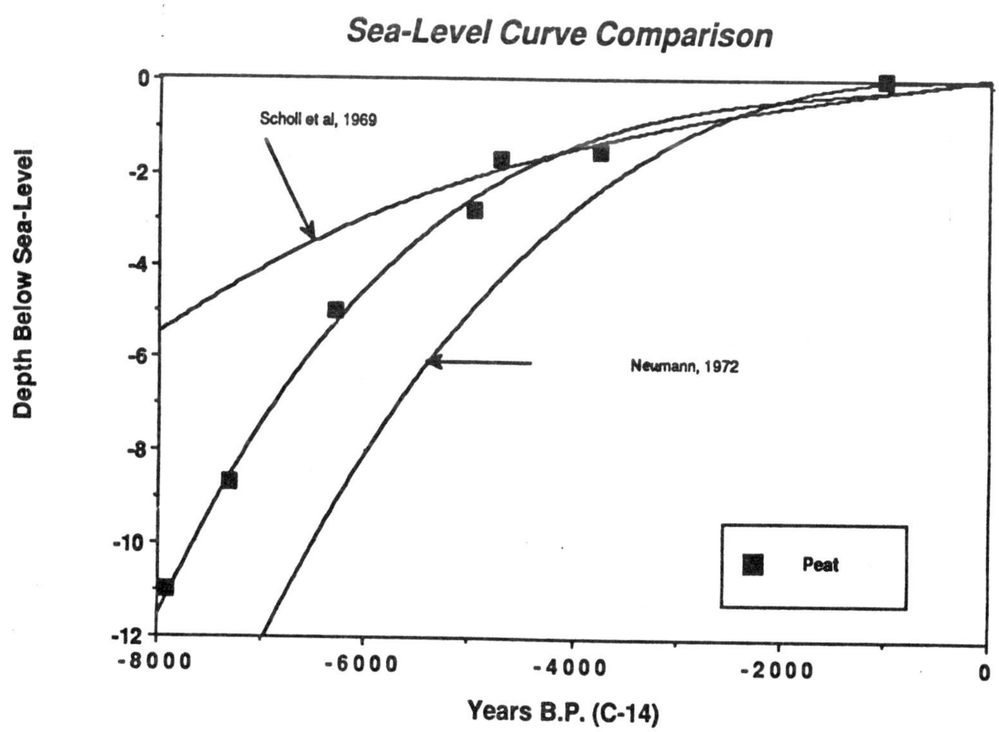

Fig. 2: A comparison of sea-level curves for Bermuda (Neumann, 1972), south Florida (Scholl et al., 1969), and the Bahamas (this paper) shows the increasing difference for dates older than 2,000 years BP.

analysis of the sediment, as well as the geology of the exposure surface. Shallowing-upwards sequences of fine-grained sediment (James, 1984), and the vertical sequences of coarse-grained sediments (Hine, 1977; Harris, 1979) and mud mounds (Bosence et al., 1985) are closely tied to sea level. In some lagoons, the depth of a sill controls the degree of restriction of the lagoon water.

In the Bahamas, sills have controlled the early deposition of Graham's Harbor, San Salvador (Colby and Boardman, 1988; Colby, 1989), and Bight of Abaco, Little Bahama Bank (Boardman, 1976; Rasmussen and Neumann, 1986). In Bight of Abaco, a prominent sill about 2 meters below present sea level acted as a barrier to water exchange and strongly controlled the early depositional history of the lagoon (Boardman, 1976). A macrofaunal assemblage is found at the base of several cores (to about 9.6 meters below present sea level) which suggests that a restricted environment with widely fluctuating salinity and little *Thalassia* was the earliest "marine" environment. A C-14 date of shells from the base of this core (-9.6 meters) yields 6,600 (plus or minus 100) YBP. Using the Bermuda curve, these shells would have to have been deposited above sea level (by about 1.5 meters) or were exhumed (from where?) and redeposited (Figure 2). Using the Bahamas sea-level curve, the molluscs could have lived in water depths up to 2.5 meters deep within a silled basin (sediment surface at 9.6 meters, sea level at 7 meters, sill depth at 2 meters).

A normal marine macrofaunal assemblage is found at a sediment depth above 130 cm in several cores, and it is suggested that this change of environments of deposition began because the sill was breached and adequate exchange with the ocean was achieved (Boardman, 1976). Using a rate of deposition of 36 cm/1000 years (Boardman, 1976), it is estimated that normal marine conditions began 3,640 years ago. Using the Bermuda sea-level curve, sea level was -2.8 meters or 0.8 meters below the estimated level of the sill. Certainly, the estimate of the sill depth could be in error by this much. However, using the Bahamas sea-level curve, sea level was 1.1 meters below present or 0.9 meters above the estimated level of the sill. This is certainly a reasonable depth and is in agreement with all of the available information.

By using the Bahamas curve rather than a sea-level curve from Bermuda, the data makes better sense. Significant doubt must be cast on the C-14 ages or the measurements of depths if the Bermuda curve is used.

Offbank Transport and the Periplatform Signal

Carbonate production by lagoon organisms is directly controlled by sea level. Lagoon production is only possible after the bank top is flooded, and it seems likely that a certain minimum depth (perhaps 0.5 meters) is necessary for normal production to occur. Lagoons can be "drowned" (Schlager, 1981) if water depths exceed some unknown depth (perhaps deeper than 20 meters). Layers of periplatform sediment which are distinct from planktonic sediments but very similar to shallow, lagoonal sediments (aragonite-rich, high Sr values) have been recorded (Droxler et al., 1983; Boardman and Neumann, 1984; Boardman et al., 1986b). It is thought that the periplatform environment records episodes of lagoon production by recording pulses of lagoon sediment transported off the platform when sea level is high (Kier and Pilkey, 1971; Boardman and Neumann, 1984; Boardman et al., 1986b). Highstand layers (lagoon-derived) are approximately 40cm thick and lowstand layers (planktonic-derived) are about 80cm thick. This interpretation has been challenged on several grounds, one of which is that the C-14 dates of periplatform sediments indicate that the pulse of sediment geochemically similar to lagoonal sediments is deposited prior to the time of bank flooding (Droxler, 1985, 1986). There may be several reasons for this apparent discrepancy, but the most important one may be that these bank-flooding estimates are based on the Bermuda sea-level curve, which would make the flooding events appear to have occurred later than they actually did.

By using the Bahamian curve instead of the Bermuda curve, the initial flooding and rate of flooding of the Bahama Banks (Figure 3) are shown to have occurred earlier than estimated by Droxler (Droxler et al., 1963; Droxler, 1985). Thus, the Holocene flooding history of the Bahamas is in good agreement with the offbank transport model (Boardman and Neumann, 1984; Boardman et al., 1986a,b).

Eolian Dune Formation

The Bahamian sea-level curve is necessary to our proper understanding of the formation of early Holocene, transgressive sand dune forma

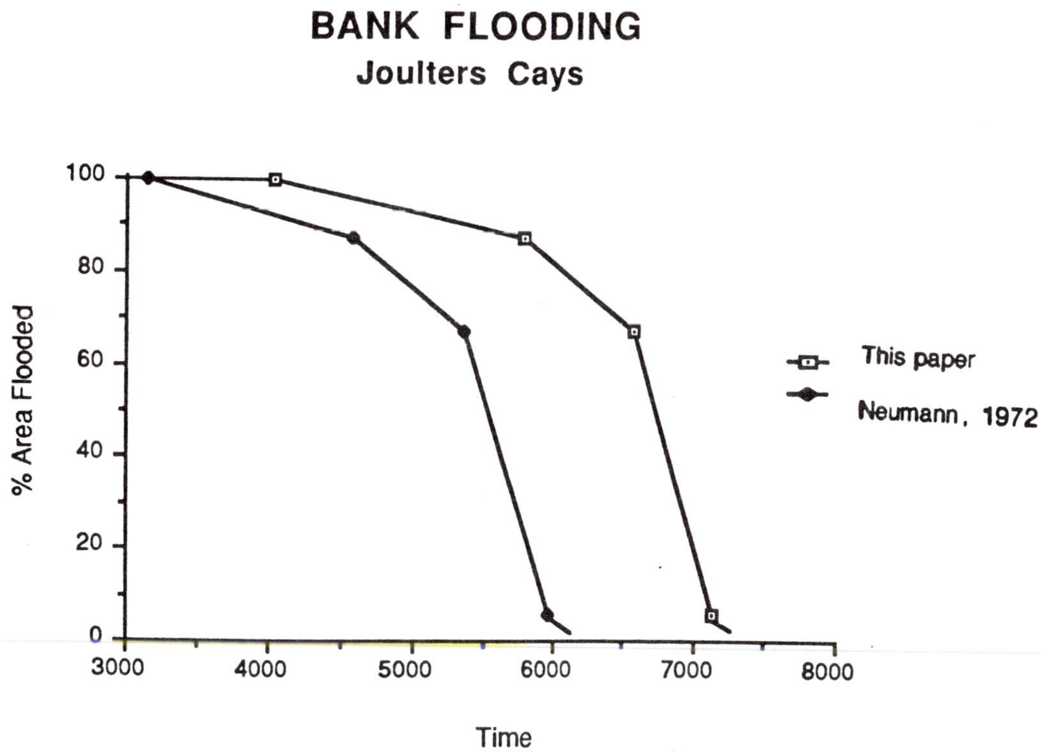

Fig. 3: Using the Bahamas curve (this paper) to determine the flooding history of Great Bahama Bank at the Joulters Cay area, it is seen that flooding is expected much earlier than predicted using the Bermuda curve (Neumann, 1972).

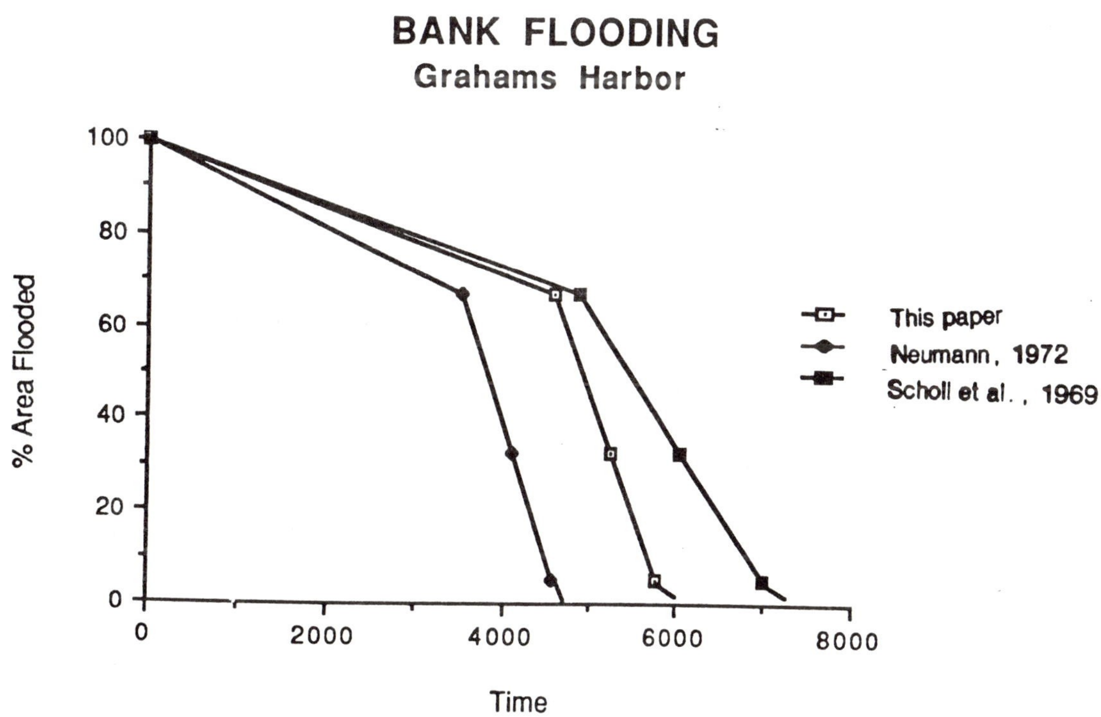

Fig. 4: The history of bank flooding of Graham's Harbor, San Salvador, is clearly dependent on the selection of a sea- level curve. Using the Bahamas curve (this paper), it is now recognized that over half of Graham's Harbor was flooded by 5,000 years BP versus 0% predicted if the Bermuda curve (Neumann, 1972) is used.

tion. C-14 dating of bioclastic sand from numerous coastal dunes of San Salvador (Boardman et al., 1987) clearly suggest that a major episode of dune building occurred between 6,000 and 4,000 years ago. Dunes require sediment, and the formation of sediment requires that the platform be flooded by 6,000 years ago. Using the Bermuda sea-level curve, the San Salvador platform was not significantly flooded until 4,000 years ago (Figure 4), which is in conflict with the measured ages of the dune sand. This conflict of timing is resolved if the Bahamian sea-level curve is used to evaluate the flooding history. If the sea-level curve from Florida applies (Scholl et al., 1969), large portions of the platforms would have been flooded earlier and older dates of the eolian dunes would be expected. These data support the general concept of a "high-energy interval" (Neumann, personal communication) for sand accumulation in many parts of the Bahamas.

CONCLUSIONS

The sea-level curve prepared from C-14 dating of peat samples from Bahamian sediment cores is significantly different from the sea-level curves from Bermuda and South Florida. Using the Bahamian curve for studies in the Bahamas is essential to the proper evaluation of late Holocene depositional events in the Bahamas.

REFERENCES CITED

Adey, W.H., 1978, Coral reef morphogenesis: a multidimensional model: Science, v. 202, p. 831-837.

Adey, W.H. and Burke, R.B., 1976, Holocene bioherms (algal ridges and bank barrier reefs) of the eastern Caribbean: Geological Society of America Bulletin, v. 87, p. 95-109.

Andersen, C.B. and Boardman, M.R., 1988, Lateral and vertical sedimentary gradients in a high-energy carbonate environment, Snow Bay, San Salvador, Bahamas: SEPM Annual Midyear Meeting, v. 5, p. 4.

Bloom, A.L., 1964, Peat accumulation and autocompaction in a Connecticut coastal marsh: Journal of Sedimentary Petrology, v. 34, p. 599-603.

Bloom, A.L., 1967, Pleistocene shoreline: a new test of isostasy: Geological Society of America Bulletin, v. 78, p. 1477-1494.

Boardman, M.R., 1976, Lime mud deposition in a tropical island lagoon, Bight of Abaco, Bahamas: Thesis, University of North Carolina, Chapel Hill, NC, 121 pp.

Boardman, M.R., Carew, J.L. and Mylroie, J.E., 1987, Holocene deposition of transgressive sand on San Salvador, Bahamas: Geological Society of America, Abstracts with Programs, v. 19, p. 593.

Boardman, M.R., Evans, L.A., and Andersen, C.B., 1986a, Sedimentary gradients of Holocene Bahamian lagoons: Controls by sills and sea level: SEPM Abstracts with Program, v. 3, p. 11.

Boardman, M.R. and Neumann, A.C., 1984, Sources of periplatform sediment to Northwest Providence Channel, Bahamas: Journal of Sedimentary Petrology, v. 54, p. 1110-1123.

Boardman, M.R. and Neumann, A.C., 1985, Sources of periplatform carbonates: Northwest Providence Channel, Bahamas, Reply: Journal of Sedimentary Petrology, v. 55, p. 929-931.

Boardman, M.R. and Neumann, A.C., 1986, Banktop responses to Quaternary fluctuations in sea level, Reply: Geology, v. 14, p. 1040-1041.

Boardman, M.R., Neumann, A.C., Baker, P.A., Dulin, L.A., Kenter, R.J., Hunter, G.E., and Kiefer, K.B., 1986b, Banktop response to Quaternary fluctuations in sea level recorded in periplatform sediments: Geology, v. 14, p. 28-31.

Bosence, D.W.J., Royston, J.R. and Quine, M.L., 1985, Sedimentary and budget of a recent carbonate mound, Florida Keys: Sedimentology, v. 32, p. 317-343.

Colby, N.D., 1989, Depositional evolution of a windward, high-energy lagoon, Graham's Harbor, San Salvador, Bahamas: Thesis, Miami University, Oxford, Ohio, 99 pp.

Colby, N.D. and Boardman, M.R., 1988, Vertical sequence of a high-energy carbonate lagoon: Graham's Harbor, San Salvador, Bahamas:

SEPM Annual Midyear Meeting, v. 5, p. 11-12.

Digerfeldt, G. and Hendry, M.D., 1987, An 8000 year Holocene sea-level record from Jamaica: implications for interpretation of Caribbean reef and coastal history: Coral Reefs, v. 5, p. 165-170.

Donn, T.F. and Boardman, M.R., 1988, Bioerosion of rocky carbonate coastlines on Andros Island, Bahamas: Journal of Coastal Research, 4, p 381-394.

Droxler, A.W., 1985, Last deglaciation in the Bahamian basins and the adjacent Atlantic Ocean - A dissolution record from variations of aragonite content, in E.T. Sundquist and W.S. Broecker, eds., The Carbonate Cycles and Atomspheric CO_2 - Natural Variations, Archean to Present, American Geophysical Union Monograph, 32, p. 195-207.

Droxler, A.W., 1986, Comment on "Banktop responses to Quaternary fluctuations in sea level recorded in periplatform sediments": Geology, v. 14, p 1039-1040.

Droxler, A.W., Schlager, W. and Whallon, C.C., 1983, Quaternary aragonite cycles and oxygen-isotope record in Bahamian carbonate ooze: Geology, v. 11, p. 235-239.

Fairbridge, R.W., 1974, The Holocene sea-level record in South Florida, in P.J. Gleason, ed., Environments of South Florida, Past and Present. Miami Geological Society Memoir, 2, p. 223-232.

Grottzinger, J.P., 1986, Cyclicity and paleoenvironmental dynamics, Rocknest platform, northwest Canada: Geological Society of America Bulletin, v. 97, p. 1208- 1231.

Harris, P.M., 1979, Facies anatomy and diagenesis of a Bahamian ooid shoal: University of Miami, Comparative Sedimentology Laboratory, Sedimenta, 7, 163 pp.

Hine, A.C., 1977, Lily Bank, Bahamas; case history of an active oolite sand shoal: Journal of Sedimentary Petrology, v. 47, p. 1554-1581.

James, N.P., 1984, Shallowing-upward sequences in carbonates, in R. Walker, ed., Facies Models, Geoscience Canada Reprint Series, v. 1, p. 213-228.

Kaye, C.S. and Barghoorn, E.S., 1964, Late Quaternary sea-level change and crustal rise at Boston, Massachusetts, with notes on the autocompaction of peat: Geological Society of America Bulletin, v. 75, p. 63-80.

Kendall, G.St.C. and Schlager, W., 1981, Carbonates and relative sea level: Marine Geology, v. 44, p. 181-212.

Kier, J.S. and Pilkey, O.H., 1971, The influence of sea-level changes on sediment carbonate mineralogy, Tongue of the Ocean, Bahamas: Marine Geology, v. 11, p. 189-200.

Lighty, R.G., Macintyre, I.G. and Stuckenrath, R., 1982, *Acropora palmata* reef framework: a reliable indicator of sea level in the western Atlantic for the past 10,000 years: Coral Reefs, v. 1, p. 125-130.

Macintyre, I.G. and Glynn, P.W., 1976, Evolution of modern Caribbean fringing reef, Galeta Point, Panama: American Association of Petroleum Geology Bulletin, v. 60, p. 1054-1072.

Matthews, R.K., 1984, Dynamic Stratigraphy, Prentice Hall, Inc., Engelwood Cliffs, NJ, 489 pp.

Mörner, N.-A., 1980, Eustasy and geoid changes as a function of core/mantle changes, in N.-A. Mörner, ed., Earth Rheology, Isostasy, and Eustasy, Wiley, Chichester, p. 535-553.

Neumann, A.C., 1972, Quaternary sea level history of Bermuda and the Bahamas: AMQUA Abstract, p. 41-44.

Pitman III, W.C., 1978, Relationship between eustacy and stratigraphic sequences of passive margins: Geological Society of America Bulletin, v. 89, p. 1389-1403.

Pratt, B.R. and James, N.D., 1986, The St. George Group (Lower Ordovician) of western Newfoundland: tidal flat island model for carbonate sedimentation in shallow epeiric seas: Sedimentology, v. 33, p. 313-343.

Rasmussen, K.A. and Neumann, A.C., 1987, Holocene overprints of Pleistocene paleokarst: Bight of Abaco, Bahamas, in N.P. James and P.W. Choquette, eds., Paleokarst, Springer-Verlag, New York, p. 132-148.

Read, J.F., 1988, Carbonate cycles, sea level curves, and computer modelling: SEPM Annual Midyear Meeting, v. 5, p. 44.

Schlager, W., 1981, The paradox of drowned reefs and carbonate platforms: Geological Society of America Bulletin.

Schlager, W. and Ginsburg, R.N., 1981, Bahama carbonate platforms - The deep and the past: Marine Geology, v. 44, p. 1-24.

Scholl, D.W., Craighead, F.C. and Stuiver, M., 1969, Florida submergence curve revised: its relation to coastal sedimentation rates. Science, v. 163, p. 562-564.

Shinn, E.A., 1986, Modern carbonate tidal flats: their diagnostic features: Colorado School of Mines Quarterly, v. 81, p. 7-35.

Shinn, E.A., Lloyd, R.M. and Ginsburg, R.N., 1969, Anatomy of a modern carbonate tidal flat, Andros Island, Bahamas: Journal of Sedimentary Petrology, v. 39, p. 1202-1228.

Strasser, A., 1988, Shallowing-upwards sequences in Purbeckian peritidal carbonates (lowermost Cretaceous, Swiss and French Jura Mountains): Sedimentology, v. 35, p. 369- 383.

Vacher, H.L. and Hearty, P., in press, History of stage-5 sea level in Bermuda: review with new evidence of a brief rise to present sea level during substage 5a: Quarternary Science Reviews.

Van de Plassche, O., 1980, Compaction and other sources of error in obtaining sea-level data: some results and consequences: Eiszeitalter Ggw, v. 30, p. 171-181.

Walcott, R.I., 1970, Flexural rigidity, thickness, and viscosity of the lithosphere: Journal of Geophysical Research, v. 75, p. 3941-3954.

Wanless, H.R., 1974, Mangrove sedimentation in geologic perspective, in P.J. Gleason, ed., Environments of South Florida, Past and Present, Miami Geological Society Memoir, 2, p. 190-200.

Wanless, H.R., 1982, Editorial: Sea level is rising-So what?: Journal of Sedimentary Petrology, v. 52, p. 1051- 1054.

GEOLOGICAL IMPLICATIONS OF DEEP WELL DISPOSAL IN THE BAHAMAS

Richard V. Cant
Chief Hydrologist
Ministry of Works and Utilities
Nassau, Bahamas

ABSTRACT

In the Bahamas a variety of liquid wastes have been disposed of via boreholes for a number of years. The most common of these wastes include storm water, raw sewage, treated effluent, heated brine, and cooling water. A smaller, but no less significant component, comprise the by-products of various industrial and commercial enterprises some of which may be toxic. All constitute a threat to valuable groundwater resources.

The boreholes used range from 6 inch to 36 inch in diameter and from 50 ft. to 650 ft. in depth. The wells may be uncased, partly cased, or multi-cased, usually with steel or PVC cemented into place.

Over the years many wells have functioned successfully though there have been significant failures, which have led to serious subsurface pollution. Most malfunctioning wells can be attributed to casing failure, particularly as a result of corrosion.

Subsurface investigations have failed to reveal any shallow aquicludes in the Bahamas that could be used to prevent mixing between the wastes disposed of at depth and shallow freshwater lenses. Instead zones of high transmissivity are used as receiving zones, and it is assumed that wastes are rapidly mixed, diluted, and dispersed at these horizons, thereby nullifying any threat that they may pose to the environment. The zones of high transmissivity correspond to cavernous horizons that developed in the subsurface carbonates in relationship to low sea level stands and stratigraphic disconformites. Several such horizons occur in the Lucayan Limestone, but these vary in depth regionally. More widespread cavern zones occur beneath the base of the Lucayan Limestone and at depths that equate to the Pleistocene sea level lows. These are the main ones used for deep well disposal.

A good deal more research is needed on the character and fluid dynamics of these receiving zones as this means of disposal is providing an essential service in a situation where all other alternatives are considered too costly.

INTRODUCTION

Associated with development and urbanization is an increase in the production of wastes. Man's proliferation is accompanied by a proliferation of his wastes and the increase appears to be a geometric one as living standards rise. The wastes produced are both solid and liquid and there is an ever increasing problem of finding places to dispose of them. Many wastes present a threat to the environment and should be treated prior to disposal. Unfortunately, proper treatment is a costly procedure and so other less costly alternatives are constantly being sought.

One such alternative is the disposal of liquid wastes by means of wells. On a comparative basis, this is a cheap option, however, it is an option that conceals developments and situations can arise with serious future implications. Subsurface pollution is not just difficult to detect, it is even harder to rectify. Before wastes are injected into the ground, therefore, it is imperative that all aspects of this practice be fully understood and the geological implications are the most important considerations of all.

WASTE PRODUCTS AND WELLS

The most common liquid wastes disposed of via boreholes include storm water, wash-down water, raw sewage, treated effluent, cooling water, heated brine, and reject water from reverse osmosis or similar such plants. The storm water is usually fresh, but the others vary from being fresh to highly saline. Besides these liquids, there are smaller volumes of waste that are generated as the by-products of a number of industrial and commercial activities and some of these can be regarded as fairly toxic: examples include photographic laboratory waste, garage waste, laundry water, and even abattoir wastes, to name but a few. Almost all of the wastes constitute a threat to valuable ground water resources, the occurrence of which have been described by Little et al. (1977) and Cant and Weech (1986).

The boreholes used range from 6 inch to 36 inch in diameter and from 50 feet to 650 feet in depth. The majority are probably about 100 to 150 feet deep and very few exceed 300 feet. In general, the deeper boreholes receive the greater volumes of waste, the maximum being about 5,000,000 gallons per day. Storm water disposal wells may receive more than this, but there is no means of quantifying their use, and often these are fairly shallow and liable to fail in very heavy downpours.

Almost all disposal wells are cased, at least for the uppermost 20 feet or thereabouts. The main exceptions to this are the storm water drainage wells used in Grand Bahama which are often uncased. A few wells contain multiple casings, particularly those built with proper specifications.

The casings usually comprise steel or PVC pipe and there is often a cement grout emplaced between the outer casing and the borehole wall. Packers and annulus monitoring devices are rare.

In New Providence it is estimated that about 300 disposal wells are presently in operation, but there is no register of wells available and many will have been installed without a permit or similar form of authorization. Disposal wells are presently only common in New Providence and Grand Bahama. The wide spread use of disposal wells in these two Islands started in the early 1970's and has accelerated since that time.

Probably the most important disposal well in New Providence is the Malcolms Park well which was installed in 1971. This well is 20 inch in diameter and cased to 400 feet and open to 640 feet. The well has received some 2 to 3 million gallons of raw sewage per day from the city sewerage system and appears to have functioned successfully for about 4 years. In 1978, however, monitor wells around Malcolms Park clearly indicated sewage pollution at shallow horizons and it was surmised that the steel casing had failed as a result of corrosion.

Other wells have been known to fail for whatever reason, but in general it is rare that there is any means of monitoring the wells, and so the failure rate is unknown. There can be no doubt, however, that large areas of New Providence have been affected by subsurface contamination as a result of the existing disposal well practice.

THE ENVIRONMENT AND SUBSURFACE GEOLOGY

General

The Bahamas consists of a series of shallow marine banks that emerge locally above sea level, but are separated from one another by deep water. The banks which are usually less than 30 feet deep, are akin to submerged plateaux with steep sides. From sea level down to a depth of about 5 miles, the geology is dominated by limestone and with anhydride, salt, and gypsum, appearing at deeper horizons. Throughout the sequence, and particularly towards the surface, there are cavities and cave systems. The most dramatic examples of which are the picturesque blue holes that occur in places on the surface of the banks. Some blue holes exceed depths of 350 feet and one has been plumbed to more than 600 feet (Little et al., 1977). All of the larger Bahama platforms appear to be honeycombed by such cave systems and it is evident that they dominate the hydraulics of the limestone plateaux (Dill 1977). In Southeast Florida cavities are reported down to 6000 feet (Puri et al., 1973), and it is likely that this depth is exceeded in the Bahamas (Tator and Hatfield 1975).

Information on the subsurface geology of the Bahamas has been gathered by geophysical methods and drilling. Geophysical evaluation suggests that the igneous basement may be as deep as 7 miles and that the carbonate sequence extends down to about 24,000 feet. It is postulated that a clastic-evaporite sequence extends from 24,000 feet down to the basement. Five deep exploratory wells have confirmed that limestone and dolomite predominate to 18,000 feet and that clastic deposits start to appear at this horizon. Except for the five deep wells, almost no other drilling has been done below 600 feet, and no continuous cores have been obtained below 550 feet. The surface down to about 100 feet, however, has been thoroughly explored in the land areas because it is at this horizon that the ground water resources occur (Little et al., 1977).

Most of what is known about the stratigraphy below 100 feet is extrapolated from knowledge of the subsurface of Southeast Florida where there has been extensive investigation (Winston 1982).

There is abundant literature available on the unconsolidated sediments of the Bahamas, but very little has been published on the subsurface geology of the area. A few significant studies, however, have been made on cores that were collected by the Bahamas Government, the University of Miami, and various oil companies. These include the work of Supko (1970), Tator and Hatfield (1975), Little et al. (1971-1977), Cant (1977), Gidman (1978), Beach and Ginsburg (1980), Winston (1982), Pierson (1982), and Williams (1985). The subsurface geology outlined below has emerged from these studies.

The Lucayan Limestone

From the rock surface exposed on the banks down to a depth that varies as follows:

Andros and Great Bahama Bank	129 ft.
Long Island	114 ft.
San Salvador	96 ft.
Inagua	84 ft.
Crooked Island and Acklins	63 ft.
Grand Bahama and Abaco	63 ft.
Mayaguana	24 ft.

there is a limestone formation that has been named the Lucayan Limestone (Beach and Ginsburg 1980). It is a dull-yellow to buff pelloidal and ooidal limestone that is irregularly cemented and mottled in appearance. Usually the rock shows evidence of marine bioturbation, but fine bedded structures occur near the bank margins. Throughout the formation there are discontinuity surfaces that are interpreted as horizons of subaerial exposure. The evidence of exposure includes laminated crusts, burnt limestone fragments, fossil soils with land snails, and karstic solution features. Since the banks are subsiding (Banks 1967), the horizons of subaerial exposure must result from low stands of sea level. The units between the discontinuity surfaces correspond to periods of sedimentation when sea level was high. Horizontal cavern systems relate to the discontinuity surfaces the presence of which is often inferred in cores when there is little or no recovery. There are at least 12 discontinuity surfaces in the Lucayan Limestone and some of these are laterally so extensive that they can be used as marker horizons to correlate between the banks. Beach and Ginsburg (1980) report that these surfaces occur on average every 6 feet in the upper 30 feet of the formation and every 10 feet below that level. Pierson (1982) has suggested that the thickness of the Lucayan Limestone is controlled by regional folding and that the great Bahama Bank lies in the center, where subsidence has been highest.

Deeper Formations

Below the Lucayan Limestone the rock changes to a poorly stratified skeletal packstone with common megafossils and thick accumulations of coral and coralline algae near the margins. Besides the change in sediment composition, which is very rapid and widespread, there is also a change in the frequency of discontinuity surfaces. On average these now occur in every 15 feet of core. Other significant changes include the random, appearance of dolomite and the coral Stylophora. Stylophora became extinct in the western Atlantic area at the end of the Pliocene, and therefore, the evidence suggest that the Lucayan Limestone is Late Pliocene to Late Pleistocene age, or, equivalent to the onset of the classic glaciation in the Northern Hemisphere (i.e. ranging from 11,000 to 3,000,000 years in age approximately).

The formations beneath the Lucayan Limestone have not been studied in any detail in the Bahamas. None-the-less the 550 feet core from San Salvador was described by Supko (1970) but it is uncertain how representative this is for any of the other banks particularly the larger ones, and the description is largely petrographic giving little attention to cavities and similar features that affect the hydraulics of the banks. In the San Salvador core there is a breeciated paleosol from 90 feet to 110 feet and from 133 feet down to 493 feet the rock is almost pure dolomite. The thick paleosal suggests a long period of surface exposure (perhaps the onset of glaciation), and the dolomite indicates large scale limestone alteration in a ground water environment that was probably fresh or brackish. A number of discontinuity surfaces can be identified in the sub-100 feet sequence, each of which may be associated with lateral cavern systems.

GROUNDWATER AND HYDRAULICS

Two significant factors regarding the geology and pre-history of the Bahamas impart fundamental controls of the hydraulics of the Bahamas. One is that the banks are made up of marine formed sediments, which are unstable under atmospheric conditions and the other is that there were dramatic fluctuations in sea level during the Pleistocene which exposed these sediments to meteoric conditions. World-wide research has revealed that sea level fell 400 to 600 feet lower than it is at present at the height of the periods of ice advance. It is also known that there were several such advances and in fact sea level fluctuated quite rapidly for more than two million years.

During the low sea levels all of the banks would have been exposed forming high limestone plateaus. Rainfall on the plateaus would have drained away internally, via solution sink holes, until the water table was met, at which point the flow would become more laterally orientated. The vertical phase of flow occurring in the vadose zone would have induced the development of such features as the blue holes, which were already referred to, and the horizontal flow of water in the phreatic zone would have aided in the creation of cavern systems that were horizontally aligned. Because these low sea stands occurred over a wide range of levels, cavern systems could have developed at many difference depths. Longer periods of stable sea level will have created larger more effective caverns particularly if the sea levels reoccurred at the same horizons.

The net effect of the above is that the Bahamas Banks contain cave systems that allow massive volumes of water to flow through them. In other words ocean currents and tides can actually pass through the rock. The movement of this water is concentrated at horizons of ancient sea levels, and the fact that the ice ages resulted in several long periods of sea level low, between depths of 400 and 600 feet, implies that there should be good water movement and caves between these depths. The surface of the Pliocene formations (i.e., the rocks beneath the Lucayan Limestones) were also exposed for a long time and so this horizon should also be cavernous.

The degree and effectiveness of hydraulic connection of a particular subsurface horizon to the ocean is easily proven. For example water in a hole drilled to a particular depth, presumably deep, will have tidal responses that are more closely attuned to the deep ocean as opposed to the local water table, or even the sea on the banks. This was demonstrated in deep holes on Long Island that gave tidal responses that were greater than, and in advance of, those of a tide gauge that was actually positioned in the sea on the west side of the island. This evidence indicates that sea water could pass through the subsurface more easily than it could over the banks. Further evidence of this phenomena is demonstrated by Blue Holes that are out of synchronization with local tides and sometimes these give rise to quite dramatic effects, for example whirlpools, or "fountains". The fact that the Bahamas appears to have an inverted geothermal gradient is also evidence that the deep subsurface is hydraulically linked to the ocean.

Effective hydraulic connection to the sea is obviously a limitation to the development and preservation of freshwater lenses, in an oceanic island, since these lenses are little more than entrapped bodies of rainwater. Easy access of saline water will flush them away. It is significant that the thickness of freshwater lenses on many Bahamian islands is strongly controlled by a particular subsurface horizon. These controls have been described in detail by Cant and Weech (1986), and good examples occur in Andros, Grand Bahama, and Abaco, where freshwater is limited by the base of the Lucayan Limestone, and New Providence where lenses are widely restricted to the first discontinuity surface known locally to drillers as the "Hard Brown Crust". No freshwater lenses are known to extend below the base of the Lucayan Limestone and therefore, this horizon indicates that there is a significant change in rock characteristics at this level. It explains why a land area like North Eleuthera can possess a thicker freshwater lens than either Grand Bahama or Abaco even though the latter are larger land masses and receive greater rainfall. In this context the Lucayan Limestone must be regarded as the freshwater aquifer of the Bahamas, though of course in many areas, particularly in the smaller islands, it may contain salt water.

WASTE DISPOSAL AND RECOGNITION OF DISPOSAL ZONES

The movement of large volumes of salt water through the subsurface, with eventual discharge to the sea, provides an ideal situation for liquid waste disposal. If the unwanted liquid wastes can be injected into a horizontal cavern system with a high hydraulic efficiency, they are rapidly diluted and dispersed, and then carried away to an open sea outfall which is well removed from human activity. The most important requirements are that the wastes be put into adequate cavern systems (disposal zones), and that important overlying aquifers are not contaminated. Obviously one important consideration must be well design, but it is equally important that the geological controls be properly evaluated.

The ideal situation arises where a cavernous formation is overlain by a totally impermeable one, but unfortunately in the Bahamas there are no impermeable formations, or at least none near to the surface. Some horizons are highly permeable and some are not, and again there is something of a wild card situation created by the widely dispersed vertical solution systems.

In Southeast Florida where subsurface waste disposal is fairly well regulated, and geological conditions are relatively similar to the Bahamas, there are recognized waste disposal zones extending down from 2950 feet in a horizon known as the "Boulder Zone", which is approximately 1000 feet thick. The Boulder Zone comprises a highly cavernous dolostone and contains water similar in chemistry and temperature to that of the bottom of the Florida Straits, which implies an effective hydraulic connection to the sea. Some of the caverns in the Boulder Zone are reported to be as much as 90 feet in depth. Above the Boulder Zone, in rocks of Miocene Age, there is a 700 foot section of clays and marls that is supposed to be confining and therefore provides a protective barrier to the Biscayne Aquifer above, from which potable supplies are obtained.

Other cavernous zones have been located in Southeast Florida at shallower and deeper depths, but these are not used for waste disposal. Some contain brackish water and are being conserved for future demineralization, where as others are intended for other purposes, such as providing a source of water for heat exchange purposes, or even storage of treated effluent for future use. Drilling evidence suggests that the caverns have a broad horizontal distribution and that there is practically no vertical connection.

In the Central Bahamas the Boulder Zone horizon probably occurs at a depth of 4000 to 5000 feet., and it is interesting to note that a large number of limestone caves were reported in the Long Island well at a depth of 4600 feet. Other than the 5 deep oil exploration wells no other penetrations have been made to this level in the Bahamas and so little is know about it - in any case the horizon is thought to be too deep to be used as a practical waste receiving zone.

Theoretical considerations suggest that there should be a well-developed horizontally aligned cavern system at depths ranging from 400 to 500 feet. These would have formed during the periods of low sea level, but unfortunately they also demarcate the base of the vertical solution system and so low density wastes could possibly migrate from them to the surface. No impermeable horizons are recognized in the subsurface of the Bahamas though drillers logs sometimes refer to them. Fine-grained dense limestones and dolostones may appear impermeable in a hand specimen, but the formation would still contain a solution network that would allow liquids to pass though. On this basis one is forced to accept the fact that there are probably no "ideal" disposal zones near the surface in the Bahamas, and economic constrains make it impractical to go to deeper levels where such horizons might occur. The only way in which the problem can be bypassed is by ensuring that the wastes are put into rock formations with very high transmissibilities so that there is rapid mixing and dispersal of the effluent and less likelihood of a stationary waste plume developing. An entrapped plume of waste could easily migrate upwards if it is less dense than the surrounding ground water and this is usually the case where sewage effluent is concerned.

Assuming that there are no effective confining layers present in the shallow subsurface it is clear that waste disposal must be made to highly cavernous zones. This means that large volumes of waste should be put down to the 400 to 600 foot level and even modest quantities should go below 100 feet. The Lucayan Limestone must be regarded as the Bahamian aquifer and should be protected in similar fashion as the Biscayne Aquifer. Unfortunately every septic tank discharges effluent directly to the water table and almost all disposal wells discharge into the Lucayan Limestone. The existing Building Code should be amended to specify minimum depth requirements, so that in time all disposal wells are cased through the Lucayan Limestones and double cased through any freshwater resources. Wells into which more than 1 million gallons per day are discharged should take the liquids below 400 feet (i.e. should be cased to 400 feet), and quantities that are large, but less than this can be proportioned between casing depths of 200 feet and 400 feet. For example, a 500,0000 gallon-a-day well should be cased to 300 feet and the range of open hole discharge should be at least half that again of the casing depth, i.e. a 300 feet cased well should have 150 feet of open hole beneath the casing.

One aspect that is important, but has not had the attention it should is the fact that subsurface saline water which is used for heat exchange purposes, and demineralizing, is usually obtained from below 100 feet, i.e. beneath the Lucayan Limestone. Disposing of large quantities of liquid waste at this horizon would make the water at this level unfit for either purpose and therefore, there is strong justification for preserving a saline horizon for future use. For this reason it is suggested that any disposal well discharging more than 100,000 gallons per day should be cased to 200 feet. The 100 foot to 200 foot level could provide saline water for whatever future needs there may be.

For the time being the above must be regarded, at best, as uncertain guidelines which should be modified as more drilling information becomes available.

OTHER CONSIDERATIONS

Earth Movements

Any geotectonic activity must constitute a threat to deep well disposal as it could result in well casings breaking or splitting, and lead, in turn, to contamination of the aquifer. Most people assume that the Bahamas Bank area is tectonically stable, however, there is a wealth of evidence that indicates that earth movements have occurred in the region in recent geological time.

Scheidegger (1977) evaluated join patterns in surface rocks of the Bahamas and obtained a stress pattern that supported the idea that the North American tectonic plate is being dragged westward from the Mid-Atlantic Ridge and is therefore in a state of internal E-W stress relief. On this basis there is a potential threat, but it is hard to evaluate, in human terms, as most of the evidence has been obtained from rocks that are 125,000 years old. It is also possible that the jointing evident is a surface phenomenon only, and therefore presents no threat to deep wells. One deep well at the Blue Hills desalination plant is known to have malfunctioned because its steel casing sheared. The actual break was photographed, but it was not possible to ascertain if it was due to earth movement or poor installation.

In general, there are no measures that can be taken to prevent casings from breaking as a result of earth movement. If the rock itself is fracturing, then plastic or steel will also break. The only solution to the problem is to ensure that the integrity of the casing is monitored so that those using the well will know when there is a casing failure.

Well Design

It is beyond the scope of this report to provide detailed specifications of well design, except with regard to recommended well depths and casing levels, which are given at the end of a previous chapter. Besides those guidelines, a few other suggestions are listed that are worth considering whenever a deep disposal well is being installed.

Casings: raw sewage and salt water can create acidic solutions that corrode metallic components. In these circumstances plastic piping should be used or similar resistant material. If the casing needs to be strengthened, then this can be achieved by means of outer casings and grouting. As a rule, all wells should be double cased through fresh or brackish water horizons. If toxic wastes are to be disposed of by means of a deep well, in an area where there is an important ground water resource, then the annulus between the casings should be used as a means of monitoring for casing failure. There are several methods by which this can be done and these are described in literature that is available.

Monitoring: as mentioned above, the annulus between two casings can be used to monitor the integrity of the casings, however, aquifer contamination can result from more than casing failure and therefore another means of monitoring is needed when it is essential that ground-water should not be polluted. In this circumstance isolated, purpose built, monitor wells are required. These should be installed to depths that will give significant meaningful information, and spaced out to provide the same. To reduce drilling costs a single well can be used for multi-level observation, if properly designed with a number of inner casings that open out to different depths. Unfortunately, karst hydrology results in erratic and unpredictable movement of ground water, and so serious pollution can occur via a fissure system that is undetected by a monitor well network. The development of a massive waste plume however, should be detected in the well sampling results.

Liquid wastes: the precise composition of the waste being disposed of is an important consideration that should play a part in the design requirements of any disposal well. For example certain waste substances may require treatment before being injected into the well, and an obvious example is heated acidic liquid which is liable to eat away the groundrock to create a variety of problems. Similarly some wastes may contain too many solids that are too large and these could lead to blockages in the disposal zone or well. In this case screening is necessary.

In contrast to the two cases given above, there may be some justification in not lining the disposal well at all. For example a storm water drainage well could be used as a means of artificial recharge, and therefore, should be open from the water table down. Casing a drainage well down to saline horizons results in the loss of potential recharge.

Disposal Well Operation

Many people live in parts of the world where a deep well will not accept injected waste, or, where the well may even eject water as in artisan situations, and so brief mention should be made as to how these wells operate. In principal it is quite simple: the wells simply link the land surface to cavern horizons that open out into the sea. On this basis they act as drains and will continue to work as long as they do not become blocked, and as long as there is enough free head available to ensure that flow is maintained towards the sea. Problems arise where there is poor hydraulic connection to the sea and where the water table periodically rises to cover the level of discharge. The fact that freshwater floats on seawater, tends to add a little confusion to the situation in Bahamian Islands because it means that the head component must be measured relative to the local water table, and not sea level itself. If head were effective relative to sea level and if the Ghyhen-Hertzberg principal were not influential then one would expect the top of the freshwater lenses to drain down into the sea via the deep blue hole system.

CONCLUSIONS

(a) Deep disposal wells can work in the hydrologeological environment of the Bahamas.

(b) The most important requirements are that acceptable liquid wastes be conveyed to adequate receiving zones, without leakage in transmission, and without blockage.

(c) If there is any chance that water resources, fresh, brackish, or saline, may be polluted then measures should be taken to see if and when this happens.

(d) At present there is not enough subsurface information available to be certain where wastes are going. Exploratory drilling is needed in conjunction with proper geophysical logs.

(e) Many disposal wells in current use are too shallow, and have not been built to adequate standards.

(f) Major disposal systems often rely on one well only and so use of the well continues even when the well has failed.

(g) The Building Code and equivalent rules are not strict enough to protect the subsurface from pollution.

RECOMMENDATIONS

(a) Studies should be made to establish recognized waste disposal zones in the subsurface of the Bahamas. These studies will require exploratory well drilling, coring, and geophysical logging.

(b) Once accepted disposal zones have been identified, a full range of specifications should be established to control all aspects of deep well disposal. These will include such items as well depths, well head installations, casing requirements, pretreatment, and subsequent monitoring. The specifications should be enforced by the Building Code, and appropriate legislation.

ACKNOWLEDGEMENTS

This paper is published with the kind permission of the Honourable Darrel E. Rolle, the Minister of Works and Utilities, Mr. Luther Smith, the Permanent Secretary, and Mr. Peter Gordon, the Director of Public Works. The opinions expressed are those of the author and not necessarily those of the Government of the Commonwealth of the Bahamas.

REFERENCES CITED

Banks, J.E., 1967, Geologic history of the Florida-Bahamas Platform in: Symposiumon the geological history of the Gulf of Mexico, Antilleon-Caribbean region Gulf Cost Assn, Geol. Society Trans 17.

Beach, D.K., and Ginsburg, R.N., 1980, Facies Succession of Pliocene-Pleistocene Carbonates, Northwestern Great Bahama Bank. A.A.P.G. Bull Vol 64.

Cant, R.V., 1977, Role of coral deposits in building the margins of the Bahama banks. in: Taylor, D.L. (Ed). Proc third Internat. Coral

Reef Symposium Miami, Florida, Vol 2.

Cant, R.V., and Weech, P.S., 1986, A review of the factors affecting the development of the Ghyhen-Hertzberg lenses in the Bahamas. Journal of Hydrology, 84, p. 333-343.

Dill, R., 1977, The Blue Holes - geologically significant submerged sink holes and caves off British Honduras and Andros, Bahama Islands, in: D. L.Taylor (Ed) Proc. Third Internat Coral Reef Symposium Miami, Florida Vol. 2.

Gidman, J., 1978, Diagenesis of cored Pleistocene carbonates, Great Abaco Island, Little Bahama Bank. Ph.D. dissertation, Univ. of Liverpool, U.K.

Little, B.G., Buckley, D.K., Cant, R., Gillis, W.T., Henry, P.W.T., Jefferies, A., Mather, J.D., Stark, J., and Young, R.N., 1977, Land Resources of the Bahamas: a Summary. Land Resources Study 27, Land Resources Division, MOD Surbiton, Surrey.

Pierson, B.J., 1982, Cyclic sedimentation, Limestone diagenesis, and dolomitization in Upper Cenozic Carbonates of the Southeastern Bahamas. Unpublished Ph.D. dissertation, Univ. of Miami, Coral Gables, Florida.

Puri, H.S., Faulkner, G.L., Winston, G.O., 1973, Hydrogeology of subsurface liquid waste storage in Florida, in: proc.2nd Internal Symposium on Underground Waste Management and Artificial Recharge, New Orleans, Louisiana.

Scheidegger, A.E., 1977, Joints in the Bahamas and their geotectonic significance, Rivista Italiana Di Geofisica E. Scienza affin V. 4.

Supko, P.R., 1970, Depositional and diagenetic features in subsurface Bahamian Rocks, Technical Report 70 - 1 Univ of Miami, Rosenstiel School of Marine and Atmospheric Sciences, Miami, Florida.

Tator, B.A., and Hatfield, L.E., 1975, Geology of the Bahamas with evaluation of petroleum potential. Consultants report produced for Min. of Development, Nassau, Bahamas. Weeks Tator Consulting Geologists Inc., Miami, Florida.

Williams, S.C., 1985, Stratigraphy, Facies Evolution, and Diagenesis of Late Cenozoic Limestone and Dolomite, Little Bahamas Bank, Bahamas, Unpublished Ph.D. dissertation, Univ. of Miami, Coral Gables, Florida.

Winston, G.O., 1982, Comparison of the stratigraphy of the Bahamas Deep Tests and Florida Coastal Wells, Consultants report produced for Bahamas Government, Tator & Associates, Miami, Florida.

SELECTIVE CEMENTATION OF EOLIAN STRATIFICATION IN PLEISTOCENE CALCARENITES, SAN SALVADOR ISLAND, BAHAMAS

Mario V. Caputo
Department of Geology and Geography
Mississippi State University
Mississippi State, Mississippi 39762

ABSTRACT

Subdivisions of the Grotto Beach Formation, such as the French Bay Member and eolian beds in the Cockburntown Member, comprise part of the Pleistocene interval of fossiliferous, peloidal oosparite or calcarenite exposed on southern San Salvador Island. Stratification is typified by complex crossbeds consisting of sandflow and local grainfall foresets and wind-ripple strata. Each stratification type is characterized by specific grain size, packing, and pore size inherited from eolian subprocesses such as sandflow, grainfall and ripple migration. These grain textures have controlled the degree of fluid adhesion and capillarity which have affected the degree of cementation within each stratum. Climbing wind-ripple strata are recognized in outcrop by thin, even laminations which locally exhibit inverse grading. Both outcrop exposures and thin sections are typified by pairs of lamina-sets consisting of alternating positive-relief "micro-ledges" and negative- relief "micro-recesses". A "micro-ledge" forms the lower part of a wind-ripple stratum and consists of moderately to tightly packed, well-cemented fine sand among which the pores have been completely cemented with interlocking equant crystals of low-magnesium calcite. The upper part of a ripple stratum forms a "micro-recess" composed of moderately to loosely packed, poorly cemented fine medium sand. Sandflow beds show a two-fold segregation of fine sand in the basal few millimeters and medium sand in the upper few centimeters in each bed in outcrop and in thin section. The basal fine layers are characterized by moderate to tight grain packing and small pores. The pores are completely filled with interlocking equant crystals of low-magnesium calcite such that the basal laminations form positive-relief "micro-ledges" in weathered outcrops. The upper part of a sandflow bed is composed of moderately to loosely packed medium sand with large pores which have been incompletely filled with isopachous and meniscus cement. Consequently, negative-relief, "micro-recess"-forming beds distinguish upper sandflow beds from other strata. Grainfall strata were not examined in this study.

INTRODUCTION

General Statement

In their pioneer studies of quartzarenites, Huntington (1907), Knight (1929), and McKee (1934a,b) distinguished physical aspects of cross-bedding which supported an origin by migration of ancient eolian dunes. Since then, the recognition of specific eolian stratification types or fine structure (Hunter and Rubin, 1985) within cross-bedded sets of quartzose sediments (Hunter, 1977a,b) and sandstones (Hunter, 1981) has greatly enhanced the ability of sedimentologists to discriminate eolian from subaqueous stratification. Later studies have defined fine structure in the interpretation of some Paleozoic and Mesozoic eolian sandstones in the Western Interior (Kocurek and Dott, 1981; Blakey and Middleton, 1983; Loope, 1984; Porter, 1987; Caputo, 1988) and other parts of the world. Despite the fact that wind-blown limestones (cf. eolian calcarenites or eolianites) have been recognized since the turn of the century (Evans, 1900) in various tropical and subtropical parts of the world (Ball, 1967; MacKenzie, 1964a,b; Ward, 1970, 1973, 1975; McKee and Ward, 1983), distinctive eolian stratification has only recently been described in Holocene (White and Curran, 1985, 1986, 1988), and Paleozoic (Loope, 1986; Hunter, 1988) eolian limestones.

Purpose

This report presents preliminary observations and interpretations of eolian bedding styles preserved in intervals of the Pleistocene Grotto Beach Formation, namely the French Bay Member, and eolian beds in the Cockburntown Member, on southern San Salvador Island. The report will attempt to bridge the gap between studies on siliciclastic eolian deposits and eolian limestones and describe a relationship between eolian processes, grain characteristics, and cementation patterns which are expressed in different weathering patterns in outcrop. The following ideas are presented in the report: 1. Specific bedform subprocesses such as sandflow, grainfall, and the migration of ripples in a general wind-blown environment do indeed operate for carbonate grains. 2. These subprocesses produce sandflow, grainfall and ripple strata in carbonate sediment which are similar to those in siliciclastic eolian deposits. 3. These strata are characterized by different grain sizes and packing which are inherited from bedform subprocesses. Furthermore, the inherited grain size and packing have controlled fluid capillarity between grains and subsequent cementation patterns. 4. The grain-cement relationships are expressed by different weathering patterns in outcrop. 5. Weathering patterns which are associated with specific eolian stratification types can be used to identify sandbody architecture in outcrops.

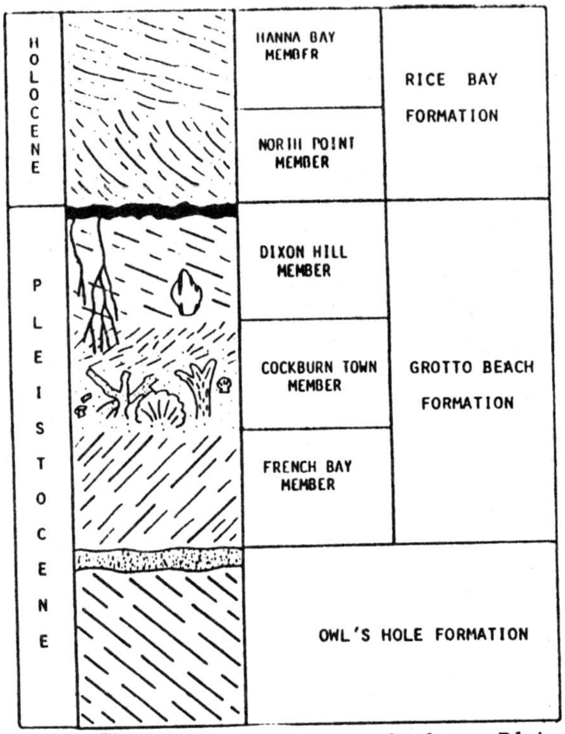

Fig. 1. Nomenclature as applied to Pleistocene and Holocene sedimentary intervals on San Salvador Island (Carew and Mylroie, 1985).

Location and Stratigraphic Setting

San Salvador is one of a series of emergent island platforms of the Bahamian Islands (Index map 1). It consists entirely of late Pleistocene and early to late Holocene, cemented, skeletal and oolitic carbonate rock.

Outcrop examples of eolian stratification and petrographic examples of grain texture, fabric, and cementation are selected from the Pleistocene Grotto Beach Formation. This unit comprises part of a well- exposed, significantly thick stratigraphic interval in the upper Neogene System of rocks on San Salvador Island (Fig. 1). The development of stratigraphic intervals and nomenclature for this locality is given in Carew and Mylroie (1985). The French Bay Member, Cockburntown Member, and Dixon Hill Member compose the Grotto Beach Formation in ascending order (Fig. 1). Physical sedimentary features and stratigraphic relations suggest an eolian origin for the lower and upper members and marginal marine and eolian affinities for the middle Cockburntown Member. The French Bay Member forms gray, wave-battered seacliffs which armor the shore south of Grotto Beach, around the Sandy Point headland, and along part of the north rim of French Bay on the southwestern part of the island. The exposure of Grotto Beach Formation is interrupted to the east in Blackwood Bay, an embayment on the northeastern side of French Bay, east of which the interval is probably represented by a facies of eolian origin in the Cockburntown Member and is exposed as far east as the Gulf and Sandy Hook areas along the southern edge of the island (Index map 1).

Previous Work

Other sedimentologists have previously noted how grain size and packing within sedimentary units affect the behavior of fluid and subsequent cementation within intergranular pores. Laminations and beds consisting of fine, tightly packed particles have a stronger capability of retaining liquid in the smaller pore spaces because of enhanced fluid surface tension and capillarity (Ward, 1975). The relationship between grain size and fabric, fluid retention, and effective cementation has been observed in marine (Halley and Harris, 1979; Strasser and Davaud, 1986; Evans and Ginsburg, 1987), and eolian (Ward, 1975; McKee and Ward, 1983; White and Curran, 1988) limestones and in unconsolidated quartzose sand of eolian origin (Fryberger and Schenk, 1988). No previous report has further drawn a relationship between eolian processes and grain characteristics and consequent cementation and weathering patterns in eolian limestones, the latter of which may be useful in distinguishing eolian stratification in carbonate rocks.

SEDIMENTARY PETROLOGY

Stratification

The French Bay Member and the eolian beds of the Cockburntown Member of the Grotto Beach Formation form a sequence of complex wedge-and tabular-shaped bedding units consisting of parallel wind-ripple strata and cross strata. These bedding units are internally punctuated by intricate second-order bounding surfaces. Thick cosets of crossbeds and wind-ripple strata are interrupted by a laterally extensive interval of terra rossa calcrete which is up to 15cm thick and forms the present topographic surface in places (Fig. 2). Individual bedding units are preserved as partial or complete dune-form bedding. Complete dune-form bedding consists of topset, brinkset, foreset, and bottomset beds, may compose one entire bedset (Fig. 3), and ranges in thickness from 0.8m to 6.0m. Subhorizontal wind-ripple strata form amalgamated topset beds 0.3m to 1.5m thick and grade downcurrent into gently dipping brinkset laminations which are transitional with sandflow and grainfall foresets (Fig. 3). Angle-of-repose foreset beds are composed of sandflow strata 2-5cm thick and local grainfall laminations 1-3mm thick. Foreset beds ultimately grade downcurrent into bottomset beds which form a complex of sandflow tongues and grainfall and wind-ripple strata.

In modern and ancient eolian quartz sand deposits, climbing translatent strata (Hunter, 1977a,b) or wind-ripple strata are recognized by the thin, even, cyclic occurrence (hence, pin stripe laminations of Fryberger and Schenk, 1988), poorly developed foresets between bounding surfaces, and a general upward-coarsening grain-size trend between lower and upper bounding surfaces (Hunter, 1977a, b; 1981; Fryberger and Schenk, 1981; Schenk, 1983). Wind-ripple strata in rocks are generally so thin that foreset laminations and inverse grading are not clearly discernible. Consequently, the thin, even cyclic nature of the bedding may be the only diagnostic criterion for recognizing wind-ripple strata (Hunter, 1977a; Fryberger and Schenk, 1981).

A wind-ripple stratum in eolian beds of the Grotto Beach Formation is 2 to 5mm thick and was formed by the migration of a wind ripple in an ancient coastal dune setting. Each stratum is cyclic and is composed of an apparent paired or two-fold layering associated with inverse grading and selective cementation of the finer-grained layer. Each layer or lamination is distinguished by color, grain size and packing, and cementation and weathering patterns (Fig. 4). Other criteria suggesting that the thin, even, parallel strata in the Grotto Beach Formation are the product of migrating ripples include the absence of parting lineation, a weathering feature attributed to deposition by upper flow regime plane beds or laminations, and the presence of high index ripple marks on bedding plane surfaces (Fig. 5).

Inverse textural grading is very well developed and is visible in oblique exposures which exaggerate the thickness of each ripple-produced stratum (Fig. 6). The lower portion of each stratum or fine-coarse couplet is characterized by fine grained, moderately well-packed and well cemented sand (next section, Petrography). It weathers to form a white, positive-relief "micro-ledge" on the outcrop. The upper part of each stratum consists of fine medium-grained, moderately to poorly packed, poorly cemented sand laminations (next section, Petrography) which weather to form gray, negative-relief "micro-recesses" on the outcrop. Collectively, the white and gray laminations form a sequence of closely spaced, parallel ledges and recesses (Fig. 4).

In general, eolian foreset bedding or cross-bedding is made of a series of individual sandflow strata or interbedded sandflow and grainfall strata (Fig. 7). Sandflow strata are lenticular, tongue-shaped units and usually form by grains slipping or avalanching then coming to rest on the steep leeside or slipface of bedforms. Gusts of wind flowing over the crest of a bedform may suspend finer grains which fall out of suspension about midway along and near the base of the leeside to produce grainfall laminations which interfinger with sandflow beds. The bottomset of foreset bedding is usually a complex of grainfall and wind-ripple strata and sandflow toes (Fig. 7). Crossbedding in the Grotto Beach Formation consists mostly of sandflow beds 2 to 5cm thick which are locally interbedded and surrounded by grainfall laminations (Fig. 8). This relationship is poorly preserved in the Grotto Beach limestone and can be locally observed on exposure surfaces which show the up-dip truncations of sandflow and grainfall strata (Fig. 9). The sandflow beds appear as gray, discontinuous lenses enclosed by white grainfall laminations of similar geometry.

Sandflow beds are inversely graded as a result of grain flow sediment mechanics (Bagnold, 1954; Middleton, 1970; Hunter, 1977a,b; 1981; 1985b) so there is a two-fold segregation of grain sizes similar to that in wind-ripple strata. Fine grained sand is concentrated near the base of a sandflow bed, exhibits improved packing, and is well cemented. The fine sand within each stratum grades vertically into medium sand which is loosely packed, and poorly cemented (next section, Petrography). Consequently in outcrop, crossbedding is characterized by a series of white positive relief "micro-ledges" of the basal sandflows and gray, negative relief "micro-recesses" of the upper sandflows spaced 2 to 5cm apart (Fig. 10). The well-cemented ledge-forming laminations can alternatively be identified as grainfall laminations only if: 1. the laminations in question can be traced down-dip into bottomsets and thicken at the expense of sandflow tongues, and 2. the laminations can be seen forming lense-shaped outlines surrounding sandflow lenses. These relationships are only very locally exposed in the French Bay Member and have not been examined in thin section.

Petrography

Details of the grain size and packing and cementation patterns associated with the eolian strata described above can be examined in oriented thin sections. Eleven samples were collected from the French Bay Member of the Grotto Beach Formation; seven oriented samples are from sandflow beds and four oriented samples are from beds of wind-ripple strata. All the limestones examined in this study can be classified as sparsely fossiliferous, peloidal oosparites or ooid grainstones, the allochemical grains of which have been cemented with low-magnesium calcite (Hutto and Carew, 1984) and partly micritized.

The two-fold subdivision of wind-ripple strata as a result of grain segregation during migration of eolian ripples is clearly visible in thin section (Fig. 11). The lower part of each stratum is characterized by fine-grained sand with an average diameter of 0.149mm and the upper layer contains fine medium-grained sand with an average diameter of 0.283mm. Because of smaller grain and pore size and improved packing similar to that observed in basal sandflow beds, the lower layer is well-cemented with equant calcite crystals which completely fill the pore (Fig. 11). Subsequently, this part of the wind-ripple stratum forms a positive-relief, "micro-ledge" weathering pattern on the outcrop (Fig. 4). The upper layer is less well-cemented because of coarser grains of relatively looser packing, larger pores, and therefore poorer fluid retention. These are features similar to those in the upper part of a sandflow bed. Consequently, the upper part of a wind ripple forms a negative-relief, "micro-recess" weathering pattern in outcrop (Fig. 4).

In thin section, sandflow beds are characterized by inverse size grading. A sharp base marks an abrupt grain size change from fine-grained to medium-grained sand between vertically adjacent sandflows. The gradual rather than abrupt change in grain size from fine- to medium-grained sand within each sandflow bed suggests that grain segregation was the result of grainflow mechanics and not from differences between sandflow and grainfall processes. Experimental studies have shown that during a grainflow, finer grains accumulate at or near the base of the flow (Bagnold, 1954; Middleton, 1970; Fryberger and Schenk, 1981). Schenk (1983) has further shown that these finer grains either remain behind on the plane of shearing or are incorporated into the base of the sandflow stratum when the flow ceases.

Fryberger and Schenk (1988) have demonstrated that decelerating gusts of wind can produce normally-graded grainfall strata and that accelerating gusts of wind are capable of producing inversely graded grainfall strata which resemble sandflow strata. However, the inversely graded foreset beds in the Grotto Beach Formation are sandflow beds because of their lenticular nature along strike and because they wedge-out into complex bottomsets of grainfall and wind ripple strata.

The lower portion of a sandflow bed is characterized by medium-grained sand with an average diameter of 0.160mm. The grains exhibit point or tangential to long contacts which suggest moderate to high compaction and packing. Entire intergranular pores are filled with a primary generation of blocky, isopachous calcite cement around each grain and a secondary generation of interlocking calcite cement crystals which filled the remainder of the pore (Fig. 12). This portion of the sandflow bed exhibits a positive-relief, "micro-ledge"-forming weathering pattern in outcrop because of the well-cemented nature of the grains.

Fig. 2. Thin interval of calcrete (arrow) which locally interrupts bedding units in eolian beds of Cockburntown Member, Grotto Beach Formation. Seacliff exposure west of the Gulf area. Scale is in inches and centimeters.

Fig. 3. Crossbed set consisting of topset (ts), brinkset (bks), foreset (fs) and bottomset (bs) beds. Eolian beds of Cockburntown Member, Grotto Beach Formation in seacliffs west of the Gulf area. Scale is in inches and centimeters.

Fig. 4. Wind-ripple strata preserved as amalgamated topset beds. Note local angular discordances (arrow) between bedsets. Scale is in inches and centimeters. Eolian beds in Cockburntown Member, Grotto Beach Formation in seacliffs west of the Gulf area.

Fig. 5. High-index ripple marks poorly preserved on bedding plane surface within amalgamated wind ripple strata. Scale is 15 cm long. Eolian beds of Cockburntown Member, Grotto Beach Formation in seacliffs west of the Gulf area.

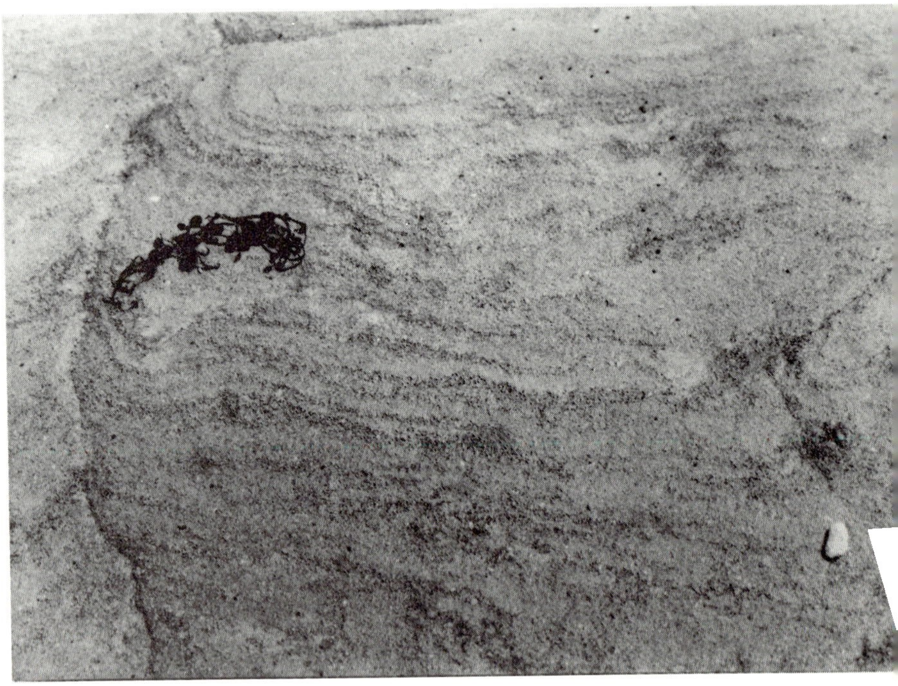

Fig. 6. Thickness of wind ripple strata exaggerated along exposure oblique to bedding. Each stratum is composed of finer-grained laminations in the lower part and coarser grained laminations in the upper part. Dark object in upper left is piece of Sargassum weed. Eolian beds in Cockburntown Member, Grotto Beach Formation in seacliffs west of the Gulf area.

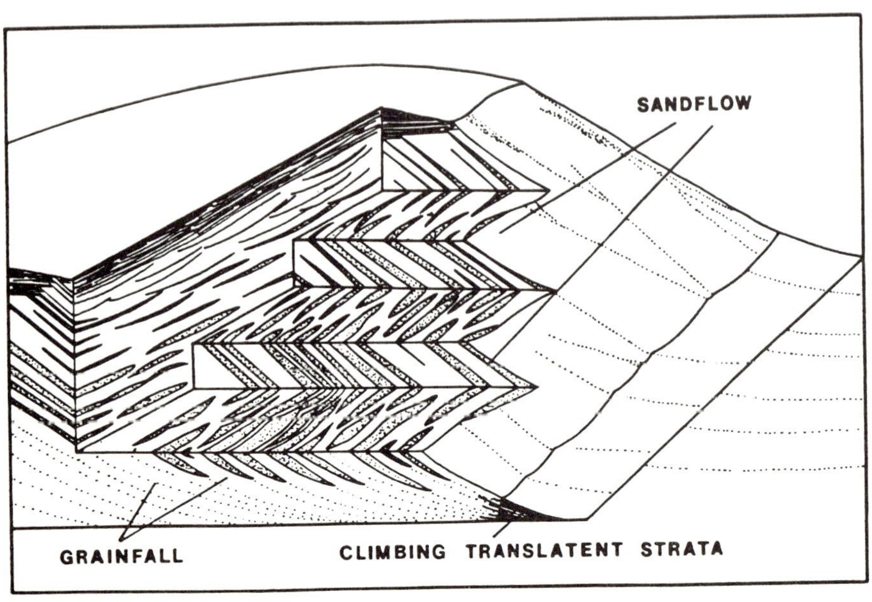

Fig. 7. Geometry and occurrence of stratification types within a general eolian dune-form (redrawn from Hunter, 1977a).

Fig. 8. Tangential bottomset of sandflow beds forming "micro-recesses" and wedging-out into a complex of grainflow and wind ripple strata which form "micro-ledges".

Fig. 9. Locally preserved grainfall laminations (arrow) encircling lenticular sandflow beds in transverse section. Eolian beds in the Cockburntown Member, Grotto Beach Formation in seacliffs west of the Gulf area. Scale is in inches and centimeters.

Fig. 10. Alternating small-scale ledge- and recess-forming strata composed of basal and upper parts of sandflow beds. Eolian beds in the Cockburntown Member, Grotto Beach Formation in seacliffs west of the Gulf area. Scale is in inches and centimeters.

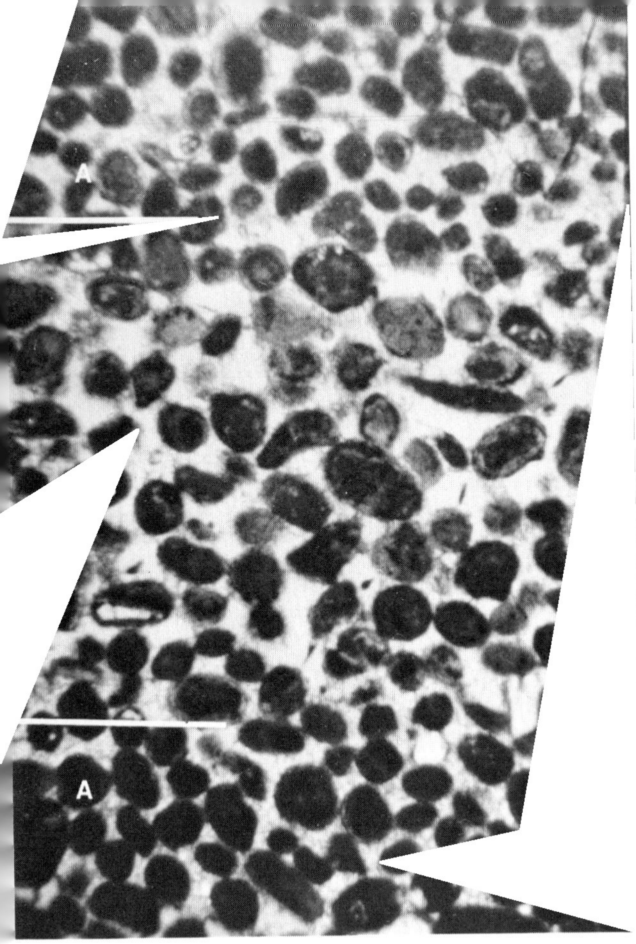

Fig. 11. Photomicrograph of a ripple stratum consisting of basal well-cemented fine grained (A) laminations which grade upward into poorly cemented fine medium grained laminations (B). Low power; uncrossed nicols.

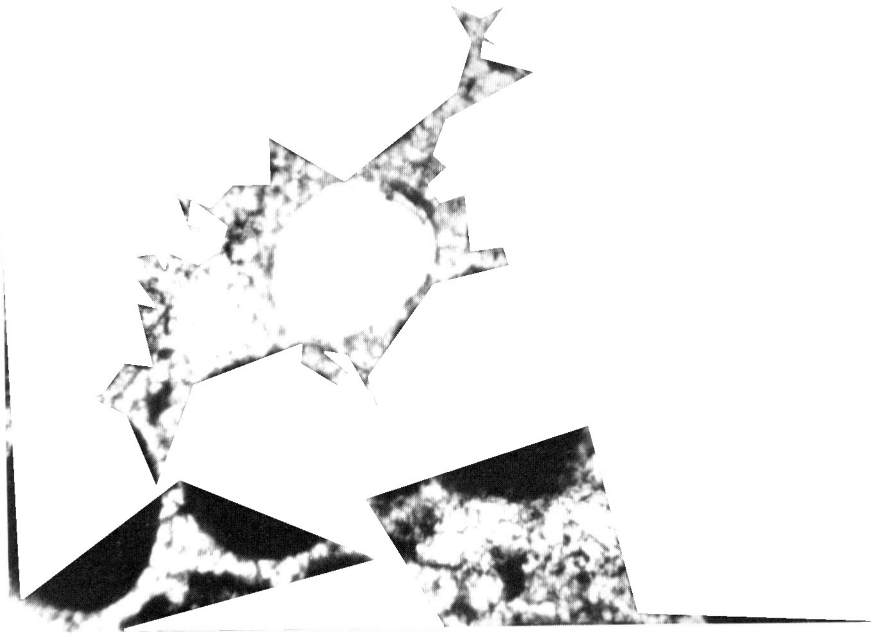

Fig. 12. Photomicrograph of basal part of sandflow bed. Carbonate sand is fine grained and is completely cemented with equant crystals of calcite. Medium power; uncrossed nicols.

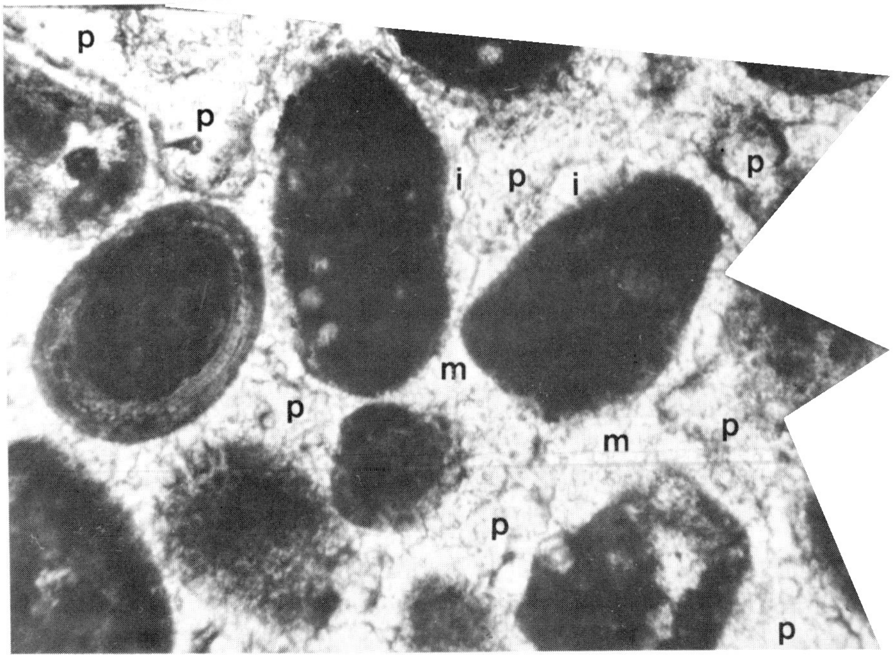

Fig. 13. Photomicrograph of upper part of sandflow bed. Carbonate sand is medium grained and partially cemented with isopachous (i) and meniscus (m) cements. Pore spaces (p). Medium power; uncrossed nicols.

In unconsolidated sedimentary beds, fluid surface tension and capillarity is enhanced by small pore space related to tight grain packing (Bagnold, 1938; Ward, 1975). These conditions create a micro-environment at the scale of a thin bed or lamination which resemble those of a phreatic diagenetic environment where the pore is filled entirely with liquid. Calcite cements formed in freshwater phreatic environments are generally characterized by interlocking equant crystals which fill entire pores (Longman, 1980).

Carbonate particles in the upper parts of sandflow beds are medium-grained sand with an average diameter of 0.330mm. They exhibit point and floating contacts which suggest moderate to low compaction and packing. The grains are held together by equant to bladed isopachous and equant meniscus cement crystals which partly fill the intergranular pores (Fig. 13). Meniscus calcite cements are usually precipitated in a freshwater vadose diagenetic environment where drops of water having meniscus or hour-glass shapes adhere between grains (Longman, 1980). However, the association of meniscus vadose cements with isopachous phreatic cements in the same bedding interval suggests that the pores were temporarily filled with fluid to form isopachous cement and later drained or evacuated by evaporation to form meniscus cement.

CONCLUSIONS AND SUMMARY

Physical subprocesses such as sandflow, grainfall, and the migration of ripples operate in a general wind-blown sedimentary environment to produce sandflow, grainfall, and wind-ripple strata in modern and ancient siliciclastic deposits. These subprocesses have generated similar eolian strata during Pleistocene time in fossiliferous, pelloidal oosparites of the French Bay Member and eolian beds in the Cockburntown Member of the Grotto Beach Formation exposed on southern San Salvador Island, Bahamas. Interpretations in this study suggest that grain textures inherited from eolian subprocesses have influenced cementation patterns among different eolian strata. Cementation patterns are expressed as ledge- and recess-forming strata in outcrop. The relationship between cementation-weathering patterns may be useful in distinguishing fine structure in eolian calcarenites.

Ripples associated with ancient eolian dunes have produced wind-ripple strata in the Grotto Beach Formation. Each stratum is composed of a lower fine sand lamination and an upper fine medium sand lamination. Differences in weathering pattern reflect cementation patterns which are controlled by grain texture. Smaller pores are related to finer grain size and tighter grain packing and favor strong fluid capillarity and retention. Consequently, laminations which have such a texture are well cemented and crop out as "micro-ledges" in weathered exposures. Conversely, larger pores are related to coarser grains and looser packing which diminish fluid retention and capillarity. Incomplete cementation of pores by meniscus calcite cement has resulted in calcarenite laminations which form "micro-recesses" in weathered exposures.

Grain flow mechanisms have controlled the geometry and distribution of sand flow beds and internal characteristics such as grain size and packing which ultimately have controlled cementation patterns in each bed. Grain segregation during grain flow produced a concentration of more tightly packed, fine-grained sand at the base and a gradual vertical change toward more loosely packed medium-grained sand in the upper part of sandflow beds. Fluid surface tension and capillarity was enhanced by small pore space related to fine grain size and greater grain compaction in the basal sandflow. Therefore, at the time of cementation, interlocking, bladed, equant crystals of low-Mg calcite, as isopachous rim and complete pore-filling cement, precipitated in the intergranular pore space. The pattern of complete cementation in the basal sandflow is expressed in outcrop as a thin positive-relief "micro-ledge". Conversely, medium-grained sand in the upper parts of sandflow beds exhibits looser packing resulting in larger pores which have been partially cemented with equant isopachous and meniscus calcite cement. This part of the sandflow is recognized as a "micro-recess" between well cemented "micro-ledges" in weathered outcrops.

ACKNOWLEDGEMENTS

I greatly appreciate the cooperation and financial support given by John Mylroie and the Department of Geology and Geography at Missisipi State University, Donald Gerace, Director of the Bahamian Field Station on San Salvador Island, and Conrad Gazzier, (Director), Chuck Peal, and Cragin Knox at the Mississippi Bureau of Geology for the initiation of this study. Helpful comments were given by Ralph Hunter, John Barwis, Brian White, Al Curran, Mark Boardman, and John Mylroie. Jim Harrington, owner and operator of The Thin Section Lab in San Diego, California made incomparable thin sections.

REFERENCES CITED

Bagnold, R.A., 1954, Experiments on a gravity-free dispersion of large solid spheres in a Newtonian fluid under shear: Proceedings Royal Society of London, 225 A, p. 49-63.

Bagnold, R.A., 1938, Grain structure of sand dunes and its relation to their water content: Nature, v. 142, p. 403-404.

Ball, M.M., 1967, Carbonate sand bodies of Florida and the Bahamas: Journal Sedimentary Petrology, v. 37, no. 2, p. 556-591.

Blakey, R.C., and Middleton, L.T., 1983, Permian shoreline eolian complex in central Arizona: dune changes in response to cyclic sealevel changes, in Brookfield, M.E., and Ahlbrandt, T.S., eds., Eolian sediments and processes: Amsterdam, Elsevier Publishers, p. 551-581.

Caputo, M.V. 1988, Origin of sedimentary facies in the upper San Rafael Group (Middle Jurassic), east-central Utah: unpublished Ph. D. dissertation, University of Cincinnati, 354 p.

Carew, J.L., and Mylroie, J.E., 1985, the Pleistocene and Holocene stratigraphy of San Salvador Island, Bahamas, with reference to marine and terrestrial lithofacies at French Bay, in Curran, H.A., ed., Pleistocene and Holocene carbonate environments on San Salvador Island, Bahamas: Geological Society of America, College Center Finger Lakes, Bahamian Field Station, p. 11-62.

Evans, C.C., and Ginsburg, R.N., 1987, Fabric-selective diagenesis in the late Pleistocene Miami Limestone: Journal Sedimentary Petrology, v. 57, no. 2, p. 311-318.

Evans, J.W., 1900, Mechanically formed limestones from Junagarh (Kathiawar) and other localities: Geological Society London Quarterly Journal, v. 56, pp. 559-583; 588-589.

Fryberger, S.G., and Schenk, C., 1981, Wind sedimentation tunnel experiments on the origins of aeolian strata: Sedimentology, v. 28, no. 6, p. 805-822.

Fryberger, S.G., and Schenk, C.J., 1988, Pin-stripe lamination: a distinctive feature of modern and ancient eolian sediments: Sedimentary Geology, v. 55, p. 1-15.

Halley, R.B., and Harris, P.M., 1979, Fresh-water cementation of a 1,000-year-old oolite: Journal Sedimentary Petrology, v. 49, no. 3, p. 969-988.

Hunter, R.E., 1977a, Basic types of stratification in small eolian dunes: Sedimentology, v. 24, p. 361-387.

Hunter, R.E., 1977b, Terminology of cross-stratified sedimentary layers and climbing-ripple structures: Journal Sedimentary Petrology, v. 47, p. 697-706.

Hunter, R.E., 1981, Stratification styles in eolian sandstones: some Pennsylvanian to Jurassic examples from the western interior U.S.A., in Ethridge, F.G., and Flores, R.M., eds., Recent and ancient nonmarine depositional environments: models for exploration: Society Economic Paleontologists and Mineralogists Special Publication 31, p. 315-330.

Hunter, R.E., 1985b, Subaqueous sand-flow cross-strata: Journal Sedimentary Petrology, v. 55, no. 6, p. 886-894.

Hunter, R.E., 1988, Eolianites in the Ste. Genevieve Limestone (Mississippian) of southern Indiana (abs.), in Abstracts: Society Economic Paleontologists and Mineralogists, v. 5, p. 26-27.

Hunter, R.E., and Rubin, D.M., 1985, Interpreting eolian cross-strata: the limits of the possible (abs.): Society Economic Paleontologists and Mineralogists Annual Midyear Meeting, Abstracts, v. 2, p. 44.

Huntington, E., 1907, Some characteristics of the glacial period in nonglaciated regions: Geological Society of America Bulletin, v. 18, p. 351-388.

Hutto, T., and Carew, J.L., 1984, Petrology of

eolian calcarenites, San Salvador Island Bahamas, in Teeter, J.W., ed., Proceedings of second symposium on geology of Bahamas: College Center of Finger Lakes, p. 197-207.

Knight, S.H., 1929, The Fountain and Casper Formations of the Laramie Basin - a study on genesis of sediments: Wyoming University Special Publication in Geology, v. 1, no. 1, p. 1-82.

Kocurek, G., and Dott, R.H., Jr., 1981, Distinctions and uses of stratification types in the interpretation of eolian sand: Journal Sedimentary Petrology, v. 51, no. 2, p. 579-595.

Longman, M.W., 1980, Carbonate diagenetic textures from nearsurface diagenetic environments: American Association of Petroleum Geologists Bulletin, v. 64, no. 4, p. 461-487.

Loope, D.B., 1984, Eolian origin of upper Paleozoic sandstones, southeastern Utah: Journal Sedimentary Petrology, v. 54, no. 2, p. 563-580.

Loope, D.B., 1986, Pennsylvanian eolian limestones, Paradox Basin, U.S.A. (abs), in Abstracts, Sediments down under: International Association of Sedimentologists, 12th International Sedimentological Congress, p. 189-190.

MacKenzie, F.T., 1964a, Geometry of Bermuda calcareous dune cross-bedding: Science, v. 144, p. 1449-1450.

MacKenzie, F.T., 1964b, Bermuda Pleistocene eolianites and paleowinds: Sedimentology, v. 3, no. 1, p. 52-64.

McKee, E.D., 1934a, An investigation of the light-colored cross-bedded sandstones of Canyon de Chelley, Arizona: American Journal Science, 5th Series, v. 28, no. 165, p. 219-233.

McKee, E.D., 1934b, the Coconino Sandstone - its history and origin: Carnegie Institute, Washington, Publication 440, p. 77-115.

McKee, E.D., and Ward, W.C., 1983, Eolian, in Scholle, P.A., Bebout, D.G., and Moore, C.H., eds., Carbonate Depositional Environments, Memoir 33: American Association of Petroleum Geologists, p. 131-170.

Middleton, G.V., 1970, Experimental studies related to the problems of flysch sedimentation, in Lajoie, J., ed., Flysch sedimentology in North America: Geological Society of Canada Special Publication 7, p. 253-272.

Porter, M.L., 1987, Sedimentology of an ancient erg margin: the Lower Jurassic Aztec Sandstone, southern Nevada and southern California: Sedimentology, v. 34, p. 661-680.

Schenk, C.J., 1983, Textural and structural characteristics of some experimentally formed eolian strata, in Brookfield, M.E., and Ahlbrandt, T.S., eds., Eolian sediments and processes: Amsterdam, Elsevier Scientific Publishers, p. 41-49.

Strasser, A., and Davaud, E., 1986, Formation of Holocene limestone sequences by progradation, cementation, and erosion: two examples from the Bahamas: Journal Sedimentary Petrology, v. 56, no. 3, p. 422-428.

Ward, W.C., 1970, Diagenesis of Quaternary eolianites of N.E. Qiuntana Rio, Mexico: Houston, Rice University, unpub. Ph.D. dissertation, 206 p.

Ward, W.C., 1973, Influence of climate on the early diagenesis of carbonate eolianites: Geology, v. 1, p. 171-174.

Ward, W.C., 1975, Petrology and diagenesis of carbonate eolianites of northwestern Yucatan Peninsula, Mexico, in Wantland, K.F., and Pusey, W.C., eds., Belize shelf carbonate sediments, clastic sediments and ecology: American Association of Petroleum Geologists, Studies in geology no. 2, p. 500-571.

White, B., and Curran, H.A., 1985, the Holocene carbonate eolianites of North Point and the modern marine environments between North Point and Cut Cay, in Curran, H.A., ed., Pleistocene and Holocene carbonate environments on San Salvador Island, Bahamas, guidebook for Geological Society of America Annual Meeting, field trip guidebook: College Center Finger Lakes, Bahamian Field Station,

p. 73-93.

White, B., and Curran, H.A., 1986, Holocene carbonate eolianites from San Salvador, Bahamas, in Abstracts, Sediments down-under: 12th International Sedimentological Congress, International Association of Sedimentologists, p. 330.

White, B., and Curran, H.A., 1988, Mesoscale physical sedimentary structures and trace fossils in Holocene carbonate eolianites from San Salvador Island, Bahamas: Sedimentary Geology, v. 55, p. 163-184.

THE GEOLOGY OF EASTERN SOUTH ANDROS ISLAND, BAHAMAS: A PRELIMINARY REPORT

James L. Carew
Department of Geology
College of Charleston
Charleston, SC 29424

John E. Mylroie
Department of Geology and Geography
Mississippi State University
Mississippi State, MS 39762

ABSTRACT

Field investigation and preliminary laboratory study of rocks exposed along the eastern margin of South Andros Island, Bahamas has revealed the following. There are essentially no lithified Holocene rocks, in contrast to many other Bahamian islands (e.g. San Salvador, New Providence, Eleuthera, Great Inagua). All surficial rocks appear related to only one high stand of sea level, most likely that of the Sangamon (Oxygen isotope stage 5). The oldest rocks are eolian calcarenites exposed on an offshore cay just north of Mars Bay, where truncated eolianites are encrusted by fossil corals and both are capped by a paleosol. There are few fossil coral localities, but extrapolation between sites suggests an eastern string of patch to bank-margin coral reefs that now outcrop as much as 2 m above sea level.

The majority of the island below 5-6 m elevation consists of ooid shoals and lagoonal deposits, that are sometimes capped by Pleistocene eolianites. The more significant Pleistocene ridges trend approximately parallel to the shore, but lie 100 to 1000 meters or more inland. At all localities the rocks are covered with a paleosol or its remnants. No underlying paleosols were observed.

All rocks are solutionally altered on a scale from small pits, to banana holes, to blue holes. The majority of blue holes are developed along a sub-parallel to bank-margin fracture zone that has the physical appearance of a graben. It is suggested that progradational collapse upward from a deep cavern system developed along that trend provides a plausible origin for those blue holes. The original fracturing may be the result of bank margin subsidence, or original Mesozoic tectonics. There are also numerous small branching dry caves that were developed along the margins of a freshwater lens perched on a past higher sea level that may or may not be coincident with the highstand discussed above.

INTRODUCTION

This study was undertaken as part of the British-organized Andros '87 Project. The Project was a multidisciplinary scientific investigation of the geology, geochemistry, and biology of the blue holes of South Andros Island, Bahamas (Palmer, 1988). The field work for our part of the project was conducted in July 1987.

We were invited to be a part of the Project for the purpose of establishing the stratigraphic, sedimentologic, and structural setting of the blue holes. Additionally, we were to determine whether the stratigraphic model established on San Salvador Island (Carew & Mylroie, 1985) (Fig. 1) could be used successfully on South Andros.

METHODS

Field investigation and outcrop sampling was conducted by walking all available coastal outcrops along the east coast of South Andros Island from Driggs Hill to Mars Bay (see Fig. 2). Use of a zodiac boat permitted us to visit and investigate exposed rock on all the offshore cays from Goulding Cay to High Cay (see Fig. 2). We also investigated all quarries and pits, road cuts, and some blue holes (down-blue hole geology was directed by Dr. Peter Smart of the University of Bristol). We studied the rocky exposures inland along the banks of Deep and Little Creeks (see Fig. 2), and hiked inland to investigate all inland paleo-dune ridges in eastern South Andros. Finally all macroscopic (enterable) subaerial caves were investigated and where possible, mapped.

Eighty-eight samples were taken for further laboratory investigation. To this date slab descriptions and some thin section analyses have been completed. Further thin-section study, U/Th dating of some *in situ* fossil corals, and S.E.M. investigations are still to be done.

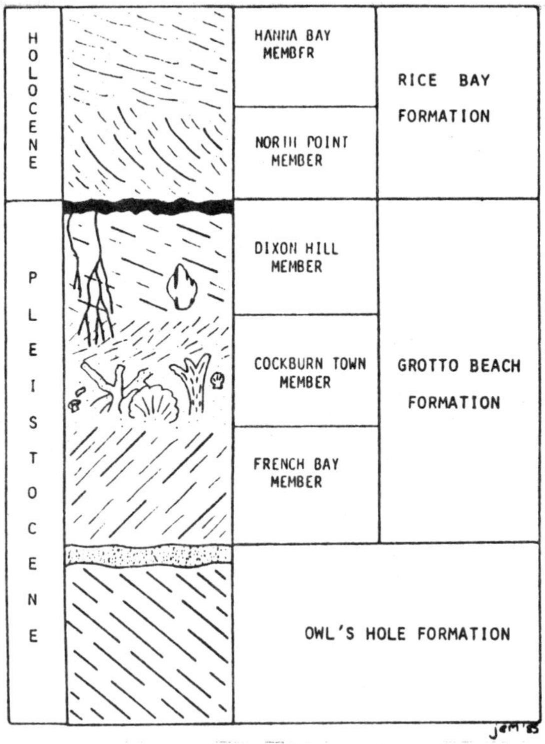

Fig. 1. Stratigraphic column for the Late Quaternary of the Bahamas. This stratigraphy was developed on San Salvador and has been successfully applied to the geology on New Providence, Great Inagua and Little San Salvador islands.

HOLOCENE

Unlike many other Bahamian islands (e.g. San Salvador, New Providence, Eleuthera, Great Inagua), there are essentially no lithified Holocene deposits, with the exception of a few scattered, weakly cemented back-beach deposits at a coastal location just N of Deep Creek. There are unlithified Holocene sediments comprising beach and back-beach, shallow subtidal lagoon, offshore shoal, creek channel, and lowland mangrove marsh and grass flat facies.

The offshore shoal that parallels much of the southern half of South Andros is composed primarily of 0.5-1mm diameter ooids. The lagoon that lies behind this shoal contains a variable mixture of ooids, bioclastic fragments, forams, peloids and aggregate grains with variable amounts of micrite. The beaches comprise a variety of calcarenites that show diminished ooid content with distance from the shoal. The creek channels, marsh, and grass flats that occupy much of the low-lying interior of this primarily flat island were not investigated.

PLEISTOCENE

Offshore Cays

The outcrops found on the offshore cays comprise subtidal ooid shoal deposits, and *in situ* reef corals and associated facies. There are a few larger cays (e.g. High Cay) that have outcrops of dune facies that extend to about 15 m above present sea level. The *in situ* corals extend upward to approximately 2 m above present sea level.

For much of the relatively flat low-lying cays it is not possible, in the field, to determine whether steeply dipping (25°-30°) cross beds are part of a shoal or are truncated eolian dune. Because there are relatively coarse shell hash and whole shell layers, and herring-bone cross bedded horizons associated with these steeply dipping beds in a moderate, but widespread number of locations on these cays, we believe that most of these deposits were part of a series of shoals that occupied this area during the Pleistocene. In a few locations there were subaerial eolian accumulations that formed islands along the eastern margin of South Andros during a time of past sea level higher than present. There is a paleosol/calcrete or its remnants that caps all outcrops.

Coastal Outcrops

Most of the coastal outcrops are low lying (0-2 m above sea level). These exhibit textures and compositions assignable to subtidal, shoal, beach, or reef. In a few notable localities (e.g. The Bluff) there are coastal cliffs that reveal subtidal facies low in the section, that grade upward into beach and then back-beach dune facies that extend up to 10 m above present sea level.

Outcrops along The Bluff settlement are particularly well developed and exposed. At the northern end of the cliffs there are sea caves along the coast that provide a window carved into the interior of a well developed coral reef complex comprising numerous taxa, conspicuous among which are *Acropora cervicornis, Montastrea annularis* and *Diploria* sp. Along the cliffs lateral to these reef deposits the rocks comprise a variety of subtidal deposits including ooid shoals and mixed bioclastic ooid subtidal sands. These subtidal facies extend to approximately 5 m above sea level where they grade upward into beach facies. These facies then pass upward into back--beach dune facies that extend inland to eolian ridges up to 20 m above sea level. The outcrops are capped by a paleosol/calcrete or its remnants.

Pits and Quarries

Depending on the elevation of, and depth to which pits and quarries penetrate, a variety of sedimentary facies are exposed. All exposures within a few meters of sea level are assignable to subtidal facies. A few exposures at higher elevations reveal beach and dune facies.

A particularly noteworthy exposure was seen at Driggs Hill where bulldozer excavation of a new harbor was underway. Excavation behind coffer dams below sea level revealed a complex subtidal facies mosaic composed of patches of reef corals surrounded and entombed by subtidal sands. These deposits extended upward to several meters above present sea level. At the present time all of this outcrop below sea level should be flooded by sea water. A paleosol/calcrete or its remnants caps all natural exposures.

Roadcuts

There is one major N-S road that is close to and parallels the eastern shoreline. There are also other minor N-S roads, and a few short and two long E-W roadway/tracks (e.g. Blister Rock Road). Road cuts on the N-S roads bisect a variety of facies depending on road elevation. Exposures representing subtidal, shoal, beach, and dune facies can be seen. At The Bluff settlement the N-S road splits the outcrops described earlier, within the beach to back-beach facies.

East-west roads penetrate into the interior of the island. They mostly reveal only exposures of eolian dune facies, but they do show that the N-S elongate eolian ridges are primarily oolitic.

Only the E-W road to Black Point settlement (just N of Deep Creek) penetrates facies other than eolian. Along that road can be found a sequence of outcrops beginning at its eastern end with subtidal facies, and progressing along the road and upsection to an oolitic eolian ridge outcrop at its western termination. All outcrops are capped by a paleosol/calcrete or its remnants.

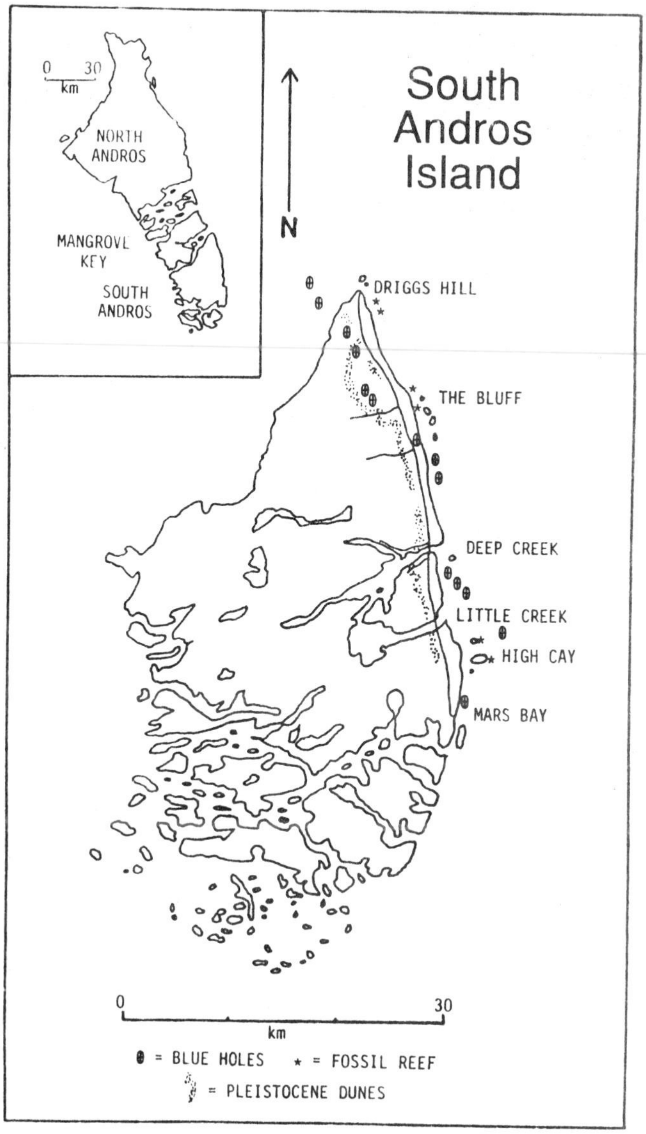

Fig. 2. Map of South Andros Island, Bahamas, showing major landmarks, fossil reefs, dune tracts and roads. The low-lying, western part of the island was not examined.

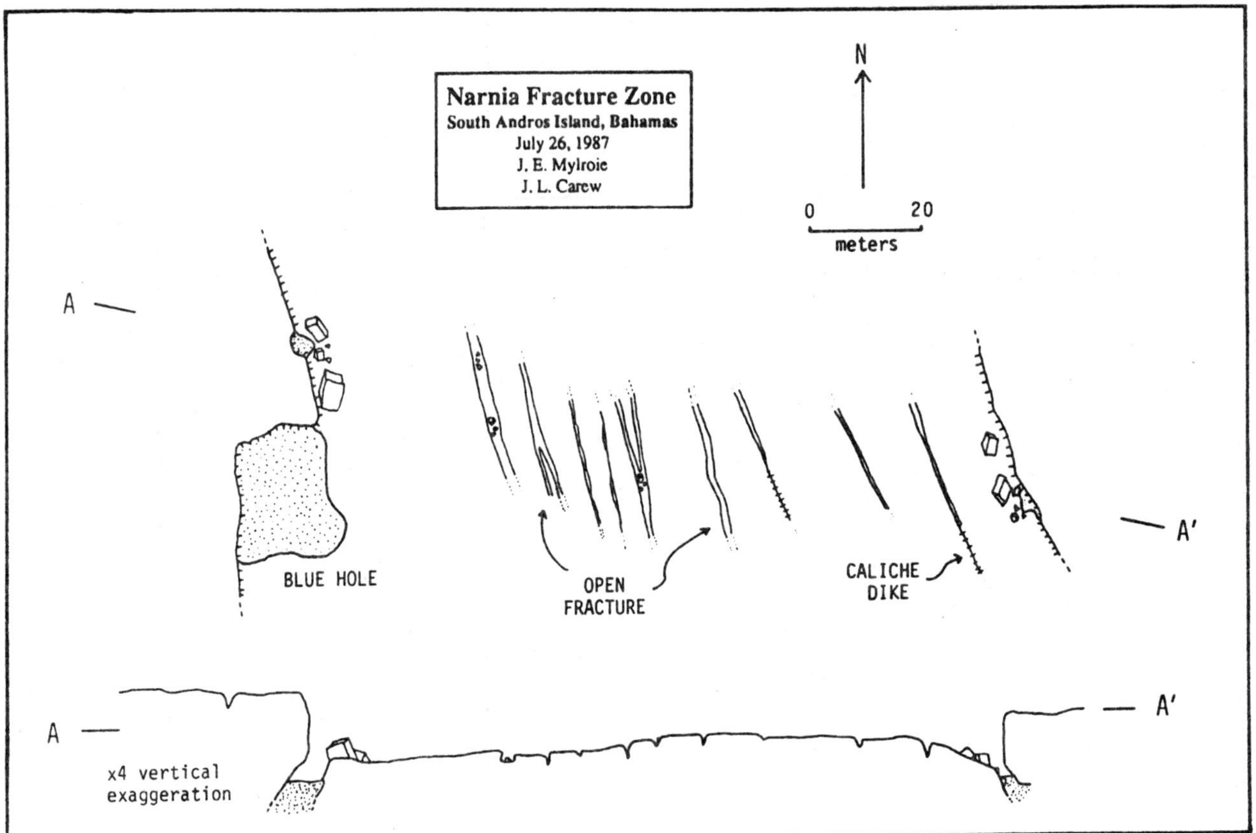

Fig. 3. Map and cross section view of the Narnia Fracture Zone, located halfway between the Bluff and Deep Creek (see Fig. 2). This graben-like feature is one of many along a zone that is subparallel to the Tongue of the Ocean to the east.

Creeks

Utilizing an inflatable boat and short inland treks at appropriate and accessible locations we investigated the margins of the channels of Deep Creek and Little Creek (see Fig. 2). Those waterways provided another avenue for exploration westward into the island's interior. We proceeded westward as far as was possible or practical. In each case outcrops seen near the eastern end revealed subtidal oolitic and oolitic-bioclastic facies. In some cases demonstrable subtidal shoal facies extend to 5 m - 6 m above present sea level.

At more westerly locations the creeks sequentially truncate the interior N-S dune ridges. At some locations subtidal deposits crop out from the water level to several meters above it, and then beach through dune facies are developed sequentially up the outcrop to a dune ridge crest 12-15 m above water level. The dune facies are oolitic. Outcrops are all capped by a paleosol/calcrete or its remnants. Most of the island's interior is a flat lowland plain covered by marsh, swamp, algal and grass flats with "islets" of rocky bedrock sprinkled in. We did not investigate those areas.

Fracture Zones

Behind the Batelco offices south of The Bluff settlement the Bahamians had cleared a large area by slash and burn. The cleared area exposed part of the fracture zone that trends subparallel to the coast, and along which the blue holes are developed, in this case Narnia Blue Hole. The cleared area provided a relatively long line-of-site view of the fracture zone which is usually buried in vegetation and very difficult to see. Figure 3 depicts this portion of the graben-like fracture zone. The cliffs on this "graben" margin reveal predominantly oolitic subtidal and shoal facies. Detailed sampling was done in this area, but the samples have not yet been analyzed. Narnia Blue Hole is found along the western margin of this fracture zone.

The fracture zones along the eastern margin of South Andros pose an interesting geologic problem. The fractures appear tectonic in origin, but occur in an area (the Bahamas platform) noted for its tectonic stability (Mullins and Lynts, 1977). The fractures are oriented subparallel to the trend of the eastern margin of the Andros portion of the platform, and therefore nearly parallel to the orientation of Tongue of the Ocean (TOTO).

Determining cause and effect is difficult based on the available data. Arguments could be made to suggest that the fractures represent a surficial expression of the crustal fractures thought by some (Mullins and Lynts, 1977) to be the source of an original horst and graben topography of the Bahamas region, that is expressed today as linear platforms with co-linear embayments. Alternatively, the proximity of the depths of TOTO to the platform margin may result in gravity sliding of the platform towards TOTO, with subsequent fracturing parallel to TOTO. The fracturing could: (1) result from gravity sliding active today; (2) result from faulting only active during Pleistocene sea level low stands when 100+ m of bouyant seawater support to the platform margin would be lost; or (3) result from faulting that has been inactive since some much earlier time.

Yet another prospect is that the grabens and associated fractures represent stoping structures from large solution conduits at depth. This idea requires a mechanism for the development of very large solutional voids subparallel to the bank margin. A possible reason can be found in the interaction between outward discharging freshwater and inward encroaching marine waters. That mixing produces a very solutionally aggressive groundwater that has been well documented (Back and others, 1986; Mylroie, 1988; Whitaker and others, 1988; Smart and others, 1988). Those conditions would place a zone of high solutional agressivity inward from the platform margin, but parallel to it. As the fractures penetrate rocks as young as 125,000 years old, regardless of their origin the final surficial expression of these features is relatively recent.

A polygenetic origin for the fractures is also possible. Given a fracture pattern either inherited from basement tectonics, or developed from gravity sliding towards TOTO, this fracture system could be expected to become a preferred flow path for groundwater in the island, causing a preferred site for solution at depth which is subsequently expressed at the surface as a graben fracture system. Study of the subsurface geology of the fascinating hydrological/structural aspects of the blue holes was conducted under the direction of Dr. Peter Smart of the University of Bristol, and preliminary results have been reported (Smart and others, 1988; Whitaker and others, 1988).

Caves

South Andros Island has a number of caves that are currently subaerial. These caves developed by solution at a past higher sea level (assuming platform tectonic stability) when the freshwater lens stood at a higher elevation than today. The caves conform to the pattern of the flank margin model of cave development (Mylroie, 1988). They consist of simple phreatic chambers with radiating tubes that developed in the zone of mixing between the discharging freshwater and landward encroaching marine waters. A map of South Deep Creek Cave, a typical flank margin cave, shows the main features of this type of cave development (see Fig. 4). The cave has a large central chamber that occurs on the margin of a dune, and which has been breached by recent surface erosion. A single phreatic tube leads inward into the dune mass, and ends abruptly. Caves of this type are common all over the Bahamas (Mylroie, 1988).

A different type of cave development is shown in figure 5. Ratbat Cave is a feature produced by stoping upward from a large void (assumed to be solutional) at depth. Instead of producing a graben feature as is common on South Andros, the fracture zone passed beneath an eolian ridge which had sufficient mechanical strength to bridge the void produced by collapse. On either side of this dune ridge, the fracture can be recognized by a series of linear blue holes. The plan and cross section of Ratbat Cave show that the cave has regions of deep water and dry areas of collapsed rock. This type of cave development is unusual in the Bahama islands, but is common on Bermuda (Mylroie, 1984).

SUMMARY AND CONCLUSIONS

All of the rocks exposed on South Andros Island appear to have been deposited during only one high stand of sea level about 5 to 6 m above present. It seems most reasonable that these rocks were deposited during the late rise, still-stand, and early fall of the high stand associated with oxygen isotope substage 5e (Sangamonian; circa 125,000 years ago) (Shackleton & Opdyke,

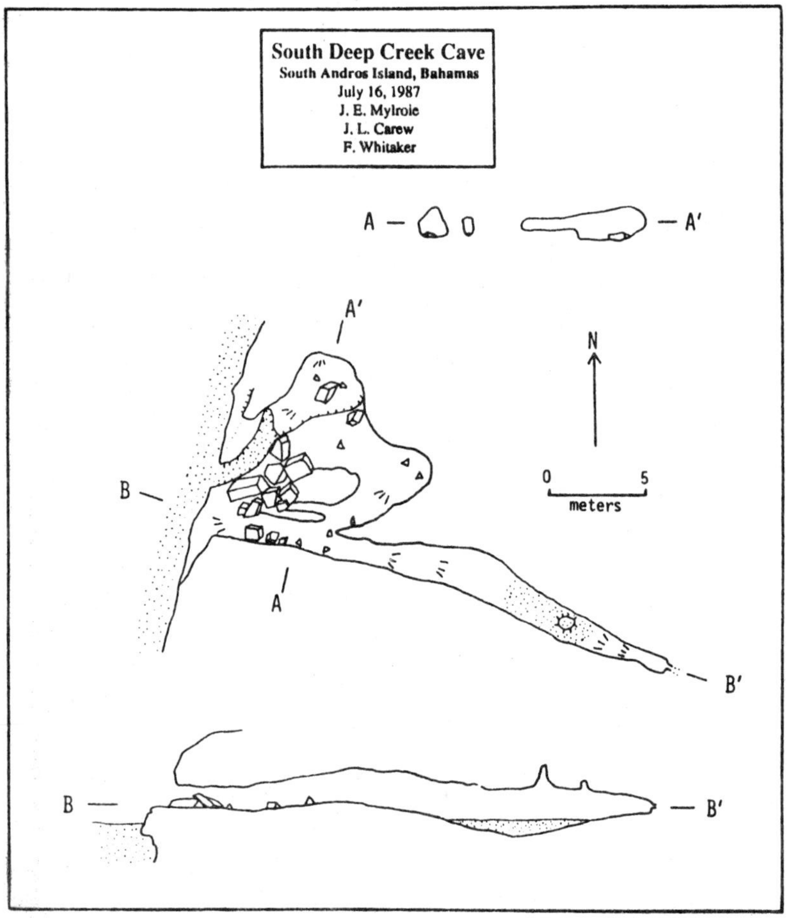

Fig. 4. Map and cross section of South Deep Creek Cave. This cave developed by solution along the margin of a freshwater lens where it mixed with sea water during a past sea level higher than at present. The rocks in which the cave is developed are believed to be 125,000 years old (oxygen isotope stage 5e). This requires that the cave developed on the regression of that sea level high, or during some later sea level high-stand that reached above current sea level. The cave is located on the south bank of Deep Creek where it cuts through the dune ridge system.

1973). All of these rocks can be assigned to the Grotto Beach Formation as defined on San Salvador (see Fig. 1), where U/Th and amino acid analyses yield ages of around 125,000 years (Carew & Mylroie, 1985, 1987). Like the Grotto Beach Formation on San Salvador the eolian facies are predominantly oolitic.

Only one paleosol is exposed on South Andros, and it is found to cap depositional as well as truncated bedding surfaces of Pleistocene rocks. The paleosol varies from a relatively thin brownish layered calcrete to better expressed paleosol and solution-filling remnants. As the rocks covered by the paleosol are believed to be Grotto Beach Formation, the paleosol represents the Pleistocene-Holocene transition, as shown on San Salvador (Carew and Mylroie, 1985).

The demonstrably oldest rocks seen on South Andros consist of an eolian ridge on High Cay (see Fig. 2) that was truncated to form a flat platform. On its eastern margin, fossil corals can be seen encrusted on this truncation surface, and both the dune and the corals are partially covered by a paleosol (Fig. 6). We suggest that this outcrop can be interpreted as follows. During the transgression of sea level associated with oxygen isotope substage 5e, oolitic dune ridges were built, among them High Cay. Continued rise in sea level truncated part of High Cay and then inundated the truncation platform on which the corals later grew. After sea level fell the paleosol developed on all exposed surfaces. The lack of a paleosol below the coral argues against an earlier origin (oxygen isotope stage 7) for the dune.

There are only a few fossil coral reef outcrops, but extrapolation among sites suggests the existence of a string of patch to bank-margin reefs developed along the eastern margin of South Andros during the Sangamon. Some of these now outcrop 2-3 m above sea level.

The bulk of the island below 5 to 6 m elevation comprises a suite of subtidal deposits consisting of ooid shoal and shallow lagoon deposits. In a few places these subtidal deposits can be seen to grade upward into beach and dune deposits.

The larger N-S trending eolian ridges are developed subparallel to the bank margin (see Fig. 2). Their relationship to other facies usually cannot be observed. We interpret these as deposits formed during transgression to the acme of the Sangamonian high stand when most of the

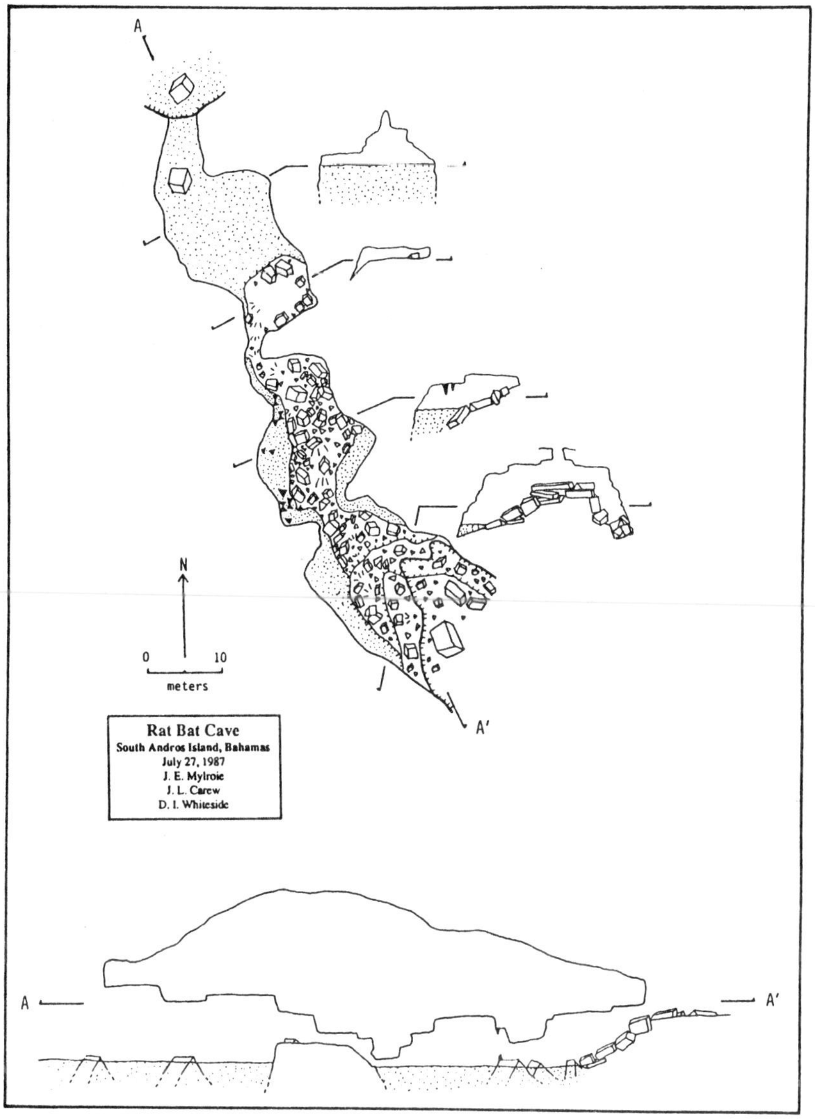

Fig. 5. Map and cross section of Ratbat Cave. This cave developed by upward collapse of a void at depth along a fracture zone. The cave exists where a dune ridge crosses the fracture zone. The dune ridge provided a roof to the fracture zone. This cave lies inland between The Bluff and Driggs Hill.

platform was flooded. These ridges may have been a series of cays at that time.

There are virtually no lithified Holocene rocks. This is in contrast to what is found on many other Bahamian islands (e.g. Great Inagua, Eleuthera, San Salvador, New Providence, etc.).

We note that our investigation of South Andros has yielded a geologic scenario that is remarkably similar to that found by Boardman & Bergstrand (1988) on North Andros.

The blue holes are primarily developed along a fracture system that lies sub-parallel to the eastern bank margin. The blue holes penetrate nearly 100 m down, and open to the surface through Pleistocene rocks assigned to the Grotto Beach Formation (circa 125,000 years old). The surface expression of the fracture zone is often graben-like.

The fractures are problematical as they occur in a supposedly tectonically stable area. Their development suggests a complex interaction between solutional process and perhaps gravity sliding of the platform edge into the Tongue of the Ocean.

The subaerial caves of South Andros fit models known from other carbonate islands. They are solution caves that developed during a higher sea level than exists at present. From their morphology and location, they fit the flank margin model of development (Mylroie, 1988). Additionally, there are stoping or collapse caves on South Andros. These were produced by upward progradation of a collapsing bedrock ceiling. These stoping caves are found associated with the graben-like fracture systems and their associated blue holes.

The flank margin solutional caves indicate that sea level had to be higher than at present on South Andros at a time after the deposition of the rock in which the caves are developed. As the

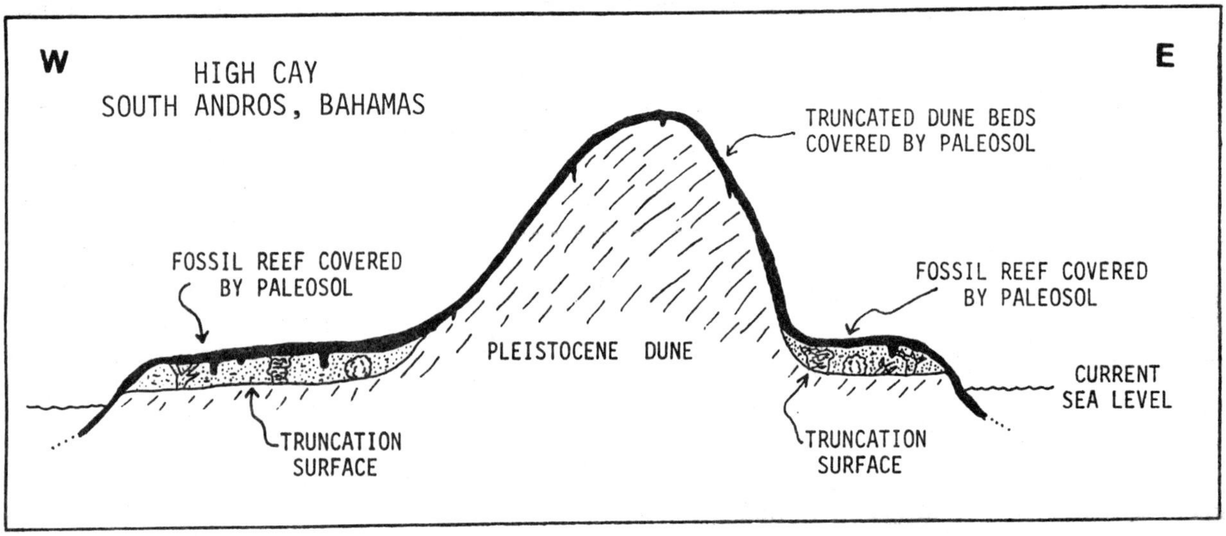

Fig. 6. Diagrammatic representation of the geology on High Cay. The truncation of the dune was followed by growth of coral at a higher sea level. All surfaces were subsequently modified by karst processes and paleosol development. The lack of a paleosol between the coral and the underlying dune suggests the dune was built on the Oxygen Isotope Stage 5e transgression, and subsequently notched by continued sea level rise. Coral then grew on the truncation surface. The truncation surface was formed by removal of some of the upper part of the dune by sea cliff retreat.

rock is believed to be 125,000 years old (oxygen isotope stage 5e), it implies that the caves developed in the dunes during the regression of the stage 5e highstand, or that a second high stand (oxygen isotope stage 5c, 5a or 3) produced the caves.

The surficial rocks of South Andros fit into the stratigraphy proposed by Carew and Mylroie (1985) for San Salvador Island (Fig. 1). Not all units exposed on San Salvador can be found exposed on South Andros. The Owl's Hole Formation and its associated paleosol were not observed on South Andros, in fact, no lower paleosol was seen. The Grotto Beach Formation is the dominant rock unit, and two members can be recognized. The French Bay Member can be seen on High Cay, as a transgressive dune cut and notched by the oxygen isotope Stage 5e highstand. The vast majority of South Andros is Cockburn Town Member subtidal and eolian carbonate facies. Cockburn Town Member facies rest on the truncated dune facies of the French Bay Member at High Cay. No Dixon Hill Member material (Stage 5a) was identifiable. The Holocene was remarkable for its paucity of lithified material, but what is present can only be generally categorized as Rice Bay Formation.

ACKNOWLEDGEMENTS

We thank the various individual and corporate sponsors of the Andros '87 Project, and the support of H.R.H. the Duke of Kent KG, for providing the funds and equipment necessary to undertake this international research expedition. Special thanks to: Ian Bishop for his whole hearted support and organization of this expedition, and his tireless good humor; Rob Palmer for leadership and drive to get the project off the ground; Neil Cave for companionship, entertainment and diving skill; and the Oxford Scientific and National Geographic film crew. A College of Charleston Faculty Research Grant provided additional funds for J. L. C. The Office of Research of Mississippi State University assisted J. E. M. Thanks to Joan Newell for word processing.

REFERENCES CITED

Back, W., Hanshaw, B.B., Herman, J.J., and Van Driel, J.N., 1986, Differential dissolution of a Pleistocene reef in the ground water mixing zone of coastal Yucatan, Mexico: Geology, v. 14, p. 137-140.

Boardman, M.R., and Bergstrand, P.K., 1988, Surficial geology and origin of northern Andros Island, Bahamas: Abstracts and Programs, Fourth Symposium on the Geology of the Bahamas (17-22 June 1988), Ft. Lauderdale, Florida, C.C.F.L. Bahamian Field Station, p. 10, and this volume.

Carew, J.L., and Mylroie, J.E., 1987, A refined geochronology for San Salvador Island, Bahamas, in Curran, H.A., ed., Proceedings of the Third Symposium on the Geology of the Bahamas: Fort Lauderdale, Florida, C.C.F.L. Bahamian Field Station, p. 35-44.

Carew, J.L., and Mylroie, J.E., 1985, The Pleistocene and Holocene stratigraphy of San Salvador Island, Bahamas, with reference to marine and terrestrial lithofacies at French Bay, in Curran, H.A., ed., Pleistocene and Holocene carbonate environments on San Salvador Island, Bahamas - Guidebook for Geological Society of America, Orlando annual meeting field trip #2: Ft. Lauderdale, Florida, C.C.F.L. Bahamian Field Station, p. 11-61.

Mullins, H.T. and Lynts, G.W., 1977, Origin of the northeastern Bahamian Platform: review and reinterpretation: Geological Society of America Bulletin, v. 88, p. 1447-1461.

Mylroie, J.E., 1984, Speleogenetic contrast between the Bermuda and Bahama islands, in Teeter, J.W., ed., Proceedings of the Second Symposium on the Geology of the Bahamas: Ft. Lauderdale, Florida, C.C.F.L. Bahamian Field Station, p. 113-128.

Mylroie, J.E., 1988, Karst of San Salvador, in Mylroie, J.E., ed., Field Guide to the karst geology of San Salvador Island, Bahamas: Mississippi State, MS, Department of Geology and Geography, Mississippi State University and C.C.F.L. Bahamian Field Station, p. 17-43.

Palmer, R., ed., 1988, Report of the 1987 International Blue Holes Project. The Andros Project, Department of Geography, University of Bristol, England, 28p.

Shackleton, N.J., and Opdyke, N.D., 1973, Oxygen isotope and paleomagnetic stratigraphy of equatorial Pacific core V28-238: oxygen isotope temperatures and ice volumes on a 10^5 to 10^6 year scale: Quaternary Research, v. 3, p. 39-55.

Smart, P.L., Palmer, R.J., Whitaker, F., and Mylroie, J.E., 1988, Processes and controls on cave development in carbonate platforms: some observations from the Blue Holes of the Bahamas: Abstracts and Programs, Fourth Symposium on the Geology of the Bahamas (17-22 June 1988), Ft. Lauderdale, Florida, C.C.F.L. Bahamian Field Station, 29-30.

Whitaker, F., Smart, P.L., Dawans, J.M., 1988, Carbonate dissolution by groundwater mixing in the Blue Holes of the Bahamas: Abstracts and Programs, Fourth Symposium on the Geology of the Bahamas (17-22 June 1988), Ft. Lauderdale, Florida C.C.F.L. Bahamian Field Station p. 33.

TEXTURE AND COMPOSITION OF HOLOCENE BEACH SEDIMENT, SAN SALVADOR ISLAND, BAHAMAS

Daryl D. Clark and John E. Mylroie
Department of Geology and Geography
Mississippi State University
Mississippi State, Mississippi 39762

James L. Carew
Department of Geology
College of Charleston
Charleston, South Carolina 29424

ABSTRACT

Sedimentation on San Salvador may be seen as a model for that of many other Bahamian islands. Because of its location on a small, isolated, carbonate bank, there are constraints on beach development and sediment supply. To determine the nature of the beach sediment, samples were collected along transects across the subaerial portions of 18 selected beach sites along the island periphery. These beaches provided a total of 71 samples that were dry sieved to determine texture. Sixty-five samples were impregnated and cut into thin sections for petrographic study.

The results of dry sieving show a majority of the beach environments (60 out of 71 sampled) are moderately well to well sorted. In a beach by beach comparison however, grain sizes are more diverse. Beaches along the eastern margin are predominantly fine to medium grained while western beaches, though mostly well sorted, show a high degree of variability in grain size. Preliminary interpretations indicated that textural findings may be related to the strength and consistency of active physical processes such as wind, wave and current activity and also to antecedent topography and platform morphology. Analysis of thin sections show a mostly bioclastic composition for San Salvador beaches. Comparisons of fine and medium grained eastern beaches with their western counterparts show that percentages of red algae, lithoclasts, peloids, foraminifera and minor constituents vary significantly. This indicates that composition is not a controlling factor for grain size and sorting or vice versa.

INTRODUCTION

General Setting

San Salvador Island is a small island situated on a shallow, isolated carbonate bank on the eastern edge of the northwest Bahamas. The island measures eleven kilometers in width (E-W) and nineteen kilometers in length (N-S). Surrounding San Salvador is a narrow shelf, which drops almost vertically from the shallow bank to deep ocean on the western and southern margins. On the eastern and northern sides, the transition from shallow bank to deep ocean is gradual and the slope more gentle. Numerous bank/barrier and fringing reefs are found offshore.

The topography is highlighted by Pleistocene and Holocene eolian ridges. Many of these ridges form headland cliffs along the shoreline, and between them beach sediments are deposited. Like other Bahamian islands, San Salvador beaches are associated with a high energy carbonate shelf environment (Inden and Moore, 1983) in which wind, waves and currents play a dominant role in the sedimentation.

Purpose

Most previous research in the Bahamas and on San Salvador has been concentrated on shallow bank margins rather than beaches. Studies of the Little Bahama Bank carbonate margins (Hine and others, 1981; Hine and Neumann, 1977; Hine, 1983) resulted in classifications and models for sedimentation onto the shallow water banks. These classifications generally indicate that platform morphology, physical energy and antecedent topography are controlling factors on sedimentation.

There have been a few general descriptions of beaches on San Salvador: at French Bay (Bain,

1986), and a grain size distribution study along selected lower beachfaces (Lee and others, 1986) around the island.

The purpose of this study was to analyze the texture (grain size and grain sorting) of modern carbonate beach sediments, identify the allochems which comprise these sediments and determine the relationship of sediment properties to energy (wind, wave and currents) and physiographic features like platform morphology, beach slope and antecedent topography.

METHODS

Field Methods

Between May 17 and June 8, 1987, eighteen beach sites on San Salvador were selected for a textural analysis (Fig. 1). At each site, sediment samples were gathered from each of the subaerial facies (lower beachface, upper beachface, back beach and dune), if all were present, along a measured transect across the beach environment (Fig. 2). Clinometer reading were taken from the lower beachface to the highest topographic feature adjacent to the shoreline in order to measure the beach slope.

Only the subaerial beach environments were included in order to allow comparison between the composition of modern beach sands and dunes, and lithified eolianites of both Pleistocene and Holocene age. A total of 71 individual samples were collected.

Laboratory Methods

Approximately 50 grams of sediment from each of the 71 samples were washed and dried in preparation for sieving. Sieve sizes -1.5, -1.0,-0.5, 0, .05, 1.0, 1.5, 2.0, 2.5, 3.0, 3.5, 4.0 were used. Samples were rotapped for 15 minutes. The contents of each sieve was weighed using a

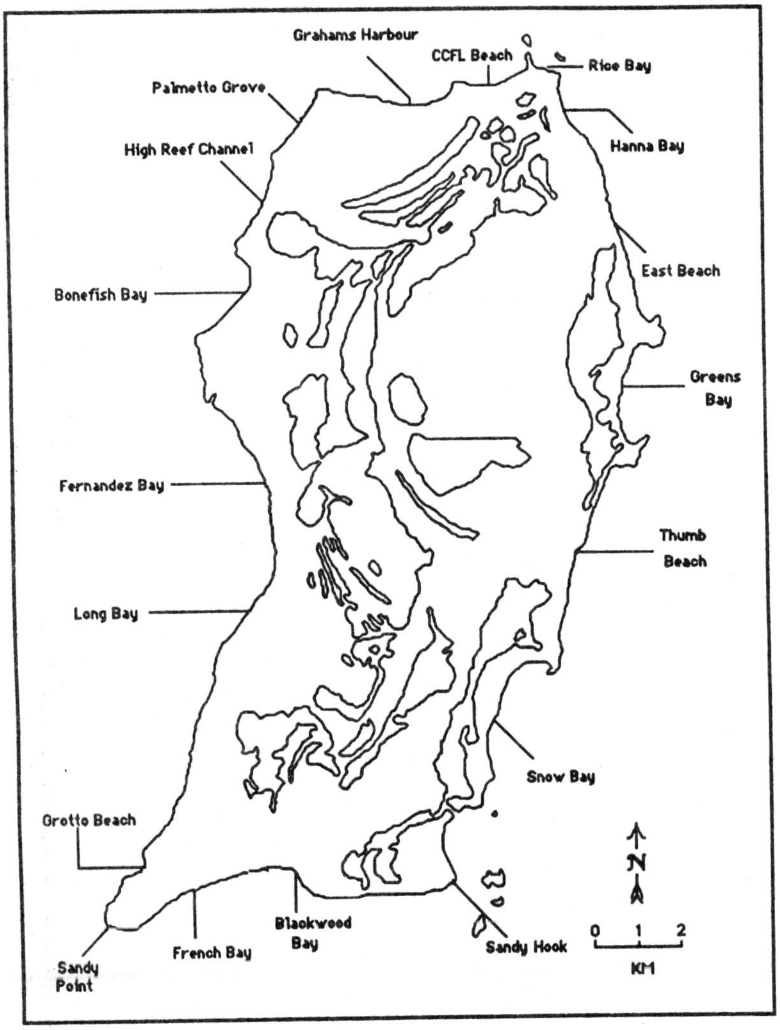

Fig. 1. Map showing transect and sampling locations of study area.

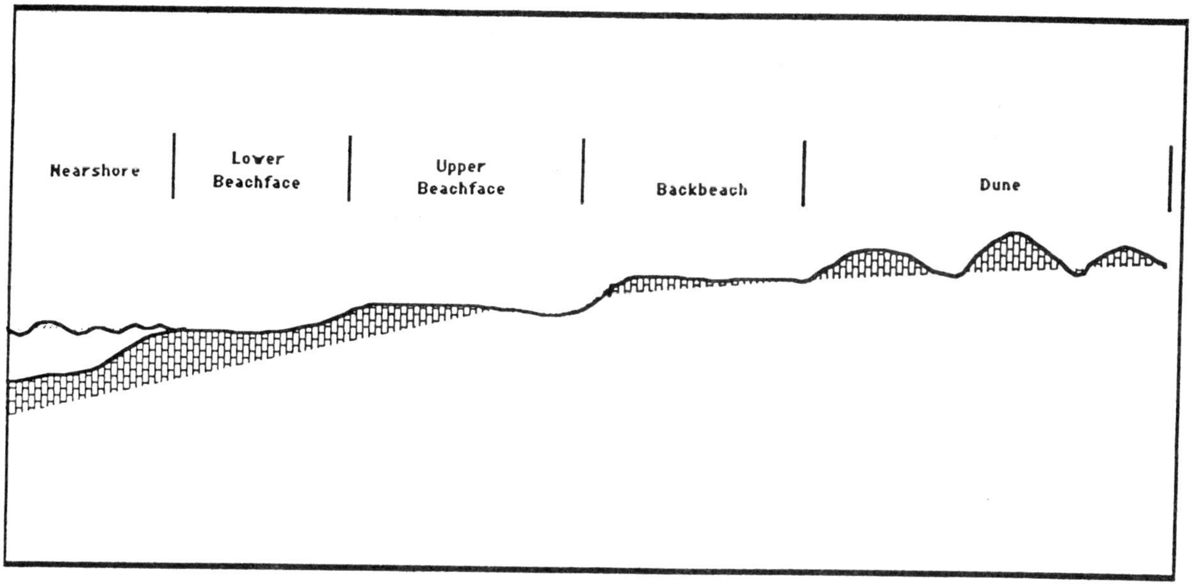

Fig. 2. General profile of beach environment transect.

Fisher scientific balance. All textural data was then calculated and determined using the Folk classification method (Folk, 1974) and the Udden-Wentworth grain size scale.

Samples from 16 of the 18 beaches (a total of 64 samples) were chosen for compositional analyses. Sediments were impregnated with clear epoxy, then mounted and cut into thin sections. These sections were studied using a petrographic microscope at 10X magnification. Allochems were identified by comparison to figures in Adams and others (1984) and Scholle (1978). To determine the percent of each allochem type, 150 points were counted on each thin section.

SIEVE RESULTS

Grain Sorting

A majority of the samples collected (60 out of 71) show a uniformity in grain sorting, ranging from moderately well to well sorted. Only the sieve results of the lower beachface and dune environments for the eastern and western beaches are presented in this paper (Figs. 3 and 4). These environments represent the transect end members and thereby offer a comparison of textural data from mean sea level and the dune area. The eastern and western shorelines, because of their length, differing platform morphology and varying magnitudes of physical energy offer interesting comparisons.

Textural data from the lower beachface were similar to those obtained by Lee and others (1986) in their study of grain size distribution at these same beaches. The consistency of the textural characteristics over a period of four years (their samples were collected in 1983 and 1984) indicates that the results of this study may be regarded as representative of the beaches and not simply anomalous examples.

The lower beachfaces along both eastern and western beaches are predominately well sorted. This indicates that the factors most influential for sorting: consistency of energy levels from wind, waves, currents and storms in the depositional area, are yielding well sorted beaches on the island. As expected, the dune sediments are well sorted as a result of the winnowing effect of consistent winds that average 10 knots in the San Salvador platform region (U.S. Naval Weather Service Command, 1974).

A combination of high wind (75% frequency and 25% calm averaging 12 knots) and wave (77% frequency and 23% calm) activity along the windward eastern side may account for the well sorted lower beachface, while on the western edge, the consistency of lower winds (9% frequency, 10 knot average) and waves (9% of all waves) are a major factor of beachface sorting (U.S. Naval Weather Service Command, 1974).

Deviations from the predominately well sorted texture may be the result of localized disruptions in the normally consistent energy processes. For example, at Grotto Beach, the steepness of the beach slope (33.8% grade) and its location in a cove may result in more varied wave and current action, causing its moderate sorting.

Snow Bay is a small, protected, steep (22% grade) pocket beach adjacent to a narrow platform edge. These physiographic features could be disrupting the consistency of wave and wind activity.

Grain Size

Average grain size is primarily a function of the magnitude of physical energy in the depositional environment, platform slope and antecedent topography. Figures 3 and 4 show grain size information for the eastern and western beaches.

East side beaches are nearly consistently fine grained sand. Western beaches, however, are markedly different. In the lower beachface, there is a high degree of variation among the beaches. The dune environments are somewhat less variable and are predominately medium sand.

Explanations for the contrast in beach sediments in the east and west must be attributed to the differing platform morphology and the presence of headlands and cays. The highest frequency and intensity of winds and waves striking San Salvador Island and its platform are along the eastern margin, based on synoptic meteorological data of this region (U.S. Naval Weather Service Command, 1974), and yet the beaches are finer. High winds and waves should produce coarser beaches. Therefore, morphology must be very important.

As shown by bathymetric data (Fig. 5), the carbonate platform is wider and has a gentler slope on the eastern margin. Fringing reefs and numerous cays are common along the eastern shoreline. All of the eastern beaches (Rice Bay, Hanna Bay, East Beach, Greens Bay, Thumb Beach, Snow Bay) are protected, because they are developed between headland cliffs and ridges. The combined factors of shelf slope, reefs and cays act as barriers to wind and wave action, thus reducing energy levels. Wave energy is further reduced at the shoreline by headland cliffs which

Location of Sample Area	Grain Size			Sorting			Skewness			Kurtosis			
	fine	med	coarse	poor	mod	well	fine	symm	coarse	m	l	sym	p
Rice Bay	X					X		X		X			
Hanna Bay	X					X			X	X			
East Beach	X					X		X		X			
Green Bay	X					X		X		X			
Thumb Beach		X				X			X	X			
Snow Bay		X			X				X				X

Location of Sample Area	Grain Size			Sorting			Skewness			Kurtosis			
	fine	med	coarse	poor	mod	well	fine	symm	coarse	m	l	sym	p
Rice Bay	X					X		X			X		
Hanna Bay	X					X			X		X		
East Beach	X					X			X	X			
Green Bay	X					X		X		X			
Thumb Beach	X					X			X		X		
Snow Bay		X				X		X		X			

Fig. 3. Sieve results of lower beachface (top) and dune environment (bottom) for eastern beaches.

Location of Sample Area	Grain Size			Sorting			Skewness			Kurtosis			
	fine	med	coarse	poor	mod	well	fine	symm	coarse	m	l	sym	p
Palmetto Grove		X				X	X				X		
High Reef Channel			X			X	X				X		
Bonefish Bay	X					X			X		X		
Fernandez Bay			X			X			X				X
Long Bay	X					X			X	X			
Grotto Beach			X		X				X		X		

Location of Sample Area	Grain Size			Sorting			Skewness			Kurtosis			
	fine	med	coarse	poor	mod	well	fine	symm	coarse	m	l	sym	p
Palmetto Grove		X				X		X		X			
High Reef Channel			X			X		X			X		
Bonefish Bay	X					X	X				X		
Fernandez Bay		X				X		X		X			
Long Bay		X				X		X		X			
Grotto Beach		X				X		X		X			

Fig. 4. Sieve results of lower beachface (top) and dune environment (bottom) for western beaches.

refract and dissipate incoming waves. The lowered energy levels are allowing only the finer particles to reach the beaches.

An exception to the fine grained characteristics of the eastern beaches are Snow Bay and Thumb Beach, which are predominately medium grained. Both have slightly narrower platform widths than the other eastern beaches. Transport distance is shorter and energy dissipation is less as a result. Discontinuous offshore reefs and headland cliffs, however, act as wave barriers, thereby allowing medium sized rather than coarse particles to reach the beaches.

On the western side, an equal amount of fine, medium and coarse grained environments are found. The frequency of high wind and waves is lower along the western margin (U.S. Naval Weather Service Command, 1974), but the variations of platform width, beach orientation, presence of headlands and storm activity may be the main contributors to the wide grain size distribution. Storms frequently blow in from the northwest side of San Salvador during the winter and spring for short durations. Beaches in this area, therefore, are subjected to high intensity short duration storm waves and winds that deposit coarse material onto the shore at the less protected sites.

At coarse beaches like Fernandez Bay and Grotto Beach, their steep grade (37% and 34% respectively) and narrow platform allow the full brunt of waves and currents to hit the shores with little wave dissipation. Offshore reefs act as only slight barriers to the waves.

The coarse and medium sediment within High Reef Channel and Palmetto Grove beaches must be related to beach orientation and storm activity since the shelf is relatively broad along these beaches. A lack of barriers (no protruding cliffs and discontinuous reef development) to reduce wave energy and northwesterly storm activity must account for the coarse sediment.

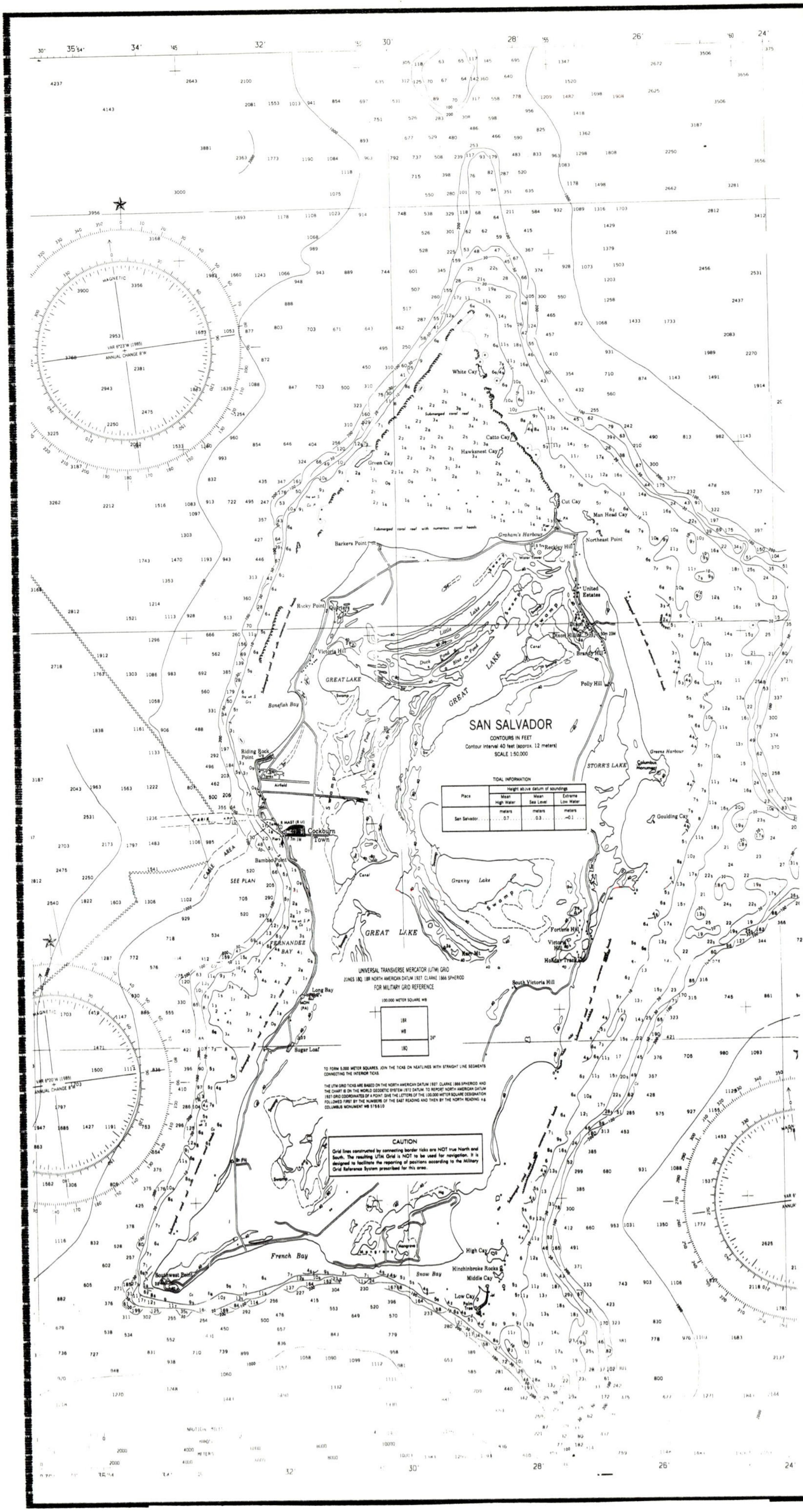

Fig. 5. Bathymetric map of San Salvador Island.

Bonefish Bay, located in the northwest part of San Salvador, is a very fine grained beach. An important distinction of this beach is its location between two headlands, unlike any other western beach studied here and a fringing reef that stretches across the bay very close to shore. Except for the narrower platform, these characteristics are similar to many of the fine-grained eastern beaches. Long Bay has the widest shelf among the western beaches and is protected by discontinuous coral reef development along the bank margin. The broader shelf dissipates much of the wave energy and results in fine and medium grained sediment reaching the shore. The medium grain size reported for the Long Bay dune is unusual as dune areas are typically finer than the other beach environments. One explanation is that erroneous calculations of grain size data were made. Another possible reason is that the dune is relict, and not in equilibrium with the current beach.

COMPOSITION RESULTS

No attempt was made to determine where the allochemical grains were created or the type of biota within the reef ecosystem. The intent was only to report the type and percent of allochemical constituents which make up the beaches. Petrographic analysis of 64 samples from 16 of the 18 beaches reveal that the sediments are predominantly bioclastic. Figures 6 through 11 show the results of the composition study. The most common constituents are foraminifera, red algae, green algae, mollusk fragments, lithoclasts and peloids. The term lithoclast is used to describe grains that have been welded or cemented together. This includes grapestones and beachrock fragments. Peloids is an inclusive term for micritized grains, pellets and grains too small to identify the shape or internal structure. Minor percentages of bryozoans, sponge spicules, coral fragments, annelid tubes, echinoderms and ostracods were also counted.

Eastern beaches were looked at to determine if allochemical amounts differed in a beach by beach comparison. Most allochemical percentages were found to be very similar for all environments. The only major exception was red algae, whose percentages varied in the lower beachface, upper beachface and backbeach environments of all eastern beaches. For western beaches, only the lithoclast and peloidal sands showed percentage variations in beach by beach comparisons. An analysis of this variation indicates that lithoclast content increases with increasing grain size, and peloid content decreases.

A comparison of the eastern versus western beach environments shows that foraminifera and red algae percentages are higher on the eastern beaches, while lithoclast and peloid sediment predominate on western beaches. Eastern beaches also contain a wider variety of minor constituents.

The fine and medium grained eastern beaches were also compared with their western counterparts. Results of the comparison show that the percentages of red algae, lithoclasts, peloids, minor constituents and to some extent foraminifera vary significantly. Therefore, composition is not seen as a controlling factor for grain size and sorting or vice versa. It is more probable

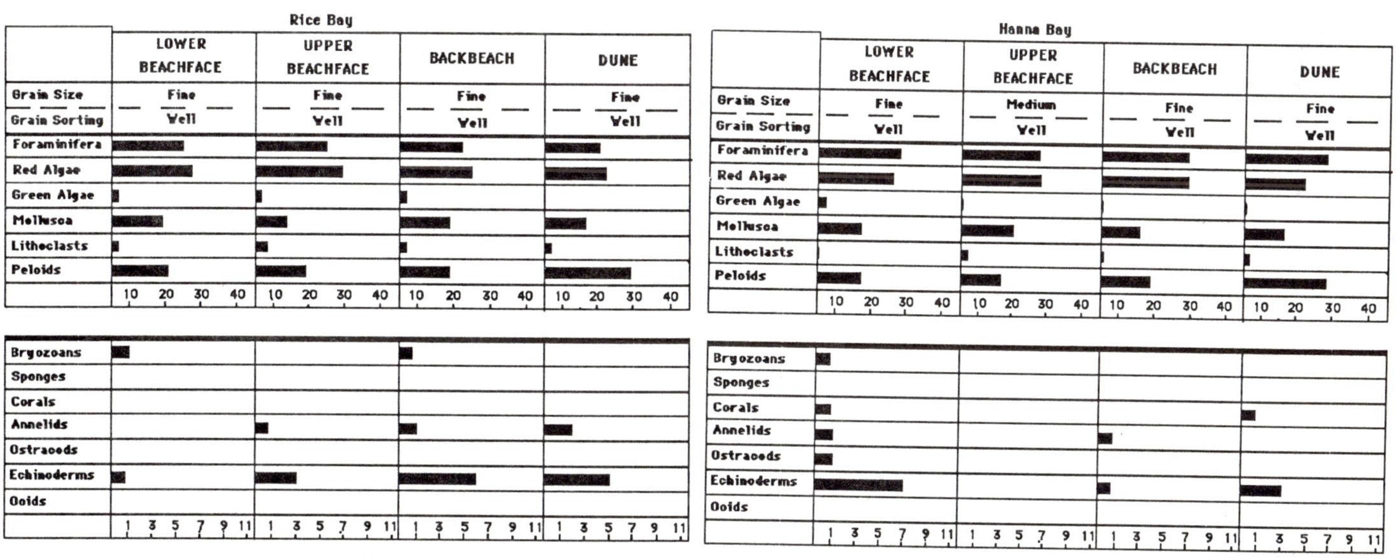

Fig. 6. Composition results for Rice Bay and Hanna Bay.

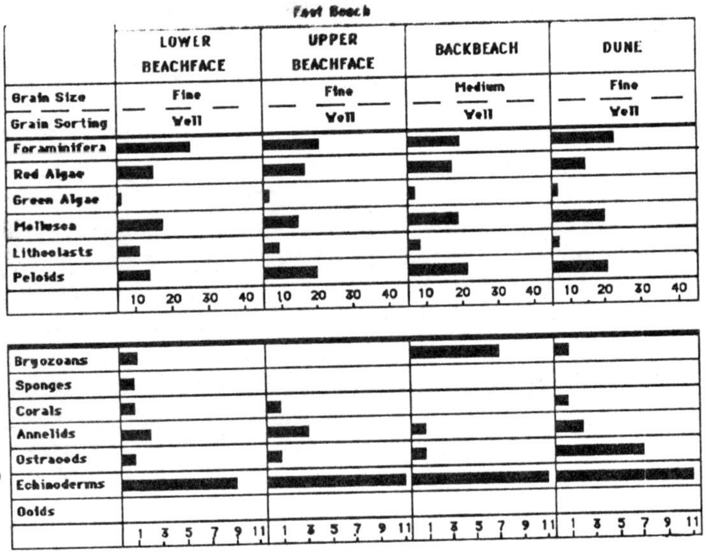

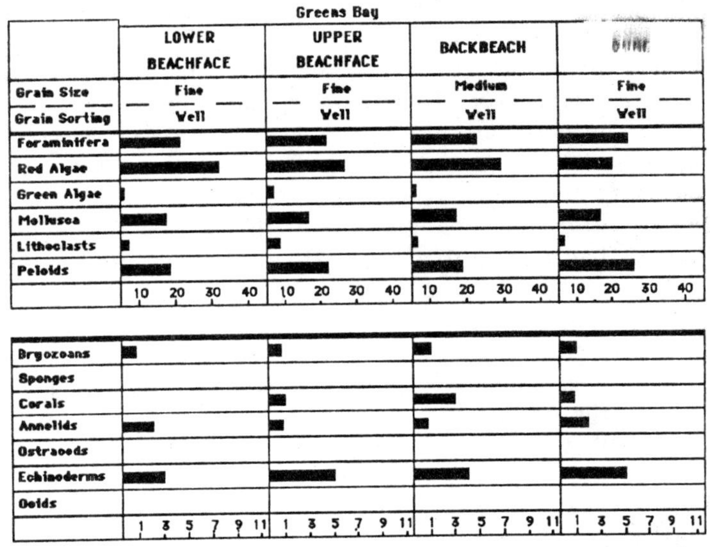

Fig. 7. Composition results for East Beach and Greens Bay.

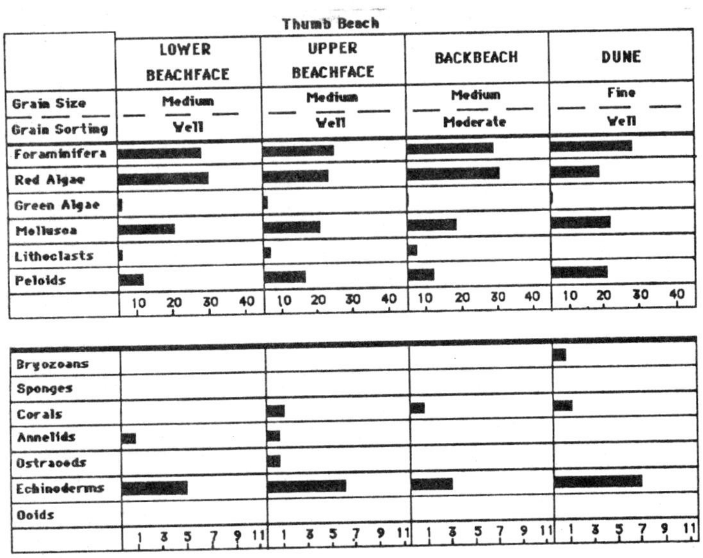

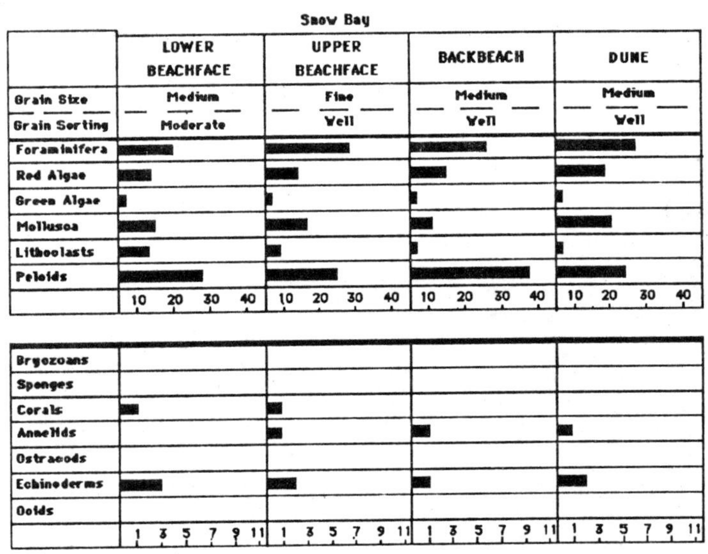

Fig. 8. Composition results for Thumb beach and Snow Bay.

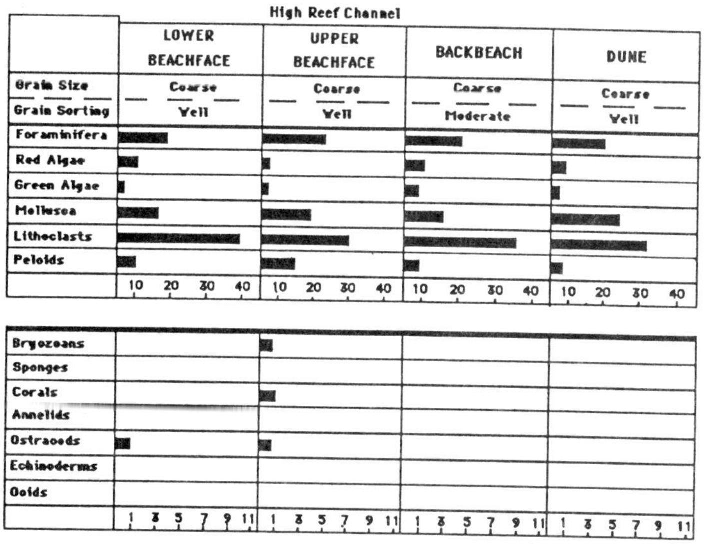

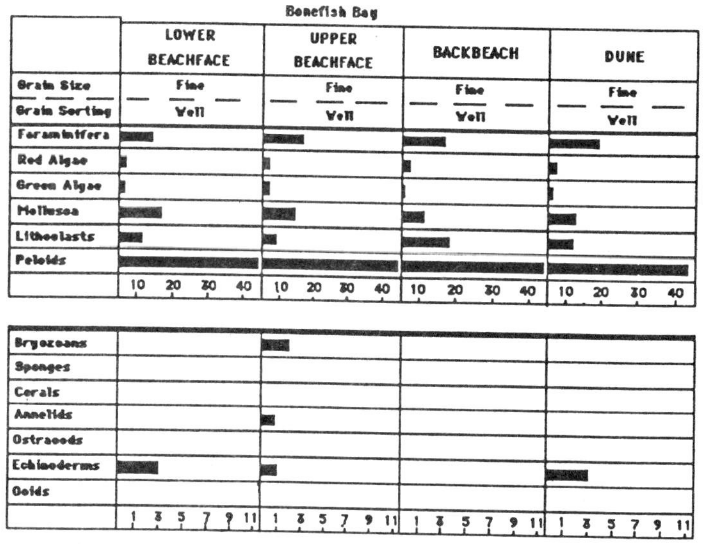

Fig. 9. Composition results for High Reef Channel and Bonefish Bay

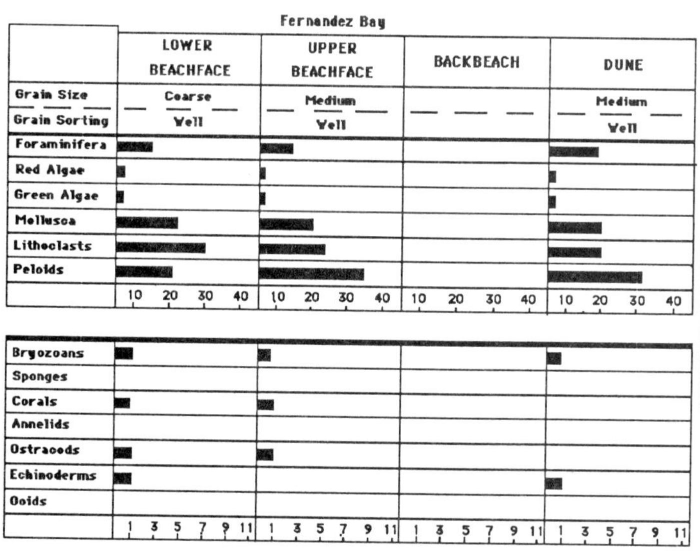

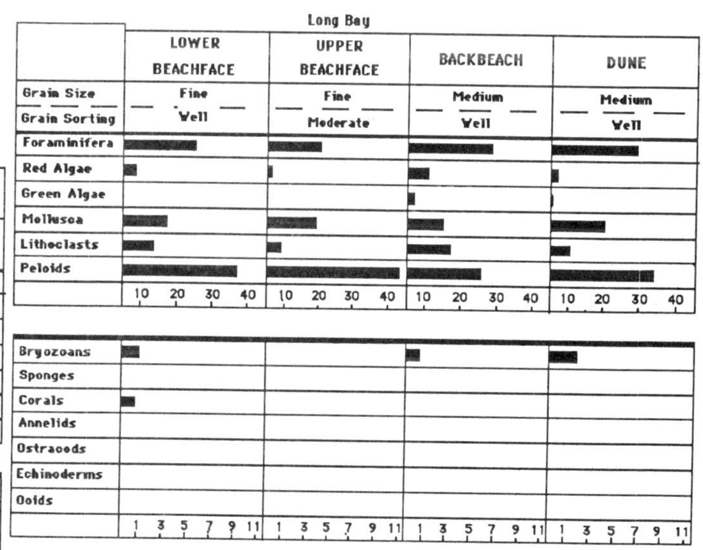

Fig. 10. Composition results for Fernandez Bay and Long Bay.

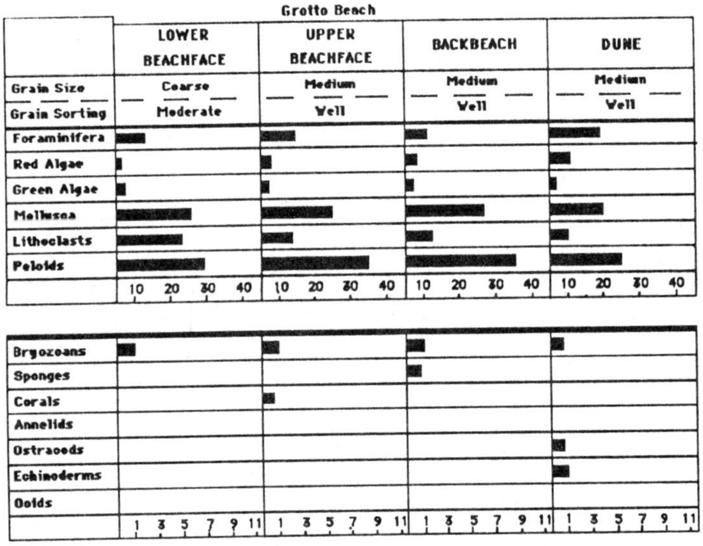

Fig. 11. Composition results for Grotto Beach.

that platform slope and the presence or absence of offshore reefs are the major reasons for the composition differences. A broader platform and the continuous nature of reefs may be allowing a larger abundance and variety of biota to exist on the eastern bank than on the narrower western bank.

SUMMARY

Sedimentation onto the beaches of San Salvador is controlled by a number of factors. Wind and wave activity, the presence of antecedent topography such as reefs, cays and headlands, beach slope and platform morphology were examined in relation to grain sorting and grain size.

Both the lower beachface and dune environments of the eastern and western beaches show a predominantly well sorted texture. This indicates that energy levels from wind and waves striking San Salvador are for the most part very consistent around the island.

The magnitude of wind and wave energy is usually considered the most influential factor in controlling grain size. Results from this study, however, show that platform width and antecedent topography (headland cliffs, reefs, cays) appear to be the dominant factors controlling grain size. Eastern beaches, subject to the highest wind and wave activity on the island, are consistently fine grained sand. This is unlike the coarse grained sand that is expected in a very high energy environment. It is concluded here that the broad, gently sloping nature of the platform, development of prominent headlands between beaches (characteristics of all eastern beaches) and the presence of continuous coral reefs act as dissipators of wave energy. The lowered energy levels only allows the finer particles to reach shore.

Western beaches, on the other hand, show wide variations in grade, platform width, orientation, development of reefs, cays and headlands. Seasonal storm activity also plays an important sedimentation role. Therefore, grain size differences exist among the beaches, which range from very fine to coarse.

Beach sediments are bioclastic sands in which the major constituents are foraminifera, red and green algae, mollusk fragments, lithoclasts and peloids. Eastern beaches show a higher percentage of forams, red algae and minor constituents, while lithoclasts and peloids are found in higher amounts on the western beaches. Comparisons of fine and medium grained eastern beaches with their western counterparts show that the percentages red algae, peloids, lithoclasts, minor allochems and to some extent forams differ significantly. This is an indication that sediment composition and grain size are not necessarily related. It is suggested that the broad platform and continuous nature of reef development along-the eastern margin is responsible for the higher percentages of red algae, forams and minor allochemical constituents by allowing a wider variety and higher abundance of sediment producing biota to exist along the eastern bank.

The similarity of the beach texture data from this study (collected in 1987) to that of Lee and others (1986) (collected in 1983 and 1984) indicates the consistency of the beach texture over a four-year time period. This consistency demonstrates that synoptic sampling of beaches can provide a meaningful measure of beach texture for relatively long time periods, despite daily to seasonal variations in conditions.

The beaches of San Salvador are conspicuous in their lack of ooids. This peloidal bioclastic beach composition is consistent with modern dune deposits and with the lithified Holocene marine and eolian rocks of the Rice Bay Formation (Carew and Mylroie, 1985). The contrast of these modern beaches with the ooid-dominated Pleistocene Grotto Beach formation is striking, indicating that the beach source areas in Grotto Beach time were very different from today. It is interesting to note that the modern beach sediments, although very different from rocks of the Grotto Beach Formation, are quite similar to those of the older Owl's Hole Formation.

ACKNOWLEDGEMENTS

Appreciation is extended to Dr. Donald Gerace, Director of the CCFL Bahamian Field Station and his staff, for the financial and logistical support of our research while on San Salvador Island.

Thanks to Rob Stowers for his field assistance in helping D. Clark to measure transects and collect sediment samples, and to Pete Vogel and his family for extending their hospitality to D. Clark during his work in Charleston, SC.

For the use of the facilities at the Mississippi Bureau of Geology, we want to thank the Director, Dr. Conrad Grazzier and also Chuck Peele for his assistance and patience in the use of the thin section equipment.

REFERENCES CITED

Adams, A.E., Mackenzie, W.S., Guilford, 1984, Atlas of Sedimentary Rocks Under the Microscope: Wiley and Sons, New York, p. 104.

Bain, R.J., 1986, Beach Rock, French Bay: in Curran, A.E., ed., Pleistocene and Holocene Carbonate Environments on San Salvador Island, Bahamas: Georgia Geological Society Field Trip Guidebook and College Center of the Finger Lakes Bahamian Field Station, p. 121-128.

Carew, J.L. and Mylroie, J.E., 1985, The Pleistocene and Holocene stratigraphy of San Salvador Island, Bahamas, with reference to marine and terrestrial lithofacies at French Bay, in Curran, H.A., ed., Pleistocene and Holocene carbonate environments on San Salvador Island, Bahamas - Guidebook for Geological Society of America, Orlando annual meeting field trip #2: Ft. Lauderdale, Florida, C.C.F.L. Bahamian Field Station, p. 11-61.

Folk, R.L., 1974, Petrology of sedimentary rocks: Hemphills Publishing Company, Austin, Texas, p. 179.

Hine, A.C. and Neumann, A.C., 1977, Shallow carbonate bank margin growth and structure, Little Bahama Banks, Bahamas: American Association of Petroleum Geologists Bulletin, v. 61, no. 3, p. 376-406.

Hine, A.C., Wilbur, R.J., and Neumann, A.C., 1981, Carbonate sand bodies along contrasting shallow bank margins facing open seaways in northern Bahamas: American Association of Petroleum Geologists Bulletin, v. 65, no. 2, p. 261-290.

Hine, A.C., 1983, Modern shallow water carbonate platform margins, in Cook, H.E., Hines, A.C., and Mullins, H.T., eds., Platform Margin and Deep Water Carbonates: Society of Economic Paleontologists and Mineralogists Short Course No. 12, p. 3-34.

Inden, R.F. and Moore, C.H., 1983, Beaches in Scholle, P.A., Bebout, D.G., and Moore, C.H., eds., Carbonate Depositional Environments: American Association of Petroleum Geologists Memoirs #33, p. 218-256.

Lee, Y.I., Lindsey, B., May, T., and Mann, C.J., 1986, Grain size distribution of calcareous beach sands, San Salvador Island, Bahamas: College Center of the Finger Lakes Bahamian Field Station Occasional Paper, no. 1, 12p.

Scholle, P. A., 1978, A color illustrated guide to carbonate rock constituents, textures, cements and porosities: American Association of Petroleum Geologists Memoir #27, 241 p.

United States Naval Weather Service Command, 1974, Summary of synoptic meteorological observations: Caribbean and nearby island coastal marine areas, vol. 3: Asheville, N.C., National Climatic Center, p. 317-395.

DEPOSITIONAL EVOLUTION OF A WINDWARD, HIGH-ENERGY CARBONATE LAGOON, SAN SALVADOR, BAHAMAS

Norman D. Colby and Mark R. Boardman
Geology Department
Miami University
Oxford, Ohio 45056

ABSTRACT

Graham's Harbor is a windward, 2x3km high-energy lagoon located at the northeast end of San Salvador. It is open to the west and rimmed to the north and east by cays and a barrier reef. Cluster analysis reveals four marine facies in Graham's Harbor: 1) abraded grain "grapestone"; 2) abraded grain "grainstone"; 3) *Halimeda* and foram-rich "packstone"; and 4) grapestone/bioclast "wacke-/packstone". These facies are distinct both laterally and vertically within the lagoon.

Studies of sediment cores reveal a coarsening upwards sedimentary sequence. This sequence differs from previous models for lagoonal deposition. Sediment probes indicate that a pronounced sill exists in Graham's Harbor. The presence of this sill has been recorded in the vertical sedimentary sequence during the Holocene sea-level rise. A muddy facies (grapestone/bioclast "wacke-/packstone") is overlain by a grainy facies (abraded grain "grapestone"). In protected, seagrass stabilized areas of the lagoon, the latter facies is overlain by a slightly muddier facies (*Halimeda* and foram-rich "packstone").

The sedimentary sequence in Graham's Harbor records the banktop response to the Holocene sea-level rise. Deposition within the silled basin allowed the accumulation of lower energy, muddy sediments followed by less muddy, higher energy sediments as sea level rose and breached the sill. Radiocarbon dates of peat and sediment from cores indicate that the sequence in Graham's Harbor was deposited at rates of 30-80 cm/1000 years. The entire sequence formed in less than 7000 years.

INTRODUCTION

Vertical and lateral facies changes in limestones are related to sea-level, energy, and production rates. Sea-level rise often results in a shallowing-upwards sequence (James, 1984) since sedimentation rates generally outpace sea-level rise and/or rates of subsidence (Schlager, 1981).

Models developed to explain carbonate deposition are based primarily on the rock record and concentrate on the formation of tidal flat and low-energy lagoonal sediments (Goodwin and Anderson, 1985; Pratt and James, 1986; Wright, 1986). James (1984) has developed a hypothetical model for carbonate deposition in a high-energy environment. Few modern analogues to this model exist, however. In addition, many of the studies of high-energy lagoons have focused on benthic communities and their relationship to the sediment record (Ginsburg and Lowenstam, 1957; Taylor and Lewis,1970; Miller, 1988). Few studies have focused on lateral and vertical facies relationships in high-energy lagoons (Ginsburg, 1956; Swinchatt, 1965; Locker, 1981; Andersen, 1988).

The purpose of this study is to examine vertical and lateral facies relationships in a high-energy lagoon in order to better understand the depositional processes and controls acting on this type of environment. Establishing a timeframe for deposition of the vertical sequence during a sea-level rise is also a primary focus of the study.

METHODS

Field

Graham's Harbor is a 2x3km high-energy lagoon located at the northeast end of San Salvador, Bahamas. It is open to the west and rimmed to the north and east by cays, eolian dunes, and a barrier reef (Fig. 1). Fourteen cores and 48 surface samples have been collected in several transects across the lagoon. Water depths, sediment thicknesses, bedforms, and floral and faunal variability have been recorded.

Laboratory

All cores were described megascopically and

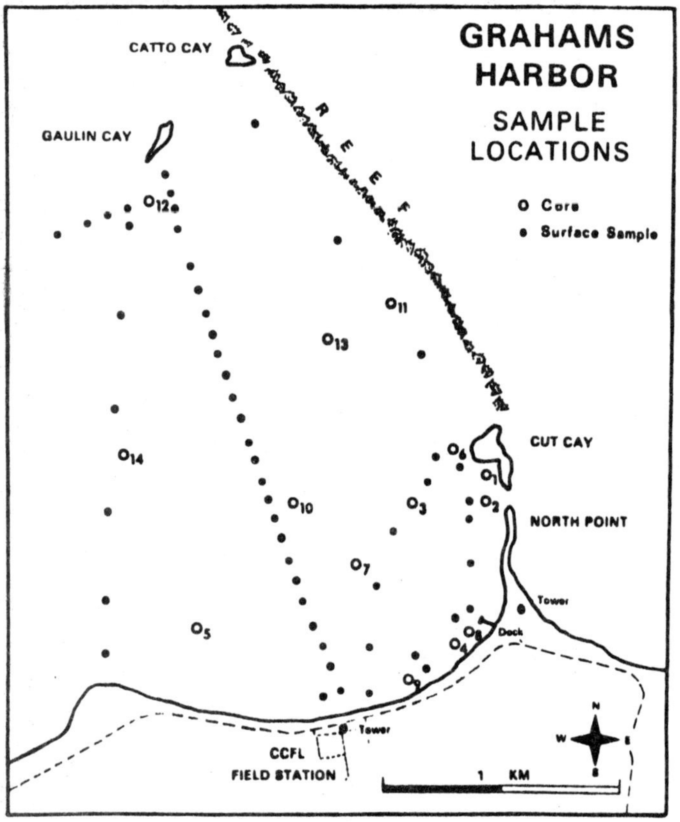

Fig. 1: Graham's Harbor sample locations. Open circles represent core locations, closed circles represent surface samples.

three cores were sampled at 10 or 20 cm intervals. Of the surface samples, 36 were chosen in order to cover the widest geographic area of the lagoon. All of these samples were analyzed with respect to texture, composition, and mud fraction mineralogy.

Texture

Standard wet sieve analysis was used on samples at one phi intervals. Pipette analysis was performed on the <63μm, <16μm, and <4μm mud fractions. All samples were rinsed in a 50% Clorox solution prior to sieving to remove organics. The method of moments was used to determine sorting (Folk, 1966; Lewis, 1984).

Composition

At least 300 grains from the 1-2mm size fraction were counted from each sample. This size fraction was chosen because it provided the greatest number of readily identifiable grains.

Mud Fraction Mineralogy

The ratio of aragonite to calcite was determined from x-ray diffraction data using the peak area method (Chave, 1954). The type of calcite (high or low magnesium) was determined using the position of the 104 peak of calcite (Goldsmith, Graf, and Joensuu, 1955).

Statistical Analysis

Cluster analysis was used to recognize the relationships between samples. The program CLAP (written by Sepkoski and Sharry, 1976; modified by A. Miller, 1987) was used to accomplish both Q-mode and R-mode analyses. The UPGMA method of clustering was used on standardized data (Imbrie and Van Andel, 1964; Harbaugh and Mirriam, 1968). The cosine theta similarity coefficient was used in combination with a total of 10 compositional, textural, and mineralogic variables per sample. A combination of Q-mode and R-mode analyses provided a two-way cluster analysis (Miller, 1988), from which the various facies are derived.

Radiometric Dating

Selected core, surface, and rock samples were chosen for carbon-14 dating. The sand (125-1000μm) and mud (<63μm) fractions were separated from three samples (sand-mud clouplets) and dated. A total of 15 samples were analyzed.

RESULTS

Depositional Environments

Graham's Harbor deepens towards the center to a maximum depth of 6m and shallows to the north, east, and south (Fig. 2). It remains open to the west, allowing excellent exchange with the Atlantic Ocean. The coast is dominated by beaches, with beachrock exposed in many areas.

Sediment thickness also increases towards the center of the lagoon, reaching a thickness of at least 4m (Fig. 3). By combining sediment thickness with bathymetry, a map of antecedent (Pleistocene) topography was developed (Fig. 4). This shows that Graham's Harbor was a silled, bowl-shaped basin with a maximum depth of about 10m below present sea level before Holocene sedimentation began.

Facies

Four marine facies are found in Graham's Harbor: 1) abraded grain "grapestone"; 2) abraded grain "grainstone"; 3) *Halimeda* and foram-rich

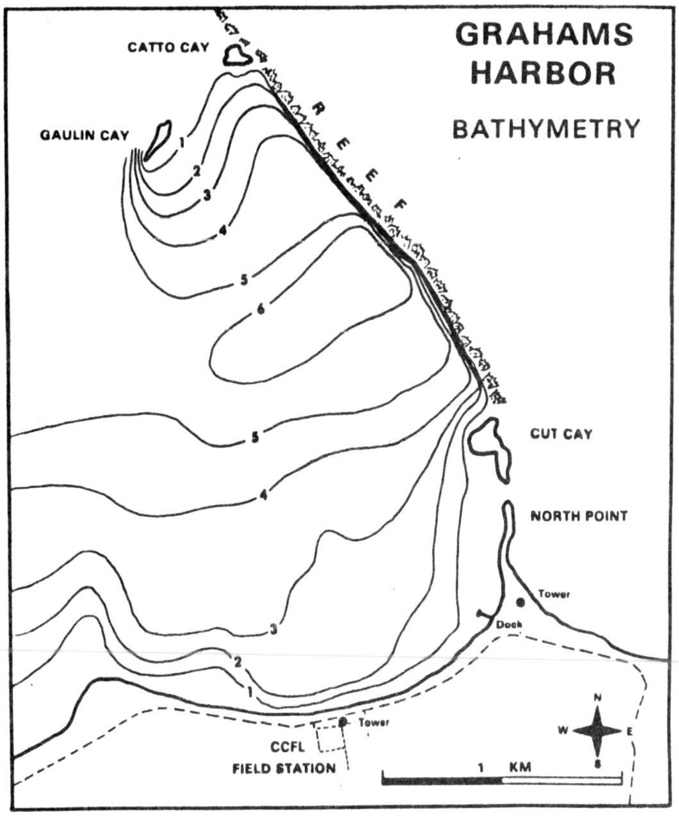

Fig. 2: Graham's Harbor bathymetry. The lagoon is deepest in the center. It is open to the west but rimmed on the north and east by a barrier reef and cays.

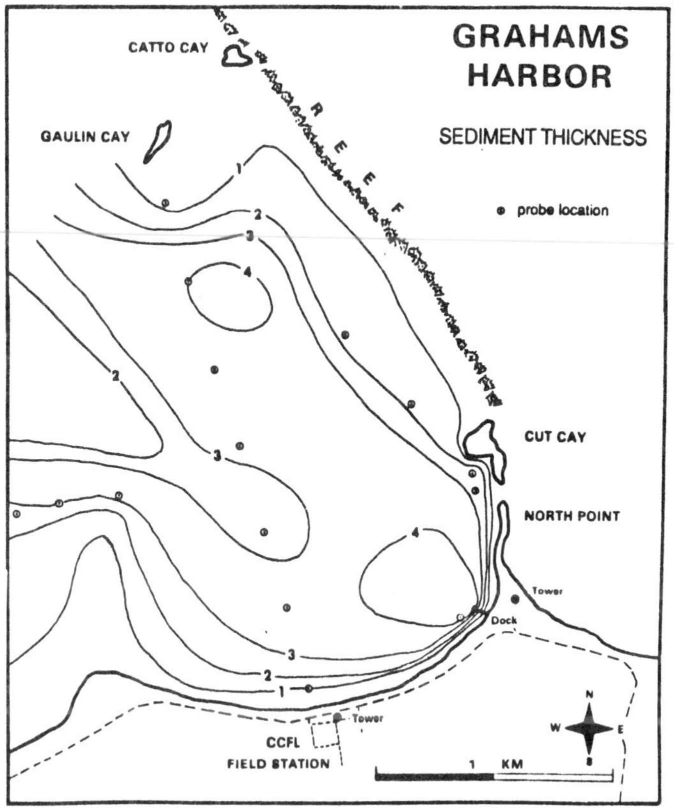

Fig. 3: Sediment thickness in Graham's harbor. Thickness increases toward the center of the lagoon. Contour intervals are in meters.

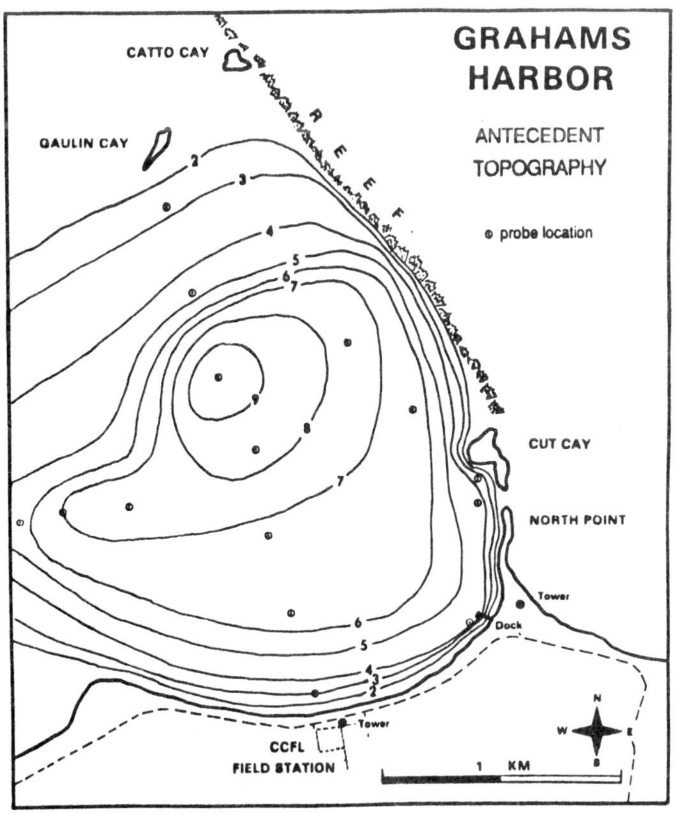

Fig. 4: Antecedent (Pleistocene) topography. Graham's Harbor was a silled basin at the time of the Holocene sea-level rise.

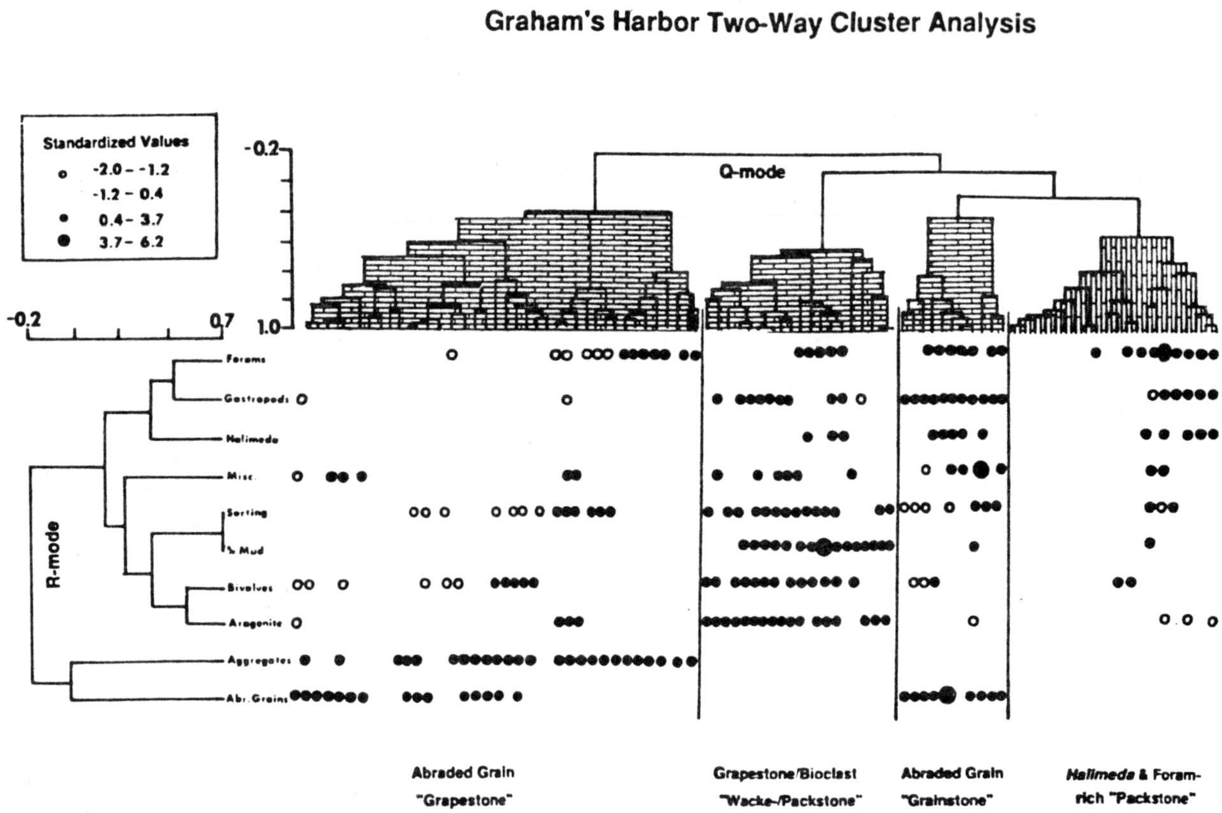

Fig. 5: Two-way cluster analysis of Graham's Harbor core and surface samples. Negative values represent variables below the mean, positive values represent variables above the mean.

"packstone"; and 4) grapestone/bioclast "wacke-/packstone". These facies are recognized by cluster analysis (Fig. 5) which groups core and surface samples together based on 10 sedimentary characteristics (Table 1). Facies are named on the basis of the most abundant allochems within each cluster group. A corresponding environment of deposition is then assigned to each facies. All cores facies were combined to form a composite vertical sequence for Graham's Harbor.

Lateral Gradients

All of the facies are found surficially in Graham's Harbor except the grapestone/bioclast "wacke-/packstone". The abraded grain "grapestone" covers much of the lagoon in areas of periodically shifting sand. The abraded grain "grainstone" is found in a high-energy zone near the barrier reef and south of Gaulin and Catto Cays. The *Halimeda* and foram-rich "packstone" is located behind North Point, in a relatively low-energy, seagrass stabilized area (Fig. 6).

Vertical Gradients

The grapestone/bioclast "wacke-/packstone" forms the basal portion of 2 of the 3 cores analyzed (Fig. 7). This facies contains the highest percent mud of the four facies as well as a high proportion of grapestone. Where this facies is present, it is abruptly overlain by the abraded grain "grapestone", a much coarser facies. The *Halimeda* and foram-rich "packstone" is found overlying the abraded grain "grapestone" and also directly overlying peat (core 1). This facies is bioclast-rich and slightly muddier than the previous facies. The vertical extent of the abraded grain "grainstone" is not yet established, but it has been found in the upper portion of core 11. This facies contains very little mud and a high proportion of abraded grains.

The mud fraction mineralogy of cores 4 and 7 changes from aragonite dominated to roughly equal proportions of aragonite and high magnesium calcite. Core 1 mud contains co-equal proportions of aragonite and high magnesium calcite throughout.

Radiometric Dating

Carbon-14 dating was performed on selected core, surface, and rock samples (Table 2).

Samples were taken from cores 1, 7, and 8

Graham's Harbor Facies

Variable Means and Standard Deviations

	Abraded Grain Grapestone	Grapestone/Bioclast Wacke/Packstone	Abraded Grain Grainstone	Halimeda & Foram-rich "packstone"
Halimeda	3.1 ± 1.9	11.2 ± 9.6	14.9 ± 8.7	33.3 ± 11.8
Bivalves	10.2 ± 5.4	22.64 ± 5.0	7.9 ± 2.5	14.6 ± 5.5
Gastropods	2.0 ± 0.9	4.8 ± 2.7	5.1 ± 1.5	3.9 ± 1.9
Forams	10.4 ± 6.5	18.9 ± 7.3	16.8 ± 5.5	22.4 ± 7.8
Aggregates	52.1 ± 14.0	25.2 ± 13.6	16.3 ± 13.0	8.7 ± 5.7
Miscellaneous	4.1 ± 2.6	8.1 ± 3.0	6.8 ± 6.2	5.8 ± 2.2
Abraded Grains	18.1 ± 11.7	9.3 ± 7.9	32.2 ± 9.0	11.2 ± 4.4
Mud	6.6 ± 4.6	32.6 ± 15.8	7.7 ± 9.7	14.7 ± 5.4
Aragonite	52.9 ± 5.9	67.6 ± 4.9	46.9 ± 4.7	48.4 ± 2.9
Sorting	1.7 ± 0.5	2.2 ± 0.3	1.59 ± 0.5	1.9 ± 0.1

Table 1: Variable means and standard deviations of sedimentary facies recognized by cluster analysis.

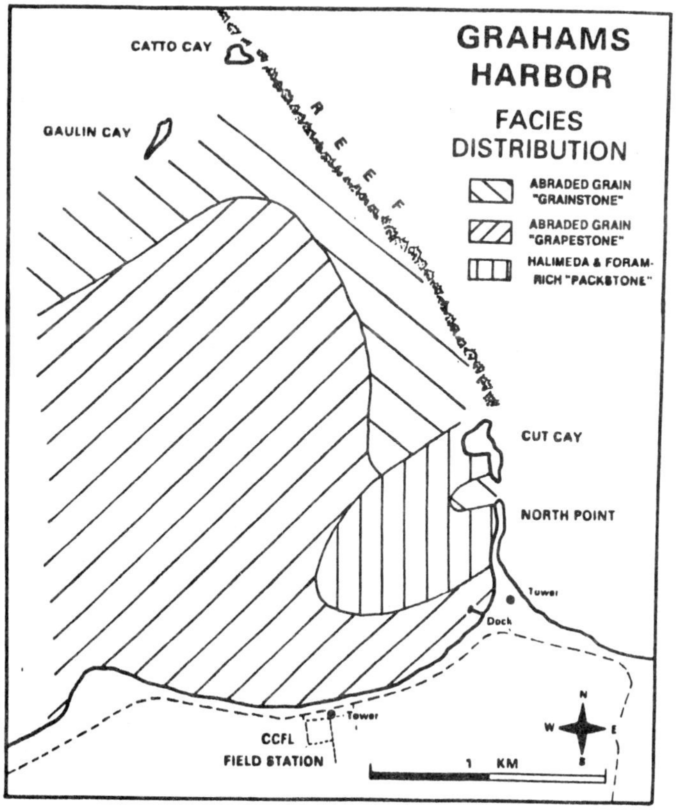

Fig. 6: Lateral facies distribution in Graham's Harbor.

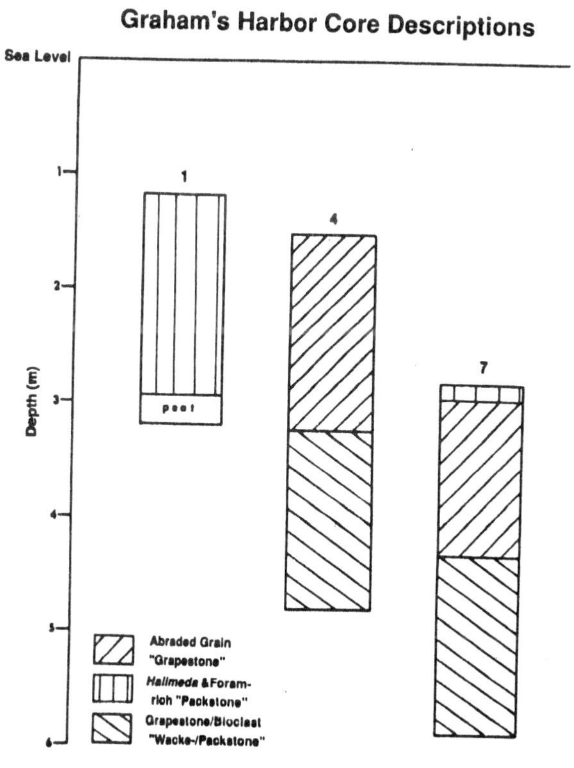

Fig. 7: Vertical distribution of facies in cores 1, 4, and 7. Cores are drawn relative to mean sea-level.

GRAHAM'S HARBOR RADIOCARBON DATES

Core Samples

Core	Sample Type	Sample Depth	Water Depth	Total Depth (cm)	Age
1	peat	198	120	318	4960 ± 70
1	shell	170	120	290	4700 ± 90
7	sand	20	300	320	3080 ± 70
7	mud	20	300	320	960 ± 90
7	sand	110	300	410	3550 ± 80
7	mud	110	300	410	2550 ± 90
7	sand	284	300	584	6480 ± 100
7	mud	284	300	606	6500 ± 100
7	shell	306	300	584	4640 ± 140
8	peat	406	120	526	7080 ± 90

Rock Samples

Sample Type	Sample Location	Age
Bulk Rock	North Point, 2m above sea level from calved off face of eolianite	5360 ± 110
Bulk Rock	(same as above)	5435 ± 125
Bulk Rock	North Point, near top of dune, +4.2m m.s.l.	5670 ± 160
Bulk Rock	North Point, near center of dune, +2.2m m.s.l.	6000 ± 100
Bulk Rock	North Point, near bottom of dune, approx. +1m m.s.l.	6190 ± 150

Table 2: Radiocarbon dates from core, rock, and surface samples in Graham's Harbor.

for radiocarbon dating. Peat samples were collected from the base of cores 1 and 8. These samples yielded ages of 4960±70 and 7080±90 ybp, respectively. A shell sample was taken at the sand/peat interface of core 1, which yielded an age of 4700±90 ybp.

Sand-mud couplets were collected from the abraded grain "grapestone" facies of core 7. Radiocarbon dates of these samples shows a wide discrepancy between sand and mud ages upward in the core. Sand and mud ages are very close in the lowest part of the core, but diverge toward the top. The largest difference is near the top of the core, where the *Halimeda* and foram-rich "packstone" overlies the abraded grain "grapestone" facies (Table 2).

DISCUSSION

Facies in Graham's Harbor

The source of sediment in Graham's Harbor is the lagoonal "carbonate factory" (Neumann and Land, 1975; James, 1984)). This in-situ accumulation of sand sized skeletal material from breakdown of molluscs, bivalves, and reef material and mud from calcareous algae and epibionts on seagrass provides the bulk of the sediment in the lagoon. Studies have shown that the production by green algae alone is more than enough to fill most lagoons and that a large percentage is probably transported off-platform (Neumann and Land, 1975). Radiocarbon dates from sand-mud couplets in sediment cores also suggests another source of sand in the lagoon. While mud dates get progressively younger upwards in the cores, the sand remains old (Table 2). This suggests that erosion of older North Point dune sand, dated at between 6200 and 5300 ybp, has contributed a large proportion of the sand sized material in Graham's Harbor. Other studies on San Salvador have yielded similar results (Boardman et al, 1987; Boardman et al, in prep.)

Lateral Distribution of Facies

The principal controls on lateral facies distribution in Graham's Harbor are wave energy and substrate modification by benthic flora. In most of the lagoon today, the substrate is open to wave energy, is not stabilized by benthic flora, and abraded grain "grapestone" is the dominant facies. Higher wave energy and lack of stabilization allows winnowing of mud and abrasion of grains. Presence of grapestone also indicates periodic shifting of sand followed by periods of stabilization and cementation (Illing, 1954, Winland and Mathews, 1974). Similar sediments are found under similar conditions today on northern Great Bahama Bank (Imbrie and Purdy, 1962; Trumbel, 1986), Bight of Abaco (Neumann et al, 1975; Boardman, 1976; Locker, 1981), and Snow Bay (Andersen and Boardman, 1987).

In protected (lower wave energy) areas of the lagoon that have been stabilized by seagrass, *Halimeda* and foram-rich "packstone" is found. North Point sand dunes provide protection from the dominant easterly waves in Graham's Harbor. The dunes refract and dampen much of the wave energy. The North Point dunes are 10m high, 15m wide, and extend about 500m to the north. A barrier reef extends from the northwest tip of Cut Cay to Catto Cay, the northern portion of the study area, and absorbs much of the wave energy entering the lagoon. In addition, seagrass baffles wave energy and provides a substrate for encrusting forams and other epibionts (Scoffin, 1970; Scoffin and Tudhope, 1985; Boscence, 1986).

Areas of greatest wave energy are found next to the barrier reef and south of Gaulin and Catto Cays, which receive uninterrupted wave energy from the west. Abraded grain "grainstone" rims the lagoon in these areas. Abundant east-west trending blowouts adjacent to the reef, the high percentage of abraded grains, and the low mud content attest to the heavy wave and current activity.

Vertical Gradients

Vertical facies gradients in Graham's Harbor are controlled by energy and substrate modification, as well as the presence of a sill. The presence of a sill strongly influenced the basal portion of the vertical sequence (Fig. 8) in Graham's Harbor by restricting wave energy and allowing a very muddy sediment to accumulate. Grapestone/bioclast "wacke-/packstone" forms the basal component of cores which penetrate the deeper portions of the lagoon (Fig. 8), representing deposition within a silled, semi-enclosed basin. Cores taken on the higher portions of the Pleistocene bedrock do not contain this facies.

Abraded grain "grapestone" abruptly overlies the grapestone/bioclast "wacke-/packstone" facies, indicating a rapid change in deposition. The low mud content and abundance of abraded grains in

Graham's Harbor

COMPOSITE VERTICAL SEQUENCE

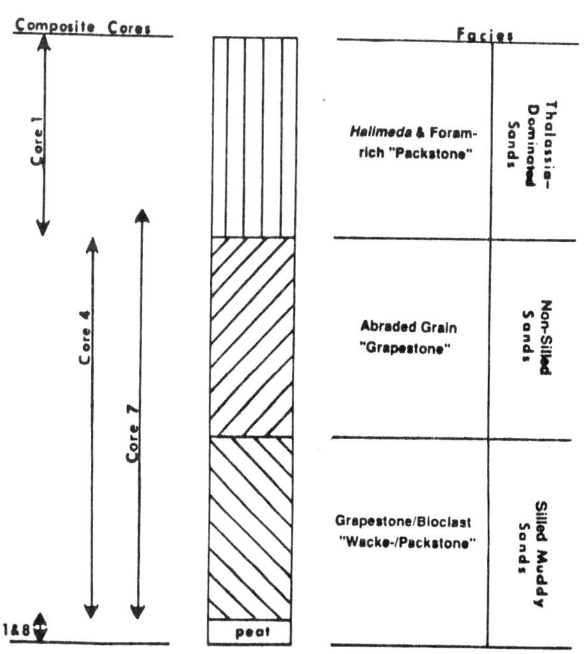

Fig. 8: Composite vertical sequence in Graham's Harbor. Composite cores represent cores used to develop each part of the vertical sequence.

this facies suggest a greater degree of agitation and winnowing by waves and currents. This facies signifies deposition after the sill had been breached by rising sea level under higher-energy, less restricted conditions.

The vertical sequence is capped by *Halimeda* and foram-rich "packstone", which comprises all of core 1 and the upper 15cm of core 7. This facies accumulated in the sheltered environment behind North Point only after seagrass was able to colonize and provide a substrate for epibionts and a stabilized environment for *Halimeda.* It therefore indicates a protected, bottom-stabilized environment. Radiocarbon dates from peat and shells at the base of core 1 indicate that this facies began forming in some areas less than 5000 years ago. The abraded grain "grainstone" forms in a zone near the barrier reef and south of the cays. This facies may represent an alternate, high-energy cap to the vertical sequence in Graham's Harbor.

The vertical sequence in Graham's Harbor is also characterized by a distinct mineralogic signature. The mud fraction mineralogy changes from about 70% aragonite to roughly equal proportions of aragonite and high-magnesium calcite upward in the core. This is due to the colonization by seagrass in the upper part of the cores and the associated production of high-magnesium calcite from epibionts. This mineralogic signature can be a more reliable indicator of seagrass stabilization than percent mud (Andersen et al, 1988).

History of Deposition

The depositional history of Graham's Harbor began about 7000 years ago with the Holocene sea level rise and flooding of the platform. Peat formation began and was preserved in low-lying, protected areas of the silled basin. Radiocarbon dates of peat are 7080$\pm$90 ybp and 4960$\pm$70 ybp, indicating that peat formation continued in Graham's Harbor as sea level rose. As sea level continued to rise, subtidal muds (grapestone/bioclast "wacke-/packstone") were deposited above the peat in the lowest lying areas of the silled basin. As the sill was breached, higher energy conditions prevailed, less mud was deposited, and these muds were covered by subtidal sands (abraded grain "grapestone"). This facies overlies the grapestone/bioclast "wacke-/packstone" wherever it is found in the lagoon. Protected areas behind the North Point dunes allowed colonization by *Thalassia* and accumulation of a somewhat muddier sediment (*Halimeda* and foram-rich "packstone"). Alternately, heavy wave and current activity in a zone near the barrier reef allowed a high-energy, sandy sediment to accumulate (abraded grain "grainstone"). Fig. 9 illustrates the depositional evolution of Graham's Harbor. Radiocarbon dates of mud indicate the sequence was deposited at rates of 30-80 cm/1000 years.

Vertical Sequence

The vertical sedimentary sequence in Graham's Harbor coarsens upwards from subtidal muds to subtidal sands and muddy sands. This sequence differs from models proposed by James (1984), Fischer (1964), and Boardman (1976) for lagoonal deposition. The sequence described here illustrates a previously undescribed set of conditions: deposition in a semi-enclosed, silled basin followed by higher-energy, open marine deposition. Standard sedimentological analysis combined with multivariate statistical techniques have permitted the recognition of this sequence.

CONCLUSIONS

Four marine facies are found in Graham's Harbor: 1) abraded grain "grapestone"; 2) abraded grain "grainstone"; 3) *Halimeda* and foram-rich "packstone"; and 4) grapestone/bioclast "wacke-/packstone". These facies are distinct both laterally and vertically in the lagoon. Very little overlap occurs. Standard sedimentological analysis combined with multivariate statistical techniques have permitted recognition of these facies.

Abraded grain "grapestone" blankets most of the lagoon today, because the substrate is open to wave energy and is not stabilized by benthic flora. *Halimeda* and foram-rich "packstone" is found today in protected (lower wave energy) areas of the lagoon that have been stabilized by seagrass. Rimming the lagoon on the east and north is the abraded grain "grainstone" facies. This facies accumulates along the barrier reef and next to the northern cays, areas which receive the greatest wave energy.

The vertical sequence in Graham's Harbor begins with the grapestone/bioclast "wacke-/packstone", which forms the basal component. This facies represents deposition within a low energy, silled, semi-enclosed basin. Abraded grain "grapestone" overlies the grapestone/bioclast "wacke-/packstone" facies. This facies results from high-energy, less restricted conditions after the sill

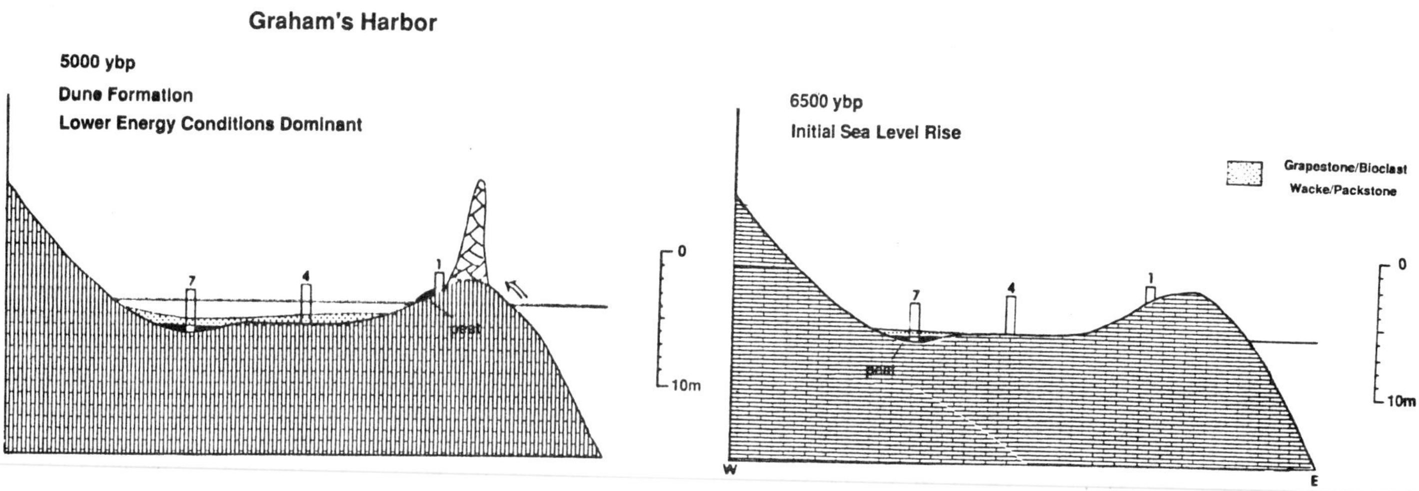

Fig. 9a: Depositional evolution of Graham's Harbor from 6500-5000 ybp. Sequence illustrates the response of the lagoon to a progressive sea-level rise.

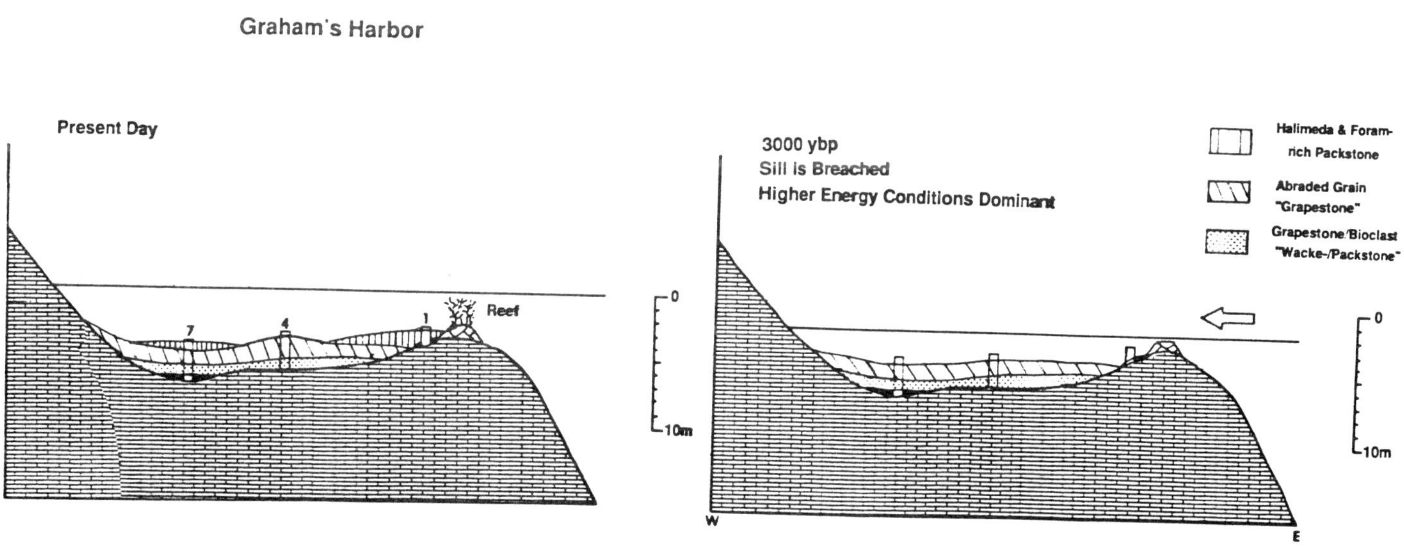

Fig. 9b: Depositional evolution of Graham's Harbor from 5000 ybp-present.

has been breached by rising sea level. The vertical sequence is capped by *Halimeda* and foram-rich "packstone", produced in a protected, bottom-stabilized environment.

The depositional history of Graham's Harbor began about 7000 years ago with platform flooding accompanying Holocene sea level rise. Peat accumulated in low-lying, poorly drained areas. Subtidal muds were then deposited within the silled basin as sea level continued to rise. These muds were covered by subtidal sands and muddy sands as the sill was breached, higher energy conditions prevailed, and less mud was deposited. Radiocarbon dates of the mud fraction indicate the sequence was deposited at rates of 30-80 cm/1000 years. Radiocarbon dates of the sand sized fraction from cores indicates that erosion of old (6200-5300 ybp) North Point sands has contributed much of the sand in the lagoon. The vertical sequence in Graham's Harbor coarsens upwards from subtidal muds to subtidal sands and muddy sands. This sequence is a significant departure from other models for lagoonal deposition.

ACKNOWLEDGEMENTS

We wish to thank Don and Kathy Gerace and the staff of the CCFL Bahamian Field Station for their generous help and use of facilities. Special thanks to Ron Pelle, Paul Bergstrand, and Brannon Andersen, who provided invaluable help with field work. Thanks also to John Mylroie for review of this paper and many helpful suggestions. This project was performed in collaboration with the annual summer workshop Modern Carbonate Deposition, directed by Mark Boardman. Additional funding was provided by two Grants-in-Aid of Research from Sigma Xi, The Scientific Research Society.

REFERENCES CITED

Andersen, C.B. and Boardman, M.R., 1987, Lateral and vertical sedimentary gradients in a high-energy carbonate environment, Snow Bay, San Salvador, Bahamas: Abstracts of Society of Economic Paleontologists and Mineralogists Annual Midyear Meeting, v. 4, p. 4.

Boardman, M.R., 1976, Lime mud deposition in a tropical island lagoon, Bight of Abaco, Bahamas: Unpublished M.S. thesis, University of North Carolina, Chapel Hill, N.C., 121 p.

Boardman, M.R., Carew, J.L., and Mylroie, J.E., 1987, Holocene deposition of transgressive sand on San Salvador, Bahamas: Geological Society of America Abstracts with Programs, v. 19, p. 593.

Boardman, M.R., Carew, J.L., Mylroie, J.E., and Andersen, C. B. (in prep.), Holocene deposition of transgressive sand on San Salvador, Bahamas.

Bosence, D.W.J., 1986, Controls on lateral gradients in carbonate grains and mounds, Florida Bay: Abstracts of Society of Economic Paleontologists and Mineralogists Annual Mid-year Meeting, v. 3, p. 13.

Chave, K.E., 1954, Aspects of the biochemistry of magnesium 1. Calcareous marine organisms: Journal of Geology, v. 62, p. 266-283.

Ginsburg, R.N. and Lowenstam, H.A., 1956, The influence of marine bottom communities on the depositional environment of sediments: Twentieth International Geological Congress, Mexico, p. 232.

Ginsburg, R.N., 1956, Environmental relationships of grain size and constituent particles in some south Florida carbonate sediments: Bulletin of the American Association of Petroleum Geologists, v. 40, p. 2384-2427.

Goldsmith, J.R., Graf, D.L., and Joensuu, O.I., 1955, The occurrence of magnesian calcites in nature: Geochimica et Cosmochimica Acta, v. 7, p. 212-230.

Goodwin, P.W. and Anderson, E.J., 1985, Punctuated aggradational cycles: a general hypothesis of episodic stratigraphic accumulation: Journal of Geology, v. 93, p. 525-533.

Harbaugh, J.W. and Merriam, D.F., 1968, Computer Applications in Stratigraphic Analysis: Wiley, New York, 282 p.

Illing, L.V., 1954, Bahamian calcareous sands: American Association of Petroleum Geologist Bulletin, v. 38, p. 1-95.

Imbrie, J. and Van Andel, T.H., 1964, Vector analysis of heavy-mineral data: Geological Society of America Bulletin, v. 75, p.

1131-1156.

James, N.P., 1984, Shallowing-upward sequences in carbonates, in Walker, R.G., ed., Facies Models, p. 213-228, Geoscience Canada, Reprint Series 1.

Locker, S.D., 1981, Origin and depositional history of a semienclosed windward lagoon off Great Abaco Island, Bahamas: Unpublished M.S. thesis, University of North Carolina, Chapel Hill, N.C., 181 pp.

Miller, A.I., 1988, Spatial resolution in subfossil molluscan remains: implications for paleobiological analyses: Paleobiology, v. 14, p. 91-103.

Neumann, A.C. and Land, L.S., 1975, Lime mud deposition and calcareous algae in the Bight of Abaco, Bahamas: a budget: Journal of Sedimentary Petrology, v. 45, p. 763-786.

Pratt, B.R., and James, N.P., 1986, The St. George Group (Lower Ordovician) of western Newfoundland: tidal flat model for carbonate sedimentation in shallow empiric seas: Sedimentology, v. 33, p. 313-343.

Purdy, E.D., 1963, Recent calcium carbonate facies of the Great Bahama Bank: Journal of Geology, v. 71, p. 334-355.

Schlager, W., 1981, The paradox of drowned reefs and carbonate platforms: Geological Society of America Bulletin, v. 92, p. 197-211.

Scoffin, T.P., 1970, The trapping and binding of subtidal carbonate sediments by marine vegetation in Bimini Lagoon, Bahamas: Journal of Sedimentary Petrology, v. 40, p. 249-273.

Scoffin, T.P. and Tudhope, A.W., 1985, Sedimentary environments of the central region of the Great Barrier Reef of Australia: Coral Reefs, v. 4, p. 81-93.

Swinchatt, J.P., 1965, Significance of constituent composition, texture, and skeletal breakdown in some recent carbonate marine sediments: Journal of Sedimentary Petrology, v. 35, p. 71-90.

Taylor, J.D. and Lewis, M.S., 1970, The flora, fauna, and sediments of the marine grass beds of Mahe Seychelles: Journal of Natural History, v. 4, p. 199-220.

Trumbel, W. and Neumann, A.C., 1986, The depositional history of an open windward marginal lagoon, northern Great Bahama Bank: Society of Economic Paleontologists and Mineralogists Annual Midyear Meeting Abstracts, v. 3, p. 110-111.

Winland, H.D. and Matthews, R.K., 1974, Origin and significance of grapestone, Bahama Islands: Journal of Sedimentary Petrology, v. 44, p. 921-927.

Wright, V.P., 1986, Facies sequences on a carbonate ramp: the Carboniferous limestone of South Wales: Sedimentology, v. 33, p. 221-241.

COMPARATIVE MORPHOLOGIC ANALYSIS AND GEOCHRONOLOGY FOR THE DEVELOPMENT AND DECLINE OF TWO PLEISTOCENE CORAL REEFS, SAN SALVADOR AND GREAT INAGUA ISLANDS, BAHAMAS

H. Allen Curran and Brian White
Department of Geology
Smith College
Northampton, Massachusetts 01063

J. H. Chen and G. J. Wasserburg
Lunatic Asylum
Division of Geological and Planetary Sciences
California Institute of Technology
Pasadena, California 91125

ABSTRACT

Sizeable bank/barrier coral reefs of Sangamon age crop out along the coast at Cockburn Town on San Salvador and at Devil's Point on Great Inagua. The Cockburn Town reef has been mapped, and the relationships between various parts of the reef, lithofacies contacts, and modern sea level have been established precisely. At Devil's Point, three detailed profiles have been surveyed relative to high tide line.

^{238}U-^{234}U-^{230}Th dates were determined for sixteen coral samples from the Cockburn Town reef and three coral samples from the Devil's Point reef using recently developed mass spectrometric techniques for the measurement of ^{230}Th abundance. Typical errors are ± 1.5 ky. This narrow error range permits detailed chronologic study with time resolution adequate to define stages of the reef's history. The initial stage of reef development is not revealed at either site, but the main parts of both reefs are well preserved. The Cockburn Town reef existed by 130 kybp and an *Acropora palmata* reef crest with adjacent patch reefs flourished between 124 to 121 kybp. By 119 kybp the reef was in decline. Total longevity of the Cockburn Town reef was about 12 ky. In the vicinity of the measured profiles at Devil's Point, an *A. cervicornis*-dominated rubblestone and corals preserved in growth position have been bevelled off, apparently representing an ancient wave-cut surface. A patch reef dominated by *Montastrea annularis* and *Diploria* sp. overlies the bevelled surface. Three coral ages from Devil's Point range from 125 to 122 kybp, corresponding to the time when the Cockburn Town reef also was flourishing.

As Wisconsinan glaciation expanded, the resulting fall in sea level led to the progressive burial of the reefal facies at both sites by subtidal, then beach, and finally dunal sands.

INTRODUCTION

Brief and usually sketchy reports of reefal facies of presumed Pleistocene age in the carbonate rocks that cap the Bahamas have been made in various geologic and geographic surveys over the years from a number of Bahamian islands. However, except for the Cockburn Town fossil coral reef on San Salvador, none of the larger fossil coral reefs in the Bahamas have been described in any detail. The well exposed patch reef complex at Sue Point on San Salvador recently has been mapped and studied by White (this volume).

The Cockburn Town reef is the largest fossil coral reef on San Salvador Island. The reef was described briefly by Moshier and others (1979). A detailed facies map of the reef and a thorough field description of its geologic and paleontologic setting based on the map and a series of five stratigraphic profiles was given by Curran and White (1984, 1985). The shallowing-upward facies associated with the reef were described and an analysis of the carbonate cements stratigraphy of the sequence was made by White and others (1984). On Great Inagua Island, the general geologic setting of the large fossil coral reef at Devil's Point was described by White and Curran (1987).

The purposes of this paper are to describe further the paleogeographic setting and history of development and decline of the Cockburn Town reef, to describe the setting and three stratigraphic profiles from the Devil's Point reef, and to integrate into the reef developmental models sixteen radiometric dates from corals of the Cockburn Town reef and three dates from corals of the Devil's Point reef. In addition, we will discuss the significance of these dates for interpretation of the developmental history of these two reefs and for the understanding of eustatic sea level change during that part of Sangamon time represented by the reefs. Length limitations for this paper prohibit extended discussion of the developmental model for the Cockburn Town reef; our ideas are succinctly outlined in the figures within. We expect that several more radiometric dates will be forthcoming from both reefs to expand and fill in remaining gaps of coverage at each site. These dates and results from some further refinement of the dates presented herein will be given in a future publication.

THE GEOLOGIC SETTING

The Cockburn Town Fossil Coral Reef

The Cockburn Town reef is located along the western coast of San Salvador Island, northwest of the center of Cockburn Town (Fig. 1). Reefal rocks extend in a northwesterly direction from the old town dock for a distance of about 650m, terminating near an abandoned cable trench cut into Pleistocene shallow subtidal and beach calcarenites that flank and overtop the reef at its northern end.

The geologic map of the reef and its associated facies (Fig. 2) and five detailed stratigraphic profiles from the reef area originally were published by Curran and White (1984, 1985). Reference starter points for map and profiles topography were nearby bench marks that were tied directly to accurately measured mean sea level (Adams, 1980). The reef ultimately was buried by a sequence of calcarenites deposited in shallow subtidal environments and grading up to a beach and then dunal environment. This shallowing-upward sequence and the diagenetic history of the reef complex with respect to sea level change were discussed by White and others (1984).

The Cockburn Town and Devil's Point reefs are excellent fossil examples of bank/barrier reefs as defined by Kaplan (1982). Such reefs are common today on the narrow, wave-cut shelves of islands in tropical western Atlantic and Caribbean waters. Modern bank/barrier reefs (such as

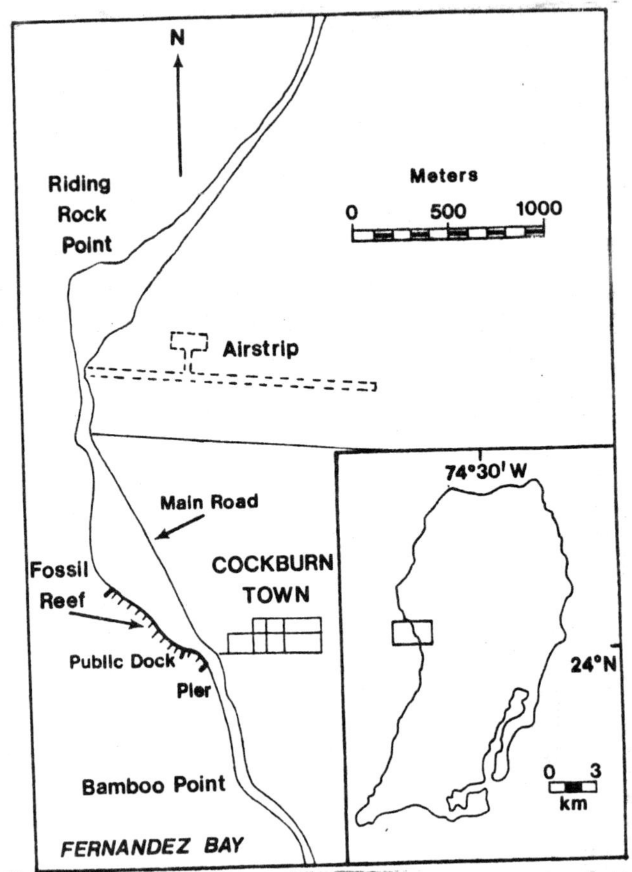

Fig. 1. Location of the Cockburn Town fossil coral reef on San Salvador Island.

the Gaulins Reef off of the north coast of San Salvador) normally are only hundreds of meters off from a land mass, not thousands of meters offshore as is the case for true barrier reefs. Bank/-barrier reefs also are much shorter in linear extent than barrier reef complexes. Our paleogeographic reconstruction of the Cockburn Town reef as a bank/barrier reef with associated patch reefs, lying about 450 m offshore and flourishing in the time period about 124,000 to 121,000 years ago is shown in Figure 3. This reconstruction is based on the map of the reef (elevations of major coral types and facies contacts), reconstruction of the growth heights of the reef crest corals (*Acropora palmata*), and the cluster of reef crest coral age dates, as will be discussed subsequently.

The Devil's Point Fossil Coral Reef

The Devil's Point reef is found along the west coast of Great Inagua Island flanking both sides of Devil's Point (Fig. 4.). This bank/barrier reef is much larger than the Cockburn Town reef; it extends for several kilometers around Devil's Point. The general geologic setting of this reef was described by White and Curran (1987). Again

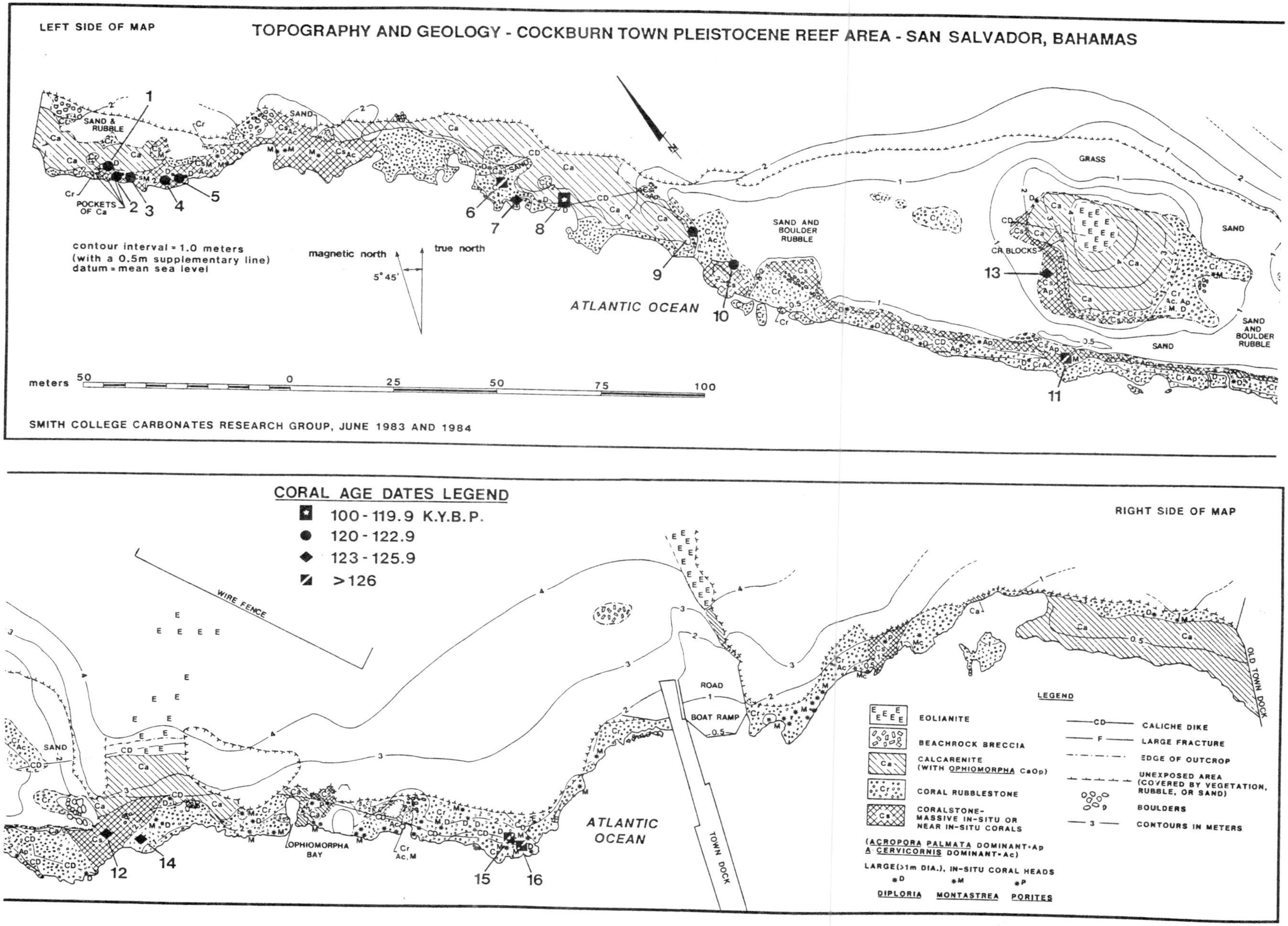

Fig. 2. Geologic map of the Cockburn Town coral reef. Numbers indicate locations of radiometrically dated coral samples as given in Table 1.

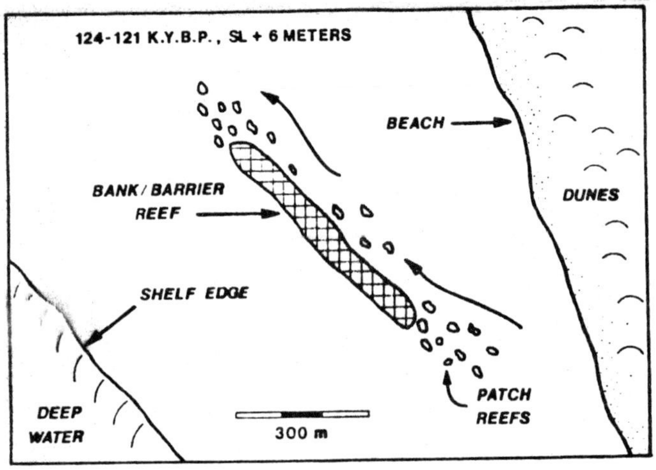

Fig. 3. Paleogeographic reconstruction of the Cockburn Town coral reef at time of its maximum development about 124-121 kybp.

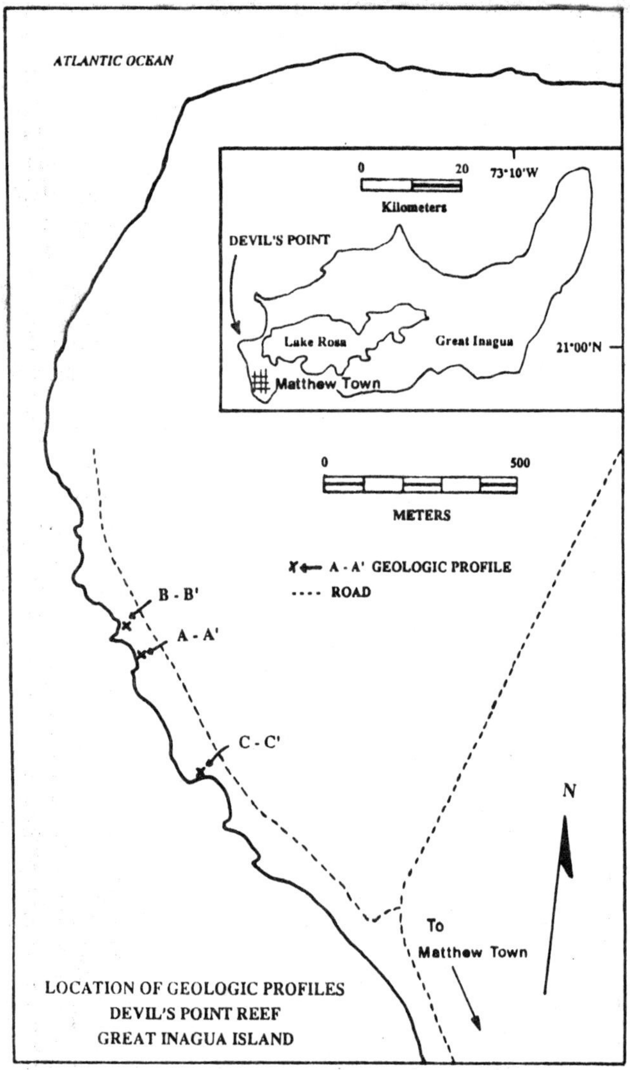

Fig. 4. Location of the Devil's Point fossil coral reef on Great Inagua Island, with locations of Profiles A-A' (Fig. 14), B-B' (Fig. 15), and C-C' (Fig. 5).

a shallowing-upward sequence of calcarenite facies caps the reefal complex.

The large size and remoteness of the Devil's Point reef have prohibited the making of a geologic map, but three detailed stratigraphic profiles as located on Figure 4 have been carefully surveyed. Probably the general history of development and decline of this reef is similar overall to that of the Cockburn Town reef, but at least one major difference in the developmental sequence has been recognized, as will be discussed later.

RADIOMETRIC DATES

Background

A total of nineteen radiometric dates, sixteen dates from the Cockburn Town reef and three dates from the Devil's Point reef, have been obtained from fossil coral samples. These ^{238}U-^{234}U-^{230}Th dates were determined by J.H. Chen and G.J. Wasserburg using their newly developed mass spectrometric technique for the measurement of ^{230}Th abundance (Edwards and others 1987a,b). Typical errors for dates from this method are $\pm$1.5 ky (2 sigma); this is about 5 to 10 times smaller than the error range for alpha counting. These low error range dates permit detailed chronologic study of a reef and, in the case of the Cockburn Town reef, with time resolution sufficient to bracket closely the timing of stages of development of the reef. A summary of data for the nineteen dated coral samples is given in Table 1.

Cockburn Town Reef Dates

The locations of dated coral samples from the Cockburn Town reef are marked on the geologic map of the reef (Fig. 2). Samples of five different coral species have been dated from the reef, and we think that the dates are fully species independent. If there is a problem coral species, it would seem to be *Diploria clivosa*, the encrusting brain coral. *D. clivosa* sample CT-6 from Cockburn Town and sample DP-3 from Devil's Point gave dates that were somewhat older than we would have predicted given that this species most commonly is a shallow water encruster.

The oldest coral dated from the Cockburn Town reef is an *Acropora palmata* specimen (sample CT-11) at 130.75$\pm$1.5 ky, and the youngest coral date is 119.2$\pm$1.5 ky from a *Diploria*

SAMPLE NUMBER	SCLERACTINIAN CORAL SPECIES	AGE IN K.Y.B.P.	PRESERVATIONAL SETTING	ELEVATION ABOVE SEA LEVEL (in meters)
COCKBURN TOWN REEF, SAN SALVADOR				
1	*Montastrea annularis*	120.0 ± 1.4	growth position	+2.25
2	*M. annularis*	120.5 ± 1.4	" "	+2.25
3	*Diploria strigosa*	122.1 ± 1.2	" "	+2.25
4	*D. clivosa*	121.8 + 1.3	encruster in growth position	+1.5
5	*M. annularis*	122.7 ± 1.3	growth position	+2.25
6	*D. clivosa*	128.7 ± 1.2	encruster in growth position	+1.5
7	*Acropora cervicornis*	124.2 ± 2.2	in coral rubblestone, close to growth location	+1.5
8	*D. strigosa*	119.2 + 1.5	growth position	+2.0
9	*A. palmata*	121.2 ± 1.4	near growth location in coralstone	+1.5
10	*A. palmata*	122.5 ± 1.5	" " " " "	+1.75
11	*A. palmata*	130.75 ± 1.5	" " " " "	+1.25
12	*A. palmata*	124.6 ± 1.3	" " " " "	+1.5
13	*A. palmata*	123.7 ± 1.6	" " " " "	+1.5
14	*A. cervicornis*	124.7 ± 1.5	in coral rubblestone, close to growth location	+1.5
15	*D. strigosa*	126.3 ± 1.2	growth position	+1.5
16	*M. annularis*	127.1 ± 2.1	" "	+1.25
DEVIL'S POINT REEF, GREAT INAGUA				
1	*M. annularis*	125.1 ± 2.1	in coral rubblestone	See Fig. 5
2	*M. annularis*	122.2 ± 1.3	growth position	" " "
3	*D. clivosa*	123.4 ± 1.5	encruster in growth position	" " "

Table 1. Compilation of data on the radiometrically dated fossil coral samples, including sample numbers as keyed to the Cockburn Town Reef geologic map (Fig. 2) or Devil's Point Profile C-C' (Fig. 5), coral species, age date and error range, coral sample preservational setting, and elevation of sample above sea level.

strigosa specimen (sample CT-8). Thus we know that the reef existed and flourished during the Sangamon interglacial, at the time of oxygen isotope substage 5e of Shackleton and Opdyke (1973) and as previously reported by Carew and Mylroie (1987). Our span of coral age dates from the Cockburn Town reef indicates a minimum longevity for the reef of 12,000 years. Furthermore, the age dates can be used to bracket the timing of reef developmental stages as discussed in the next section.

The patch reef corals at the northwest end of the Cockburn town reef yielded generally younger age dates than the middle section of the reef, the *Acropora palmata* reef crest area, and the patch reefs at the southeast end of the reef (Fig. 2). We think that this largely is the result of topographic control, with the samples from the northwest end of the reef having a higher topographic position. It also could be that the reef grew in a northwesterly direction. We favor the former interpretation, because we think that the shallow water and encruster corals that are common in the northwest area of the reef (see later discussion) and the overlying calcarenite beds truly are higher up in the reef sequence. One must remember that a modern bank/barrier reef develops with significant topographic relief (at least 6 m at Gaulins reef). In a fossil reef complex, the preservation of corals would not have occurred in a layer-cake fashion; instead considerable complexity with regard to age relationships between adjacent fossil corals should be expected.

Carew and Mylroie (1987, Fig. 1) reported three U-series dates from corals of the Cockburn Town reef. Two of these dates (error range of 12 ky) fit well into our range of dates. A third date of 171±15 ky from a *Diploria strigosa* rubble chunk is much older than any date we obtained. Carew and Mylroie postulated that the old coral was washed as rubble into the Cockburn Town reef complex and preserved there, and that its age suggests an earlier flooding event for the San Salvador shelf. This may have been the case, but we cannot confirm the hypothesis with any of our age dates.

Devil's Point Reef Dates

Three U-series age dates have been determined from coral samples taken along Profile C-C' at Devil's Point (Fig. 5, see also Table 1). In the area of Profile C-C', a well preserved, *in situ* patch reef lies on top of a wave-cut surface with underlying coral rubblestone. A *Montastrea annularis* specimen from below the wave-cut surface yielded an age of 125.1±2.1 ky. Samples from above the wave-cut surface gave age determinations of 122.2±1.3 ky and 123.4±1.5 ky.

The aggregate dates indicate that the Devil's Point reef existed contemporaneously with the Cockburn Town reef and flourished during Sangamon stage 5e time. The vertical spread of dates suggests that the wave-cut surface may reflect a short-lived sea level low-stand during the time interval of stage 5e. However, caution is warranted at this time because the error bars of the three samples somewhat overlap. Clearly more age determinations are needed before a complete history of the Devil's Point reef can be related.

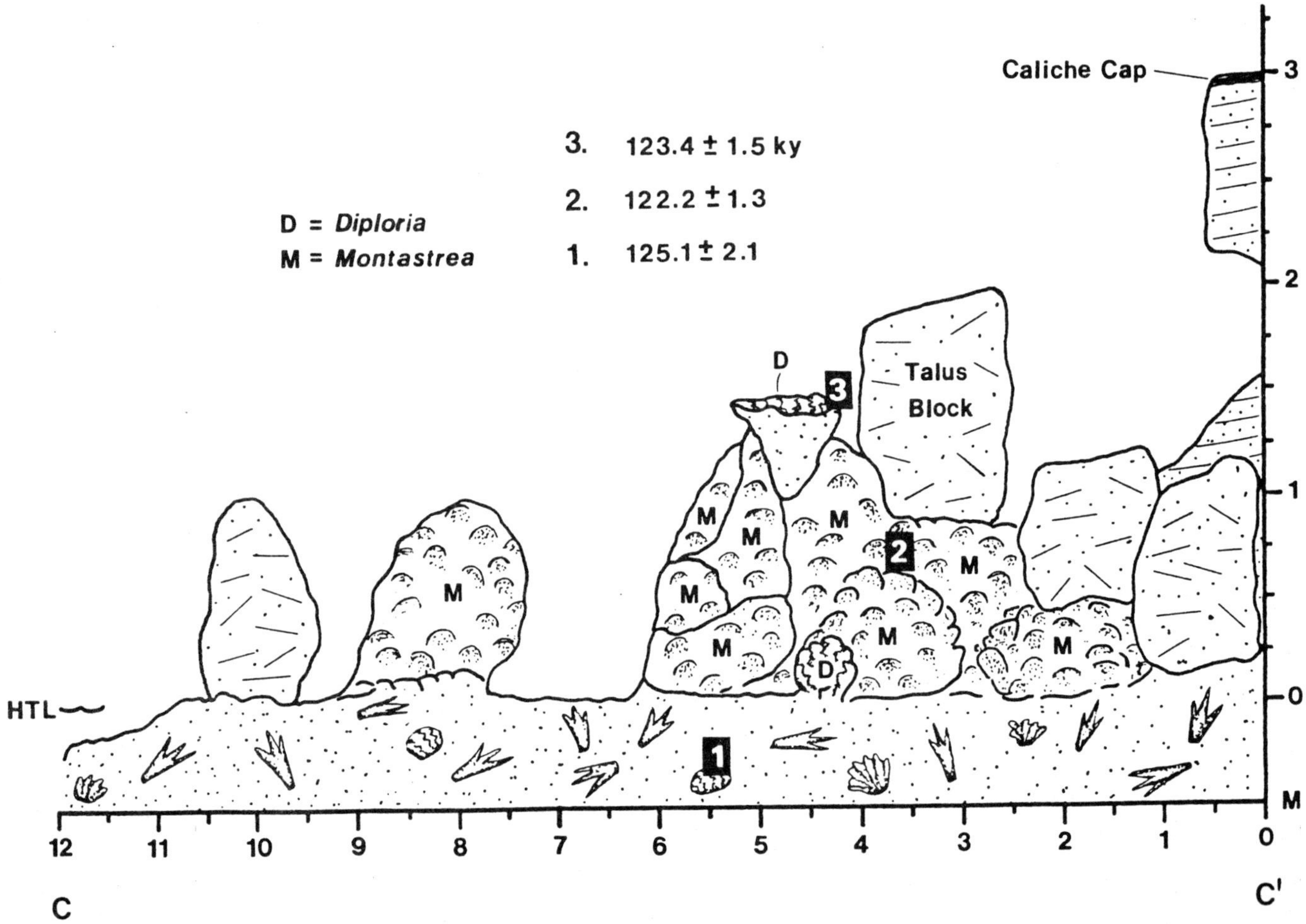

Fig. 5 Profile C-C' from the Devil's Point coral reef. The wave-cut bench occurs at top of the coral rubblestone (basal unit of the profile) and below the in situ patch reef. Locations also are shown for the three radiometrically dated coral samples as given in Table 1.

REEF DEVELOPMENT AND DECLINE

The Cockburn Town Reef

Colonization Stage

The initial stage of reef development at the Cockburn Town site is not revealed in outcrop, but the reef probably began with the colonization of hardground areas by *Acropora palmata* and other corals (Fig. 6) as rising seas encroached upon the San Salvador shelf more than 130 ky before present. An *Acropora palmata* sample (CT-11) yielded an age of 130.75±1.5 ky, indicating that the reef was established by that time. Patch reefs at the flanks of the developing bank/barrier reef were growing by 127 kybp (*Montastrea annularis* sample, CT-16).

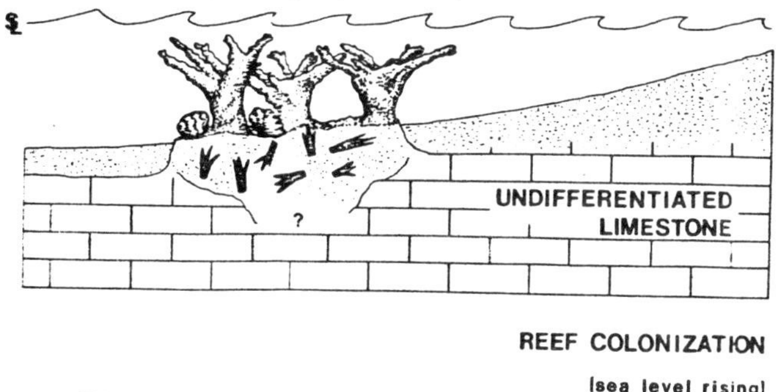

Fig. 6. Colonization stage of development of the Cockburn Town coral reef, greater than or about 130 kybp.

Maturity Stage

The crest of Bahamian bank/barrier reefs is formed by the frame-building coral *Acropora palmata*. Four of the five *A. palmata* samples from the Cockburn Town reef (samples CT-9, 10, 12, 13) fall in the time range of 124.6 to 121 kybp, indicating that the reef crest and adjacent patch reefs flourished at this time (Fig. 7). *A. palmata* coral heads grew upward to mean low sea level, which we think was about 6 m above present mean low sea level (based on our geologic/topographic map of the reef area, with no correction for tectonic subsidence). At that time, the reef lay about 450 m offshore (Fig. 3).

Reef Decline and Partial Burial

The sedimentary sequence of the reef and its overlying beds indicates that the reef began to decline (Fig. 8) as trough cross bedded, subtidal sands transported by longshore currents encroached on the reef (Fig. 9). Living patch reef

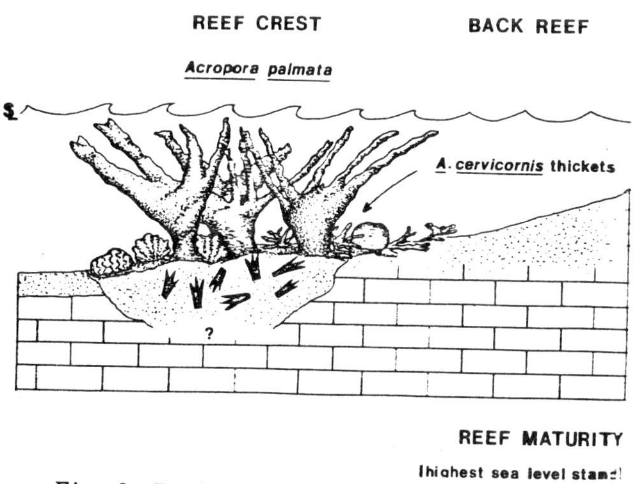

Fig. 8. Decline stage of the Cockburn Town coral reef, about 120 kybp.

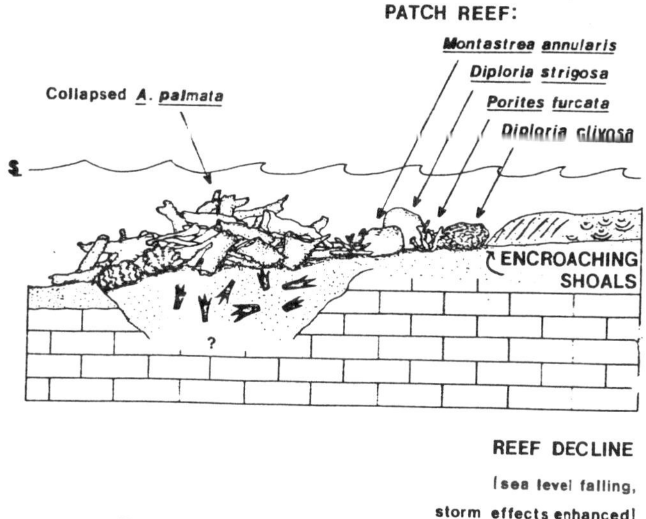

Fig. 7. Full maturity stage of development for the Cockburn Town coral reef, between 124-121 kybp.

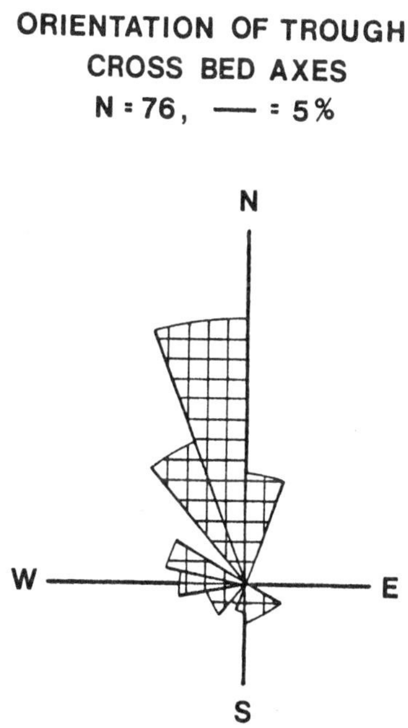

Fig. 9. Rose diagram plot of orientations of trough cross bed axes in calcarenites of subtidal beds topping the Cockburn Town coral reef.

corals were present in at least the northwest area of the reef until 120 to 119 kybp (samples CT-1, 2, 8). Particularly prominent in this area of the reef are the shallow-water corals *Porites furcata* and *Diploria clivosa* (Curran and White, 1984, 1985). At about this time, a thick bed of tabular cross bedded sand was deposited by a storm, very possibly a hurricane, and this sand layer covered much of the reef, terminating active growth (Fig. 10, 11, and White and others, 1984).

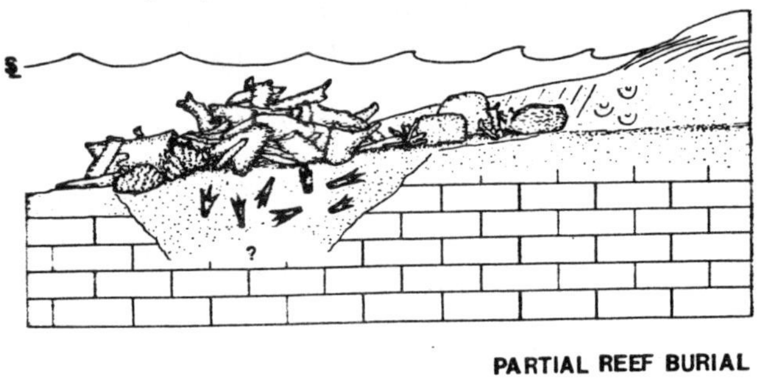

Fig. 10. Partial burial stage of the Cockburn Town coral reef, about 119 kybp.

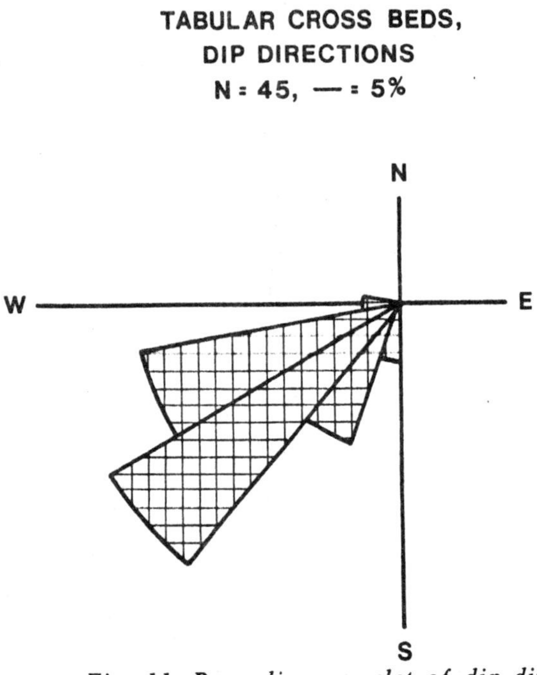

Fig. 11. Rose diagram plot of dip directions of the set of tabular cross beds in the calcarenites that cap the Cockburn Town coral reef.

Complete Burial

With the onset of Wisconsinan glaciation and continued falling sea level, sands packed around and ultimately over all remaining coral heads of the reef. The shallowing-upward sequence continued with the deposition of beach sediments and was completed by the deposition of dunal sands (Fig. 12, and White and others, 1984).

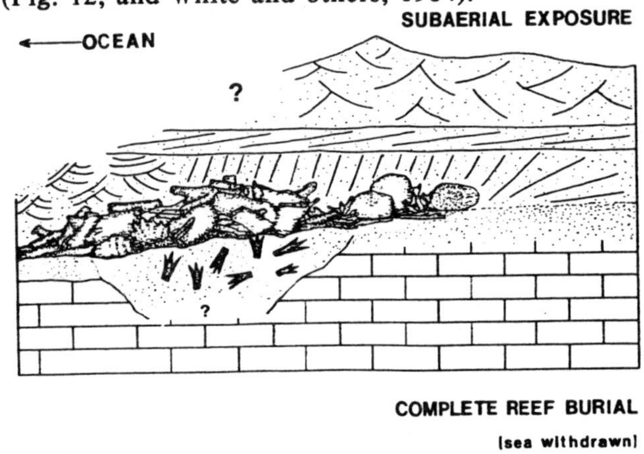

Fig. 12. Complete burial of the Cockburn Town coral reef, shortly after 119 kybp.

Summation

The major events from full maturity to burial of the Cockburn Town reef are summarized in the profiles of Figure 13. The development and

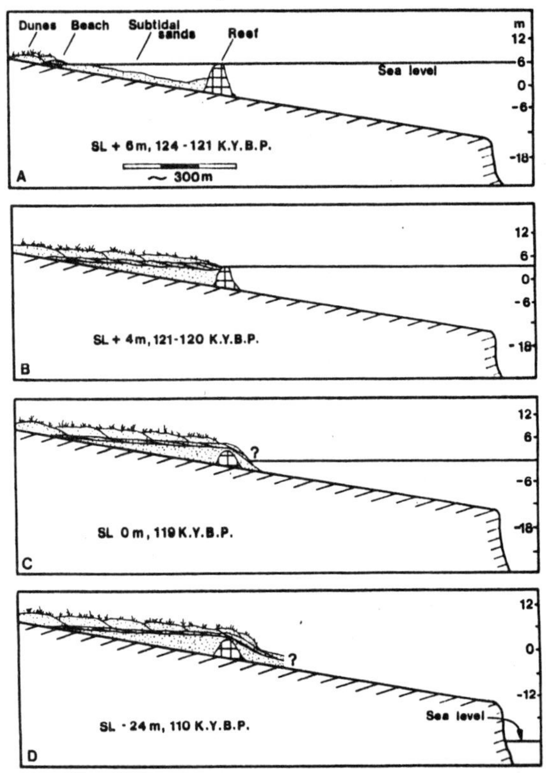

Fig. 13. Profile reconstructions of the Cockburn Town coral reef from time of its maximum development (A) to complete burial (D). Sea level positions for profiles A-C from our reconstruction of the Cockburn Town reef; sea level position for profile D from Harmon and others (1983) curve for Bermuda.

eventual decline and burial of the reef occurred over a minimum period of about 12,000 years, with full maturity at about 124 kybp (Fig. 13A). Subsequently sea level began to fall (Fig. 13B), and by shortly after 119 kybp the reef was buried (Fig. 13C). By 110 kybp, sea level had fallen to a point below the shelf/slope break off San Salvador (Fig. 13D, and Harmon and others, 1983).

Devil's Point Reef

The extensive bank/barrier fossil reef at Devil's Point on Great Inagua Island has not been studied in the detail of the Cockburn Town reef, but three surveyed profiles have been made across the reef (Figs. 5, 14, 15). The major facies of these profiles is coral rubblestone (Facies 1, Figs. 14, 15) dominated by chunks of *Acropora cervicornis* and *A. palmata* with large heads of *in situ Montastrea annularis* and *Diploria strigosa.* The top of this facies has been bevelled off and apparently represents an ancient wave-cut platform.

In the area of Profile C-C' (Fig. 5), an extremely well-preserved, *in situ* fossil patch reef lies on top of the wave-cut surface. This patch reef assemblage is dominated by *Montastrea annularis* with subordinate *Diploria strigosa* and

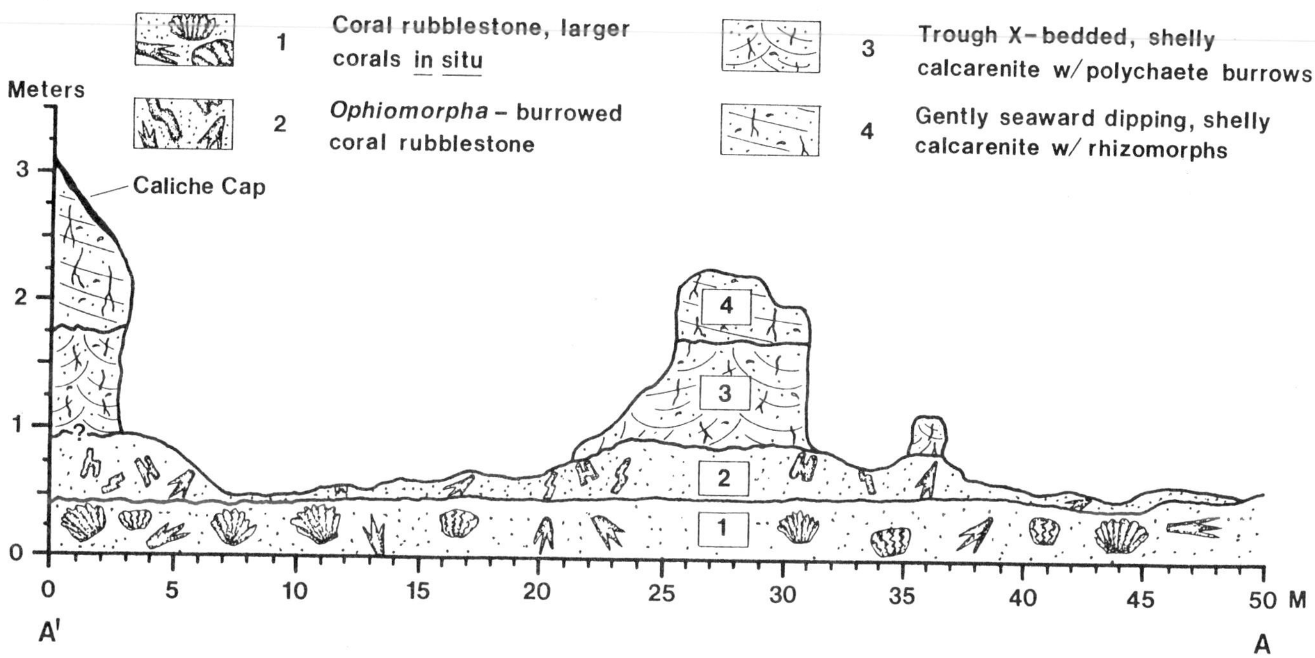

Fig. 14. Profile A-A' from the Devil's Point coral reef. A wave-cut bench occurs at the top of Facies 1 on this profile and Profile B-B' (Fig. 15).

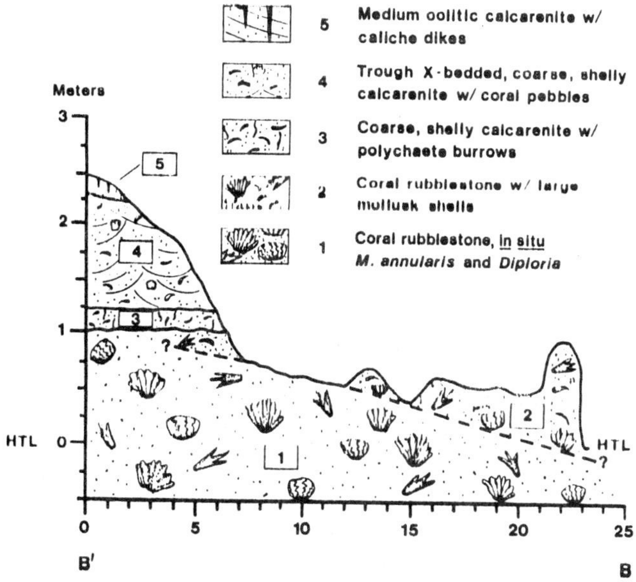

Fig. 15. Profile B-B' from the Devil's Point coral reef.

D. clivosa. The physical relationships suggest that the wave-cut surface may reflect a short-lived sea level low-stand during the time interval of stage 5e, and, as discussed previously, this is at least somewhat supported by the vertical sequence of age dates from coral samples taken along Profile C-C'. However, further work will be needed to verify this interpretation. As with the Cockburn Town reef, the stratigraphic sequence at Devil's Point reveals ultimate termination of the reef by burial during sea regression (White and Curran, 1987).

CONCLUSIONS

1. Based on sixteen age dates derived by mass spectrometric analysis of ^{230}Th abundance in coral specimens, the Cockburn Town reef on San Salvador Island had a *minimum* longevity of 12,000 years, from about 131 ky to 119 kybp. The *Acropora palmata* reef crest of this bank/barrier reef flourished in the period 124-121 kybp. Exposed patch reef corals at the southeast end of the reef are relatively old, whereas those at the northwest end are relatively young. This probably is a topographic-erosional effect.

2. The combination of paleontologic, stratigraphic, and age dating using the new mass spectrometric analysis techniques permits the recognition and time bracketing of several major stages in the development of the Cockburn Town reef, from achievement of full maturity, to decline with shallowing conditions, and finally complete burial.

3. The coral age dates from the Devil's Point reef on Great Inagua range from 125 ky to 122 kybp, corresponding to the time when the Cockburn Town reef also flourished. The wave-cut surface within the Devil's Point reef may reflect a short-lived sea level low-stand during the time interval of Sangamon stage 5e.

4. As Wisconsinan glaciation expanded, the resulting fall in sea level led to progressive burial of the reefal facies at both sites, by subtidal, then beach, and finally dunal sands after about 119 kybp.

ACKNOWLEDGEMENTS

We thank the staff of The Bahamian Field Station for full logistical support of our field work, and we are particularly grateful to Donald T. Gerace, Director, and Kathy Gerace, Associate Director, for their continuing encouragement and support of our research efforts on San Salvador. Curran and White extend their thanks to Jimmy Nixon of Matthew Town and The Bahamas National Trust for help with logistics and transportation on Great Inagua Island. Smith College students Elaine Kotler and Molly Stark assisted with field work on Great Inagua. Curran and White also thank Jim Chen and Jerry Wasserburg for their interest in our fossil coral reefs work and for their willingness to use their age-dating technique on the coral samples. Kathy Bartus, Smith College, performed the word processing for this paper with care, patience, and good humor. Acknowledgment is made to the donors of the Petroleum Research Fund, administered by the American Chemical Society, for partial support of our research through separate grants to Curran and White.

REFERENCES CITED

Adams, R.W., 1980, General guide to the geological features of San Salvador, in Gerace, D.T., ed., Field guide to the geology of San Salvador (1st ed.): San Salvador, Bahamas, CCFL Bahamian Field Station, p. 1-66.

Carew, J.L., and Mylroie, J.E., 1987, A refined geochronology for San Salvador Island, Bahamas, in Curran, H.A., ed., Proceedings of the third symposium on the geology of the Bahamas: San Salvador, Bahamas, CCFL Bahamian Field Station, p. 35-44.

Curran, H.A., and White, B., 1984, Field guide to the Cockburn Town fossil coral reef, San Salvador, Bahamas, in Teeter, J.W., ed., Proceedings of the second symposium on the geology of the Bahamas: San Salvador, Bahamas, CCFL Bahamian Field Station, p. 71-96.

Curran, H.A., and White, B., 1985, The Cockburn Town fossil coral reef, in Curran, H.A., ed., Pleistocene and Holocene carbonate environments on San Salvador Island, Bahamas-Guidebook for Geological Society of America, Orlando annual meeting field trip #2: San Salvador, Bahamas, CCFL Bahamian Field Station, p. 95-120.

Edwards, R.L., Chen, J.H., and Wasserburg, G.J., 1987a, ^{238}U-^{234}U-^{230}Th-^{234}Th systematics and the precise measurement of time over the past 500,000 years: Earth and Planetary Science Letters, v. 81, p. 175-192.

Edwards, R.L., Chen, J.H., Ku, T.-L., and Wasserburg, G.J., 1987b, Precise timing of the last interglacial period from mass spectrometric determination of thorium-230 in corals: Science, v. 236, p. 1547-1553

Kaplan, E.H., 1982, A field guide to coral reefs: Boston, Houghton Mifflin Co., 289 p.

Harmon, R.S., and others, 1983, U-series and amino-acid racemization geochronology of Bermuda: Implications for eustatic sea-level fluctuation over the past 250,000 years: Paleogeography, Paleoclimatology, Paleoecology, v. 44, p. 41-70.

Moshier, S.O., and others, 1979, Stratigraphy and facies of a Pleistocene fringing reef at Cockburn Town, San Salvador, Bahamas, Geological Society of America, Abstracts with Programs, v. 11, p. 206.

Shackleton, N.J., and Opdyke, N.D., 1973, Oxygen isotope and paleomagnetic stratigraphy of equatorial Pacific core V-28-238: Oxygen isotope temperatures and ice volumes on a 10^5 year and 10^6 year scale: Quaternary Research, v. 3, p. 39-55.

White, B., this volume, Field guide to the Sue Point fossil coral reef, San Salvador Island, Bahamas.

White, B., and Curran, H.A., 1987, Coral reef to eolianite transition in the Pleistocene rocks of Great Inagua Island, Bahamas, in Curran, H.A., Proceedings of the third symposium on the geology of the Bahamas: San Salvador, Bahamas, CCFL Bahamian Field Station, p. 165-179.

White, B., Kurkjy, K.A., and Curran, H.A., 1984, A shallowing-upward sequence in a Pleistocene coral reef and associated facies, San Salvador, Bahamas, in Teeter, J.W., ed., Proceedings of the second symposium on the geology of the Bahamas: San Salvador, Bahamas, CCFL Bahamian Field Station, pp. 53-70.

KARST HYDROLOGY OF SAN SALVADOR

R. Laurence Davis
New Hampshire Department of Environmental Services
Groundwater Protection Bureau
6 Hazen Drive, Box 95
Concord, New Hampshire 03301

Calvin R. Johnson, Jr.
Department of Geology
Indiana University
Bloomington, Indiana 47405

ABSTRACT

San Salvador is a typical carbonate platform island in the eastern Bahamas. Its major features include arcuate, consolidated dune ridges up to 40 meters high, and shallow, hypersaline lakes. These occupy the low areas between the dunes and cover a substantial portion of the island's interior. The subsurface contains large numbers of caves, conduits and vertical shafts. Despite the nearly 120 cm of annual rainfall, there is no flowing surface water. The ground water system is poorly defined and fresh water lenses are discontinuous and sporadic. Data collected during two years of field work has led to the construction of a general model of the island's hydrologic regime.

The system consists of six elements: 1) Precipitation; 2) Evapotranspiration; 3) Ground Water; 4) Inland Lakes; 5) Conduits and Blue Holes; and 6) Ocean tides.

Investigations included an examination of rock and water geochemistry, rock permeability, and physical hydrology. Over 430 water samples and 16 rock samples were analyzed. Ca^{++}/Mg^{++} and Mg^{++}/Cl^{-} ratios were used to trace water sources. Water chemistry was used to pinpoint the locations of conduits and seeps on the inland lakes. Ground water elevations were mapped in several locations and conduit flows were measured. Pump and laboratory tests were performed in order to estimate rock permeabilities. Tide gauges were used to measure ocean tides and water level fluctuations in caves and in the hypersaline inland lakes. This field data was used to construct the general model.

Groundwater in the interior of the island is found in discontinuous, isolated, fresh water lenses beneath the consolidated carbonate dunes. The lenses, which are fed by infiltrating rain water, drain into the hypersaline lakes where the water evaporates. Near the coast there are similar lenses, but these are disrupted in many places by flow through caves and conduits. Some of the interior lakes are connected to the sea by conduits, but most appear to be fed directly by ground water seeps and precipitation. The height of the tides varies considerably from day to day. The elevation of many of the inland lakes is above that of the lower high-tides. Therefore, conduit flow into them only occurs during the higher high-tides. At these times, the flow is especially vigorous and can contribute to the disruption of the fresh water lenses in the vicinity of the conduit. The rock has relatively low primary permeability and the direct influence of ocean tides on the fresh water lenses near the shore appears to be small. This model can be used to locate new freshwater resources for San Salvador and other similar islands. It may also be useful in examining the paleohydrology of the island and its impact on cave formation.

INTRODUCTION

Since January, 1987, we have been conducting research on the hydrology of San Salvador. Investigations have included water sampling and chemical analysis, electrical resistivity measurements, exploration, dye tracing, surveying, and current and tidal fluctuation measurements. Based on the field work, a preliminary model of the island's hydrologic system has been constructed. This paper examines both the data and the model.

BACKGROUND

Geography

San Salvador is located in the eastern Bahamas, approximately 600 km east of Miami (Fig. 1).

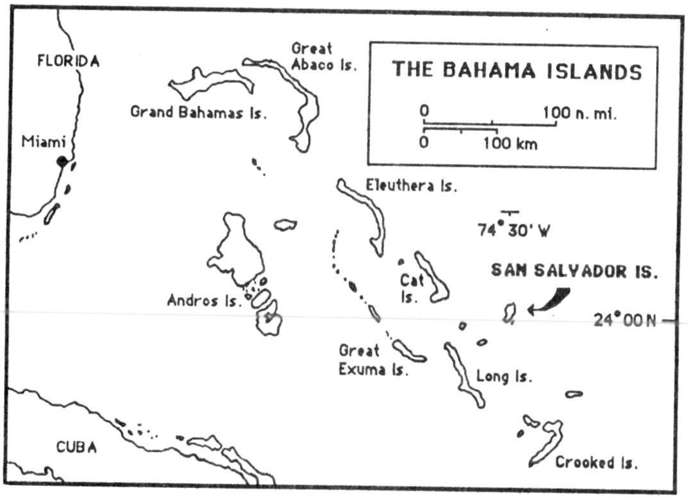

Fig. 1. Map of the Caribbean area showing the location of San Salvador, Bahamas (after Adams, 1983).

The island is part of the Bahamas platform, although it is separated from the main part of the platform by deep ocean water. The open North Atlantic Ocean lies to the east.

The island is approximately 14 kilometers North to South and 8 kilometers East to West. Major settlements are located at Cockburntown, United Estates, North Victoria Hill, and Long Bay. The CCFL Field Station is located at Grahams Harbor in the northeastern portion of the island. There is a group of vacation homes located in the Columbus Landings Subdivision along the southern coast of the island at Sandy Point and Sandy Hook (Fig. 2).

The topography of the island is dominated by consolidated carbonate dune ridges. These rise to elevations of over 40 meters above sea level, but most are in the 10-20 meter range. The low areas between the dune ridges are occupied by hypersaline lakes. Most of the lakes are about 1 to 2 meters deep and have no direct surface connection to the ocean. Many of the lakes, however, are interconnected either naturally or by man-made canals that were constructed as part of an intra-island transportation system dating from the early 1800's.

Off-shore there is a shelf, 0.5 to 25 kilometers wide, where water depths reach 15 meters. It is bordered by a fringing reef. The shelf ends with a sharp break in slope and the ocean floor plunges to a depth of over 2000 meters forming a feature known as the "wall".

The rock beneath the island contains many small caves and conduits. The caves are mostly dry and seem to be located about 2 meters above present sea level. The conduits transport water between the ocean and the inland lakes. Flooded cave openings have been observed along the wall at depths of more than 100 meters (Mylroie, 1988).

Geology

The geology of the island is similar to that found on other carbonate platforms. The rock is

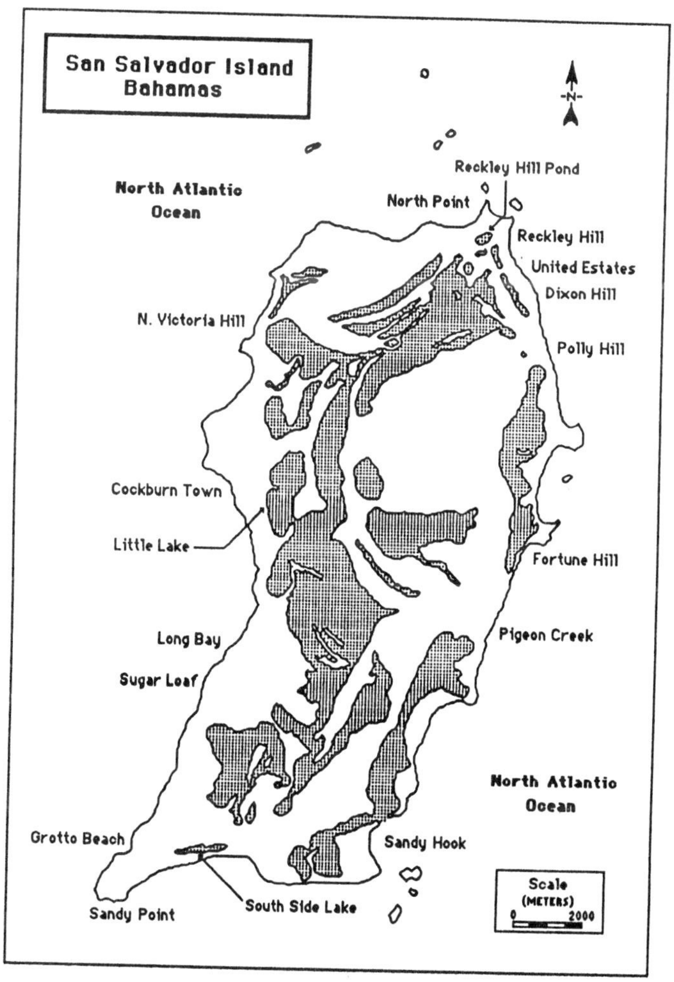

Fig. 2. Map of San Salvador Island.

no older than Pleistocene, and it is entirely carbonates (Carew and Mylroie, 1985). There are three major rock types: eolianites, beachrock, and reef rock (Adams, 1983). Eolianites make up the carbonate dune ridges found almost everywhere on the island. They also occur in the subsurface. These cross-stratified rocks consist of ooliths and bioclasts. They are often very porous and have distinct bedding planes.

Beachrock occurs on modern beaches where it is found in large tabular masses. It also may cap older rocks near the shore. The rock consists of cemented shell fragments and ooliths. It is usually porous and poorly cemented, and the units tend to be rather thin (less than 1 meter).

The final major rock type is reef rock, which is found at several localities along the shore and in the interior. This rock consists of fossilized reef materials; corals, molluscs, sponges, algae, etc. Its occurrence is spotty and the units tend to be thick and massive (Adams, 1983).

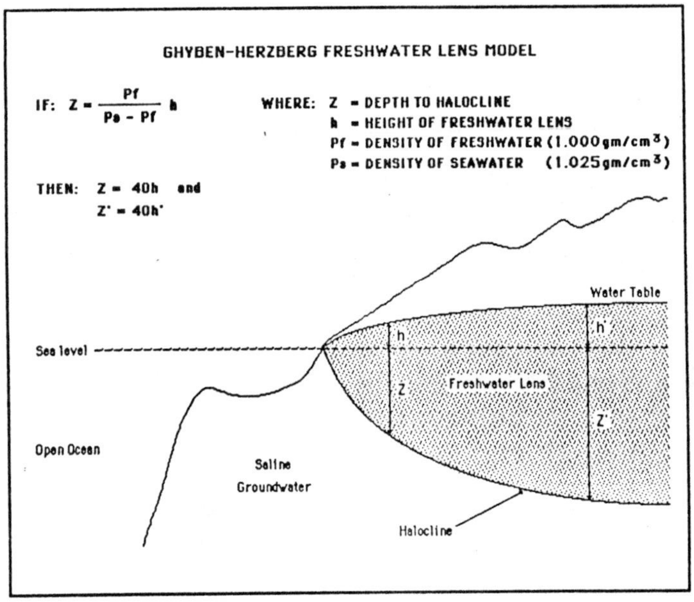

Fig. 3. Idealized figure of the Ghyben-Herzberg Freshwater Lens Model (after Mylroie, 1988).

ISLAND HYDROLOGY

The purpose of our work was to model the hydrologic system on San Salvador. The classic model of island hydrologic regimes is the Ghyben-Herzberg model, (Fig. 3). It shows a fresh ground water lens underlain by saline water from the ocean. The formation of the lens is due to a density difference between the fresh and saline water. The boundary between them is known as the halocline. The model predicts that it will be located at a depth below sea level of 40 times the height of the freshwater table above sea level. Thus, if the water table is at an elevation of 1 meter, then the halocline should be at an elevation of minus 40 meters (Fetter, 1980). Preliminary field work and sampling suggested that this model did not always fit the actual conditions on San Salvador. In many places, freshwater is not found where the model predicts it, and in other places where the model says that there should be freshwater, there is none.

The San Salvador Hydrologic System

The hydrologic system on San Salvador has two very striking aspects. The first is the total absence of surface water streams. This is typical of most karst areas. The second is the important role played by ocean tides, which is typical of most small islands. It is the combination of these two aspects that makes the island's hydrology so complex.

The system has six elements. These are shown in Table 1 and are discussed individually in the following paragraphs.

Precipitation
Evapotranspiration
Ground Water
Inland Lakes
Conduits and Blue Holes
Ocean (Tides)

Table 1: The Elements of the San Salvador Hydrologic System

<u>Precipitation</u>

Precipitation on San Salvador averages between 1000 and 1250 mm per year. The majority of it occurs during the rainy season which lasts from late May to January (Sealy, 1985).

<u>Evaporation</u>

Potential evaporation averages between 1250 and 1375 mm per year (Sealy, 1985). Because this exceeds annual precipitation, there is a yearly deficit. During the rainy season, however, precipitation exceeds potential evaporation, so there is a temporary surplus of freshwater. During the dry season, potential evaporation is greater than precipitation, so that conditions then, are arid. The seasonal alternation of surplus and deficit conditions means that large variations are possible in the water chemistry of the inland lakes and ground water.

Ground Water

Ground water on the island may be fresh or brackish. The distribution of fresh ground waters is one of the issues that our work attempted to address. Unfortunately, ground water is accessible only at wells and at a few widely scattered caves and banana holes in the interior. The wells are located predominantly along the coast. The lack of data from the island's interior means that its ground water system is difficult to study. For the most part, water table elevations at accessible locations, are less than 1 meter above sea level.

Inland Lakes

Inland lakes are located in the low areas between the dune ridges. They cover a large percentage of the interior of the island. Depths range from less than 1 meter to almost 3, with most being 1 to 2 meters deep. The salinity of water in the lakes can be slightly lower than sea water, to more than 4 times that of sea water. This varies with the seasons and from lake to lake.

Conduits and Blue Holes

In several places on the island, there are vertical holes that connect, via conduits, directly to the ocean. These are known as blue holes. Other conduits connect some of the lakes to the ocean, or carry fresh ground water to the lakes or the sea.

Ocean

Hydrologically, the most important aspect of the ocean is the tides. These range from 0.3 to 1 meter, depending on lunar phases, the season, and the time of day. It is important to note that there may be a difference of more than 0.5 meters between the heights of the two high tides on a given day.

FIELD AND LABORATORY INVESTIGATIONS

Field work was conducted during six trips to the island, encompassing more than 90 days. We looked at each of the six components of the hydrologic system, and performed investigations involving water sampling, drilling, topographic leveling, flow measurements, exploration, tide measurements, rock chemistry analysis, electrical resistivity and dye tracing. Chemical analyses were performed with a Hach Digital Titrator using standard methods. Conductivity and pH were measured with the appropriate equipment.

Rock Chemistry

We examined a total of 16 rock samples. An attempt was made to sample all of the various rock types from a number of different places around the island ("+" symbols-Fig. 4). Locations were also chosen to reflect a variety of hydrogeologic conditions. For example, samples were taken from parts of the island that have never been below sea level. Some were taken in places such as Lighthouse Cave, where they are covered by sea water at every high tide. Others were

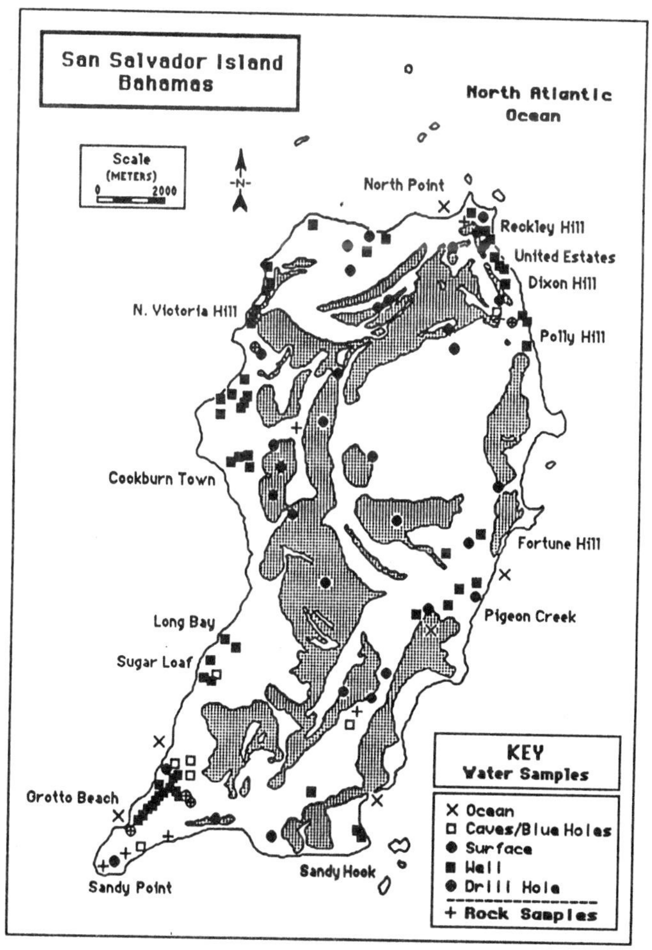

Fig. 4. Map of water sampling, rock sampling, and drilling locations.

taken from localities which are dry now, but which may have been submerged in the geologic past.

The rock chemistry was determined by dissolving a known quantity of powdered sample in hydrochloric acid and then analyzing the resulting solution.

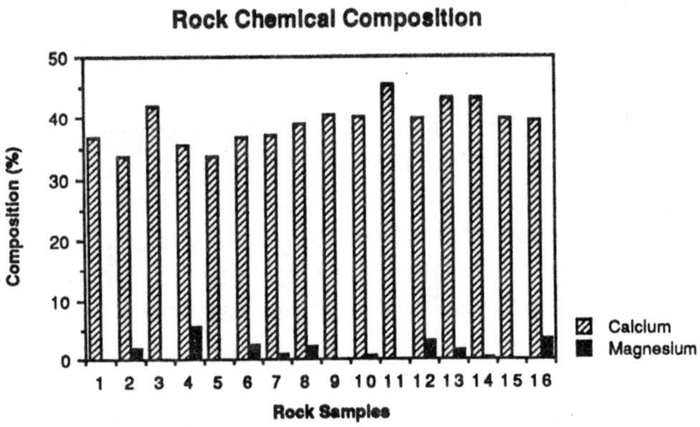

Fig. 5. Graph of chemical composition of rock samples.

Figure 5 shows the bulk percentage of calcium and magnesium for each sample. These elements occur in the Calcites that make up about 45% of the limestone rock. The remainder of the rock (not shown on the graph) is carbonate. It is important to note that none of the samples contain more than 6% Magnesium, and that several contain virtually none. Magnesium/calcium ratios range from 0.00 to 0.16 (Sample 4). Ground water derived only from precipitation which has percolated through the rock, should reflect these same magnesium/calcium ratios.

Precipitation

Precipitation was measured with a recording rain gauge, and water samples were collected for chemical analysis. Rain water was found to contain about 30 mg/l chlorides, about 12 mg/l calcium and virtually no magnesium (Table 2).

Water Chemistry

Figure 4 also shows the locations of water samples taken for analysis. In all, there were over 430, from approximately 150 different locations. Many of these were taken at wells, shown by solid squares. Others were taken as grab samples in the lakes, and some at casual surface ponds. Three lakes, Little Lake, Reckley Hill Pond, and South Side Lake, were sampled systematically. All lake samples and other samples of surface waters are shown by solid circles. Ocean water samples are shown by crosses. Samples from inside caves and conduits are shown by open squares.

Most of the water samples were analyzed for conductivity and pH plus chloride, calcium, magnesium and bicarbonate ion concentrations. Studies by other workers have shown that, with the exception of sodium (from sea water), and nitrates

DATE	CONDUCTIVITY (umhos/cm)	Cl^- (mg/l)	Ca^{++} (mg/l)	Mg^{++} (mg/l)	pH
12/30/86		17	39	3	-
01/02/87	215	26	30	1	-
01/02/87	540	44	54	3	-
01/04/87	240	22	20	3	-
05/22/87	315	59	-	-	7.85
06/14/87	346	54.8	12	-	7.3

Table 2: Chemistry of Rain Water. Dashes indicate data that is missing due to equipment failure or because the small volume of the samples did not allow for complete analysis.

(from sewage), these are the only ions of any importance in the water (Davis, 1975; Quarrier, 1975; Touchette, 1975; Countryman and Poole, 1981).

We found that Mg/Ca and Mg/Cl ratios were good indicators of the source of ground water (Johnson and Davis, 1987). As stated before, ground water with Mg/Ca ratios of less than 0.16 originates from precipitation that has flowed through the rock. Water from the ocean will have Mg, Cl, and Ca ions present. The Mg/Ca ratio of sea water is about 3.65, and therefore, ground water with ratios above 0.16 will be a mixture of water derived from both the ocean and precipitation. Mg/Cl ratios can also be used to discriminate between precipitation and ocean sources. Ocean water has a ratio of about 0.1. Fresh waters have generally higher ratios. We used a combination of these two ratios to determine the amount of seawater/freshwater mixing in any given sample.

Rock Permeability

Cores taken from the 16 rock samples (discussed under "rock chemistry") were tested for permeability using a constant-head permeameter. Based on published data for porous and oolitic limestones, values of permeability were expected to fall into the 10^{-6} to 10^{-4} centimeters/second range (Davis and DeWiest, 1966).

Actual values, however, were between 10^{-9} and 10^{-5} cm/sec with most lower than 10^{-6}. Hence, the rocks were considerably less permeable than expected.

Other Tests

Five wells were drilled and sampled with an auger-type drill rig provided by the Bahamian Government. Three were located at Sandy Point, one at North Victoria Hill, and one in United Estates at Dixon Hill (Figure 4).

Because of the small differences in well and ground water elevations on the island, careful topographic control was needed in order to interpret flow regimes. Leveling was done at Sandy Point, N. Victoria Hill, United Estates and on the east side near Fortune Hill. Leveling was also done to establish the elevations of Reckley Hill Pond (See Figure 2 for locations).

Tide gauges were installed on the ocean and at several locations inland. The timing and height of the tides were found to be irregular. In addition, some unexpected patterns were observed. For example, in Lighthouse Cave, far inland, the tidal range is greater (sometimes by as much as 1/2 m) than that in the ocean, whereas in Reckley Hill Pond, only 400 meters from the coast, the tidal range is just a few centimeters.

Pump tests were performed on some of the wells at Sandy Point. Drawdowns of almost 0.5 meters were found at pumping rates of only 1 liter/minute. This suggests that the permeability of the rock is significantly less than might be expected based on the nature of the rock. These conclusions are supported by laboratory tests on the rocks (see the previous sections).

Dye tracing was attempted in order to locate the ocean outlets for some of the conduits feeding the inland lakes, but the results proved inconclusive. Conduit flow was measured at Reckley Hill Pond, and electrical resistivity tests were conducted at United Estates, North Victoria Hill, and Sandy Point.

Work of Others

Scientific work provided by other researchers was invaluable to the success of this study. Gerhard Kunze of the University of Akron, supplied information on resistivity studies conducted at Sandy Point and Sandy Hook (Kunze and Burke, 1984; Kunze and Weir, 1987). James Teeter, also from the University of Akron shared information on the limnology of the inland lakes. John Winter of Molloy College, guided us to many sites in the island's interior and provided historical information on water sources on the island. Finally, John Mylroie of Mississippi State University and James Carew of the College of Charleston, supplied us with unpublished maps of caves and helped with information on Pleistocene sea levels, rock ages, etc.

RESULTS

The large amount of data that was collected helped us to gain an understanding of the nature of San Salvador's ground water, inland lakes, and conduits. The data obtained from each of these systems will be discussed separately in the sections that follow.

Ground Water

Chemical analyses of water samples, pump tests, permeability measurements, and topographic

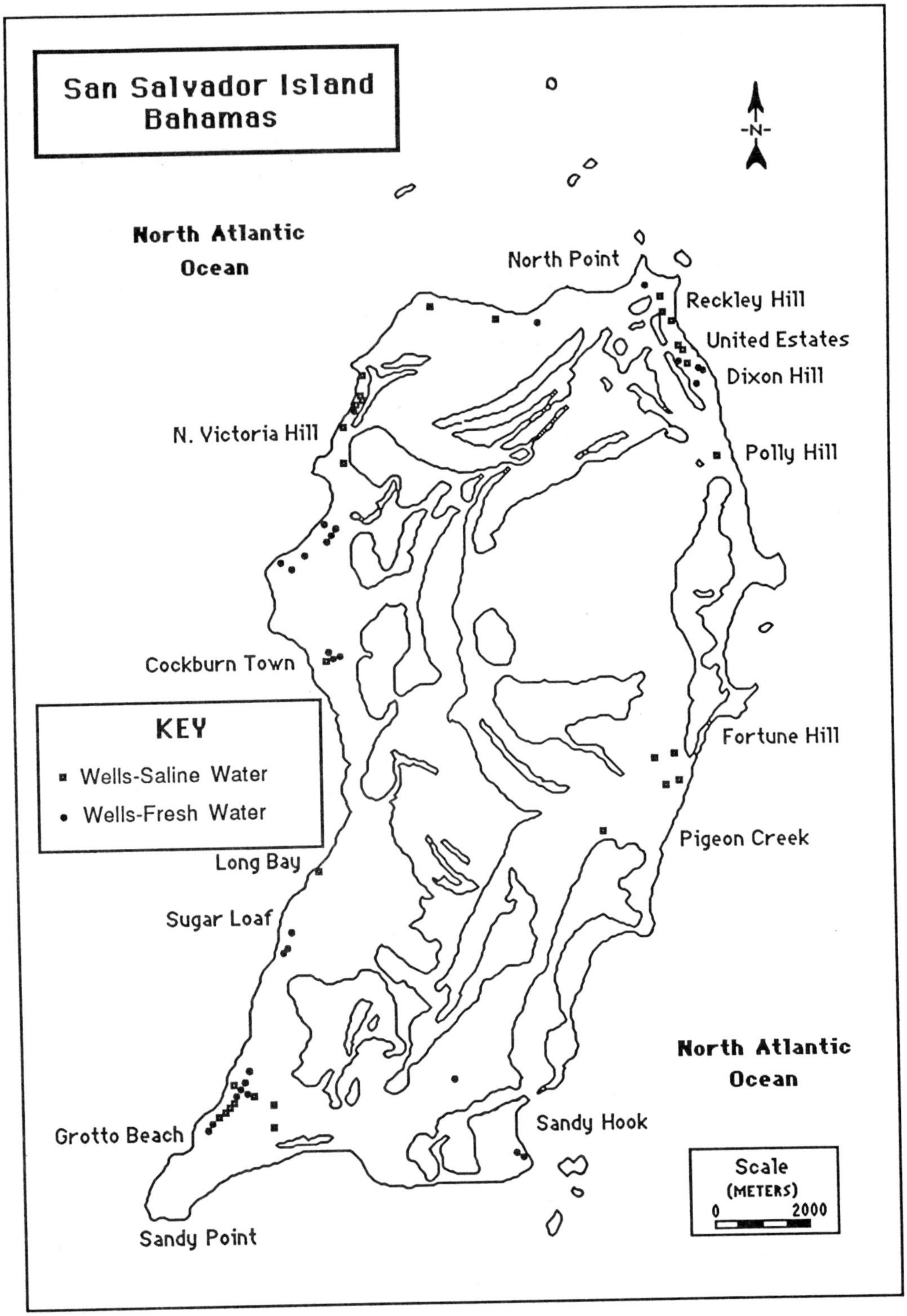

Fig. 6. Map of saline and freshwater well locations on San Salvador.

control were combined to produce insights into the nature of the ground water system.

Based on chemical analyses, Mg/Ca and Mg/Cl ratios were used to estimate the amount of mixing of saline and fresh ground waters. Figure 6 shows the chemical nature (fresh or saline) of the ground water in all of the wells that were sampled. It was found that some wells very near the coast had Mg/Ca ratios less than 0.16 and Mg/Cl ratios greater than 0.1, indicating that the ground water was derived only from precipitation. Other ratios indicated the addition of considerable amounts of saline water. Inland, most wells had fresh ground water, but in a few there was mixing. In general, mixing appeared to be related to the presence of conduits carrying sea water into and out of the inland lakes and caves.

Pump and laboratory tests have indicated that rock permeability is low. Work by Vascher (1988) suggests that under these conditions, water tables should be relatively high and freshwater lenses should be relatively thick. Resistivity work by Kunze and Burke, (1984); and Kunze and Weir (1987) supports the existence of these thick lenses. Low permeabilities would also account for the fact that freshwater lenses can exist very close to the coast, despite the potentially disruptive influence of the tides.

Drilling and resistivity work carried out during this study, however, suggest that in many areas, water tables are not as high and lenses are not as thick as expected. Furthermore, at Sandy Point, a small amount of pumping (<1 liter/min) produces salt-water intrusion, again suggesting relatively thin lenses.

In an attempt to resolve this apparent conflict, further work was done in the Sandy point area. A combination of water chemistry, pumping information, and topographic control was used to gain a more detailed understanding of the local ground water regime. Figure 7 shows the results of studies in this area. Note that in this figure, English units are used in order to conform to the topographic map of the island published by the Bahamian Government. The highest water table elevations are found under the topographic ridge. They are, however, only 1.5 (0.45m) feet above sea level. Based on this, the Ghyben-Herzberg model predicts a 60 foot (18m) thick freshwater lens. Resistivity (Kunze and Weir, 1987) and drilling suggest that a lens as thick as 35 feet (10.75m) does exist in some areas. This is well

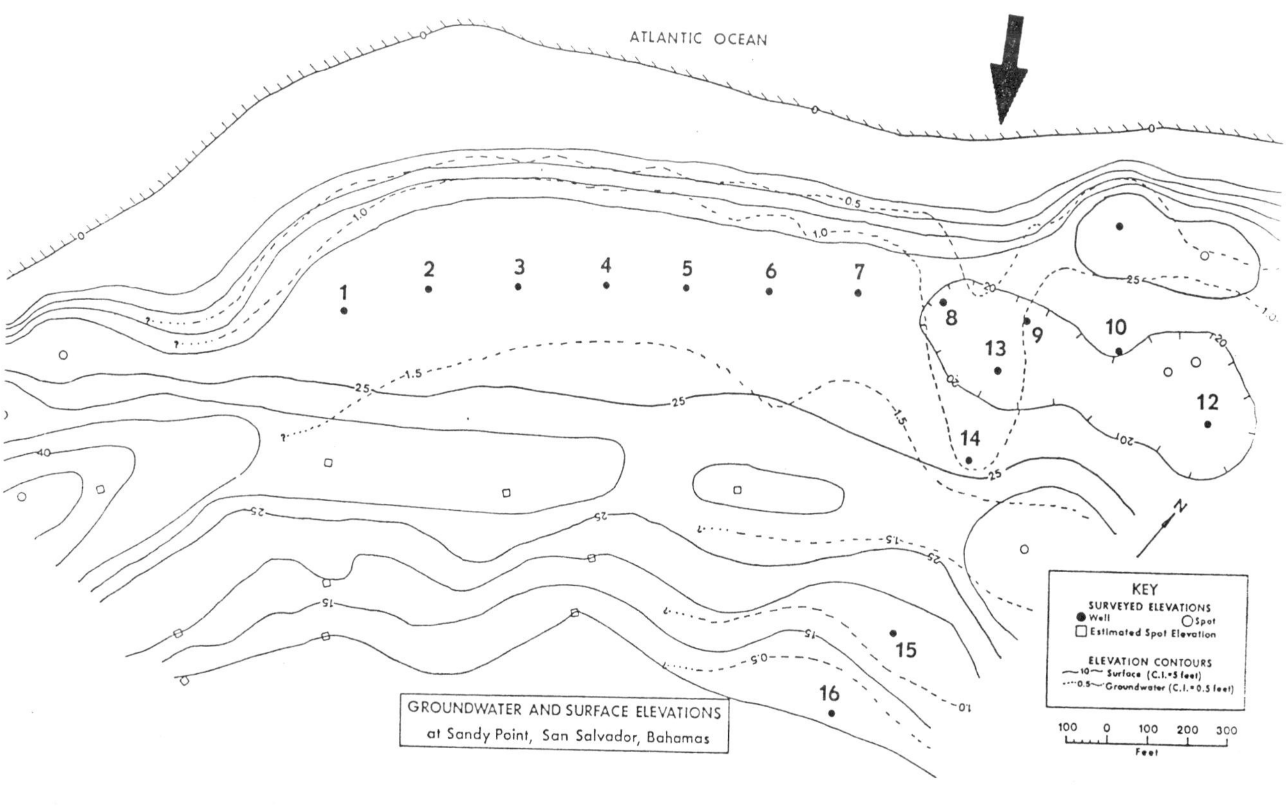

Fig. 7. Map showing surface and ground water topography at the Sandy Point well field.

Location	Number of Samples	Cl^- mg/l	Ca^{++} mg/l	Mg^{++} mg/l	Mg/Ca	pH
SP #8	8	246	104	15	0.14	7.61
SP #9	5	38	97	3	0.03	7.53
SP #10	4	145	73	7	0.10	7.79
SP #12	5	22	62	3	0.05	7.95
SP #13	6	145	96	10	0.10	7.56
SP #14	4	604	116	41	0.34	7.60
Ocean	9	20508	479	1714	3.63	8.35

Table 3: Water Chemistry (average values) of the Wells Near the Sandy Point Ground Water "Channel".

within the error range of the data.

Of particular interest, is the area on the map marked by an arrow (Figure 7). Here the contours indicate a ground water "channel". All of these wells not only have lower than expected water table elevations, but they also show higher than expected mixing of sea water (Table 3). It is important to note that none of these wells have ever been pumped, nor is there any indication that pumping in nearby wells has affected them. Consequently, this seems to be a natural phenomena. Vasher (1988) discusses what might happen to freshwater lenses and water tables under varying conditions of permeability. He states that a high permeability "basement" results in lower water tables and hence, thinner freshwater lenses. Nearby, at Grotto Beach, a conduit carrying freshwater out to the ocean has been found. If such a conduit were located beneath the ground water "channel" shown on Figure 7, then the permeability of the rocks should be very high. This would result in lower water tables, such as those shown on the map. In addition, at high tides, sea water would flow along the conduit and into the rock where it would mix with the fresh ground water. This would account for the relatively high chloride concentrations in wells 10, 13 and 14.

Topographic control and water analyses performed in the United Estates area suggests that a ground water "channel" may also exist here. Its location is directly over the inferred path of a conduit that connects Reckley Hill Pond to the sea. Careful measurements of water table elevations may reveal other locations where this phenomena is present.

Conclusions

In summary, the ground water results demonstrate that the Ghyben-Herzberg model is applicable in areas that are under the ridges and far from any conduits. In these locations, fresh water lenses, of some thickness, are found. In addition, the existence of low permeabilities in the rocks over most of the island, means that, in these areas, the direct influence of the ocean and tides on the ground water system is limited. Consequently, fresh water lenses can, and do, occur very close to the shore.

Conversely, where conduits and areas of high permeability are present, they exert a strong influence on ground water chemistry and flow. In these areas, fresh water lenses may be entirely absent, and, if they are present, they will be thinner than predicted by the Ghyben-Herzberg model. The water in the lenses will also be brackish due to the tide-driven mixing of fresh and sea waters within the permeable zones.

Inland Lakes

Information provided by John Winter (Personal Communication), and continued exploration, revealed the presence of many freshwater seeps along the shores of the inland lakes. The water coming from these seeps is considerably fresher and cooler than that in the adjacent lake (Table 4). These seeps always seem to be associated with a highly karstified type of rock, that is called moon rock locally. A similar feature is found in caves world-wide. It is the result of mixing corrosion. This occurs when two solutions, saturated at different calcium ion concentrations, mix. The resulting mixture, paradoxically, is unsaturated and hence, aggressive (White, 1988). This aggressive water dissolves the rock where the two solutions come into contact, producing "spongework" and other highly karstified features. This same phenomena would occur where the Ca-

Location	Cl^- mg/l	Ca^{++} mg/l	Mg^{++} mg/l	Mg/Ca	pH	Temperature °C
Duck Pond Seeps (samp.1)	15020	344	1233	3.58	8.31	24
Duck Pond Seeps (samp.2)	15440	408	1382	3.39	7.81	24
Duck Pond Surface	77800	1408	6431	4.57	7.95	30
Granny Lake Seep	5680	282	474	1.68	6.73	
Granny Lake Surface	20850	584	1636	2.80	7.34	
S. Side Lake Seeps	9700	493	627	1.27	7.30	
S. Side Lake Surface	139700	536	17378	32.42	7.25	
Ocean	20507	479	1714	3.63	8.35	

Table 4: Water Chemistry of Seeps and Surface Water at Several Lakes

saturated water from the seeps mixes with the lake water, which is saturated at a different Ca concentration. The mixing would result in the formation of moon rock.

In some locations, moon rock is present, even though there are no discernible seeps. These may be areas where seeps existed once but are now no longer active.

Our study of the lakes revealed two general types. The first is very muddy and contains substantial colonies of blue-green algae. These often appear red due to nutrient deprivation. The salinity of the lakes ranges from 36 to 90 parts per thousand (ocean salinity is 35 ppt), and water temperatures are usually very high (exceeding 30^{0} C). Conduits are not associated with this type of lake.

The second type of lake is clear, and salinities are similar to those of ocean water (Table 5). Water temperature also resembles that of ocean water. Conduits are always found associated with these lakes. Both types of lakes may or may not have seeps and moon rock.

It is likely that the first type of lake receives water from three sources: direct precipitation, seeps, and saline ground water flow. During the dry season, evaporation concentrates the salts in these lakes and they may become highly saline. During the wet season, direct precipitation becomes important and salinities drop, sometimes below that of sea water (Countryman and Poole, 1981). A comparison of the water chemistry of South Side Lake during the dry season (June) and during the wet season (January) illustrates this point well. In June, the chloride concentration was over 139,000 mg/l. In January, it averaged only 36,000 mg/l (see Table 5). Ocean water chloride concentration is about 20,000 mg/l. We have named this type of lake "seep-fed".

The second type of lake receives water from the same sources as the seep-fed lakes, but in addition, it receives a "slug" of sea water through the conduit during most high tides. We have named this type of lake "conduit-fed". The water chemistry in conduit-fed lakes is close to that of sea water and it remains fairly constant throughout the year. For example, the average chloride concentration in Reckley Hill Pond during the dry season is about 30,000 mg/l, while it is about 21,000 mg/l during the wet season. The fact that concentrations are higher than sea water indicates that evaporation can play a role in the water chemistry of these lakes, but it is not nearly as important as it is for the seep-fed lakes.

Three lakes, Little Lake, South Side Lake, and Reckley Hill Pond (see Figure 2), were sampled systematically during January, 1988. Little Lake is medium sized (~300 hectares), conduit-fed, and there are several freshwater seeps. Its depth averages about 2 meters. South Side Lake is small (~18 hectares) and seep-fed. It also has several freshwater seeps. It averages less than 1 meter in depth. Reckley Hill Pond is small (~14 hectares) and conduit-fed. There are no freshwater seeps. It averages about 1.25 meters in depth.

<u>Little Lake</u>

Forty-two samples were taken from the northern three-quarters of Little Lake during a six hour period on 8 January 1988. The tide was low. Figure 8 shows the locations of the samples, indicated by closed circles, and chloride ion

Location	Number of Samples	Type	Cl^- mg/l	Ca^{++} mg/l	Mg^{++} mg/l	Mg/Ca	pH
		Averages By Lake					
Miller's Pond	1	C	31200	748	2756	3.68	8.65
Reckley Hill Pond	89	C	21457	484	1729	3.57	8.12
Little Lake	46	C	25721	588	2134	3.63	8.36
S. Granny Lake	1	C	20850	584	1636	2.80	7.34
Pond N of Quarry E	1	C	24340	608	2709	4.46	
Clear Lake	1	C	21840	456	1780	3.90	7.85
N Little Lake @N End	1	S	40350	848	3266	3.85	8.38
Storr's Lake @Narrows	1	S	44100	1028	3787	3.68	8.38
Pond - Bridge (Dixon Hill)	1	S	47080	560	3660	6.54	
Duck Pond Surf	1	S	77800	1408	6431	4.57	7.95
Great Lake	6	S	40741	839	3326	3.99	7.98
Stout's Lake	2	S	44600	908	3434	3.78	
Sm. Pond W of Stout's	1	S	42220	624	3289	5.27	
S Side Lake/June	1	S	139700	536	17378	32.42	7.25
South Side Lake (January)	34	S	36622	820	2774	3.39	8.16
		Averages By Type					
Conduit-fed	139		22957	522	1877	3.59	8.20
Seep-fed	48		41043	828	3311	4.23	8.13
Ocean	9		20508	479	1714	3.63	8.35

Table 5: Average Water Chemistry in Inland Lakes. "C" in types column indicates conduit-fed, "S" indicates seep-fed.

concentrations, contoured at a 250 mg/l interval. Of particular interest is the area of low chloride ion concentration surrounding point 23 in the central portion of the lake. This is known to be the location of the conduit which feeds the lake. At times of high tide, water can be observed "boiling up" here (Winter, Personal Communication). The lake is about 2 meters deep at this point, but the conduit is not visible due to turbidity. Nevertheless, the water chemistry clearly shows the presence of this conduit, even though the tide was not flowing in at the time of the sampling.

Two other points of interest are number 5 (east-central part of the lake) and number 18 (west-central). Moon rock is present at both of these locations. The water chemistry at point 5 suggests the presence of a seep, and a small amount of water was observed flowing from the rock. The water chemistry at point 18 does not suggest a seep. However, the lake is deep (1.5 meters) here and, on the day that sampling was carried out, there was a brisk northeast wind. This would promote mixing, which could, at the scale of our sampling grid, mask the effect of a freshwater seep.

A final point of interest is number 29, located in the northwest corner of the lake. Here the chloride concentration is relatively high. The most likely explanation for this is the presence of the Cockburntown Dump along the shore between points 29 and 30. The northeast wind/ might be pushing water, bearing leachate, from the dump into the vicinity of point 29. Again, the density of the sampling grid did not permit confirmation of this hypothesis.

South Side Lake

Thirty-four samples were taken from South Side Lake over a three hour period on 4 January 1988. The wind was calm. Sample locations are shown by solid circles on Figure 9. Chloride ion concentrations are contoured at a 500 mg/l interval.

Seeps and moon rock are located on the northeast shore of the lake near points 17, 19, and 21. These are clearly shown by the chloride contours. Other areas of interest include the west end of the lake, which has high chloride concentrations. The lake is very shallow (less than

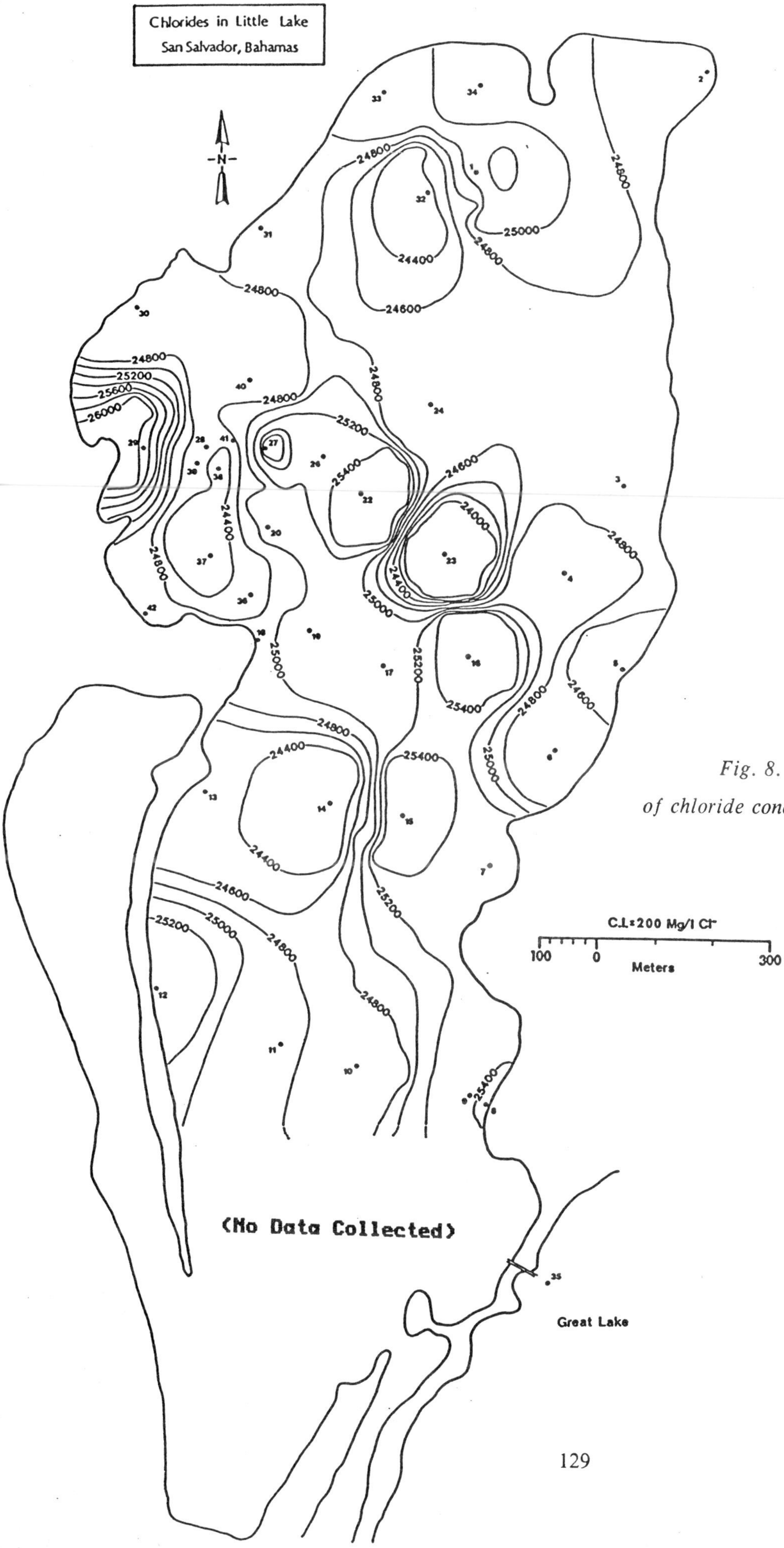

Fig. 8. Sampling locations and contours of chloride concentrations, Little Lake.

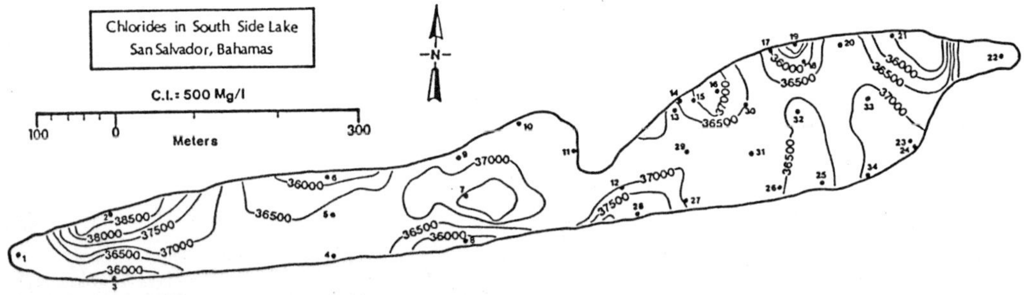

Fig. 9. Sampling locations and contours of chloride concentrations, South Side Lake.

0.3 m) and water temperatures are 1-2° C higher than in the deeper parts of the lake. Hence, the higher concentrations are probably caused by increased evaporation. The higher concentrations near points 13, 14, 15, and 16 (north shore) are probably also due to shallow water and higher temperatures.

Chloride concentrations are very high relative to both ocean water and the water in the other two lakes (Table 5). This is true even though the January samples were taken at the end of a very wet Autumn. Sampling done in June, 1987, showed chloride concentrations as high as 132,000 mg/l.

<u>Reckley Hill Pond</u>

Thirty-nine samples were taken from Reckley Hill Pond on 10 January 1988. Their locations are shown by solid circles on Figure 10. Samples were collected over a three hour period. High tide occurred while the sampling was being conducted, so water was flowing into the lake from the conduit during the entire time. On the figure, pH has been contoured at an interval of 0.04 pH units. It was not possible to contour chlorides, since the concentrations in the lake water were very close to that coming in through the conduit.

The contours show a plume of water, with lower pH, flowing from the conduit in the SE corner of the lake. It spreads out towards the

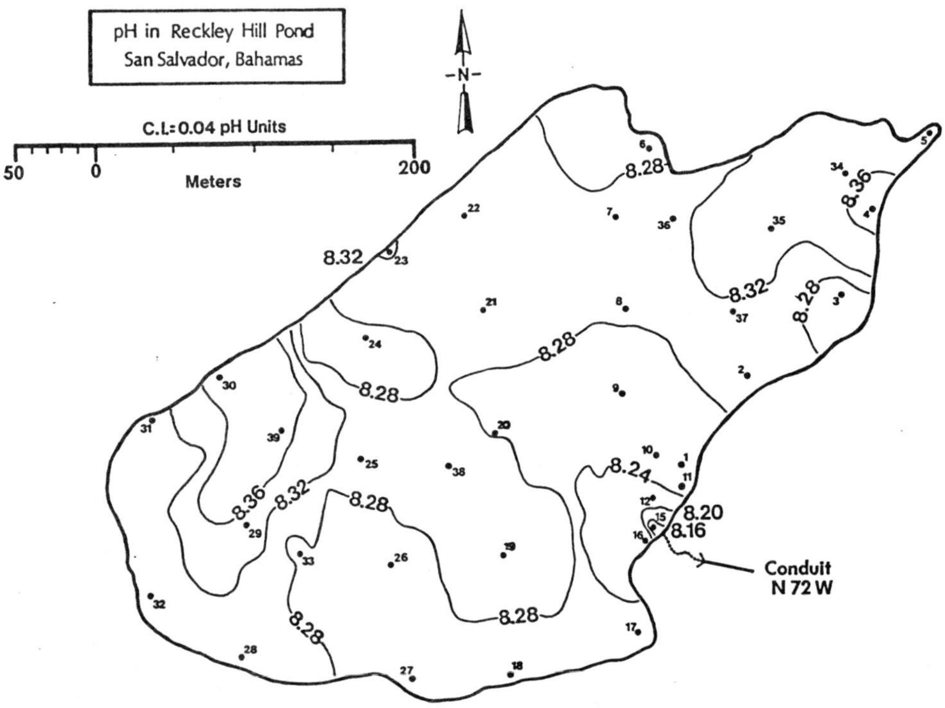

Fig. 10. Sampling locations and contours of pH, Reckley Hill Settlement Pond.

center of the lake for almost 100 meters. No seeps or moon rock were found along the lake shore, and the water chemistry also did not show any areas of incoming fresh water.

Conclusions

Based on these observations, we conclude that the influence of conduits and seeps can be seen in the water chemistry of most lakes. The nature of the water sources controls the lake's chemical condition. Those fed only by seeps, precipitation, and saline ground water (such as South Side Lake) are strongly influenced by evaporation, which results in high and variable salinities. During the wet season the salinities may be twice that of ocean water. During dry seasons, they may be as much as 6 times as high. Under these conditions, seeps are particularly noticeable when water chemistry is contoured. The difference in the calcium concentrations between the lake and the seep waters allows for mixing solution and the formation of moon rock.

In those fed by conduits, such as Reckley Hill Pond and Little Lake, waters are mixed, resulting in salinities that are relatively constant through out the year. Seeps are not as obvious under these conditions, although water chemistry differences between the lake and seep waters are still large enough to produce mixing corrosion and the formation of moon rock.

Conduits and Tides

Extensive studies of conduit flow were performed at Reckley Hill Pond. The conduit appears as an opening in the limestone at the foot of a ridge. A rectangular channel approximately 60 centimeters square, leads through the mangroves to the pond, which is approximately 25 meters away. The tides force water in and out of the conduit. Flow measurements and water chemistry were studied at the conduit on three separate occasions: June, 1987 (for 12 hours), December, 1987 (for 13 hours) and January, 1988 (for 7 hours). The results of these measurements are shown in Figures 11A and B. Figure 11A shows flow on 10 June 1987. Combined flows on December 31, 1987 and January 11, 1988, are shown on Figure 11B giving an idea of what a complete tidal cycle might look like. On both figures, positive discharges indicate flow into the lake (tide coming in) and negative discharges indicate flow out of the lake (tide going out).

Since flow is clearly driven by the tides, we expected that water would flow into the pond for 6 hours and out for 6 hours of each 12 hour tidal cycle. However, during the first observation period, on 10 June 1987, water flowed in for only 4 hours and out for 8 (Figure 11A). During the second observation period, water flowed out during the entire time that data was being collected (13 hours). During the third observation, on 11 January, 1988, water flowed in for 3 hours and out for 9 (Figure 11B). Careful measurement of the pond's elevation, and an examination of tide tables and our own records, provided an explanation (Figure 12). The elevation of the pond is about 70 cm above mean low tide. This is higher

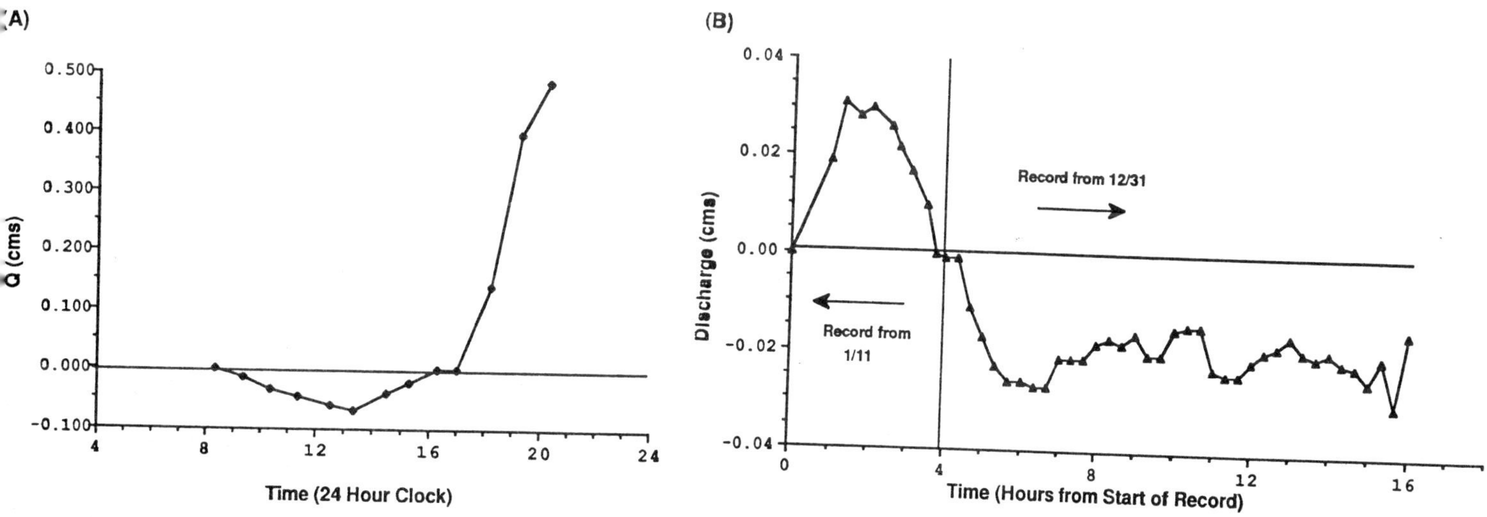

Fig. 11. Hydrographs for Reckley Hill Settlement Pond. (A) 10 June 1987 (B) 31 December 1987 and 11 January 1988.

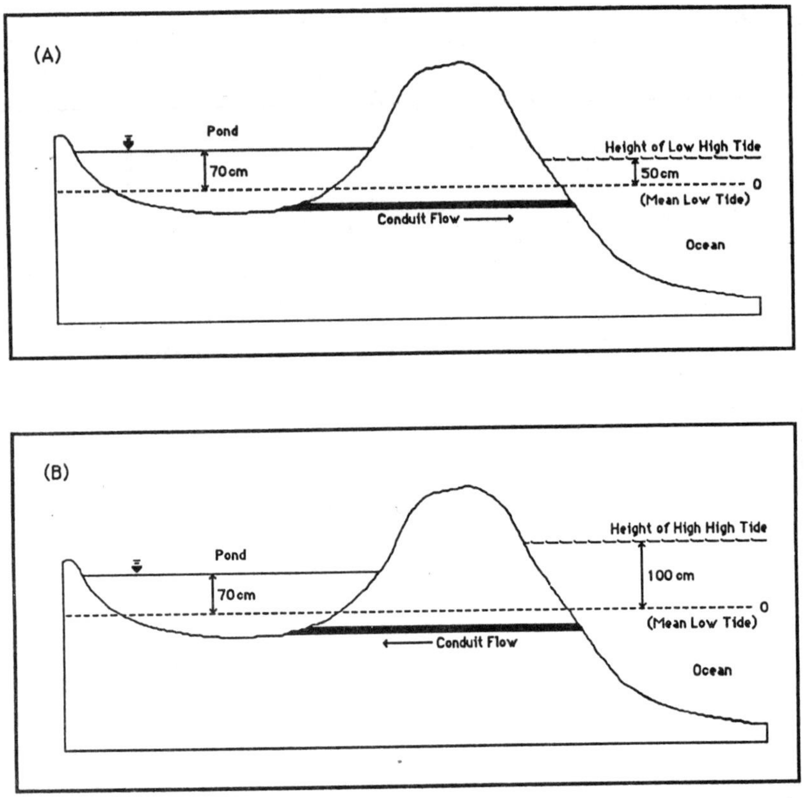

Fig. 12. General model of tidal fluctuations and conduit flow. (Based on conditions at Reckley Hill Settlement Pond).

than the elevation of many high tides. At times of low high-tide, the elevation of the water in the ocean never reaches the surface of the pond. Consequently, water will flow out during the entire tidal cycle. At times of higher high-tides, water will only flow into the pond during that part of the cycle when the tide is higher than the pond's elevation. High tides on June 10 were very high (~1.1m). Those on December 31 were much lower (~0.70m), and those on January 11 were low (~0.90m), but were about 20 cm higher than on December 31.

Mass balance calculations were carried out to see if the water coming into the pond balanced that going out. Water enters the pond through the conduit at relatively high discharges (see Figure 11). This means that even though the duration of incoming flow may be short, the volume is fairly large. Outgoing discharges are considerably smaller. In fact, on 10 June, the volume of water coming into the pond was larger than that flowing out (by about 2 times). The difference can be accounted for by evaporation (Davis, et. al., 1988). Conduits at other ponds appear to behave in a similar fashion (Teeter, Winter, and Gerace, Personal Communication), although direct observations will have to be made to confirm this.

When the water does flow into the pond, the flow is very forceful and turbulent. This would promote mixing in the ground water zone adjacent to the conduit, and may produce the kind of ground water "channel" and chemistry features that were observed at Sandy Point.

CONCLUSIONS

Based on the field work, we have been able to reach some preliminary conclusions about the nature of the hydrologic system on San Salvador. Figure 13 shows how the elements of our model interact. Freshwater lenses of varying thicknesses exist under the dune ridges both along the coast and in the interior. If there are no conduits running under the ridges, then the ground water will be fresh and potable. Where conduits cross the ground water zones, the quality of the water is diminished. Salt water from the ocean mixes with the fresh ground water producing brackish water, and water tables in these areas are usually lower.

The water in the inland lakes is supplied by

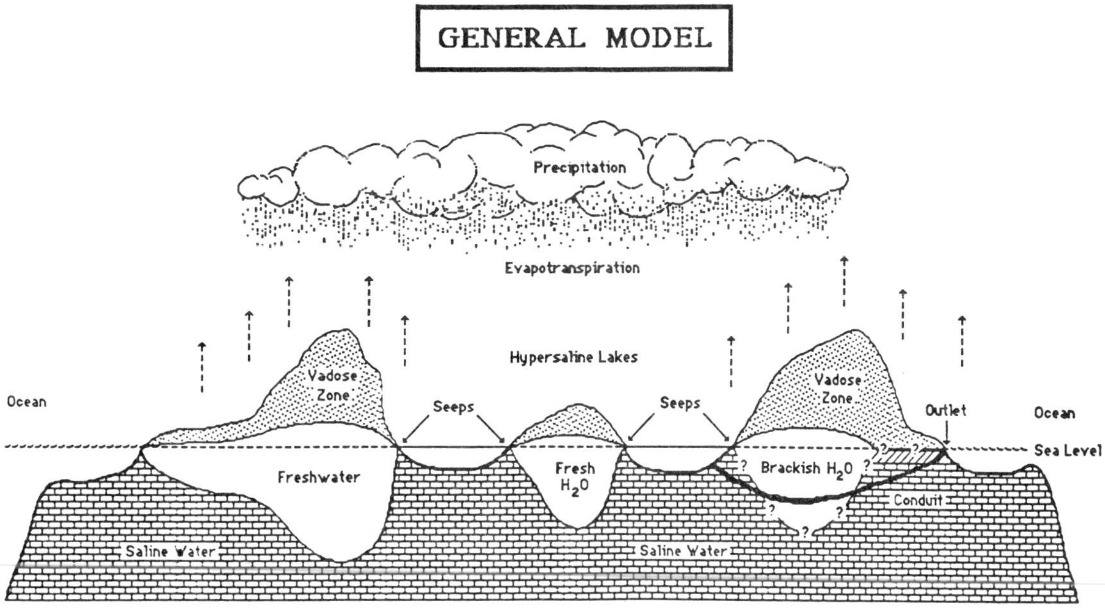

Fig. 13. General model of the hydrologic system on San Salvador.

precipitation, seeps from the lenses in the dunes, saline ground water infiltration, or conduits. Those lakes that have conduits, maintain a salinity that is close to sea water throughout the year. These lakes are also clear and free of algae.

Lakes that are not fed primarily by conduits are dominated by evaporation. During the rainy season, the water may be brackish, but during the dry season, the water is hypersaline. These lakes are muddy and full of algae. In either case, the locations of seeps and conduits can be determined by a systematic chemical analysis of the lake water.

Flow in the conduits is controlled by tides and lake elevations. Most of the lakes are higher than low high-tides, and water will flow into them only at higher high-tides. During the dry season, overall lake levels are lower, so inflow should be more sustained than during the rainy season.

Future Work

This preliminary model could be made more specific by conducting a variety of other field tests. Important ones might include: 1) deep holes drilled on the ridges, especially in the interior, 2) the examination of the coast north of Grotto Beach to see if there really is a conduit located where the ground water map suggests, and 3) an analysis of the ground water topography at other locations to see if it matches that at Sandy Point, and if it reflects the presence of known conduits. Also, other lakes should be systematically sampled, so that the locations of seeps and conduits can be mapped.

Implications

There are many important implications of our work. We have already used the model to locate one potential water source at N. Victoria Hill Settlement and it may be possible to locate others. The model can also be used to help with management of the water supply. More water sources may, in turn, bring some development to the island, providing new jobs for the residents. Pumping, however, must be done with great care, since the permeability of the rocks is low and the possibility for salt water intrusion is high.

This preliminary understanding of the hydrologic regime could also lead to speculation on how the system looked at higher and lower sea level stands. Were there more or less conduits? Was there a greater or lesser quantity of freshwater? Were freshwater lenses more continuous than they are today and did they conform more closely to the Ghyben-Herzberg model? The answers to these questions can lead to an understanding of the origin of the island's caves and karst, and, by extension, the origin of caves and karst on other similar islands world-wide. They may also provide important indicators as to how caves and conduits can be used to determine the history of

Pleistocene sea level changes.

ACKNOWLEDGEMENTS

We would like to acknowledge the financial and logistical assistance of the CCFL Bahamian Field Station and the continuous support provided by Don and Kathy Gerace, the directors. We would also like to acknowledge the financial support of the National Speleological Society, who provided us with two grants and C.I.D. Landfill, Incorporated of Chaffee, New York whose grants financed three trips to the island and allowed us to purchase much needed equipment. In the field, we were assisted by John Winter, John Mylroie, Jim Carew, Pete Vogel, John Rapp, Todd Seibel, and many others. We are grateful to them for their time and interest. Clifford Fernander of the Bahamas Ministry of Works provided equipment, logistical support, transportation and encouragement. Finally, we would like to thank the people of San Salvador for their warm welcome and their willingness to share their knowledge and ideas with us.

REFERENCES CITED

Adams, R.W., 1983, General guide to the geological features of San Salvador, in Gerace, D.T., ed. Field Guide to the Geology of San Salvador. 3rd ed. San Salvador, Bahamas: CCFL Bahamian Field Station.

Carew, J.L. and Mylroie, J.E., 1985, The Pleistocene and Holocene stratigraphy of San Salvador Island, Bahamas with reference to marine and terrestrial lithofacies at French Bay, in Curran, H.A., ed. Pleistocene and Holocene Carbonate Environments on San Salvador Island, Bahamas, Geological Society of America Annual Meeting (Orlando, Florida) Guidebook for Field Trips #2. Fort Lauderdale: CCFL Bahamian Field Station, pp. 11-61.

Countryman, K. and Poole, M., 1981, Physical and chemical characteristics of Merrill's Pond, San Salvador Island, Bahamas, Unpublished Paper, Elmira, N.Y.: Elmira College Dept. of Geology, (on file at CCFL Bahamian Field Station, San Salvador, Bahamas).

Davis, M.R., 1975, Iron in wells, in College Center of the Finger Lakes, Island Environmental Studies Report-Environmental Chemistry, San Salvador, Bahamas: College Center of the Finger Lakes.

Davis, R.L., Johnson, C.R., Jr., Rapp, J.C., Jr., and Seibel, T.D., 1988, Conduit flow and water chemistry in Reckley Hill Settlement Pond, San Salvador, Bahamas (abs), Program of the 10th Annual Friend of the Karst Meeting, San Salvador, Bahamas, February 11-15, 1988. Mississippi State, Ms: Dept. of Geology and Geography, Mississippi State University.

Davis, S,N. and DeWiest, R.J.M., 1966, Hydrogeology, New York: John Wiley and Sons.

Fetter, C.W., Jr., 1980, Applied Hydrogeology, Columbus: Charles E. Merrill.

Gerace, D.T., Director, CCFL Bahamian Field Station. Personal Communication.

Johnson, C.R., Jr. and Davis, R.L., 1987, The use of Ca^{++} and Mg^{++} ions as indicators of groundwater mixing on San Salvador, Bahamas, Abstracts with Programs Annual meeting, Southeast Section-Geological Society of America, Norfolk, VA Pg. 91.

Kunze, A.W.G. and Burke, J.C., 1984, A resistivity survey of Sandy Point, San Salvador, Bahamas, in Teeter, J.W., ed. Proceedings of the 2nd Symposium on the Geology of the Bahamas. June 16-20, 1984. San Salvador, Bahamas: CCFL Bahamian Field Station, pp. 97-112.

Kunze, A.W.G. and Weir, W.G., 1987, Geoelectrical ground-water survey of the Sandy Hook Area, San Salvador, Bahamas, in Curran, H.A., ed. Proceedings of the 3rd Symposium on the Geology of the Bahamas, June 6-10, 1986, Fort Lauderdale: CCFL Bahamian Field Station. pp. 81-89

Mylroie, J.E., 1988, Karst of San Salvador, in Mylroie, J.E., ed. Field Guide to the Karst Geology of San Salvador Island, Bahamas. Mississippi State, MS: Dept. of Geology and Geography, Mississippi State University.

Quarrier, N., 1975, Sulfate in pond and well water, in College Center of the Finger

Lakes, Island Environmental Studies Report-Environmental Chemistry, San Salvador, Bahamas: College Center of the Finger Lakes.

Sealey, N.E., 1985, Bahamian Landscapes: London Collins Caribbean.

Teeter, J.W., University of Akron, Department of Geology. Personal Communication.

Touchette, N., 1975, Nitrates in well water, in College Center of the Finger Lakes, Island Environmental Studies Report-Environmental Chemistry. San Salvador, Bahamas: College Center of the Finger Lakes.

White, W.B., 1988, Geomorphology and Hydrology of Karst Terrains. New York: Oxford University Press.

Winter, J.D., Molloy College, Division of Social Sciences. Personal Communication.

TOPOGRAPHIC EFFECTS ON ALUMINOUS LATERITIC SOIL DEVELOPMENT, ELEUTHERA ISLAND, BAHAMAS

Annabelle Foos
Department of Geology
University of Akron
Akron, Ohio 44325

ABSTRACT

Aluminous lateritic soils from Governor's Harbour, Eleuthera Island, Bahamas were investigated to evaluate the relative importance of topography on the mineralogy of soils in carbonate environments. Samples were collected from depressions, crests of hills and on slopes up to 6°. The mineralogy and major element geochemistry was determined. The soils are concentrated in solution pits on a karst surface and range in thickness from 0 to 70 cm. The mineralogy of all the samples was similar consisting of varying amounts of boehmite, hematite, goethite, dioctahedral vermiculite, calcite, aragonite and quartz. The calcite plus aragonite content ranges from 2 to 64% and averages 21%. The SiO_2/R_2O_3 ratio is low (x = 0.58) indicating the soils have undergone extensive leaching and enrichment in Al, Fe, and Ti. No statistically significant relationships were observed between topographic setting and chemical composition or mineralogy. These results suggest that topography does not play a major role in determining the mineralogy of these soils.

INTRODUCTION

Periods of subaerial exposure are recorded in paleosols and they have the potential of being important paleoenvironmental indicators of these time periods. In order to unlock the secrets stored in paleosols it is necessary to understand modern soil forming processes. Factors controlling the development of soils are parent material, topography, climate, biota and time. It is necessary to evaluate the relative importance of these factors before paleosols can be used as paleoenvironmental indicators. This study evaluates the effect of topography on the mineralogy and chemistry of soils in carbonate environments. Aluminous lateritic soils from Governor's Harbour, Eleuthera were investigated. Isolated patches of aluminous lateritic soils occur on Eleuthera, Cat Island, Abaco, Exuma, Little Exuma, Long Island, Mayaguana, and New Providence and are always associated with hard bedrock (Little and others, 1977; Ahmad and Jones, 1969). Little and others (1977) state that these soils are restricted to low spots or "valley" bottoms between dune and beach ridges. Therefore, it was believed that this would be an ideal area to study the effects of topography on soil formation.

The parent material of these soils consists of lithified eolian limestone (Little and others, 1976) plus air borne dust from North Africa (Glaccum and Prospero, 1980; Eaton and Boardman, 1985). The climate of Eleuthera is characterized by warm rainy summers and cooler drier winters with an average temperature of ($25^{o}C$) $77^{o}F$ and an annual precipitation of (114 cm) 45 in.(Little and others,1976). The Köppen classification of the Eleuthera climate would be humid tropical savanna (Gabler and others, 1977). The vegetation in the area is dense mixed broad leaf coppice. An exact age for the bedrock in the study area is unknown due to a lack of radiometric dates, but, it has been mapped as Middle Pleistocene by Little and others (1976).

METHODS

The topography was measured along a transect crossing a prominent ridge south of Governor's Harbour, Eleuthera (Fig. 1). Samples were collected at approximately 50 m intervals and the topographic setting classified as either slope, depression or crest (Fig. 2). The mineralogy of the samples was determined with X-ray diffraction on a Philips APD 3720 instrument with CuKα radiation. Organic matter was removed with Na-hypochlorite buffered to a pH of 9.5 (Jackson, 1969). The untreated sample, insoluble residue and clays after free iron oxide removal were X-rayed. Free iron oxides were removed using the Na-citrate-bicarbonate-dithionite method of Jackson (1969). The clays were also treated with ethylene glycol and heated to 400 and $550^{o}C$ to aid in their identification (Starkey and others, 1984).

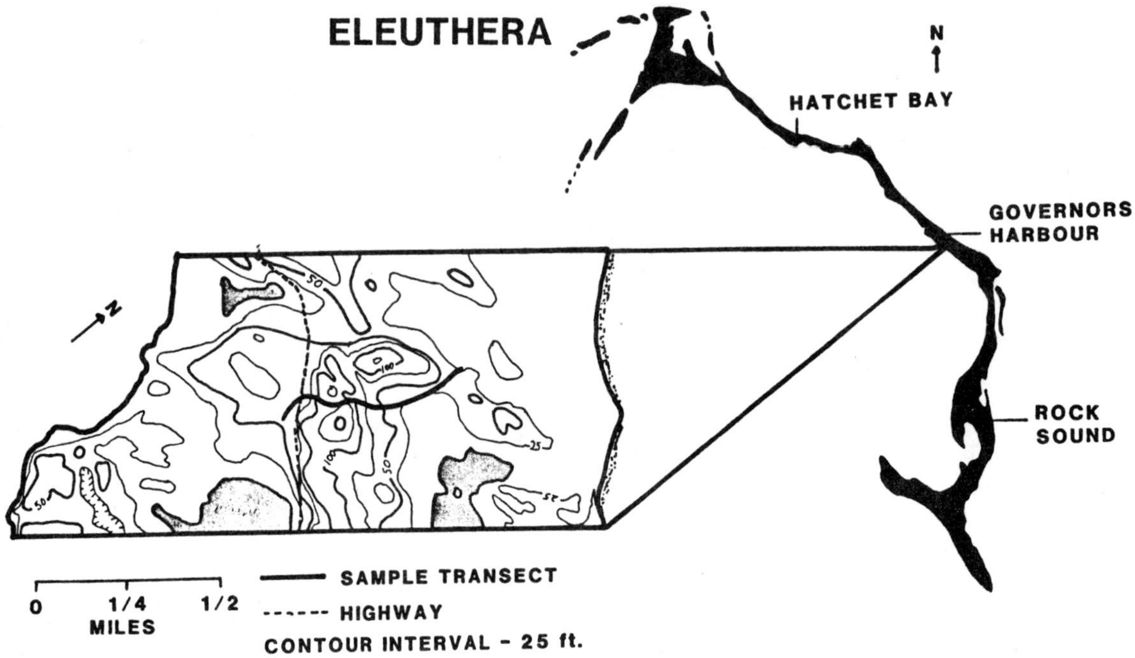

Fig. 1. Location of sample transect.

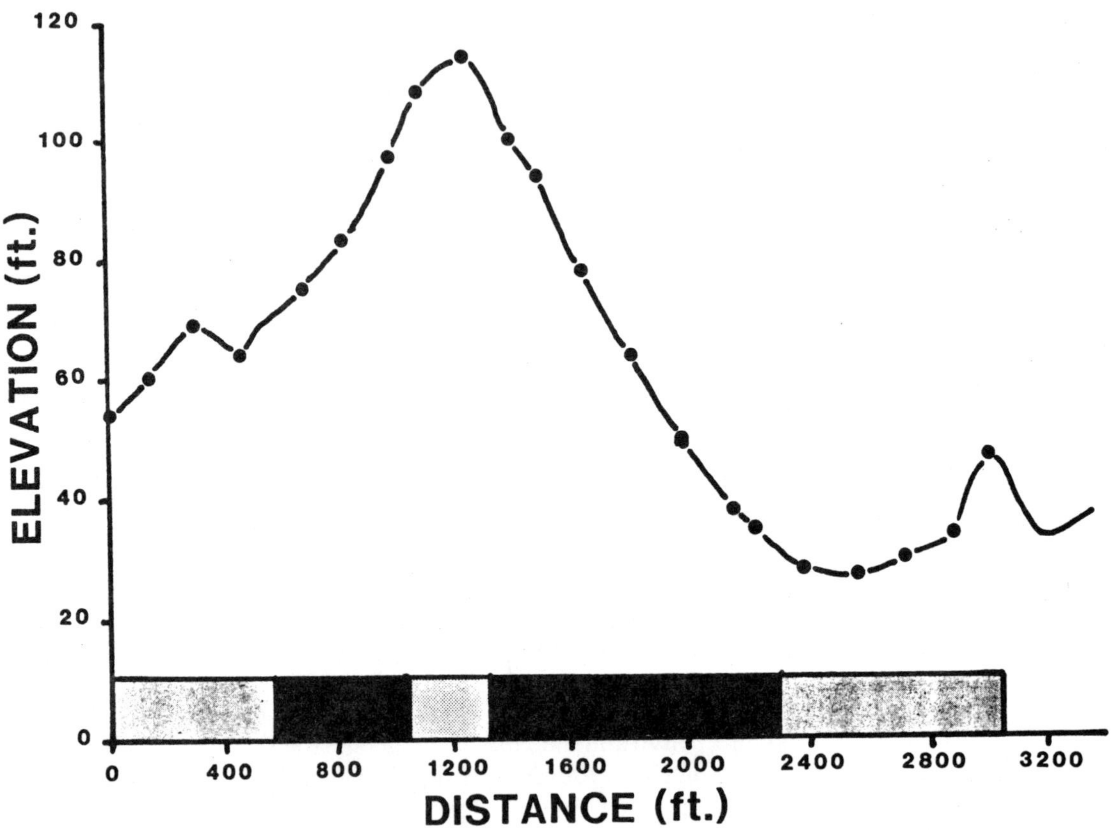

Fig. 2. Vertical profile of sample transect. (see Fig. 5 for key)

The chemical composition of the less than 2 mm fraction was determined by ICP.

RESULTS

The soils are concentrated in solution pits on a karst surface and have a maximum thickness of 70 cm. Bedrock is exposed at all sample localities. The soil profile consist of a layer of leaf litter followed by a uniform B horizon which exhibits a fine to coarse (0.1 to 1.0 cm) crumb texture. Below the B horizon is a C horizon of altered limestone followed by unaltered limestone. Textural analysis prior to removal of organic matter indicates the soils are sandy loams to loamy sands. The soil color ranged from yellow-red (2.5YR 4/8) to red (10R 2/2). Soils in depressions and on the crest of hills tend to be redder than soils occurring on slopes. The soil pH ranges from 7.3 to 8.0.

The mineralogy of all the samples is similar, consisting of varying amounts of calcite, aragonite, boehmite, hematite, goethite, dioctahedral vermiculite and quartz (Fig. 3). Boehmite has XRD peaks at 6.11 and 3.16Å. The dioctahedral vermiculite has a basal reflection at 14.1 and an 060 reflection at 1.50Å. The basal reflection did not shift after solvation with ethylene glycol, collapsed to 13.2Å after heating to 400°C and 12.4Å after heating to 550°C (Fig. 4). Differential XRD indicates that the hematite and goethite are aluminum rich containing up to 16 and 28 mole percent Al respectively.

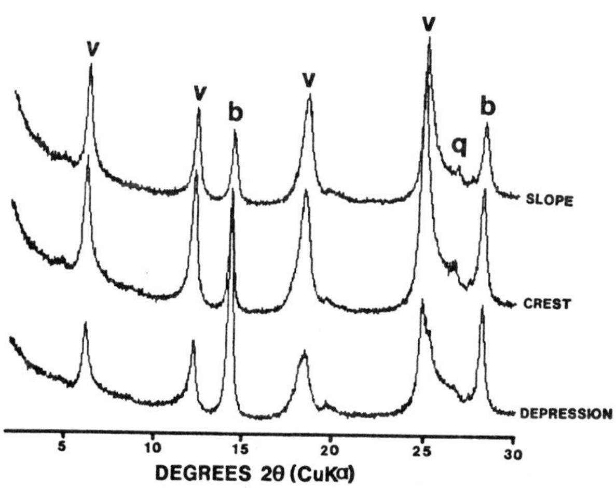

Fig. 3. X-ray diffraction pattern of clay after removal of free iron oxide for representative samples from a slope, crest and depression. (v-vermiculite, b-boehmite, q-quartz.)

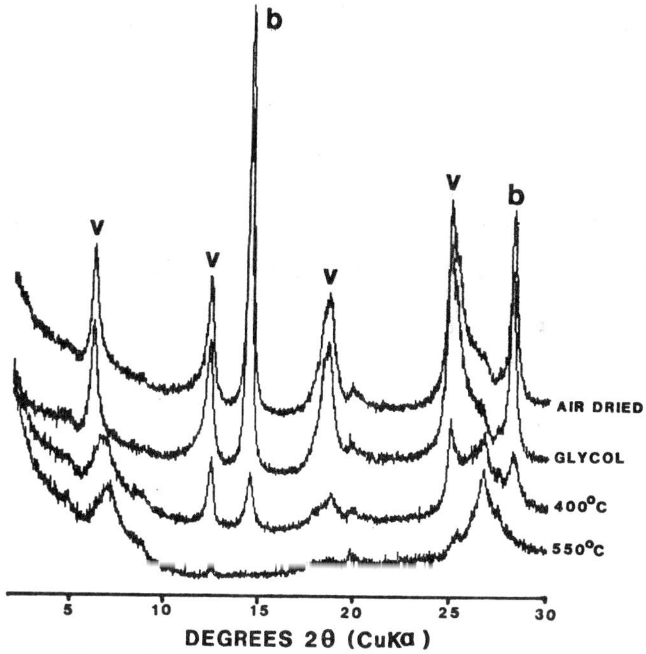

Fig. 4. X-ray diffraction pattern of clay after removal of free iron oxides. (v-vermiculite, b-boehmite)

The average chemical composition is presented in Table 1. Due to the presence of organic matter, carbonate and hydrous phases the loss on ignition (LOI) values were high. In order to compare variations in the chemical composition the oxides were normalized to 100%. The CaO concentration reflects the calcite plus aragonite content which ranges from 2 to 64% with an average of 21%. There is a negative correlation between CaO and all the other oxides reflecting dilution of the insoluble residue with carbonate material (Table 2). The chemical composition of the insoluble residue can be determined by normalizing the data to a carbonate free basis. After normalizing the data a positive correlation was observed between Al_2O_3, Fe_2O_3 and TiO_2. These elements were negatively correlated with SiO_2 and K_2O. These data suggest that there is an inverse relationship between the abundance of boehmite, goethite and hematite with dioctahedral vermiculite and quartz. As the intensity of weathering increases Fe, Al and Ti oxides, hydroxides and oxyhydroxides would accumulate at the expense of vermiculite. The SiO_2/R_2O_3 ratio is low (.58) indicating that the soils have undergone extensive leaching and enrichment in aluminum and iron.

No statistically significant relationships were observed between the topographic setting and chemical or mineralogical composition of the

Table 1. Chemical analysis of soils from Governor's Harbour, Eleuthera. (A. average analysis; B. oxides normalized to 100%)

	A	B
SiO_2	16.57	27.09
Al_2O_3	20.08	33.21
Fe_2O_3	8.56	13.66
MgO	1.14	1.97
CaO	11.87	20.64
K_2O	0.65	1.11
TiO_2	1.04	1.65
P_2O_5	0.26	0.42
MnO	0.14	0.22
LOI	37.75	-----
SiO_2/R_2O_3		0.58

Table 2. Correlation coefficients of chemical analysies of soil samples.

	CaO
Al_2O_3	-.94
Fe_2O_3	-.96
SiO_2	-.84
MgO	-.42
K_2O	-.52
TiO_2	-.94
P_2O_5	-.77
MnO	-.87

Normalized*	Fe_2O_3	TiO_2	SiO_2	K_2O
Al_2O_3	.95	.98	-.98	-.87
Fe_2O_3		.91	-.96	-.80
TiO_2			-.94	-.51
SiO_2				.82

* Normalized to a carbonate free basis.

soil (Fig. 5). There is as much variation in the chemistry of a soil within a topographic setting as there is between topographic settings. The only significant topographic trend observed was the increase in redness of soils in depressions and on crests relative to soils on slopes. The degree of redness reflects the relative ratio of hematite and goethite, with redness increasing with increasing hematite content (Taylor, 1982). The distribution of these minerals reflects the moisture regime of soils with dryer conditions favoring the formation of hematite over goethite (Schwertmann, 1985; Trolard and Tardy, 1987).

The crust of altered limestone below the soils (Ck horizon) contains megascopic and petrographic features that have been previously described in paleosols from the Bahamas (Hale and Ettensohn, 1984; Foos, 1987; McCartney and Boardman, 1987) including, laminated micrite, alveolar textures, soil breccia, root cast and iron rich glaebules. Circular voids surrounded by altered limestone are often occupied by plant roots. The mineralogy of the insoluble residue of the crust is similar to the overlying soil (Fig. 6). These data suggest that the soil and crust are contemporaneous and the crust is actually a Ck horizon of the soil profile.

DISCUSSION

The mineralogy of aluminous lateritic soils from Governor's Harbour was not significantly effected by topographic setting and they are not restricted to low spots between ridges. The lateral changes in soils on a hillslope are defined as a catena. They often result from transportation of fine material down slope by the processes of overland and throughflow which may result in an accumulation of clay at the base of the slope (Fig. 7). At Governor's Harbour the irregular karst surface and the relatively impermeable Ck horizon prevent throughflow and limit the amount of overland flow. As a result the net water movement is the same on all parts of the slope and the soils are similar (Fig. 8).

The results of this study indicate that a unique topographic setting is not necessary for the formation of aluminous lateritic soils. The climate, biota and parent material where these soils occur is similar to other areas in the Bahamas where aluminous laterites are not found. Time is most likely the important factor controlling the distribution of aluminous lateritic soils in the Bahamas. The mineralogy of Bahamian soils oc-

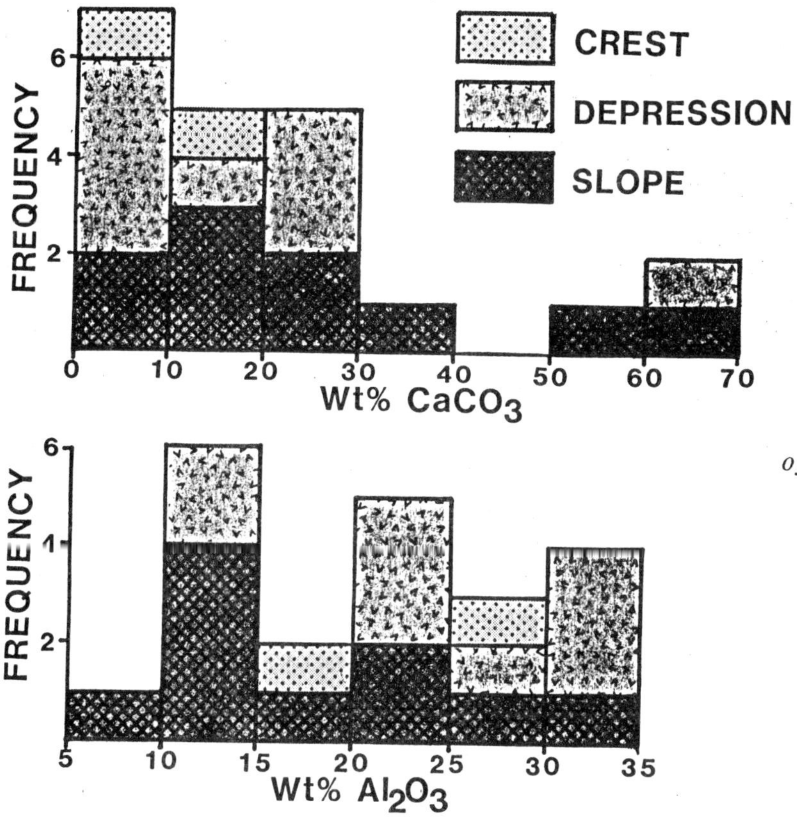

Fig. 5. Frequency distribution of chemical composition.

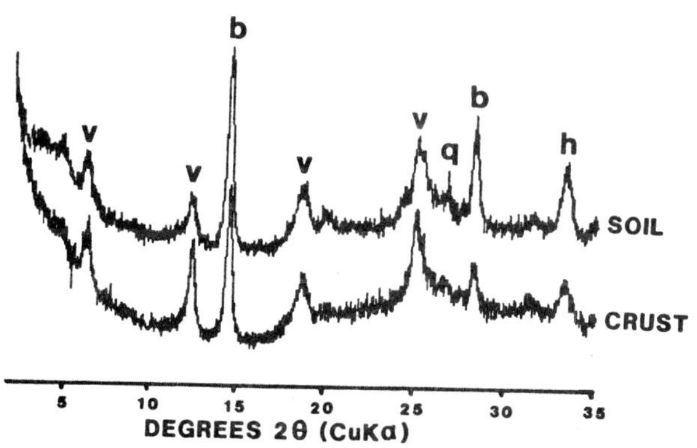

Fig. 6. Representative X-ray diffraction pattern of insoluble residue from a soil and it's underlying crust. (v-vermiculite, b-boehmite, h-hematite, q-quartz)

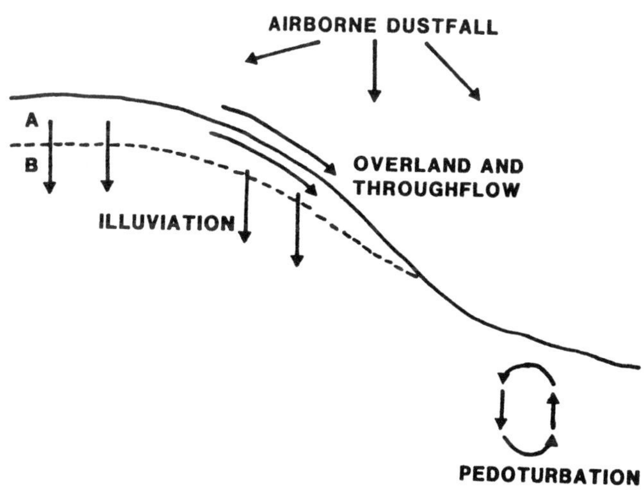

Fig. 7. An idealized example of the effect of topography on soil formation. (After Muhs, 1982)

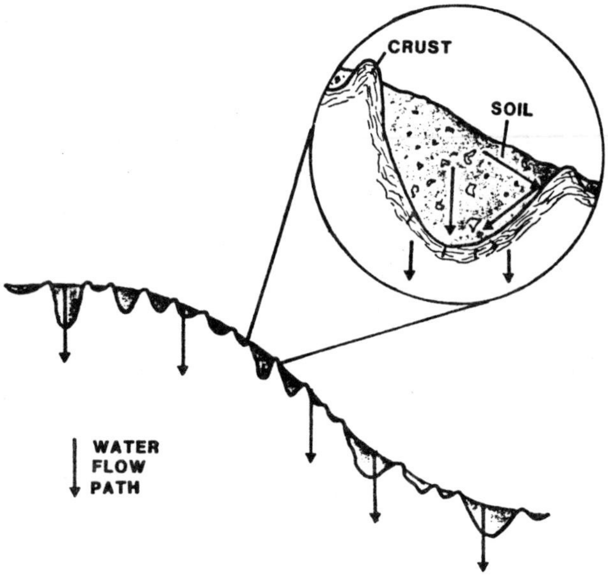

Fig. 8. An idealized view of water movement on a slope overlain by aluminous laterite.

curring on indurated bedrock may be an indicator of the soils maturity. Paleosols with a low SiO_2/R_2O_3 ratio and containing the minerals boehmite and dioctahedral vermiculite would represent more mature soils and long periods of subaerial exposure.

ACKNOWLEDGEMENTS

This research was funded by the American Philosophical Society and a University of Akron Summer Research Fellowship. I am very grateful to Ted Mowers for his assistance in the field.

REFERENCES CITED

Ahmad, N. and R.L. Jones, 1969, Occurrence of aluminous laterite soils (Bauxites) in the Bahamas and Cayman Islands: Economic Geology, V. 64, p. 804-808.

Eaton, M.R. and M.R. Boardman, 1985, North African dust and its relation to paleoclimate recorded in a sediment core from Northwest Providence Channel, Bahamas: Geological Society of America Abstracts with Programs, V. 17, p. 572.

Foos, A., 1987, Paleoclimatic interpretation of paleosols on San Salvador Island, Bahamas, in Curran, H.A., ed., Proceedings of the third symposium on the geology of the Bahamas: CCLF Bahamian Field Station, p. 67-72.

Gabler, R.E., R.J. Sanger, S. Brazier, and J. Pourciau, 1977, Essentials of physical geography: Holt, Rinehart and Winston, New York, 496 p.

Glaccum, R.A. and J.M. Prospero, 1980, Saharan aerosols over the tropical North Atlantic-Mineralogy: Marine Geology, V. 37, p. 295-321.

Hale, A.P. and F.R. Ettensohn, 1984, Micromorphological features observed in pedogenic carbonates, in Teeter, J. W., ed., Proceedings of the second symposium on the geology of the Bahamas: CCLF Bahamian Field Station, p. 265-278.

Jackson, M.L., 1969, Soil chemical analysis - Advanced course: Published by Author, Department of Soil Science, University of Wisconsin, Madison, Wis., 895 p.

Little, B.G., A. Jefferiss, J.D. Mather, J. Stark, and R.N. Young, 1976, Land Resources of the Commonwealth of the Bahamas, Volume 2A, Eleuthera Island: Land Resource Division, Ministry of Overseas Development, Surrey, England, 132 p.

Little, B.G., D.K. Buckley, R. Cant, P.W.T. Henry, A. Jefferiss, J.D. Mather, J. Stark and R.N. Young, 1977, Land resources of the Bahamas: A summary: Land Resources Division, Ministry of Overseas Development, Surrey, England, 133 p.

McCartney, R.F. and M.R. Boardman, 1987, Bahamian paleosols: Implications for stratigraphic correlation, in Curran, H.A., ed., Proceedings of the third symposium on the geology of the Bahamas: CCLF Bahamian Field Station, p. 99-108.

Muhs, D.R., 1982, The influence of topography on the spatial variability of soils in Mediterranean climates, in Thorn, C.E., ed., Space and time in geomorphology: George, Allen and Unwin, London, p. 269-284.

Schwertmann, U., 1985, The effect of pedogenic environments on iron oxide minerals, in

Stewart, B.A., ed., Advances in soil science, Volume 1: Springer Verlag, N. Y., p. 171-200.

Starkey, H.C., P.D. Blackmon and P.L. Hauff, 1984, The routine mineralogical analysis of clay-bearing samples: U. S. Geological Society Bulletin 1563, 32 p.

Taylor, R.M., 1982, Colour in soils and sediments-A review, in Van Olphen, H. and Vaniale F., eds., Proceedings of the international clay conference, Bologna and Pavia, 1981: Elsevier, Amsterdam, p. 749-761.

Trolard, F. and Y. Tardy, 1987, The stabilities of gibbsite, boehmite, aluminous goethite and aluminous hematites in bauxites, ferricretes and laterites as a function of water activity, temperature and particle size: Geochimica et Cosmochimica Acta, V. 51, p. 945-957.

WATER AND ORGANICS IN SKELETAL CARBONATES: EARLY DIAGENETIC EFFECTS - A PROGRESS REPORT

Susan J. Gaffey
Department of Geology
Rensselaer Polytechnic Institute
Troy, New York 12181

ABSTRACT

Water (liquid and bound H_2O and OH^-) and organic compounds are important primary constituents of all carbonate skeletons incorporated during growth. These phases may persist for long periods of time in the sediment column, either in their original form or as by-products of chemical or biological breakdown. Despite the fact that water and organic compounds play a central role in diagenesis of carbonate sediments, there are at present no techniques routinely in use by carbonate petrologists to detect, identify, or determine the abundances of these phases. Visible and near infrared (VNIR) (0.3 to 2.7 um) spectroscopy can potentially provide such a tool, affording a rapid, inexpensive, and non-destructive method for detecting the presence of many of the fluid and amorphous phases not accessible by X-ray diffraction, and for measuring the abundances of lighter elements such as H, C, O, and N not easily studied by more commonly used techniques such as microprobe, X-ray fluorescence, or AA. Spectral data show that there is considerable variation in the water and organic content of skeletons of different taxonomic groups, that water and organic compounds are found in fossil as well as modern skeletons, and that these phases are only partially removed by routine sample preparation techniques such as bleaching, grinding and heating. Routinely prepared carbonate samples are not representative of skeletal materials as they occur in nature, nor are they pure mineral samples of aragonite and calcite produced by organisms.

INTRODUCTION

The chemical and mineralogical composition of carbonate skeletons is more varied and complex than the usual terms "aragonite" and "high" or "low Mg calcite" imply. Water and organics, both an intrinsic part of all carbonate skeletons, can play a major role in diagenesis of carbonate sediments. Liquid water provides the medium for all diagenetic reactions (Bathurst, 1971). Presence of hydrated phases containing H_2O or OH^- can affect the stability and solubility of carbonate skeletons (e.g. Mackenzie and others, 1983; Mucci and Morse, 1984; Turner and others, 1986). Organic compounds themselves can facilitate precipitation of carbonates, and can affect the mineralogy, chemical composition, and crystal form of the precipitate (e.g. Loreau, 1982; Mitterer and Cunningham, 1985; Suess, 1973). Organic films on carbonate grains may also retard precipitation of carbonates (Suess, 1970). Organic compounds can form complexes with metal cations, affecting their mobility (Kharaka and others, 1986; Mitterer and Cunningham, 1985). Breakdown of organic compounds either by direct physicochemical oxidation or by bacteria can influence pore water chemistry, affecting Eh and pH as well as concentration of dissolved constituents such as NH_3^+, SO_4^{-2}, PO_4^{-3}, and CO_2 (Gautier, 1985; Kharaka and others, 1986) which can in turn affect the solubility and stability of carbonate phases (Walter, 1986). Although reefs occur in high energy environments where large volumes of water are circulated through the reef framework by waves and tides, anaerobic diagenesis can occur throughout reefs (Sansone, 1985; Sansone and others, 1988; Trichet and others, 1985).

Despite the importance of water and organic compounds in diagenetic reactions, and although water in various forms (liquid and bound H_2O and OH^-) and organic compounds are known to be important constituents not only of skeletal but also of non-skeletal grains such as ooids and peloids (e.g. Busenberg and Plummer, 1985; Crenshaw, 1982; Gaffey, 1983, 1984, 1985, 1988; Green and others, 1980; Mackenzie and others, 1983; Milliman, 1974; Mitterer and Cunningham, 1985; Schmalz, 1965; Weiner and others, 1983) there are at present no analytical techniques routinely in use by carbonate petrologists for identification and determination of abundances of these components. Organic contents of carbonate skeletons reported the literature vary widely (see e.g. Milliman, 1974) and data on water contents are virtually non-existent. Although considerable work has been done on diagenesis of organics in clastic sequences (e.g. Gautier, 1986; Gautier and others, 1985) organics in carbonate sediments have received much less attention. The long term goals of this work are to identify and determine abundances of water-containing phases and of organic compounds in carbonate skeletons, and to relate changes in the types and abundances of these components to diagenetic processes within modern and Pleistocene reef sediments. An important first step in such a study is the development of analytical tools for studying water and organic compounds in geologic materials. VNIR reflectance spectroscopy can potentially provide one of these tools, and this paper summarizes the advances made to date in applying reflectance spectroscopy in studies of modern and ancient carbonate sediments. Application of this tool in a specific diagenetic study can be found in Zabielski and Gaffey (this volume).

MATERIALS AND METHODS

Modern and fossil material used in this study was collected on Oahu, Hawaii, between 1980 and 1984, and from San Salvador Island, Bahamas in March, 1986 and January and June, 1988. Calcareous green algae such as *Halimeda* and *Penicillus*, and gorgonians were dried in an oven at a temperature of about $40^{o}C$. Coralline red algae, scleractinian corals, hydrozoans, and echinoids which were collected live were bleached in 5 % sodium hypochlorite solution before returning them to the laboratory for analysis. Sands from both beaches and the marine environment were collected so that samples of forams, plates of *Halimeda*, and segments of coralline red algae could be separated from them. *Homotrema rubrum* were removed from coral heads or rock slabs they encrusted. Some tests of live Foraminifera were also removed from *Penicillus* and *Halimeda*. Skeletons of dead echinoids, algae, and corals were collected from bottom sediments and from beaches.

Mineralogy of samples was determined using X-ray diffraction. All modern samples were bleached in H_2O_2 buffered to a neutral pH with NaOH or in sodium hypochlorite solution. Fossil specimens were examined in thin section and/or with SEM. VNIR (0.3 - 2.7 um) reflectance spectra were obtained both from whole samples and from bleached and unbleached powders. Spectra were obtained using the spectrophotometers at RELAB at Brown University, described by Pieters (1983) and at the U.S.G.S. in Denver, described by Clark and others (1988). Spectra of several samples were obtained on both instruments to verify subtle features.

BACKGROUND

VNIR spectroscopy provides a rapid, inexpensive, non-destructive technique for characterizing significant aspects of the mineralogical and chemical composition of geologic samples. Like X-ray diffraction it is non-selective, giving information on all optically active phases present in a sample. Unlike X-ray diffraction, however, it can be used to study many liquid and amorphous phases, and phases which occur in solid solution rather than as discrete mineral phases with an ordered lattice, providing a useful compliment to X-ray diffraction analyses. VNIR spectroscopy can also be used to study the lighter elements such as C, O, H, and N, which are not amenable to analysis by techniques such as microprobe or XRF which are more commonly employed by geologists.

Reflectance and transmission spectroscopy are not new techniques, but have been used routinely by chemists and physicists for about 60 years. All the basic principles used in interpretation of the spectra presented in this article and in Zabielski and Gaffey (this volume) are well established and are taken from the chemical literature cited below or in Gaffey (1988).

When light interacts with a compound certain wavelengths are preferentially absorbed. The number, positions, widths, and intensities of these absorptions are diagnostic of the mineralogical and chemical composition of the sample. Spectra of a sample may be obtained by reflecting light from its surface, or by transmitting light through it. The light is separated into its component wavelengths by a prism, diffraction grating or circular variable filter either before or after it has interacted with the sample, and the intensity of light reflected or transmitted at each wavelength is recorded. In the reflectance mode, the intensity of light reflected from the sample at each wavelength is ratioed to the intensity of that reflected from a standard, usually Halon in this spectral region. In the transmission mode, the intensity at each wavelength of a light beam which has passed through the sample is ratioed to the intensity of a beam which has not. In both cases, spectra like those in Figure 1 are obtained, which show the percentage or proportion of light reflected from or transmitted through the sample as a function of wavelength. A perfect reflector would produce a straight line at 1.0 reflectance.

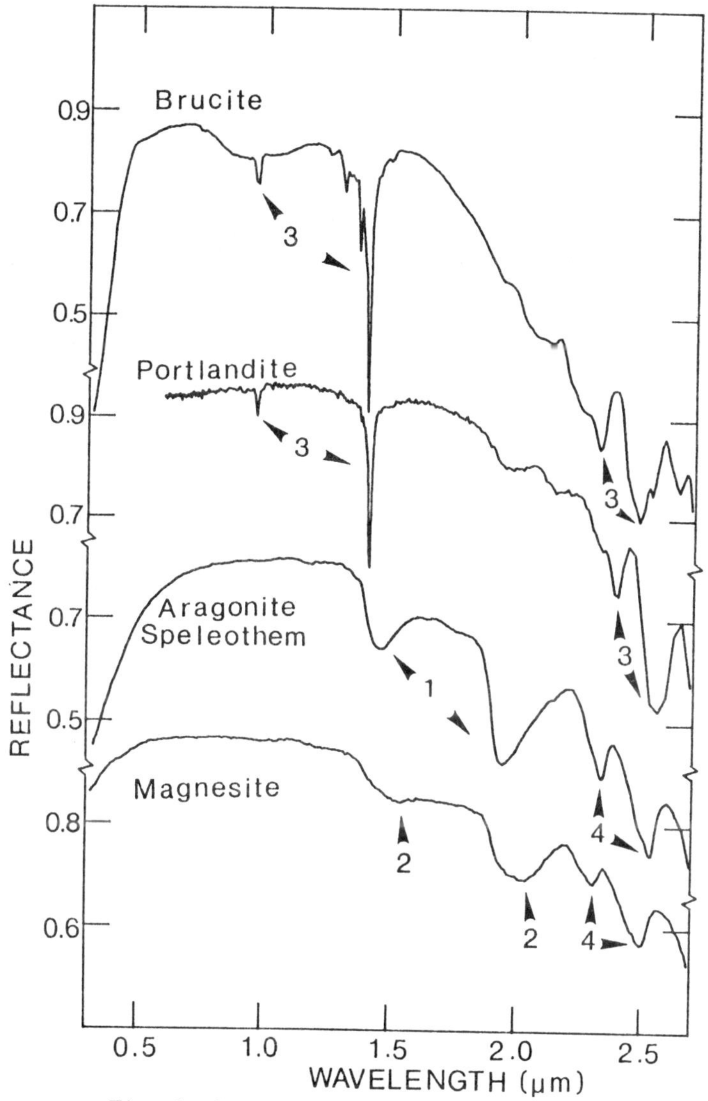

Fig. 1. Spectra of nonbiogenic mineral samples containing liquid and bound H_2O and OH^-. Arrows indicate absorption features due to: 1 - Liquid H_2O, 2 - Bound H_2O, 3 - OH^-.

Several different processes in compounds and crystals can cause absorption of light: (1) electronic transitions in molecular orbitals in compounds; (2) electronic transitions between unfilled d-shells in transition metal ions such as Fe^{2+} and Fe^{3+}; (3) internal vibrational transitions in molecules and molecular ions such as H_2O or NH_3^+ in which bond lengths and bond angles change; (4) external vibrational transitions in which the molecule or molecular ion moves as a unit. Fundamental internal vibrational modes are those in which the atoms in a molecule move in straight line paths causing a change in bond length (stretching modes) or bond angle (bending modes). External vibrational modes in crystals such as calcite or aragonite consist of librational or rocking motions of the carbonate ion, or translational motions analogous to P and S waves. Overtones and combination tones are analogous to the overtones of a vibrating violin string. In these vibrational modes the atoms do not follow straight line paths. However, these more complex motions can all be resolved into their component fundamentals. More may be learned about spectroscopy from excellent references such as those of Farmer (1974), Harris and Bertolucci (1978), Herzberg (1945), Wheatley (1968), and Wilson and others (1955).

The VNIR region of the spectrum employed in the present study and in Zabielski and Gaffey (this volume) has a number of advantages over the more commonly used mid-infrared (MIR) (3 to 20 um) and far infrared (FIR) (>20 um) regions. (1) Because the types of particle size effects which hamper work in the MIR and FIR (Estep-Barnes, 1977; Farmer and Russell, 1966; Russell, 1974; Tuddenham and Lyon, 1960) are absent in the VNIR, spectra can be obtained from samples in any form (Gaffey, 1986). Unstable or soluble phases such as organics and hydrated mineral phases can be studied *in situ*, without subjecting materials to grinding, heating, bleaching, or extraction techniques which may alter, remove, or destroy the phases of interest. (2) The VNIR region contains absorptions due to all four types of transitions discussed above, unlike the MIR and FIR which only contain features due to vibrational transitions. (3) A great deal of overlap between absorption bands occurs in the 3 um region which contains absorptions due to the fundamental O-H, C-H, and N-H stretching modes and the first overtones of the H-O-H and H-N-H bending modes. However, in the VNIR energy differences between different vibrational modes are amplified, and a useful separation occurs between absorptions due to overtones and combination tones of these modes.

Spectral Properties of Carbonates

All anhydrous carbonate minerals produce a series of sharp, narrow absorptions at wavelengths

>1.6 um, whose precise positions and widths vary with mineralogy and chemical composition. All but the two strongest bands near 2.3 and 2.5 um are usually masked by water bands in spectra of skeletal carbonates, although in spectra of some samples the carbonate band near 2.17 um can also be seen. See Gaffey (1984, 1985, 1986, 1987) for more complete discussions of the VNIR spectral properties of anhydrous carbonate minerals.

Spectral Properties of Water

As used here, "water" refers to all inorganic liquid and bound H_2O and OH^-. Spectra of some H_2O- and OH^--containing phases are shown in Figures 1 and 2. The brucite, aragonite speleothem, and magnesite are naturally occurring mineral samples. The portlandite, hydrocerussite, hydrozincite, and hydromagnesite samples are reagent grade chemicals.

H_2O has three strong absorption features near 1.4 (the first overtone of the O-H stretch), 1.9 (a combination of the H-O-H bend and O-H stretch) and 2.5 um (a combination of the O-H stretch and an external mode). H_2O also produces weaker absorptions near 0.9, 1.2, and 1.8 um (Bayly et al., 1963). The aragonite speleothem spectrum (Fig. 1) illustrates absorption features due to $CaCO_3$ and liquid H_2O.

When H_2O is incorporated into a crystal lattice, hydrogen bonding causes absorption bands to broaden, intensify, and shift to longer wavelengths (Falk and Knop, 1973; Hamilton and Ibers, 1968). OH^--containing mineral phases produce absorptions near 0.9, 1.4 and 2.2 to 2.3 um, but do not produce a features in the 1.2- and 1.9-um regions (Chapman and Nacey, 1957; Miller and Willis, 1956). OH^- features in mineral spectra tend to be sharper and narrower than H_2O features.

Spectra in Figures 1 and 2 illustrate the usefulness of VNIR spectroscopy for characterizing water in geologic samples. For example, although X-ray diffraction indicates the magnesite sample (Fig. 1) contains only magnesite, its spectrum indicates the presence of bound H_2O, in addition to water in fluid inclusions which occurs in nearly all mineral samples. The hydrozincite and hydrocerussite samples of Figure 2 are reagent grade $ZnCO_3$ and $PbCO_3$, respectively. X-ray diffraction shows the "$PbCO_3$" sample is ~50% hydrocerussite, chemical formula $Pb_3(CO_3)_2(OH)_6$ (Fleischer, 1980), and its spectrum shows the expected narrow sharp OH^- absorptions near 0.9, 1.4, and 2.3 um. However, although X-ray diffraction analysis of the "$ZnCO_3$" sample shows it is composed of hydrozincite, formula $Zn_5(CO_3)_2(OH)_6$ (Fleischer, 1980), the strong 1.9-um band and positions of both the 1.4- and 1.9-um bands indicate the presence of bound H_2O. The chemical composition of many hydrated carbonate mineral phases, including hydrated $MgCO_3$ minerals such as nesquehonite and hydromagnesite, are poorly known (e.g. White, 1971) and additional data is needed before hydrated species in carbonate skeletons can be related to specific mineral phases.

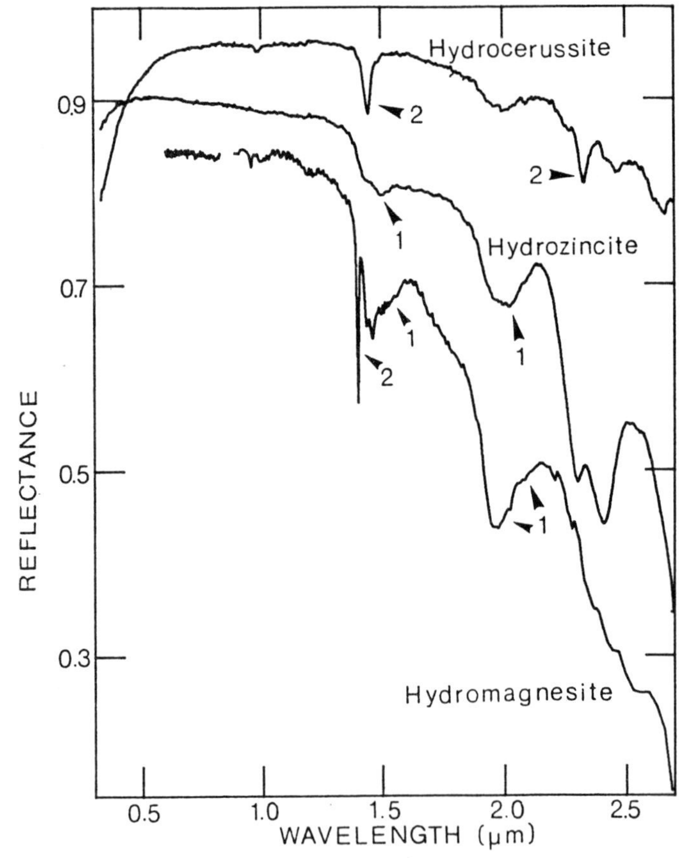

Fig. 2. Spectra of reagent grade chemicals showing features due to bound H_2O and OH^-. Note that although the chemical formula for hydrozincite indicates this mineral contains only OH^-, bound H_2O is actually the form of hydrogen present. Arrows indicate absorption features due to: 1- Bound H_2O, 2- OH^-.

Spectral Properties of Organics

A variety of organic compounds have been identified in carbonate skeletons. Mollusks have been most intensely studied, but information is also available on organics in skeletons of other organisms. The following discussion is not intended to give a complete analysis of the spectral properties of all organic compounds, but to briefly outline the spectral characteristics of broad classes of compounds known to occur in skeletal carbonates. Spectra of some organic materials are shown in Figure 3.

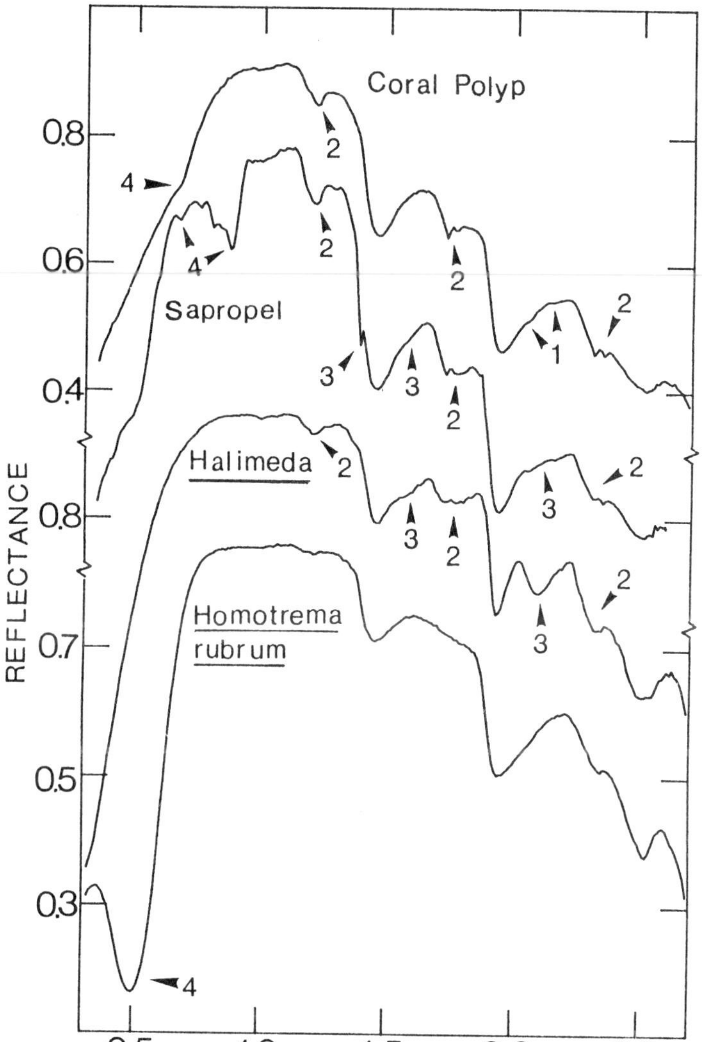

Fig. 3. Spectra of organic samples. Arrows indicate absorption features due to: 1- NH_3^+, 2- CH_2, CH_3, 3- OH, 4- organic pigments.

The near infrared portion of the spectrum is dominated by overtones and combination tones of the N-H, C-H, and O-H stretching fundamentals of functional groups in organic compounds. Despite the large size and structural complexity of organic molecules, small groups of atoms such as NH_3^+, OH, or CH_2 within those molecules will consistently produce absorption bands of similar intensities in approximately the same spectral region, called characteristic group frequencies. Precise positions of these absorption features will vary somewhat depending on the way the atomic group is bonded to the larger molecule (Colthup and others, 1964; Kaye, 1954; Wheatley, 1968).

While assignments of a single absorption feature to a given compound can't be made with confidence, suites of features considered together reduce the ambiguity in identification of absorbing species (Adams, 1975; Huguenin, 1987). Although the MIR is the wavelength region of choice for chemists, there is some data available in the literature on the NIR spectral properties of organics. In addition, because absorptions in the NIR are overtones and combination tones of fundamental modes, extrapolations can be made from the more readily available data on fundamental modes in the MIR (Herzberg, 1945).

Proteins

Proteins rich in aspartic acid are a characteristic component of many carbonate skeletons. Glutamic acid, glycine, and alanine may also be important constituents (e.g. Crenshaw, 1982; Mitterer and Cunningham, 1985; Weiner and others, 1983; Wilbur and Simkiss, 1979). NH_3^+ and CH_2 are the principle functional groups in proteins producing absorptions in the NIR (Colthup and others, 1964; Davies and Littlewood, 1979; Weber and Teale, 1965; Whetsel and others, 1958).

NH_3^+ groups produce two or more broad, rather diffuse absorptions between 2.0 and 2.2 um and broad bands near 1.5 and 1.1 um (Colthup and others, 1964; Miller and Willis, 1956; Weber and Teale, 1965; Whetsel and others, 1958). Sharp, narrow bands due to CH_2 occur in the 2.3 to 2.4-um region and near 1.7 um. Narrow bands also occur near 1.2 um (Colthup and others, 1964; Kaye, 1954; Miller and Willis, 1956; Weber and Teale, 1965).

Hydrogen bonding causes absorption bands due to functional groups to intensify and shift to longer wavelengths. NH_3^+ and CH_2 absorptions are several orders of magnitude weaker than those produced by an equal concentration of H_2O or OH^- (Bayly and others, 1963; Russell and Thompson, 1955a,b). Thus absorptions due to NH_3^+, CH_2, and CH_3 may be partially or completely masked by absorptions due to liquid H_2O in fluid inclusions (e.g. Hotta, 1969), or in polysaccharides. Absorptions due to weakly absorbing species will be detected most readily in the more transparent regions of the spectrum where absorptions due to H_2O, OH^-, and carbonate are weak or lacking.

Polysaccharides

Polysaccharides also occur in many carbonate skeletons (Borowitzka, 1986; Cabioch and Giraud, 1986; Krumbein, 1979; Weiner et al., 1983). OH^- groups in polysaccharides produce broad bands between 2.0 and 2.3 um and near 1.5 and 1.0 um (Perlin and Casu, 1982; Zhbankov, 1966). Sharp, narrow bands near 0.9 and 1.4 um indicate the presence of a weakly bound OH^- group (Zhbankov, 1966), illustrated in the sapropel spectrum in Figure 3. CH_2 groups produce narrow, sharp absorption features near 2.3 and 1.7 um, and narrow absorptions near 1.4 and 1.2 um (Colthup and others, 1964; Evans and Hibbard, 1951; Zhbankov, 1966).

Lipids

Lipids have been identified in some carbonate skeletons (Johnston, 1979, 1980). Spectra of lipids are dominated by absorptions of CH_2 and CH_3 groups near 1.2, 1.4, 1.7 and 2.3 - 2.4 um (Chapman, 1965; Colthup and others, 1964; Davenport, 1971).

Organic Pigments

Organic pigments such as chlorophyll and carotene produce very intense absorption features in the visible region due to electronic transitions in molecular orbitals. These pigments and/or the products resulting from their breakdown can persist in sediments for geologically long periods of time, and absorptions due to these pigments are found in oils and organic-rich sediments in ancient rocks (Baker and Louda, 1986).

PHASES IN MODERN SKELETONS

Inorganic H_2O and OH^-

Spectra of all skeletal carbonates studied to date show absorption features due to liquid H_2O in fluid inclusions (Gaffey, 1983, 1984, 1985, 1988). Chemical analyses of the organic component of mollusk shells also indicate the presence of H_2O in fluid inclusions (Hudson, 1967). SEM and TEM studies of a variety of types of skeletal material have shown that rapid growth of crystals results in abundant twinning and growth dislocations, and produces a polycrystalline material with numerous inter- and intracrystalline voids (Bruni and Wenk, 1985; Constantz, 1986a,b) filled with organic material and/or liquid water (e.g. Bruni and Wenk, 1985; Conger and others, 1977; Green and others, 1980; Gvirtzman and Friedman, 1977; Hudson, 1967; James, 1974; Towe and Thompson, 1972).

Spectral and other properties of skeletal carbonates indicate the presence of bound H_2O- and OH^--containing phases in many skeletons, particularly the high Mg calcites (e.g. Busenberg and Plummer, 1985; Gaffey, 1984, 1985, 1988; Mackenzie and others, 1983; Mucci and Morse, 1984; Schmalz, 1965; Turner and others, 1986; Weber and Kaufman, 1965). Lowenstam (1981) also reports the presence of hydrated amorphous $CaCO_3$ and monohydrocalcite in some mollusks and arthropods.

Lippmann (1973) notes that Mg has a marked tendency to form hydrous carbonate mineral phases, and attributes this to strong bonds which exist between Mg^{2+} and water molecules in aqueous solutions. Rapid precipitation of mineral phases from concentrated solutions, both in nature and in chemical laboratories, apparently produces these unstable, hydrated phases. During precipitation of biominerals, an amorphous or hydrated phase is often precipitated first and later reverts to a more stable anhydrous phase (Mann, 1986). It may be that some of these unstable, intermediate phases remain in skeletons.

Skeletons of *Lytechinus variegatus*, like those of most echinoids, contain very little organic matter, organics comprising $< 0.1\%$ of the skeleton by weight (Swift and others, 1986). All absorption features in the *Lytechinus* spectrum in Figure 4a can be attributed to liquid and bound H_2O.

The *Halimeda* spectrum in Figure 4a provides another example. The narrow feature near 2.3 um, superimposed on the 2.3-um carbonate band, could be attributed to CH_2. However, the more easily detected CH_2 absorptions near 1.7 and 1.2 um are lacking (compare to *Homotrema* spectrum, Fig. 4a), indicating that this absorption is best attributed to $Ca(OH)_2$. Broadening and apparent shift to longer wavelengths of the 1.4- and 1.9-um bands may be attributed to bound H_2O.

Organic Compounds in Skeletons

Organic compounds in skeletons may be part of an organic matrix used to construct the

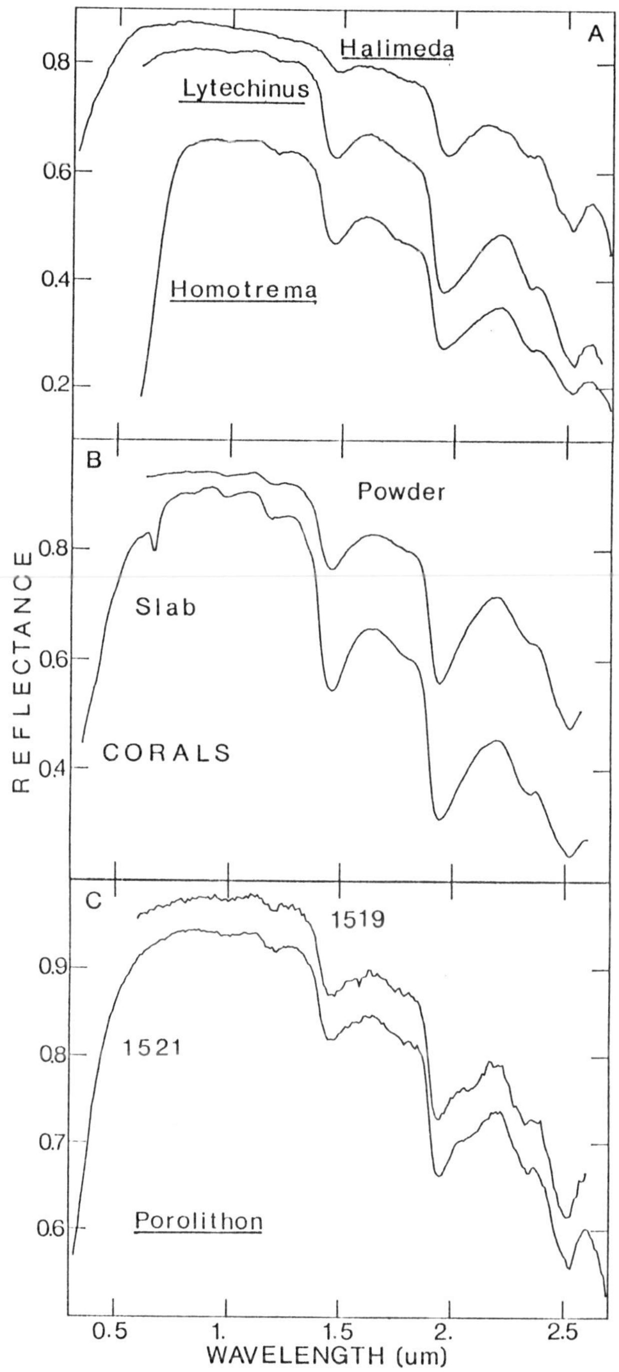

Fig. 4. Spectra of powdered and whole samples of skeletal carbonates showing absorption features due to water and organics. A. Spectra of whole plates of Lytechinus variegatus, *whole tests of* Homotrema rubrum *and bleached powder of* Halimeda. *Spectra offset vertically:* Lytechinus *0.05,* Homotrema *- 0.05. B. Spectra of bleached slab and bleached powder of scleractinian coral skeletons. C. Spectra of bleached powders of coralline red algae* Porolithon. *Spectra offset vertically: 1521 - 0.03, 1519 - 0.02.*

skeleton and perhaps to give it added strength (e.g. Crenshaw, 1982; Lowenstam, 1986; Weiner and others, 1983), soft tissues and body fluid of the organism entrapped during growth (Constantz, 1986a,b; Johnston, 1979, 1980) or material introduced by endoliths (Johnston, 1980). All skeletons contain some organic material, even after bleaching and grinding (e.g. Young, 1971). However, because many of the functional groups in organic compounds are weak absorbers, spectra of carbonate skeletons may not always indicate the presence of organics. For example, Figure 4b shows spectra obtained from a powdered, bleached coral sample and from a slab of a coral head. The spectrum of the slab shows some weak absorptions in the 2.2 - 2.3-um region due NH_3^+, a weak shoulder on the 2.3-um carbonate band and weak features near 1.7 and 1.2 um due to CH_2, as well as features near 0.67 and 0.40 um due to organic pigments introduced by endolithic algae (Baker and Louda, 1986; Lukas, 1974). However, in the powdered, bleached sample, only features due to liquid H_2O and carbonate can be seen. Additional work is needed to determine the detection limits for various organic compounds in spectra of intimate mixtures with water and carbonates.

Spectra of other samples, however, do show absorptions attributable to organic compounds even after grinding and bleaching. Comparison of the spectrum of whole *Homotrema* tests (Fig. 4a) with the spectrum of powdered, bleached *Homotrema* (Fig. 3) shows that in the spectrum of the powdered sample the C-H features near 2.3 and 1.7 um are not evident. However, the C-H feature near 1.2 um is still present, although weakened, in the spectrum of the powdered sample, as well as the strong band near 0.5 um caused by the organic pigments which give this foram its distinctive red color.

Spectra of calcareous red algae in Figure 4c may also show some features due to organic compounds. Calcification in the coralline red algae results from impregnation of the cell wall with high Mg calcite (HMC), a process controlled by a polysaccharide matrix (Cabioch and Giraud, 1986). Comparison of spectra of coralline red algae with spectra of organic materials (Fig. 3) and nonbiogenic mineral samples (Figs. 1 and 2) show that features in these spectra could be assigned either to organic and or to inorganic phases. It has been known for some time that skeletons of coralline red algae contain brucite (Busenberg and Plummer, 1985; Schmalz, 1965; Weber and Kaufman, 1965). Grinding and bleaching aid in distinguishing between absorption features due to organics and those due to mineral phases. Grinding and bleaching will weaken features due to liquid water and organics, making those due to mineral phases more prominent. Some features originally attributed to hydrated mineral phases (Gaffey, 1988), now appear to be better assigned to organic compounds. For example, the feature near 2.2 um in the spectrum of sample 1521 in Figure 4c is probably due to OH in polysaccharides. These features are weakened in the spectrum of 1519 which was bleached more thoroughly, and features due to liquid and possibly to bound H_2O and OH^- become more evident.

Spectra clearly indicate that routine grinding and bleaching do not remove all the water and organic materials from skeletons. Routinely prepared samples are not pure specimens of aragonite and calcite produced by organisms.

Water and organics are present in Pleistocene and even older samples (e.g. Gaffey, 1983, 1984, 1985, 1988; Wyckoff, 1972; Zabielski and Gaffey, this volume), and fluid inclusions may still be abundant in fossils even after alteration of aragonite and high Mg calcite to low Mg calcite (Gaffey, 1984, 1985, 1988; Sandberg, 1984). Thus, organic compounds and liquid water incorporated in skeletons during their formation may remain to play important roles in carbonate diagenesis for geologically long periods of time. Organic pigments which produce absorption features in the visible region of the spectrum also persist in geologic samples even after considerable breakdown and heating of organic materials, providing biomarkers for determining the history and origin of organic materials (e.g. Baker and Louda, 1986).

UNRESOLVED QUESTIONS

The complexity and variability in composition of skeletal carbonates and the sensitivity of VNIR reflectance spectroscopy to these variations have combined to slow progress in this diagenetic study. The water and organics in skeletons are strongly affected by routine sample preparation such as bleaching, grinding, and heating. However, despite routine bleaching and grinding, samples still contain significant quantities of water and organics. Spectra indicate that routinely processed samples are not representative of the original materials as they occur in nature, nor are they pure samples of aragonite and calcite produced by various organisms. This is true even of fossil specimens from which much organic material is presumably lost.

Additional information is needed on the effects of grinding, bleaching, and heating on the composition of carbonate samples. In addition, because spectral data available in the literature of physics and chemistry are generally inadequate for studies of geologic materials, additional spectra of organic compounds and hydrated mineral species must be obtained. Additional work is needed employing other analytical techniques (e.g. NMR, ESR, Raman spectroscopy, FTIR, coulomtery) to completely characterize the water and organic content of skeletons. These techniques should provide a quantitative measure of the water and organics in skeletons, and should make it possible to develop calibration curves allowing abundances of these phases to be determined from spectra. More work is needed using reflectance spectra in the visible region, where absorptions due to electronic transitions in molecular orbitals occur.

In addition, all phases present in skeletons must be characterized. Much of the previous work on organics in carbonate skeletons has focused on those compounds believed to make up the organic matrix used by organisms to construct their skeletons. However, organic compounds may affect diagenesis of carbonates regardless of their source, and skeletons may also contain organics derived from other tissues or body fluids of plants and animals which build the skeletons, as well as from endoliths which infest skeletons after death.

ACKNOWLEDGEMENTS

I would first like to thank Brian White and Allen Curran for introducing me to San Salvador, and Don and Kathy Gerace and the staff at the Bahamian Field Station for their logistical support. I thank Carle Pieters at Brown University and Roger Clark at the USGS in Denver for the use of their spectrometers. Thanks also to Steven Pratt at RELAB at Brown University for his assistance in obtaining spectral data. I would also like to thank Victor Zabielski for his assistance in the field. Funding for this work was provided by NSF Grant Number EAR-8721094 and by Gaylord, Leonard and Edna Cobeen and Edith Jenks.

REFERENCES CITED

Adams, J.B., 1975, Interpretation of visible and near-infrared diffuse reflectance spectra of pyroxenes and other rock-forming minerals, in Karr, C., ed., Infrared and raman spectroscopy of lunar and terrestrial minerals: New York, Academic Press, p. 91-116.

Baker, E.W., and Louda, J.W., 1986, Porphyrins in the geological record, in Johns, R.B., ed., Biological markers in the sedimentary record: New York, Elsevier p. 125-225.

Bathurst, R.G.C., 1971, Carbonate sediments and their diagenesis: New York, Elsevier Publishing Company, 620 p.

Bayly, J.G., Kartha, V.B., and Stevens, W.H., 1963, The absorption spectra of liquid phase H_2O, HDO, and D_2O from 0.7 to 10 um: Infrared Physics, v. 3, p. 211-223.

Borowitzka, M.A., 1986, Physiology and biochemistry of calcification in the Chlorophyceae, in Leadbeater, B.S.C. and Riding, R., 1986, Biomineralization in the lower plants and animals: Oxford, Clarendon Press, p. 107-124.

Bruni S.F., and Wenk, H.R., 1985, Replacement of aragonite by calcite in sediments from the San Cassiano Formation (Italy): Journal of Sedimentary Petrology, v. 55, p. 159-170.

Busenberg, E., and Plummer, N.L., 1985, Kinetic and thermodynamic factors controlling the distribution of SO_4^{2-} and Na^+ in calcites and selected aragonites: Geochimica et Cosmochimica Acta, v. 49, p. 713-725.

Cabioch, J., and Giraud, G., 1986, Structural aspects of biomineralization in the coralline algae (calcified Rhodophyceae), in Leadbeater, B. S. C., and Riding, R., Biomineralization in lower plants and animals: Oxford, Clarendon Press, p. 141

Chapman, D., 1965, The structure of lipids: New York, John Wiley and Sons, 323 p.

Chapman, D., and Nacey, J.F., 1957, A rapid spectroscopic method for determination of water in glycerols: The Analyst, v. 83, p. 377-379.

Clark, R.N., King, T., Klejwa, M., Swayze, G.A., and Vergo, N., 1988, High spectral resolution reflectance spectroscopy of minerals: Journal of Geophysical Research, in press.

Colthup, N.B., Daly, L.H., and Wiberly, S.E., 1964, Introduction to infrared and Raman spectroscopy: New York, Academic Press, 511 p.

Conger, S.D., Green, H.W., and Lipps, J.H., 1977, Test ultrastructure of some calcareous Foraminifera: Journal of Foraminiferal Research, v. 7, p. 278-296.

Constantz, B.R., 1986a, Coral skeleton construction: A physiochemically dominated process: Palaios, v. 1, p. 152-157.

Constantz, B.R., 1986b, The primary surface area of corals and variations in their susceptibility to diagenesis, in Schroeder, J.H., and Purser, B.H., eds., Reef diagenesis: New York, Springer-Verlag, p. 53-76.

Crenshaw, M.A., 1982, Mechanisms of normal biological mineralization of calcium carbonates, in Nancollas, G.H., ed., Biological mineralization and demineralization: New York, Springer-Verlag, p. 243-258.

Davenport, J.B., 1971, Infrared spectroscopy of lipids, in Johnson, A.R., and Davenport, J. B., eds., Biochemistry and methodology of lipids: New York, Wiley Interscience, p. 231-242.

Davies, J., and Littlewood, B.S., 1979, Elementary biochemistry: Englewood Cliffs, New Jersey, Prentice-Hall, 346 p.

Estep-Barnes, P.A., 1977, Infrared spectroscopy, in Zussman, J., ed., Physical methods in determinative mineralogy, Second edition: New York, Academic Press, p. 529-603.

Evans, A., Hibbard, R.R., and Powell, A.S., 1951, Determination of carbon-hydrogen groups in high molecular weight hydrocarbons: Analytical Chemistry, v. 23, p. 1604-1610.

Falk, M., and Knop, O., 1973, Water in stoichiometric hydrates, in Franks, F., ed., Water, a comprehensive treatise, Volume 2: Water in crystalline hydrates, aqueous solutions of simple nonelectrolytes: New York, Plenum Press, p. 55-113.

Farmer, V.C., 1974, The infrared spectra of minerals: London, Mineralogical Society Monograph 4, 539 p.

Farmer, V.C., and Russell, J.E., 1966, Effects of particle size and structure on the vibrational frequencies of layer silicates: Spectrochimica Acta, v. 22, p. 389-398.

Fleischer, M., 1980, Glossary of mineral species: Tucson, Arizona, Mineralogical Record, 192p.

Gaffey, S.J., 1983, Spectral reflectance of carbonate rocks and minerals [abstr.]: American Association of Petroleum Geologists Bulletin, v. 67, p. 465.

Gaffey, S.J., 1984, Spectral reflectance of carbonate minerals in the visible and near infrared (0.35-2.55 um) and its applications in carbonate petrology [Ph.D. thesis]: Honolulu, Hawaii, University of Hawaii at Manoa, 236 p.

Gaffey, S.J., 1985, Reflectance spectroscopy in the visible and near-infrared (0.35-2.55 um): Applications in carbonate petrology: Geology, v. 13, p. 270-273.

Gaffey, S.J., 1986, Spectral reflectance of carbonate minerals in the visible and near infrared (0.35-2.55 microns): Calcite, aragonite, and dolomite: American Mineralogist, v. 71, p. 151-162.

Gaffey, S.J., 1987, Spectral reflectance of carbonate minerals in the visible and near infrared (0.35-2.55 um): Anhydrous carbonate minerals: Journal of Geophysical Research, v. 92, no. B2, p. 1429-1440.

Gaffey, S.J., 1988, Water in skeletal carbonates: Journal of Sedimentary Petrology, v. 58, p. 397-414.

Gautier, D.K., 1985, Interpretation of early diagenesis in ancient marine sediments, in Gautier, D.L., Kharaka,Y.K., and Surdam, R.C., Relationship of organic matter and mineral diagenesis: Society of Economic Paleontologists and Mineralogists Short Course Notes Number 17, p. 6-78.

Gautier, D.K., 1986, Roles of organic matter in sediment diagenesis: Society of Economic Paleontologists and Mineralogists Special Publication No. 38, 203 p.

Gautier, D.L., Kharaka, Y.K., and Surdam, R.C., 1985, Relationship of organic matter and mineral diagenesis: Society of Economic Paleontologists and Mineralogists Short Course Notes No. 17, 279 p.

Green, H.W., Lipps, J.H., and Showers, W.J., 1980, Test ultrastructure of fusulinid Foraminifera: Nature, v. 283, p. 853-855.

Gvirtzman, G., and Firedman, G.M., 1977, Sequence of progressive diagenesis in coral reefs, in Frost, S.H., Weiss, M.P., and Saunders, J.B., Reefs and related carbonates-ecology and sedimentology: American Association of Petroelum Geologists Studies in Geology No. 4, p. 357-380.

Hamilton, W.C., and Ibers, J.A., 1968, Hydrogen bonding in solids: New York, W.A. Benjamin, 284 p.

Harris, D.C., and Bertolucci, M.D., 1978, Symmetry and spectroscopy: New York, Oxford University Press, 550 p.

Herzberg, G., 1945, Molecular spectra and molecular structure: New York, Van Nostrand Reinhold, 632 p.

Hotta, S., 1969, Infrared spectra and conformation of protein constituting the nacreous layer of molluscan shell: Earth Science, v. 23, p. 133-140.

Hudson, J.D., 1967, The elemental composition of the organic fraction, and the water content of some Recent and fossil mollusc shells: Geochimica et Cosmochimica Acta, v. 31, p. 2361-2378.

Huguenin, R.L., 1987, The silicate component of Martian dust: Icarus, v. 70, p. 162-188.

James, N.P., 1974, Diagenesis of scleractinian corals in the subaerial vadose environment: Journal of Paleontology, v. 48, p. 785-799.

Johnston, I.S., 1979, The organization of a structural organic matrix within the skeleton of a reef-building coral: Scanning Electron Microscopy, v. 2, p. 421-431.

Johnston, I.S., 1980, The ultrastructure of skeletogenesis in hermatypic corals: International Review of Cytology, v. 67, p. 171-214.

Kaye, W., 1954, Near infrared spectroscopy, A review, I. Spectral identification and analytical applications: Spectrochimica Acta, v. 6, p. 257-287.

Kharaka, Y.K., LeRoy, M.L., Carothers, W.W., and Goerlitz, D.F., 1986, Role of organic species dissolved in formation waters from sedimentary basins in mineral diagenesis, in Gautier, D.K., ed., Roles of organic matter in sediment diagenesis: Society of Economic Paleontologists and Mineralogists Special Publication 38, p. 111-122.

Krumbein, W.E., 1979, Calcification by bacteria and algae, in Trudinger, P.A., and Swaine, D.J., Biogeochemical cycling of mineral-forming elements: New York, Elsevier, p. 47-68.

Lippmann, F., 1973, Sedimentary carbonate minerals: New York, Springer-Verlag, 228 p.

Loreau, J.P., 1982, Sediments aragonitiques et leur genese: Memoires du Museum National d'Histoire Naturelle, Serie C, Geologie, V. 47, 312 p.

Lowenstam, H.A., 1981, Minerals formed by organisms: Science, v. 211, p. 1126-1131.

Lowenstam, H.A., 1986, Mineralization processes in monerans and protoctists, in Leadbeater, B.S.C., and Riding, R., eds., Biomineralization in lower plants and animals: Oxford, Clarendon Press, p. 1-17.

Lukas, K.J., 1974, Two species of the chlorophyte genus *Ostreobium* from skeletons of Atlantic and Caribbean reef corals: Journal of Phycology, v. 10, p. 331-335.

Mackenzie, F.T., Bischoff, W.D., Bishop, F.C., Loijens, M., Schoonmaker, J., and Wollast, R., 1983, Magnesian calcites: Low-temperature occurrence, solubility and solid-solution behavior, in Reeder, R.J., ed., Carbonates: Mineralogy and chemistry: Mineralogical Society of America Reviews in Mineralogy Volume 11, p. 97-144.

Mann, S., 1986, Biomineralization in lower plants and animals - chemical perspectives, in Leadbeater, B.S.C., and Riding, R., eds., Biomineralization in the lower plants and animals: Oxford, Clarendon Press, p. 39-54.

Miller, R.G.J., and Willis, H.A., 1956, Quantitative analysis in the 2-u region applied to synthetic polymers: Journal of Applied Chemistry, v. 6, p. 385-391.

Milliman, J.D., 1974, Marine carbonates: New York, Springer-Verlag, 375 p.

Mitterer, R.M., and Cunningham, R., 1985, The interaction of natural organic matter with grain surfaces: Implications for calcium carbonate precipitation, in Schneidermann, N., and Harris, P. M., eds., Carbonate cements: Society of Economic Paleontologists and Mineralogists Special Publication 36, p. 17-32.

Mucci, A., and Morse, J.W., 1984, The solubility of calcite in seawater solutions of various magnesium concentration, I_t=0.697 m at 25°C and one atmosphere total pressure: Geochimica et Cosmochimica Acta, v. 48, p. 815-822.

Perlin, A.S., and Casu, B., 1982, Spectroscopic methods, in Aspinall, G. O., ed., The polysaccharides: New York, Academic Press, p. 133- 193.

Pieters, C.M., 1983, Strength of mineral absorption features in the transmitted component of near-infrared reflected light: First results from RELAB: Journal of Geophysical Research, v. 88, no. B11, p. 9534-9544.

Russell, J.D., 1974, Instrumentation and techniques, in Farmer, V.C., ed., The infrared spectra of minerals: London, Mineralogical Society Monograph 4, p. 11-25.

Russell, R.A., and Thompson, H.W., 1955a, Intensities of vibration bands. Part VI. Alkyl esters: Journal of the Chemical Society, p. 479-482.

Russell, R.A., and Thompson, H.W., 1955b, Intensities of vibration bands. Part VII. The NH group: Journal of the Chemical Society, p. 483-486.

Sansone, F.J., 1985, Methane in the reef flat porewaters of Davies Reef, the Great Barrier Reef (Australia): Proceedings of the fifth international coral reef congress, v. 3, p. 415-419.

Sansone, F.J., Tribble, G.W., Andrews, C.C., and Buddemeier, R.W., 1988, Spatial and temporal variability of anaerobic diagenesis within a coral reef [abs], in Choat, J.H., and Bellwood, O., eds., Abstracts for the Sixth International Coral Reef Symposium, p. 90.

Schmalz, R.F., 1965, Brucite in carbonate secreted by the red alga *Goniolithon* sp.: Science, v. 149, p. 993-996.

Suess, E., 1970, Interaction of organic compounds with calcium carbonate - I. Association phenomena and geochemical implications: Geochimica et Cosmochimica Acta, v. 34, p. 157-168.

Suess, E., 1973, Interaction of organic compounds with calcium carbonate - II. Organo-carbonate association in Recent sediments: Geochimica et Cosmochimica Acta, v. 37, p. 2435-2447.

Swift, D.B., Sikes, C.S., and Wheeler, A.P., 1986, Analysis and function of organic matrix from sea urchin tests: The Journal of Experimental Zoology, v. 240, p. 65-73.

Towe, K.M., and Thompson, G.R., 1972, The structure of some bivalve shell carbonates prepared by ion-beam thinning: Calcareous Tissue Research, v. 10., p. 38-48.

Trichet, J., Chambers, L.A., and Wilkinson, C.R., 1985, Biogeochemical and microbial processes in reef environments - role of micro- organisms in coral reef ecosystems: Proceedings of the fifth international coral reef congress, v. 3, p. 471-475.

Tuddenham, W.M., and Lyon, R.J.P., 1960, Infrared techniques in the identification and measurement of minerals: Analytical Chemistry, v. 32, p. 1630-1634.

Turner, J.V., Anderson, T.F., Sandberg, P.A., and Goldstein, S.J., 1986, Isotopic, chemical and textural relations during the experimental alteration of biogenic high-magnesium calcites: Geochimica et Cosmochimica Acta, v. 50, p. 495-506.

Walter, L.M., 1986, Relative efficiency of carbonate dissolution and precipitation during diagenesis: A progress report on the role of solution chemistry, in Gautier, D.L., ed., Roles of organic matter in sediment diagenesis: Society of Economic Paleontologists and Mineralogists Special Publication 38, p. 1-12.

Weber, G., and Teale, F.W.J., 1965, The interaction of proteins with radiation, in Neurath, H., ed., The proteins: Composition, structure, and function, Volume 3, Second edition: New York, Academic Press, p. 445-516.

Weber, J.N., and Kaufman, J.W., 1965, Brucite in the calcareous alga *Goniolithon*: Science, v. 149, p. 996-997.

Weiner, S., Traub, W., and Lowenstam, H.A., 1983, Organic matrix in calcified exoskeletons, in Westbroek, P., and de Jong, E.W., Biomineralization and biological metal accumulation: D. Reidel Publishing Company, p. 205-224.

Wheatley, P.J., 1968, The determination of molecular structure: New York, Dover Publications, 264 p.

Whetsel, K.B., Roberson, W.E., and Krell, M.W., 1958, Near-infrared spectra of primary aromatic amines: Analytical Chemistry, v. 30, p. 1598-1604.

White, W.B., 1971, Infrared characterization of water and hydroxyl ion in the basic magnesium carbonate minerals: American Mineralogist, v. 56, p. 46-53.

Wilbur, K.M., and Simkiss, K., 1979, Carbonate turnover and deposition by metazoa, in Trudinger, P.A. and Swaine, D.J., eds., Biogeochemical cycling of mineral-forming elements: New York, Elsevier, p. 69-106.

Wilson, E.B., Decius, J.C. and Cross, P.C., 1955, Molecular vibrations: New York, Dover, 388 p.

Wyckoff, R.W.G., 1972, The biochemistry of animal fossils: Bristol, Scientechnica Limited, 152 p.

Young, S.D., 1971, Organic material from scleractinian coral skeletons-I. Variation in composition between several species: Comparative Biochemistry and Physiology, v. 40B, p. 113-120.

Zhbankov, R.G., 1966, Infrared spectra of cellulose and its derivatives: New York, Consultants Bureau, 333 p.

MASS MORTALITY OF THE WEST INDIAN ECHINOID *DIADEMA ANTILLARUM* ADJACENT TO ANDROS ISLAND, BAHAMAS: A NATURAL EXPERIMENT IN TAPHONOMY

Benjamin J. Greenstein & David L. Meyer
San Salvador Echinoderm Paleobiology Research Group
Dept. of Geology, University of Cincinnati
Cincinnati, Ohio 45221-0013

ABSTRACT

During 1983, populations of the common long-spined echinoid *Diadema antillarum* were decimated by disease throughout the Caribbean and as far north as Bermuda. The role of *Diadema* as a grazer of algae and agent of bioerosion is well documented. The slow rate of recovery of *Diadema* populations has been demonstrated to have had profound effects on the coral reef ecosystem. The hypothesis that the sudden incorporation of tests and spines of innumerable urchins into reef sediments would leave a clear record was tested at Andros Island, Bahamas, where the mortality was reported to have occurred in October, 1983. Sediments collected from three reef subenvironments in May, 1984 were analyzed for total echinoderm content. The echinoderm fraction shows a slight increase over pre-mortality levels reported from Bonaire, Netherlands Antilles, but it is not clear that this increase resulted from an influx of *Diadema* skeletal elements. The fragility of the *Diadema* test and post-mortem processes are suggested to be responsible for the lack of signature. The lack of a strong signal of the mass mortality event demonstrates the inadequacy of the reef sedimentary record to preserve many short-term although ecologically significant events.

INTRODUCTION

Mass Mortality

The long-spined echinoid *Diadema antillarum* Philippi is often the dominant herbivore in tropical Western Atlantic reef environments, reaching densities of 71 individuals per square meter (Sammarco, 1980; Hawkins & Lewis, 1982). As such it is responsible for large amounts of bioerosion as it grazes algae from reef surfaces (Scoffin et al., 1980; Hunter, 1977). *Diadema* also preys on living coral (Bak & van Eys, 1975; Carpenter, 1981) contributing to the control of the coral community as predators on coral recruits (Sammarco, 1980, 1982).

Beginning in January, 1983, *Diadema* suffered a widespread mass mortality (Lessios et al., 1984). The mortality was first observed at Galeta Point, Panama and by January, 1984, had spread to most of the Caribbean and to Bermuda (Figure 1). The sequence of mortality events indicates that the causative agent was generally dispersed by major surface currents over large distances until October, 1983. After that month, mortality events no longer followed a regular pattern. The causative agent is suspected to be a species-specific pathogen because the mortality spread over such a wide area without any dissipation of its severity, and no other species of sea urchin was affected (Lessios, 1984). Whatever the exact nature of the disease, its effect was catastrophic on *Diadema* populations, reducing them 98- 100% in Curacao (Bak et al., 1984), 94-99% in Panama (Lessios et al., 1985) and 99% in Jamaica (Hughes et al., 1985) where a sharp increase in bottom cover by non-crustose algae occurred within two weeks of the urchins' demise (Figure 2) (Liddell & Ohlhorst, 1986).

Urchins inhabiting the backreef, reef tract and fore-reef environments adjacent to Andros Island, Bahamas were reported dying in great numbers by the end of October, 1983 (Lessios, 1984). By May, 1984, only broken, disarticulated spines and test plates were evident in surficial reef sediments, and limited recovery of the live population had begun.

Mass mortality events affecting echinoid populations are commonly the result of prolonged exposure to elevated temperature (Glynn, 1968; Hendler, 1977), storms (Schafer, 1972) or disease (Pearse et al., 1977; Boudouresque et al., 1981; Scheibling & Stephenson, 1984). Such events occur in specific environments and on a local scale. If the current mass mortality is indeed the result of a species-specific pathogen, it represents

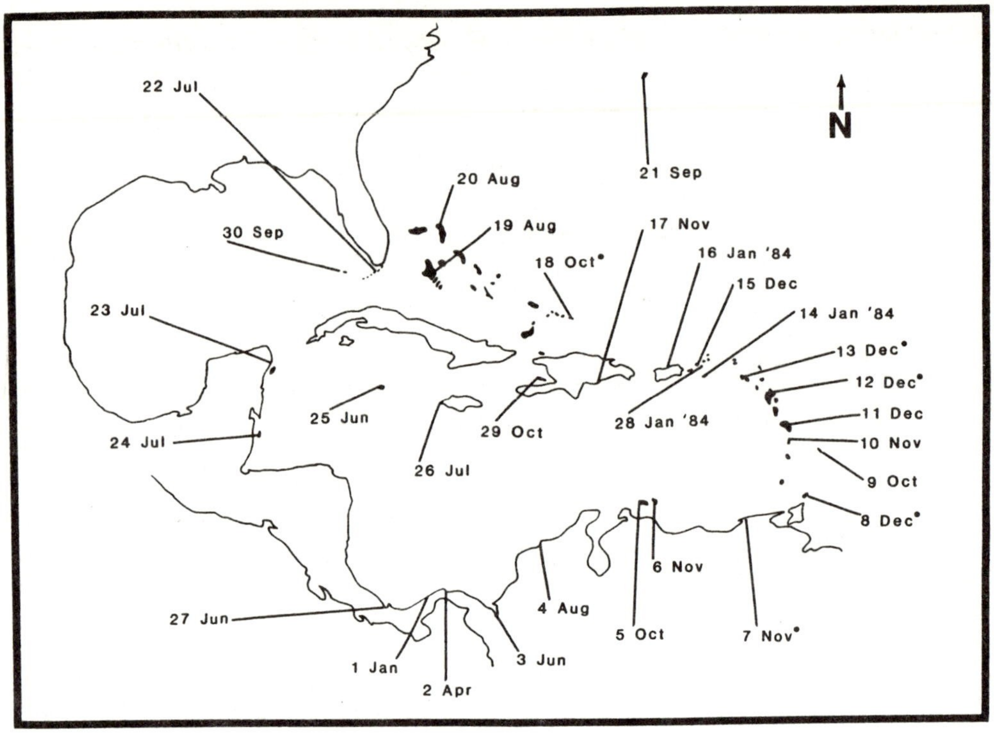

Fig. 1. Spread of Diadema *mass mortality through the Caribbean and Western Atlantic. Dates indicate the first time mortality was noted at each locality. Dates marked with a dot indicate the last time lack of mortality was verified in unaffected areas. Unless otherwise indicated, dates refer to 1983. Locality numbers: 1) Galeta Point, Panama. 2) San Blas Archipelago. 3) Puerto Obaldia. 4) Santa Marta, Columbia. 5) Curacao. 6) Bonaire. 7) Mochima, Venezuela. 8) Tobago. 9) Barbados. 10) Saint Lucia. 11) Martinique. 12) Guadeloupe. 13) Saint Kitts. 14) Saint Croix. 15) Saint Thomas and St. John. 16) Puerto Rico. 17) Santo Domingo. 18) Grand Turk. 19) Andros and New Providence Islands. 20) Grand Bahama. 21) Bermuda. 22) Florida Keys. 23) Cancun, Mexico. 24) Belize. 25) Grand Cayman. 26) Jamaica. 27) Cahuita, Costa Rica. 28) Tortola, Virgin Gorda and Salt Island. 29) Gulf of Gonave, Haiti. 30) Dry Tortugas. (after Lessios et al., 1984).*

the most extensive epidemic ever documented for a marine invertebrate (Lessios et al., 1984).

Taphonomy

A mass mortality affecting entire populations of a particular species provides a natural experiment in taphonomy, the systematic study of fossil preservation and preservational processes (see Muller, 1979, for an excellent review). More specifically, observing the result of the incorporation of large numbers of individuals in Recent reef sediments can yield information on their biostratinomy. The study of biostratinomy begins with the death struggle of an organism and ends with the final burial and arrangement of the dead or dying organism or its disarticulated remains (Muller, 1979) (Fig. 3).

The incorporation of innumerable echinoids into reef sediments suggests that sediment composition may be altered by an increase in the amount of echinoderm material present. Mass outbreaks and subsequent mortalities of the crown-of-thorns starfish *Acanthaster* have been reported preserved as layers enriched in their ossicles on the Great Barrier Reef (Frankel, 1978). Well-preserved echinoid assemblages are reported rarely in the literature (e.g Aslin, 1968; Rosenkranz, 1971; and Bloos, 1973), as are sediments composed primarily of their skeletal debris (e.g. Cherns, 1983). The mass mortality of *Diadema* provides an opportunity to assess the liklihood that an ecologically significant event will be recorded as a recognizable change in the composition of reef sediments.

STUDY AREA

Andros Island lies on the eastern margin of the Great Bahama Bank (Figure 4). The reef

Fig. 2. Overgrowth of algae occurring on the reef tract of the Florida Keys. Picture was taken in June, 1984, 11 months after the mass mortality was reported to have occurred (Photo by D. L. Meyer).

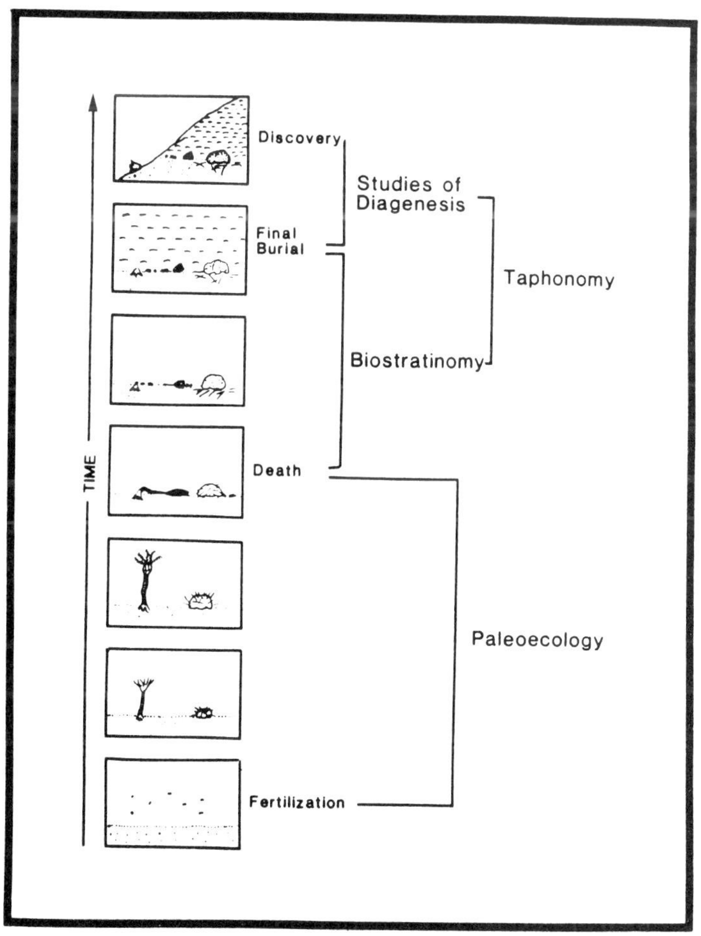

Fig. 3. Schematic illustration of the disciplines of taphonomy and biostratinomy (after Lewis, 1981).

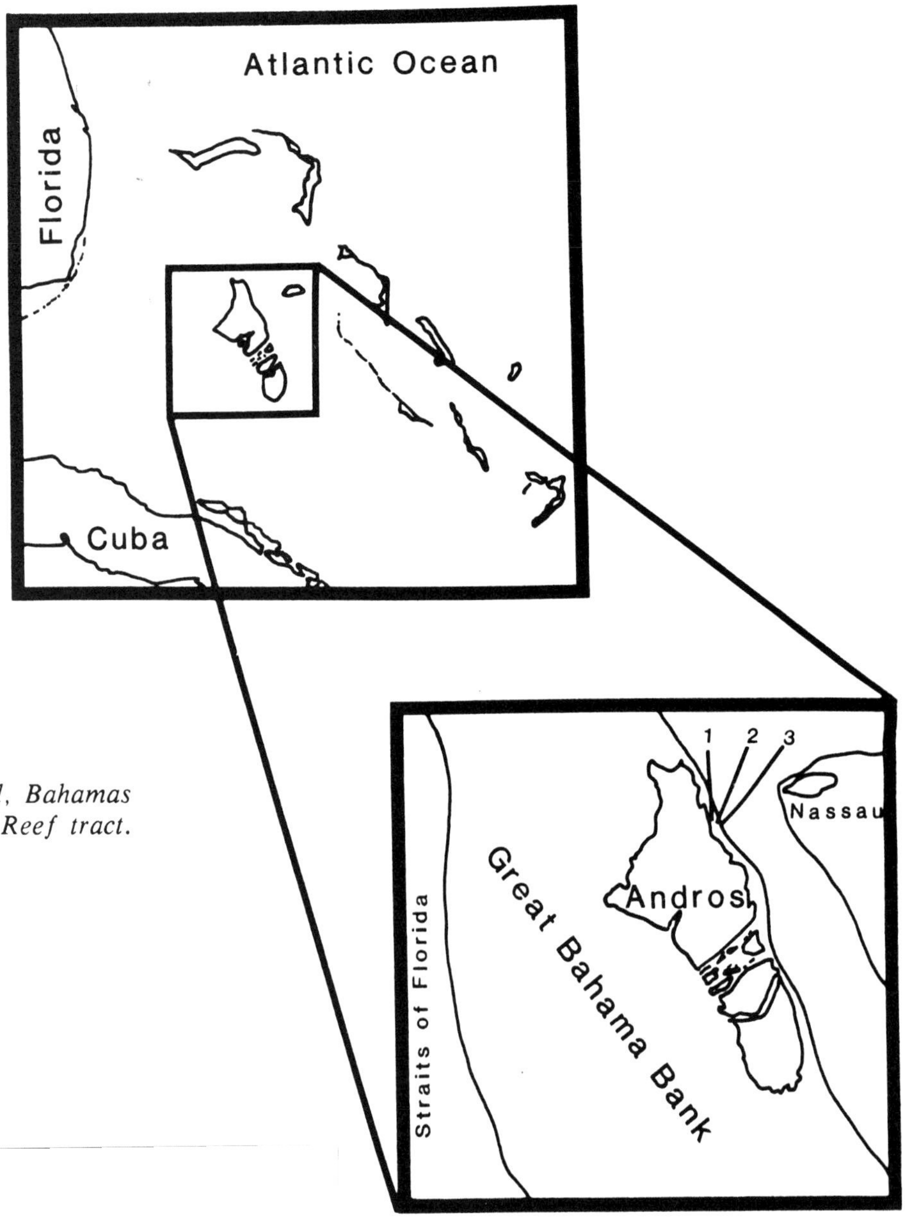

Fig. 4. Location of Andros Island, Bahamas and sample localities: 1) Backreef, 2) Reef tract. 3) Fore-reef.

Fig. 5. Echinoid skeletal elements possessing characteristic structures include spines, parts of the Aristotle's Lantern, teeth, and test plates.

tract on the east side of Andros is quite narrow and contains islands of Pleistocene carbonate in some places. Landward of the reef tract lies the backreef environment. It is shallow (water depth less than 6 m) and subject to varying amounts of environmental energy. Seaward of the reef tract the fore-reef slopes moderately steeply to a depth of 50 m where it reaches the drop-off into the Tongue-of-the-Ocean. The three subenvironments sampled are quite narrow as a result of the islands' proximity to the drop-off.

METHODS

Samples from the three subenvironments were collected while on SCUBA (Table 1). Surficial sediments were obtained as well as sediments 5-6 cm below the sediment-water interface using a scoop. Samples were air-dried, sieved and the 500-1000 micron size fraction of each was isolated for analysis. The size fractions were split on an Otto micro-splitter until sample sizes of 400-600 grains were obtained. The samples were then analyzed under a Bausch & Lomb stereozoom microscope and all echinoderm grains identified were counted and separated from the rest of the sample. Additional quantities of the 500-1000 micron size fraction of each sample were impregnated with epoxy and ground into standard thin sections. The thin sections were then point-counted following the method of Ginsburg (1956).

Recognition of Echinoderm Grains

Echinoderm grains can be recognized under a stereo microscope on the basis of their characteristic structures (Fig. 5). Skeletal elements of less common echinoderms such as ophiuroids, crinoids and starfish can also be recognized in this manner. Thin sections were made in order to take advantage of the distinctive appearance of the stereomic microstructure and the unit extinction exhibited by echinoderm calcite in cross-polarized light. It was therefore possible to identify echinoderm skeletal elements that had been mechanically, chemically or biologically reduced to grains of high-Mg calcite lacking any characteristic shape features.

RESULTS

Echinoderms are minor contributors to reef sediments ranging from .58-3.9% of each sample (Table 2). Thin section point counts and grain counts reveal similar trends (Figure 6) and echinoderm percentages obtained by point counting are consistently higher than those obtained by counting grains. Comparison of the samples obtained by digging 5-6 cm below the sediment-water interface with those obtained from the sediment surface reveals that the amount of echinoderm material remains essentially unchanged at the backreef locality, decreases slightly at the

TABLE 1, Sample Locations

Sample	Environment	Location
1SA 1D	Backreef	1 km N. of Pigeon Key and 100 m W. of Reef tract. Water depth 3 m.
6S 6D	Reef Tract	Reef tract E. of Forfar Field Station Water depth 3 m.
3S 3D	Forereef	E. of reef tract and E. of Forfar Field Station. Water depth 20 m.
4S 4D	Forereef	Staniard Rock, SE of Pigeon Key. Water depth, 20 m.

TABLE 2

THIN SECTION POINT COUNTS

Sample	# Echinoderm	Total Grains	% Echinoderm
1SA	1	172	0.58
1D	2	119	0.52
6S	6	196	3
6D	2	137	1.4
3S	3	191	1.5
3D	5	169	2.9
4S	6	153	3.9

GRAIN COUNTS

# Echinoderm	Total Grains	% Echinoderm
4	406	1
5	418	1.2
5	409	1.2
3	300	1
9	473	1.9
12	627	1.9
7	556	1.3

reef tract locality and increases slightly at the fore-reef locality.

DISCUSSION

Higher echinoderm percentages revealed in thin section as compared to grain counts result from better identification of echinoderm grains in polarized light. Under reflected light, the stereomic microstructure of echinoderm grains cannot be reliably recognized. Moreover, all sediments analyzed were obtained from the active marine phreatic zone where precipitation of aragonite and Mg-calcite occurs (Longman, 1980). Grains coated by these cements lack the characteristic appearance of stereom, making their identification under reflected light unlikely. The actions of boring organisms in the active marine phreatic may further alter a grain's appearance. The unit extinction exhibited by echinoderm calcite in thin section provides the most precise means of identification. All echinoderm skeletal elements, regardless of the degree of their destruction by mechanical, chemical or biological processes, can be reliably identified.

Frankel (1978) demonstrated that mass outbreaks and subsequent mortalities of the crown-of-thorns starfish *Acanthaster* on the Great Barrier Reef, Australia, are being preserved as layers in the sediment enriched in their ossicles.

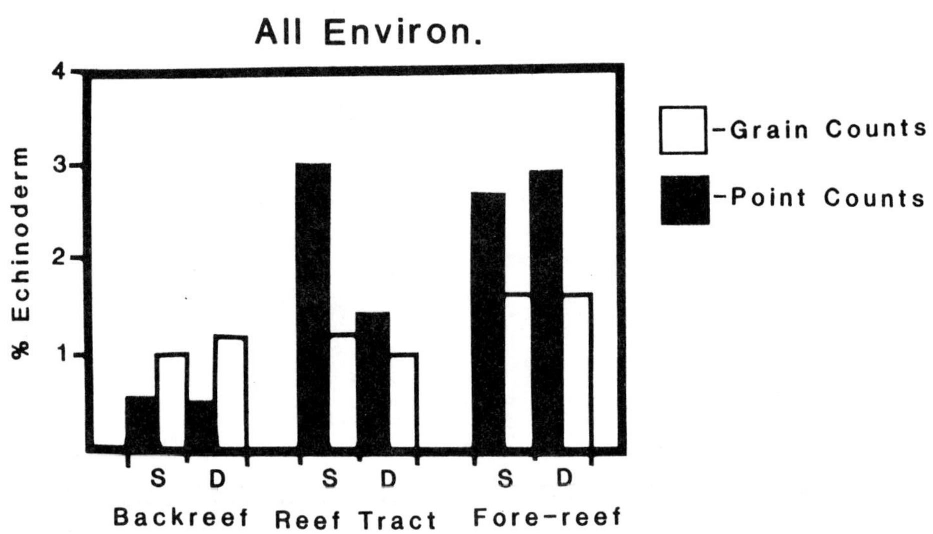

Fig. 6. Data obtained from the three reef subenvironments. In all histograms, S refers to surficial sediments and D refers to sediments collected 5-6 cm below sediment-water interface.

Comparison of surficial sediments with those 5-6 cm beneath the sediment surface does not yield a consistent pattern of "echinoderm enrichment". Sampling technique may, in part, be responsible for the lack of pattern, since sediment easily slides down the side of a hole excavated to 5 cm depth. However, recent work in the Netherlands Antilles (Greenstein & Meyer, 1985) with a sediment lift and encasing pipe also show a lack of layering of echinoderm skeletal elements suggesting that the results from Andros are accurate. The depth sampled is well within the range of burrowing organisms such as marine worms (Jacobsen, 1967; Rhoads, 1967), thalassinid shrimp (Pemberton et al., 1976) and holothurians (Mosher, 1980). Thus any layering that had occurred was rapidly disseminated as a result of bioturbation.

Comparison to Previous Work

Although no layering occurred as a result of the mass mortality, the results of this study were compared to those of constituent particle analyses conducted in the area prior to the mortality event to determine if the overall echinoderm content has increased. Because of the minor contribution echinoderm material makes to reef sediments, it is rarely reported as a separate constituent. Rather, it is included with other minor constituents in a more general category. Only if the echinoderm percentages reported in this study are greater than those of the general categories constructed by previous workers may an overall increase in echinoderm skeletal elements be suggested. Comparisons of the results of this study to those of the classic constituent particle analyses conducted in the area reveal that this is not the case (Table 3).

Kobluk & Lysenko (1984) included echinoderms as a distinct constituent in their analysis of surficial sediments collected at various depths from a fringing reef adjacent to Bonaire, Netherlands Antilles. In addition, the samples analyzed were collected between 1978-82, prior to the mass mortality event. Comparison of the results of this study to the pre-mortality values reported by these authors reveals that a very slight increase in the percentages of echinoderm material has occurred (Figure 7). A slight increase in overall echinoderm content was also reported when samples collected from the same area after the mortality event were analyzed and compared to those obtained by Kobluk & Lysenko (Greenstein & Meyer, 1985).

Recognition of the Event

It is unlikely that the slight increase of echinoderm material will constitute a recognizable signature of the mass mortality event. Liddell & Ohlhorst (1986) have suggested that the rate of recovery of the *Diadema* population will dictate the extent of the impact the mass mortality will have on the reef biota. The same may also be said with respect to the effect on reef sediments. A rapid recovery of the *Diadema* population with subsequent continued incorporation of their skeletal elements into the sediments may have the effect of masking the event entirely. If the opposite occurs, and the *Diadema* population recovers very slowly, the slight increase in echinoderm material documented by this study may then be followed by an interval of lower than normal levels as a result of the absence of *Diadema* ossicles. Thus it may not be the increase in echinoderm material resulting from the mass

Table 3. Comparison to previous work in similar environments (after Ginsburg, 1956).

THIS STUDY (2 Samples)	FLORIDA (Ginsburg, 1952) (25 Samples)	BAHAMAS (Thorp, 1937) (10 Samples)	BAHAMAS (Illing, 1954) (5 Samples)
2.2 %	9.0 %	8.0 %	4.0 %

Value reported from this study is the average percent of echinoderms obtained from sediments collected from the reef tract environment. The other values are for the percent "miscellaneous" categories of these workers. Values obtained by point counting (this study, Ginsburg, 1956) and visual estimates of percentages (Thorp, 1936, Illing, 1954).

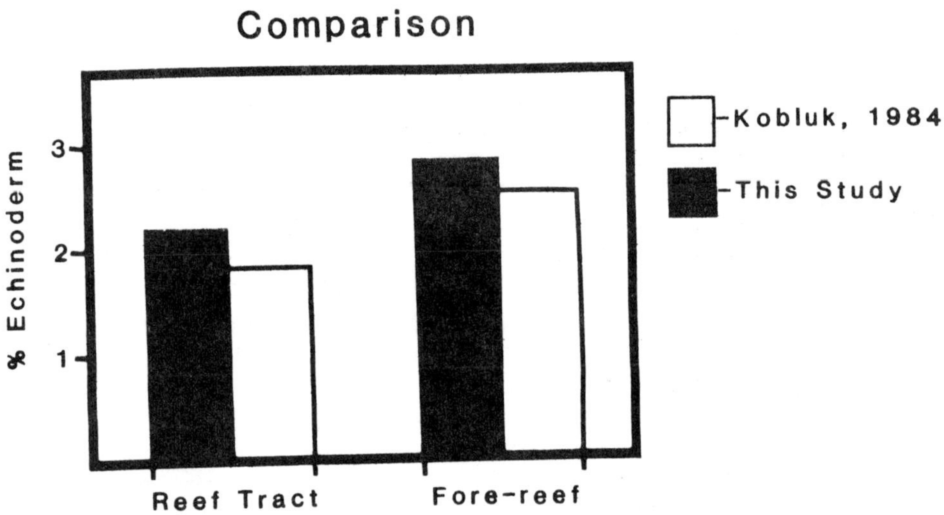

Fig. 7. Comparisons of the results of this study to those reported by Kobluk and Lysenko (1984).

mortality, but rather the duration of the interval before the population returns to normal that will determine the signature of the event in reef sediments. It should be emphasized, however, that the bioturbation processes responsible for destroying any layering of echinoderm material in the sediments are likely to be at work disseminating any "echinoderm-poor" sediments into "normal" sediments.

Application to Echinoid Fossil Record

The fact that there has been no clear signature of the mass mortality event in reef sediments underscores the rapidity with which populations of *Diadema* have been reduced to essentially unrecognizable carbonate sand grains. Although less than one year elapsed between the time the mortality was reported (October, 1983) and the reef subenvironments were sampled (May, 1984), the potential record of the event has been lost. A year is infinitesimally short geologically speaking yet there is likely to be no record of the event nor recognizable tests of *Diadema* remaining: the latter fact emphasizing that occurrences of well-preserved diadematids record truly extraordinary burial conditions.

Test plates of genera belonging to the Family Diadematidae are compounded in a manner that makes their test extremely fragile (Smith, 1984). The fragile test, coupled with an epifaunal mode of life in high energy reef environments makes preservation of these animals an exceptional event. Well-preserved echinoid assemblages must therefore be interpreted with respect to the source of the mortality as well as the processes responsible for preserving it. There are many sources of mass mortality affecting echinoids. These include disease, salinity changes, temperature changes, desiccation and storms. All of these sources will generate many dead echinoids at any given instant of geologic time. However, the mortality will only be recorded if it coincides with or is caused by an event enhancing its preservation potential. The long evolutionary history of echinoids as delineated by their fossil record should be interpreted with respect to the extraordinary conditions their preservation requires. Further work is necessary to document the taphonomic and sedimentologic aspect of fossil echinoid assemblages to ascertain the conditions of their preservation.

CONCLUSIONS

1. Thin section point counts are a more accurate means of identifying echinoderms in Recent carbonate sediments than are grain counts.
2. Echinoderms are a minor constituent of reef sediments and are therefore rarely specifically included in constituent particle analyses by recent authors. Comparison of the results of this study with those of previous workers is therefore difficult.
3. Any layering of echinoderm sediments in the three reef subenvironments sampled has been destroyed by bioturbation and/or environmental energy in less than one year.
4. The mass mortality of *Diadema antillarum* is recorded in the reef sediments adjacent to Andros by a slight increase in echinoderm content as compared to pre-mortality levels reported from Bonaire, Netherlands Antilles.
5. Preservation of the mortality event may

be contingent on the rate of recovery of the *Diadema* population, although it is unlikely that the slight increase in the echinoderm fraction documented by this study could be observed in a stratigraphic succession.

6. The reef sedimentary record may be inadequate to record this short-term but ecologically significant event.

7. Echinoid mass mortalities observed in ancient rocks are the result of extraordinary preservational events that remain poorly understood.

ACKNOWLEDGEMENTS

This study was conducted while participating in a modern carbonate field course conducted by Dr. Mark Boardman of Miami University, Ohio. We are grateful for his support. We would also like to thank Dr. Donald Gerace, Dr. John Mylroie and the staff of the CCFL Bahamian Field Station for their work organizing the symposium. Greenstein's trip to Andros was partially subsidized by the Caster Fund of the University of Cincinnati.

REFERENCES CITED

Aslin, C.J., 1968, Echinoid preservation in the Upper Estaurine Limestone of Bilsworth, Northamptonshire. Geological Magazine, 105: p. 506-518.

Bak, R.P. and van Eys, G., 1975, Predation of the sea urchin *Diadema antillarum* Philippi on living corals. Oecologia (Berlin), 20: p. 111-115.

Bak, R.P., van Eys, G., Carpay, M.J.E., and de Ruyter van Steveninck, E.D., 1984, Densities of the sea urchin *Diadema antillarum* before and after mass mortalities on coral reefs in Curacao. Mar. Ecol. Prog. Ser., 17: p. 105.

Bloos, G., 1973, Ein Fund von Seeigeln der Gattung *Diademopsis* aus dem Hettangium Worttembergs und ihr Lebenstraum. Stuttgarter Beitr. Naturk., (B) 5: p. 1-25.

Boudouresque, C.F., Nedelec, N., & Shepard, S.A., 1981, The decline of a population of the sea urchin *Paracentrotus lividus* in the bay of Port-Cros (Var, France), Rapp. Comm. Int. Medit., 27: p. 233-244.

Carpenter, R.C., 1981, Grazing by *Diadema antillarum* (Philippi) and its effects on the benthic algal community. Journal of Marine Research, 39: p. 749-765.

Cherns, L., 1983, The Hemse-elke Boundary. Facies Relationships in the Ludlow Series of Gotland Sweden. Sveriges Geologiska Undersokning, 77(2): p. 1-17.

Frankel, E., 1978, Evidence from the Great Barrier Reef of ancient *Acanthaster* aggregations in Smith, S.V., Atoll Research Bulletin, 220, Coral Reef Ecosystems. Proceedings of papers presented at the 13th Pacific Science Congress, Vancouver, p. 75-93.

Ginsburg, R.N., 1956, Environmental relationships of grain size and constituent particles in some south Florida sediments. AAPG Bulletin 40(10): p. 2384-2427.

Glynn, P.W., 1968, Mass mortalities of echinoids and other reef flat organisms coincident with midday low water exposures in Puerto Rico. Marine Biology, 1: p. 226-243.

Greenstein, Benjamin J., & Meyer, David L., 1985, Mass mortality of the West Indian echinoid *Diadema antillarum*: A natural experiment in taphonomy. Geological Society of America Abstracts with Programs 17 (7): p. 598.

Hawkins, C.M., and Lewis, J.B., 1982, Ecological energetics of the tropical sea urchin *Diadema antillarum* Philippi in Barbados, West Indies. Estuarine Coastal and Shelf Science, 15: p. 645-669.

Hendler, G., 1977, The differential effects of seasonal stress and predation on the stability of reef-flat echinoid populations. Proceedings of the 3rd International Coral Reef Symposium, Miami: p. 217-223.

Hughes, T.P., Keller, B.P., Jackson, J.B.C., and Boyle, M.J., 1985, Mass mortality of the echinoid *Diadema antillarum* Philippi in Jamaica. Bulletin of Marine Science, 36: p. 377-384.

Illing, L.V., 1954, Bahamian calcareous sands. AAPG Bulletin, 38(1): P. 1-95.

Hunter, I.G., 1977, Sediment production by *Diadema antillarum* on a Barbados fringing reef. Proceedings of the 3rd International Coral Reef Symposium. Rosenstiel School, U. Miami, Florida, 2: p. 105-110.

Jacobsen, V.H., 1967, The feeding of the lugworm *Arenicola marina*: Quantitative studies. Ophelia, 4: p. 91-109.

Kobluk, D.R., and Lysenko, M.A., 1984, Carbonate rocks and coral reefs Bonaire, Netherlands Antilles. Geological Association of Canada, Mineralogical Association of Canada Joint Annual Meeting Field Trip 13. 67 p.

Lessios, H.R., Robertson, D.R., and Cubit, J.D., 1984, Spread of *Diadema* mass mortality through the Caribbean. Science, 226: p. 335-337.

Liddell, W.D. and Ohlhorst, S.L., 1985, Changes in benthic community composition following the mass mortality of *Diadema* at Jamaica. Journal of Experimental Marine Biology and Ecology, 95: p. 271-278.

Longman, M.W., 1980, Carbonate diagenetic textures from nearsurface diagenetic environments. AAPG Bulletin, 64(4): p. 461-487.

Mosher, C., 1980, Distribution of *Holothuria arenicola* Semper in the Bahamas with observations on habitat, behavior, and feeding activity. Bulletin of Marine Science, 30: p. 112.

Muller, A.H., 1979, Fossilization (Taphonomy) <u>in</u> Robison, R.A. and Teichert, C. (eds.) Treatise on Invertebrate Paleontology A: Introduction: A2-A78.

Pearse, J.A., Costa, D.P., Yellin, M.B., and Aegian, C.R., 1977, Localized mass mortality of red sea urchins, *Strongylocentrotus franciscanus*, near Santa Cruz, California. Fisheries Bulletin of the United States, 75: p. 645-648.

Pemberton, G., Risk, M., and Buckley, D., 1976, Super shrimp: Deep bioturbation in the Strait of Canso, Nova Scotia, Science, 192: p. 790-791.

Rhoads, D.C., 1967, Biogenic reworking of intertidal and subtidal sediments in Barnstable Harbor and Buzzards Bay, Massachusetts Journal of Geology, 75: p. 461-476.

Ruhrmann, G., 1971, Riffe-nahe sedimentation palaozoischer Krinoiden-Fragmente. Neus Jahrbuch fur Geologie und Palaontologie Abhandlungen, 138: p. 56-100.

Sammarco, P.W., 1980. *Diadema* and its relation to coral spat mortality: grazing, competition, and biological disturbance. Journal of Experimental Marine Biology and Ecology, 45: p. 245-272.

Sammarco, P.W., 1982, Echinoid grazing as a structuring force in coral communities: whole reef manipulations. Journal of Experimental Marine Biology and Ecology, 61: p. 31-55.

Schafer, Wilhelm, 1972, Ecology and Paleoecology of Marine Environments. Oliver and Boyd, London, 568 p.

Scheibling, R.E., and Stephenson, R.L., 1984, Mass mortality of *Stronglylocentrotus droebachiensis* (Echinodermata: Echinoidea) off Nova Scotia Canada. Marine Biology, 78: p. 153-164.

Scoffin et al., 1980, Calcium carbonate budget of a fringing reef on the west coast of Barbados, Bull. Mar. Sci. 30, p. 475-508.

Smith, A., 1984, Echinoid Palaeobiology, Special Topics in Palaeontology; 1, George Allen & Unwin, London, 190 p.

Thorp, E.M., 1936, Calcareous shallow water marine deposits of Florida and the Bahamas. Carnegie Institute of Washington Publishing 452. Papers of the Tortugas Laboratory, 29: p. 37-119.

GIANT BAHAMIAN STROMATOLITES: A MODERN ANALOG FOR WHAT?

K. M. Griffin
S. M. Awramik
Department of Geological Sciences
Preston Cloud Research Laboratory
University of California
Santa Barbara, California 93106

ABSTRACT

With the discovery of giant subtidal stromatolites in ooid shoals of the Bahamas, a new set of environmental and biological factors important for the interpretations of stromatolites became apparent. The entirely subtidal nature of the Bahamian stromatolites along with their coarse-grained sediment, laminated and somewhat disrupted internal structure, construction by an eukaryote-prokaryote community, impressive size and variable shape along with similar subtidal stromatolites from Shark Bay, Western Australia further underscores the importance of the subtidal stromatolites for the understanding of the ancient stromatolite record. When this information is coupled with other sites where stromatolites are forming (Awramik, 1984), it becomes clear that a great variety of environmental and biological factors operate during the formation of stromatolites. Before new information on Bahamian stromatolites can be used for generalizations on fossil stromatolites, one fundamental question must be addressed: To what extent are modern marine subtidal stromatolites analogs for their ancient counterparts?

Pre-Phanerozoic stromatolites are very different from the Bahamian stromatolites. These ancient stromatolites are primarily composed of fine-grained sediment (<50 um) and often have well-defined, undisrupted laminae. Ancient stromatolites also have a greater diversity of forms than is seen in modern marine stromatolites and in many cases the forms are within definable limits of variability and hence are amenable to crude taxonomic treatment. PrePhanerozoic stromatolites were much more widespread, occupying a greater variety of environments and were not restricted by metazoans as many modern examples commonly are.

In the search for ancient counterparts to the Bahamian stromatolites, Phanerozoic stromatolites are in many ways better analogs. Paleozoic stromatolites, in particular Cambrian and Ordovician stromatolites and thrombolites are the best counterparts since they commonly have laminated to unlaminated internal fabrics, similar shapes and often occur in current swept, subtidal environments reminiscent of the Bahamian stromatolites. Most of these early Phanerozoic stromatolites, however, differ from the modern counterparts because of their micritic composition.

Therefore, the giant Bahamian stromatolites, as well as the subtidal stromatolites from Shark Bay, remain modern analogs in search of an ancient counterpart. However, by studying the processes of stromatolite formation in the Bahamian examples, and by acknowledging the similarities and differences between modern and ancient stromatolites, a better understanding of the processes important in stromatolite formation will be achieved.

INTRODUCTION

This paper is a preliminary report on some of the fundamental similarities and differences of modern and ancient marine stromatolites. It is an outgrowth of a series of discussions, reference searches, field work, hand sample and thin-section analyses of pre-Phanerozoic, Cambrian and modern stromatolites. In particular, stromatolites from the Upper Cambrian Nopah Formation in eastern California, and the modern stromatolites from Shark Bay, Western Australia and Lee Stocking Island, Bahamas were studied in great detail. The main goal is to illustrate that modern marine stromatolites are an important key to a better understanding of early Paleozoic stromatolites; however, they are not directly analogous structures. The modern stromatolites are compared to two groups of ancient stromatolites: 1)Proterozoic stromatolites and, 2) Phanerozoic stromatolites and thrombolites from the Cambrian and

	Modern calcified marine stromatolites	Pre-Phanerozoic stromatolites	M. Cambrian to E. Ordovician stromatolites and thrombolites
Environmental Occurrence	-Hypersaline bay -Ooid shoals	-Intertidal, supratidal -Subtidal platforms -Lagoonal -Reefal -Lakes, hotsprings * generally very widespread	-Patch reefs associated with active peloid and ooid shoals -Lagoonal areas in the lee of active ooid shoals -Hypersaline bay -Marginal reefs of a rimmed shelf
External Morphology	-Columns -Mounds -Irregular shaped	-Columns -Branching columns -Coniform -Mounds -Irregular shaped	-Columns -Mounds -Irregular shaped
Internal Structure	-Laminated -Unlaminated	-Well to irregularly laminated	-Laminated -Clotted (unlaminated)
Composition	-Medium-grained ooid and peloidal sand	-Very fine-grained carbonate and rare chert	-Very fine-grained carbonate and very rare chert

Fig. 1. A chart comparing several characteristics of modern, pre-Phanerozoic and early Paleozoic stromatolites.

Ordovician. The stromatolite groups are compared in terms of their environmental occurrence, external morphology, internal structure and composition (see Fig. 1).

The following definitions of the words stromatolite and thrombolite are used throughout this paper. According to Awramik and Margulis' definition, a stromatolite is an "organosedimentary structure produced by sediment trapping, binding and/or precipitation as a result of the growth and metabolic activity of microorganisms, principally cyanophytes" (in Walter, 1976, pg. 1). Aitken (1967) introduced and defined the term thrombolite as a cryptalgal structure related to stromatolites but lacking lamination and characterized by a macroscopic clotted fabric.

As stated in the first paragraph, this is only a preliminary report. The data are presented in a generalized format and future work will help to refine and test some of the broad generalizations made. Basically, we believe that by making the comparison outlined in this paper and by asking the question, "to what extent are modern stromatolites <u>different</u> from ancient stromatolites?", a better understanding of ancient stromatolites will be achieved.

PRE-PHANEROZOIC STROMATOLITES

Pre-Phanerozoic stromatolites are known to have formed under a wide variety of environmental conditions. In the marine realm, intertidal (Truswell and Eriksson, 1974), subtidal platform (Bertrand-Sarfati and Moussine-Pouchkine, 1988), deep water (Hoffman, 1974), lagoonal (Walter, et al., 1988), reefal (Hoffman, 1974), as well as supratidal environments (Grotzinger, 1986) had stromatolites. Non-marine stromatolites, although less well known and apparently less common, occurred in ephemeral lakes (Hoffman, 1976) and even in thermal springs (Walter, 1972). This is not unlike their environmental distribution today; however, subtidal marine stromatolites were most dominant in the Proterozoic.

The external morphology of pre-Phanerozoic stromatolites is extremely varied. Many of the stromatolite forms, in particular those with columnar and columnar-branching morphology, are within definable limits of variability and have therefore been assigned latinized names using an artificial binomial nomenclature. Even if the binomial nomenclature of stromatolitic forms is not widely accepted as comparable to the nomenclature of animals and plants, these "taxonomic" names are useful in discussions on the diversity in pre-Phanerozoic stromatolites. An early study on the "taxonomic" diversity of pre-Phanerozoic and early Phanerozoic stromatolites (Awramik, 1971) showed the diversity of columnar pre-Phanerozoic stromatolite "taxa" to be much higher than Phanerozoic stromatolite diversity.

The internal structure and composition of most pre-Phanerozoic stromatolites is that of well-defined thin laminae (roughly 100 um thick) composed of fine-grained (<50 um) carbonate at times replaced silica. These well- defined thin laminae of pre-Phanerozoic stromatolites differ from most early Phanerozoic and modern stromatolites.

CAMBRIAN AND ORDOVICIAN STROMATOLITES AND THROMBOLITES

As with pre-Phanerozoic stromatolites, Cambrian and Ordovician stromatolites and thrombolites are known to have formed under a variety of environmental conditions. They are often described as having formed in high energy environments such as patch reefs associated with active peloid and ooid shoals (Kennard, 1981; Moshier, 1986; Rees, 1986; and Griffin, 1988), and as marginal reefs on a rimmed shelf (Demicco, 1985). Thrombolites and stromatolites have also been described as forming in intertidal to very shallow subtidal waters (Sepkoski, 1977) and in the lagoonal areas behind active ooid shoals (Owen, 1973). Generally these Upper Cambrian and Lower Ordovician microbial structures were very widespread in terms of their paleoenvironmental and geographic distribution much like the pre--Phanerozoic stromatolites, however, at reduced numbers of occurrences.

The external morphology of early Paleozoic stromatolites and thrombolites does not show the diversity of form seen in pre-Phanerozoic stromatolites. Proterozoic stromatolites range in shape from domical to columnar, with columnar-branching and conical stromatolites being abundant and distinctive. The Cambrian and Ordovician stromatolite forms range from domical, to columns with minor branching. The branching, however, is not as pronounced or elaborate as found in Proterozoic stromatolites. The thrombolites are also columnar to mound-shaped in external form with no apparent branching. There are also numerous thrombolite forms which are extremely irregular and highly variable in shape. They range from small loaf-shaped lumps to large elongated Omega shaped structures in plan view, that preclude a generalized description.

The internal structure of Cambrian and Ordovician stromatolites ranges from coarse to irregularly laminated and the thrombolites have the diagnostic clotted, non-laminated fabric. In contrast to pre-Phanerozoic stromatolites which have well defined, thin laminae, the Upper Cambrian Nopah Formation stromatolites have a mixture of laminae that are generally >1mm thick with some that are not well defined or laminated. Preservation of the clotted thrombolitic internal structure is extremely variable in the Nopah Formation in part due to extensive dolomite recrystallization. Better preservation occurs in the Upper Cambrian of west Texas, in the San Saba Member of the Wilberns Formation where feeding-burrow traces can be observed which disrupt the laminar microstructure (Ruppel and Kerans, 1987). No burrowed laminae were observed in the Nopah Formation stromatolites.

Many Cambrian and Ordovician stromatolites and thrombolites are composed of micrite. In both the Nopah and the Wilberns stromatolites up to 85% of the structures is micrite in contrast to the coarse grainstones and packstones of the adjacent channel-fill sediments. Upper Cambrian stromatolites examined from the Hoyt Limestone, New York, have a significant component of ooids and peloids.

MODERN MARINE CALCIFIED STROMATOLITES

In contrast to the ancient stromatolite examples, large modern subtidal stromatolites presently are known from only three locations. Dill et al (1986) described the giant subtidal marine stromatolites in an active ooid shoal of the Exuma Keys. Similar stromatolites have been described by Dravis (1983) from Eleuthera in the Bahamas. The classic locality of modern marine stromatolites is the hypersaline marine inlet, Hamelin Pool, of Shark Bay in Western Australia (Playford and Cockbain, 1976). Although it is probable that more sites will be found in the future, it is safe to conclude that modern subtidal stromatolites are rare in today's oceans compared to the early Paleozoic and older times.

The external morphology of the modern stromatolites include columns, mounds and irregular "arabesque-shaped" structures (Playford and Cockbain, 1976; Dill, et al., 1986). The overall mound-shaped morphology is often elongated parallel to and over-steepened toward the current direction. Similar over steepened shapes can be found in the ancient record.

The internal structure and composition of the stromatolites from Lee Stocking Island and Shark Bay are similar. Greater than 80% of the stromatolite sediment is 0.25mm in diameter at Lee Stocking Island. The large stromatolites studied thus far are composed of small laminated columns up to 3cm wide, that are often enveloped by single laminae that cover the entire stromatolite. The laminae in certain areas of the stromatolite are disrupted by boring clams, sponges and various other escavating organisms.

The overall internal structure of these modern stromatolites is more similar to early Phanerozoic forms primarily based on their similarity of disrupted internal fabrics as exemplified by the Wilberns Formation stromatolites (Ruppel and Kerans, 1987). One important difference, however, is that some early Phanerozoic stromatolites are composed of micrite and others coarse ooid and peloid sand whereas all the presently known large modern subtidal stromatolites are composed of coarse-grained ooid sands.

DISCUSSION

The stromatolites of ooid shoals in the Bahamas and those in Shark Bay, Western Australia, are modern stromatolites that differ from pre-Phanerozoic and early Phanerozoic stromatolites in the following ways:

1. The modern stromatolites are more restricted in their environmental occurrence than the ancient stromatolites. This restriction is probably due to the destructive effects of grazing and burrowing organisms (Garrett, 1970; Awramik, 1971) that inhabit shallow water marine environments and/or the competitive exclusion by reef-building and skeleton-forming organisms that have diversified since the early Phanerozoic (Monty, 1973). The near total exclusion of modern marine stromatolites from shallow environments is not surprising when viewed from an evolutionary perspective. Dramatic changes in the ecology of shallow marine communities has occurred since the early Phanerozoic due to the diversification of marine invertebrates though time (Bambach, 1983, 1985; Sepkoski, 1982; and Sepkoski and Miller, 1985).

2. An external morphologic similarity exists between modern and early Phanerozoic stromatolites as well as with some pre-Phanerozoic stromatolites. What this actually means about the biological or environmental similarities among the stromatolite groups is not understood. The debate on whether stromatolite external morphology reflects primarily environmental or biological differences is still unresolved (see Golubic, in press; for a recent review of the subject) and no doubt both factors play significant roles. Insight into the answers to many of the questions could be resolved through careful study of the Lee Stocking Island stromatolites. By documenting: 1) the microbiology, 2) the growth and cementation rate of the living stromatolites, 3) the degree of encrustation, 4) the physical and biological erosion, 5) the sediments deposited, and 6) the environmental setting, some important parameters that influence stromatolite morphogenesis may be resolved.

3. The internal structure of modern stromatolites is superficially more similar to early Phanerozoic stromatolites than to those from the pre-Phanerozoic. This similarity lies in the coarse often disrupted laminae and in the unlaminated areas of modern and Phanerozoic stromatolites. The voids or fenestrae in the modern Bahamian stromatolites are interpreted to be caused by the degradation of organic matter (Dill et al., 1986). Other unlaminated regions are due to escavation by metazoans and sponges which is reminiscent of the Upper Cambrian Wilberns Formation where extensive feeding-burrow traces disrupted the laminar microstructure.

4. The internal sedimentological composition of the modern stromatolites is very different from pre-Phanerozoic examples and also from many of the early Phanerozoic stromatolites and thrombolites. The coarse grained nature of the modern stromatolites may reflect a fundamental difference in the sediment available and in the microbial community responsible for the accretion of the the stromatolites (Awramik and Riding, 1988). Awramik and Riding (1988) postulated that the modern stromatolites are able to accrete coarse-grained sediment because the organisms that are trapping the sediment are larger; modern stromatolites have a greater component of eukaryotic algae which are larger than the prokaryote-dominated communities of the

pre-Phanerozoic and therefore modern stromatolite communities can trap a larger size class of sediment.

In summary, this comparison between modern and ancient stromatolites demonstrates that modern marine stromatolites are not perfect analogs for their ancient counterparts. Some of the differences are due to the evolution and diversification of grazing and burrowing metazoans (Garrett, 1970; Awramik, 1971) into shallow marine communities where pre-Phanerozoic stromatolites and early Phanerozoic stromatolites and thrombolites flourished. Preliminary investigations on the differences between the internal structure and composition of modern and pre-Phanerozoic stromatolites and some early Phanerozoic stromatolites indicate a sedimentological difference and hence a possible difference in the microbial communities responsible for their construction (Awramik and Riding, 1988). Other similarities and differences between modern and ancient stromatolites such as external morphology remain unresolved. Continued research on the modern marine stromatolites from the Bahamas and Shark Bay, Western Australia may lead to a better understanding of the physical and biological processes important in stromatolite accretion which will undoubtedly bring insight to fundamental questions such as stromatolite morphogenesis. Also continued work on modern lake stromatolites such as those found on San Salvador Island, (see the paper by Neumann et al., this volume) will add to a better understanding of modern and ancient stromatolites.

CONCLUSIONS

Unlike pre-Phanerozoic or early Phanerozoic stromatolites and thrombolites, modern calcified marine stromatolites are rare in today's oceans and are restricted in their environmental occurrence.

Modern marine stromatolites have external morphologies similar to Phanerozoic stromatolites from Cambrian to Ordovician time and do not show the morphological complexity of pre-Phanerozoic stromatolites.

The internal structure of modern calcified stromatolites is coarsely laminated to unlaminated which is similar to Cambrian to Ordovician stromatolites' internal structure but differs from the pre-Phanerozoic stromatolites which lack disrupted to clotted internal fabrics.

Modern calcified marine stromatolites are composed of coarse-grained sand, dominantly ooid sand, whereas most pre-Phanerozoic and early Phanerozoic stromatolites are composed dominantly of fine-grained material.

REFERENCES CITED

Aitken, J.D. 1967, Classification and environmental significance of cryptalgal limestones and dolomites with illustrations from the Cambrian and Ordovician of Southwestern Alberta: Journal of Sedimentary Petrology, v. 37, n. 4, p. 1163-1178.

Awramik, S.M., 1971, Precambrian columnar stromatolite diversity: Reflection of metazoan appearance: Science, v. 174, p. 825-827.

Awramik, S.M., 1984, Ancient stromatolites and microbial mats, in Cohen, Y., Castenholtz R.W., and Halvorson, H.O., ed., Microbial Mats, Stromatolites: Alan Liss, Inc., New York, pg. 1-22.

Awramik, S.M., and Riding, R., 1988, Role of algal eukaryotes in subtidal columnar stromatolite formation: Proceedings of the National Academy of Science USA, v. 85, pg. 1327- 1329.

Bambach, R.K., 1983, Ecospace utilization and guilds in marine communities through the Phanerozoic, in Tevesz, M.J.S., and McCall, P.M., eds., Biotic Interactions in Recent and Fossil Benthic Communities: Plenum, New York, pg. 719- 746.

Bambach, R.K., 1985, Diversity and adaptive variety: The ecology of diversification in marine faunas through the Phanerozoic, in Valentine, J.W., ed., Phanerozoic Diversity Patterns: Profiles in Macroevolution: Princeton University Press, Princeton, New Jersey, pg. 191-254.

Bertrand-Sarfati, J., and Moussine-Pouchkine, A., 1988, Is cratonic sedimentation consistent with available models: An example from the Upper Proterozoic of the West African craton: Sedimentary Geology, v. 58, pg. 255-276.

Dill, R.F., Shinn, E.A., Jones, A.T., Kelly, K., and Steinen, R.P., 1986, Giant subtidal stromatolites forming in normal salinity water: Nature, v. 234, p. 55-58.

Dravis, J.J., 1983, Hardened subtidal stromatolites, Bahamas: Science, v. 219, p. 385-386.

Garrett, P., 1970, Phanerozoic stromatolites: Noncompetitive ecologic restriction by grazing and burrowing animals: Science, v. 169, p. 171-173.

Golubic, S., Modern Stromatolites-A progress report, in Riding, R., ed., Fossil Algae, Second International Symposium, in press.

Griffin, K.M., 1988, Sedimentology and paleontology of thrombolites and stromatolites of the Upper Cambrian, Nopah Formation and their modern analog on Lee Stocking Island, Bahamas: Unpublished Master's Thesis, University of California, Santa Barbara, 147 pg.

Grotzinger, J.P., 1986, Evolution of Early Proterozoic passive-margin carbonate platform, Rocknest Formation, Wopmay Orogen, Northwest Territories, Canada: Journal of Sedimentary Petrology, v. 56, n. 6, pg. 831-847.

Hoffman, P.F., 1974, Shallow and deepwater stromatolites in Lower Proterozoic platform-to basin facies changes, Great Slave Lake, Canada: American Association of Petroleum Geologists, v. 58, pg. 856-867.

Hoffman, P.F., 1976, Environmental diversity of Middle Precambrian stromatolites: in Walter, M.R., ed., Stromatolites: Elsevier, Amsterdam, pg. 261-271.

Kennard, J.M., 1981, The Arrinthrunga Formation: Upper Cambrian epeiric carbonates in the Georgina Basin, central Australia: Bureau of Mineral Resources, Geology Geophysics Bulletin, n. 211, p. 1-61.

Moshier, S.O., 1986, Carbonate platform sedimentology, Upper Cambrian Richland Formation, Lebanon Valley, Pennsylvania: Journal Sedimentary Petrology, v. 56, n. 2, p. 204-216.

Monty, C.L.V., 1973, Precambrian background and Phanerozoic history of stromatolitic communities, an overview: Ann. Soc. Geol. Belgium, v. 96, pg. 585-624.

Owen, R.W., 1973, Red Sea algal sediments and the Hoyt Limestone of New York: A comparison of Recent and Cabrian algal deposition (Master's thesis): Rensselaer Polytechnic Institute, Troy, NY, 121 p.

Rees, M.N., 1986, A fault-controlled trough through a carbonate platform: The Middle Cambrian House Range embayment: Geological Society America Bulletin, v. 97, p. 1054-1069.

Ruppel, S.C., and Kerans, C., 1987, Paleozoic buildups and associated facies, Llano Uplift, central Texas: Society of Economic Paleontologists and Mineralogists, Midyear Meeting Field Trip Guidebook, Austin Geological Society Guidebook n. 10, 33 pg.

Playford, P.E., and Cockbain, A.E., 1976, Modern algal stromatolites at Hamelin Pool, a hypersaline barred basin in Shark Bay, Western Australia, in Walter, M.R., ed., Stromatolites: Elsevier, Amsterdam, pg. 389-411.

Sepkoski, J.J., 1975, Depositional environments and fossil assemblages on the Cambrian shelf: An example from the Dresbachian of the Northern Rocky Mountains: Geological Society of America, Abstracts with Programs, v. 7, n. 7, pg. 1264-1265.

Sepkoski, J.J., 1982, A compendium of fossil marine families: Contributions in Biology and Geology, Milwaukee Public Museum, v. 51, 25 pg.

Sepkoski, J.J., and Miller, A.I., 1985, Evolutionary faunas and the distribution of Paleozoic benthic communities in space and time, in Valentine, J.W., ed., Phanerozoic Diversity Patterns: Profiles in Macroevolution: Princeton University Press, Princeton, New Jersey, pg. 153-190.

Truswell, J.F., and Eriksson, K.A, 1974, Stromatolitic associations and their palaeo-environmental significance; a re-appraisal of a lower Proterozoic locality from the northern Cape Province, South Africa: Sedimentary Geology, v. 10, n. 1., pg. 1-23.

Walter, M.R., 1972, A hot-spring analog for the depositional environment of Precambrian Iron Formations of the Lake Superior Region:

Economic Geology, v. 67, pg. 965-980.

Walter, M.R., 1976, Stromatolites: Elsevier, New York, 790 pg.

Walter, M.R., Krylov, I.N., Muir, M.D., 1988, Stromatolites from Middles and Late Proterozoic sequences in the MacArthur and Georgina Basins and the Mount Isa Province, Australia: Alcheringa, v. 12, pg. 79-109.

PRELIMINARY REPORT ON EVOLUTION OF THE COAST ALONG BLACKWOOD BAY, SAN SALVADOR ISLAND, BAHAMAS

Daniel E. Kramer
and
Mario V. Caputo
Department of Geology and Geography
Mississippi State University
Mississippi State, Mississippi 39762

ABSTRACT

The coast of Blackwood Bay on the south side of San Salvador Island is a low-energy carbonate shoreline, the history of which can be reconstructed from five laterally adjacent sedimentologic and geomorphic shore zones. These zones are recognized in a seaward succession: coastal ponds; gravel ridge; sand bench; upper intertidal mangrove trees and gravel; and Pleistocene bedrock. The bay is a platform margin, lagoon-tidal flat complex which is underlain by Pleistocene bedrock. The bay is restricted to the south by a partially destroyed fringing reef with a reef rubble rampart and to the southeast by cliffs and terraces of the Pleistocene Grotto Beach Formation. Abundant cobble-and boulder-sized gravel, composed of coral and rock fragments located in the upper intertidal-shore zone and backshore may represent the cumulative effects of storms striking the coast during the Holocene. The sand bench and mangrove fringing the coast may be the product of quiet interstorm periods resulting in the progradation of the shoreline and filling of the bay.

INTRODUCTION

Blackwood Bay is located on the southern side of San Salvador Island and is part of a small embayment on the eastern side of a larger bay known as French Bay (Index Map 1). Physiographically, the coast of Blackwood Bay is situated between seacliffs to the southeast and a beach-dune complex to the northwest (Fig. 1). The bay is enclosed to the south by a fringing reef and a reef rubble rampart (Pace, 1987a; 1987b; Pace and others, 1988). Pleistocene bedrock, capped by a paleosol, crops out as a rocky shoal and dampens wave energy approaching from the Atlantic Ocean to the south-southwest. The southern section of the coast is a rocky

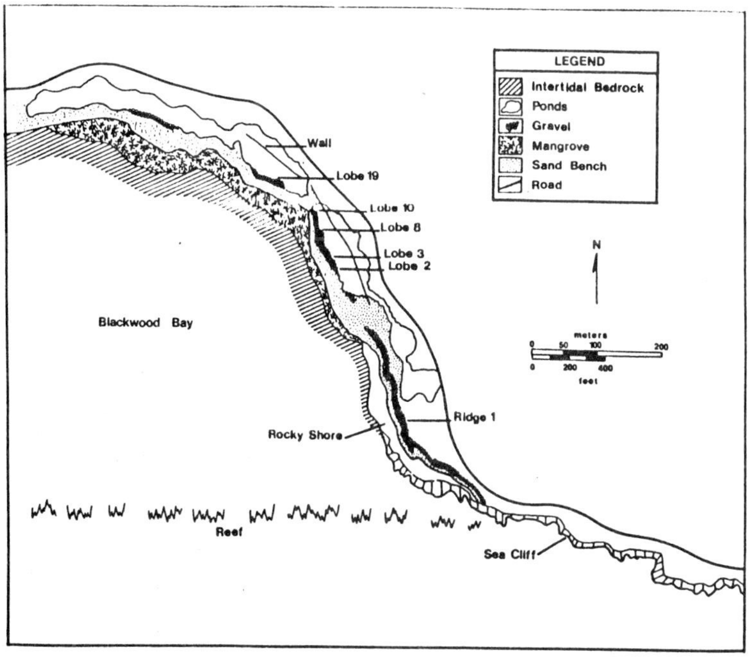

Fig. 1. Map of Blackwood Bay coastline showing shore zones, morphologic features, and sample locations discussed in Fig. 6.

shoreline without mangrove vegetation while the northern section is a partial sandy beach overgrown by red mangrove (*Rhizophora mangle*), and black mangrove trees (*Avicennia germinans*) as well as a variety of terrestrial plants.

The idea for this study evolved from descriptions of lobe-shaped carbonate gravel deposits in Pace (1987a; 1987b) and Pace and others (1988). Since the completion of field reconnaissance and mapping, the present study has grown to include laterally adjacent shore zones and sedimentary deposits. The purpose of this report is to describe the physical and biogenic sedimentary features, show the areal distribution, and interpret the sedimentary and geomorphic history of the shore zones and deposits as they relate to the overall geologic evolution of San Salvador Island.

SEDIMENTARY GEOMORPHOLOGY

Coastal Features

Field work was conducted in January, June and early July, 1988, and included mapping of coastal features and deposits along the coast of Blackwood Bay. Mapping has shown that in the present intertidal to supratidal environments there are five distinct coastal zones characterized by the following features: 1. a shoal of Pleistocene bedrock exposed in the intertidal zone; 2. a thin deposit of gravel associated with a belt of mangrove trees in the upper intertidal zone; 3. a sand body representing a beach ridge and forming a treeless sand corridor and an elevated backshore bench; 4. a discontinuous backshore gravel ridge; and 5. a series of ponds or low-lying areas in the far backshore with associated mangrove vegetation (Fig. 1).

Pleistocene Bedrock

The outcrop pattern of the Pleistocene bedrock merits discussion because it may have affected the sedimentologic and geomorphic development of Blackwood Bay. The extreme eastern edge of Blackwood Bay is characterized by low bedrock cliffs and terraces of the Pleistocene Grotto Beach Formation (Carew and Mylroie, 1985). These cliffs and terraces have been modified by erosion into small embayments and partially collapsed seacaves, the floors of which are lined with pebble-to cobble-size clasts of coral, bedrock, and shells of the mollusc, *Strombus gigas*. The erosional pattern grades into the high seacliffs which comprise the Gulf area on the southern side of the island.

Bedrock underlies the floor of the bay itself and emerges as a mantle of elevated rock shoals extending 2 to 65 meters out from the shoreline. An irregular bedrock surface is exposed in the intertidal zone and has been modified by bioerosion, wave erosion, and phytokarst processes (Fig. 2). A series of joints has disrupted the upper surface of bedrock in a pattern that is parallel and perpendicular to bedding. Mangrove roots and pneumatophores have further wedged apart square- and rectangular-shaped slabs of bedrock along joints. Consequently, rock slabs of similar shape inherited from the joint patterns have been transported higher onto the upper intertidal surface of barren bedrock and mangrove trees. Other gravel reworked with the slabs include smaller clasts of coral, bedrock fragments, and *Strombus gigas* shells.

Intertidal Mangroves and Gravel

Shoreward and adjacent to the zone of exposed bedrock is a tract of mangrove trees characterized by the black mangrove (*Avicennia germinans*) mixed with the red mangrove (*Rhizophora mangle*). They locally occupy upper intertidal and supratidal levels and may extend slightly into the sand bench zone. The mangrove swamp subcommunity on San Salvador Island consists of red mangrove, black mangrove, white mangrove (*Laguncularia racemosa*), and buttonwood (*Conocarpus erectus*) (Smith, 1985). Red and black mangrove trees are dominant along the coastal portions of Blackwood Bay, while buttonwood and poisonwood characterize the landward portions of the sand bench.

Square-and rectangular-shaped slabs of bedrock reworked from the exposures of jointed bedrock in the lower intertidal zone have been emplaced among the mangroves (Fig. 3). The slabs and other pieces of coral, shell, and bedrock debris form a layer of locally imbricated cobble- and boulder-size clasts resting on a surface of pneumatophores, sand, and bedrock. The shoot-like pneumatophores of black mangrove trees have grown out from under and in between gravel clasts in some places, partially prop-up local gravel slabs, or appear deformed and crushed underneath gravel slabs in other places. Networks of red mangrove prop roots encage local gravel slabs or bluntly terminate against them in the

Fig. 2. Intertidal bedrock shoal bordering the coast. Note joints, bedrock slabs, and black mangrove pneumatophores.

Fig. 3. Intertidal mangrove trees and gravel. Black mangrove trees are growing in front of and upon sand bench.

Fig. 4. Sand bench showing seaward most swash ridge. Swash debris is composed primarily of Thalassia *blades.*

Fig. 5. Gravel ridge with pronounced lobate edge. Dry coastal pond in background.

attempt to emplant themselves in the substrate.

Two large clusters of red mangrove up to 30 meters in width are present along the northern section of the coast. Dense branches and prop roots act as a baffle which has entrapped a layer of dark brown, organic-rich mud up to 30 centimeters thick as well as flotsam brought in by wave and storm currents. A whole assortment of drift and litter has been washed into the mangrove tract. Ship roping and netting is found entangled among mangrove pneumatophores, prop roots, and branches. Plastic, paper, styrofoam, wood and glass articles have been likewise trapped within the mangrove forest. Local plastic wrapping and bottle labels show dates which are within five years old.

Sand Bench

Landward of the mangrove trees, a tract of sand has been built up to form a backshore berm or bench 2 to 35 meters wide. It is overgrown with grass and locally with non-halophytic terrestrial trees. The surface of the bench is characterized by ridge and swale topography where the ridges are composed of linear heaps of *Thalassia*, sargassum, other vegetation and swash debris, which extend parallel along the shoreline (Fig. 4). The youngest, highest swash ridge is 0.5 meters in amplitude and has been built up at the seaward edge of the sand bench; at least 3 to 4 lower amplitude, parallel ridges had been built up behind the youngest one. The crest of the youngest, seaward-most berm is overgrown predominantly by black mangrove trees.

The width of the sand bench may be related to the intertidal bedrock exposures. The bench is wider where the Pleistocene bedrock shoal projects farthest into the bay and narrower where bedrock reentrants occur. Sediment-probing of the most recent, seaward-facing beach ridge indicates that the sediment is 130 to 55 centimenters thick. Thickness of the remainder of the sand bench ranges from 115 to 20 centimeters. In addition to the swash debris, the beach ridges are composed of fine-to medium- grained bioclastic sand which is similar to the sand in the swales. The bench forms a treeless corridor between mangroves in the intertidal zone and mangroves and other trees bordering the ponds and can be seen from the air.

Gravel Ridge

The sand bench terminates landward against gravel deposits that form a topographically elevated ridge which is discontinuous but traceable along the entire shoreline (Fig. 5). The ridge runs parallel to the coast approximately 5 to 20 meters inland of the high tide mark. The gravel is compositionally similar to the gravel located in the upper intertidal zone along the shore. The ridge runs parallel to the coast in a northern direction as a semi-continuous feature for 825 meters. Areas where the ridge is absent are covered by sand bench deposits. The ridge continues to the south for an additional 300 meters as a feature that extends to the seacliffs. The southern section of the gravel ridge is 20 meters wide and 3 to 4 meters high. The southern ridge has steep seaward and landward faces and a subhorizontal top. Along the northern section of the coast, the gravel deposit is characterized by a ridge slightly lower in elevation (1 to 2 meters) compared to the southern ridge; and by a pronounced lobate landward edge. The landward edge forms steep-sided noses or leeside surfaces facing northeast. The lobate edges of the ridge coalesce to form a mega-lobe (Pace, 1987b). The seaward edge of the ridge terminates against the sand bench along a relatively straight margin that follows the shape of the bay. The individual lobes are 4 to 15 meters in length (parallel to shore) and 4 to 9 meters wide (perpendicular to shore). In plan view the lobes are tongue-shaped and elongated landward. The gravel deposit is composed of coral clasts, rock fragments, *Strombus gigas* shells, and local bivalve shells (Fig. 6). The predominant shape of the rock fragments is subangular blocks and slabs. Size of rock fragments of the ridge varies between 5 and 27 centimeters along the longest dimension. The length of the short axis varies between 0.5 and 8 centimeters and reflects the inherited slab or blocky shape of the fragment. Isolated blocks greater that 1 meter in size are present, but are limited to the surface of the ridge. Surface rock fragments are slightly to moderately weathered and are dark grey to black. Gravel underneath the surface layer is yellowish white. The coral clasts on the surface of the ridge have retained their original biologic shape as palmate, branched, rounded-, and finger-shaped clasts. Surface coral clasts range in size from 5 to 23 centimeters. These coral clasts are slightly weathered and identifiable at the species level.

Based on the analysis of 253 surface and subsurface gravel clasts from various ridge locations, the coral and rock fraction out of the total gravel may be divided compositionally into the following (Fig. 6). Extremely abraded, unidentified

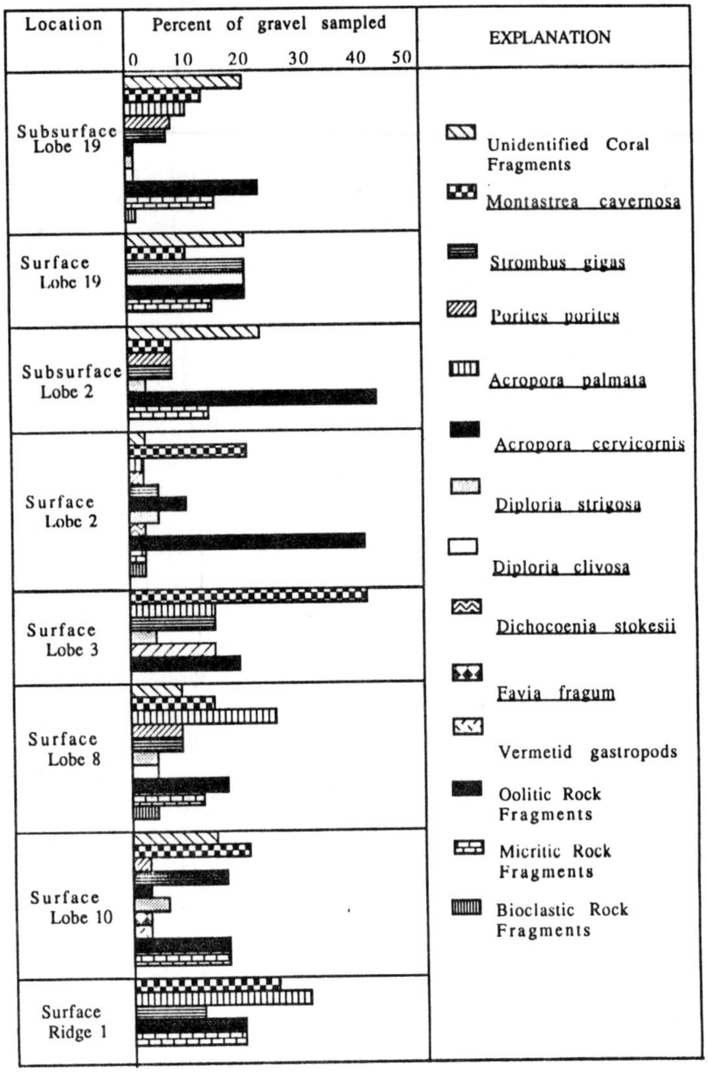

Fig. 6. Composition of gravel ridge deposits based on analysis of 253 surface and subsurface gravel clasts. Sample locations shown in Fig. 1.

coral fragments are present and generally constitute 20% of the gravel material sampled. The percentage of coral clasts ranges between 23 and 68%. The abundant coral species sampled from the ridge surface include: *Montastrea cavernosa*, and *Acropora palmata; Acropora cervicornis, Porites porites*, were fairly common while *Diploria strigosa, Diploria clivosa, Favium fragum*, and *Dichocoenia stokesii* were rare. *Strombus gigas* shells ranged between 30% to none. The occurrence of rock fragments varies between 18 and 55%.

The gravel deposits are characterized by three to four sets of pebble- and small cobble-size material (3-15 cm, long axis), alternating with beds of cobble size material (20 to 30 cm, long axis). Contacts between upper and lower beds are abrupt. The lower pebbly and small cobble beds are 10 to 20 centimeters thick while the upper cobbly beds are 20 to 30 centimeters thick. Lower beds are compositionally similar to the upper beds, although abraded coral fragments are more abundant in the lower beds. Variation in gravel size is most extreme between the uppermost pebble-to small cobble-size bed and the surface material. The gravel on the surface of the ridge is the largest material observed. Total thickness of the gravel deposit is between 90 and 120 centimeters. The gravel is clast supported with sand as a matrix in all but the upper 50 centimeters of material.

Three *Strombus gigas* shells were found 40 to 60 centimeters below the surface of a lobate portion of the gravel ridge. The shells have abraded spires and small circular openings made near the apex of the shell. These features resemble those found on *Strombus gigas* shells that were opened by early inhabitants of the island (Hoffman, personal communication). They result from the use of one shell as a tool to open another to gain access to the flesh. The practice is still occasionally used on San Salvador Island, but is more commonly associated with the early pre-Columbian inhabitants.

A small rectangular structure 4x2x1 meters in size was discovered on top of the southern gravel ridge. The upper layer of cobble- to boulder-size gravel of the ridge was removed and used to build the walls of the structure which rest on an underlying finer gravel bed. The structure is similar to buildings and walls built during an interval of San Salvador Island history known as the Loyalist Period (Sandra Riley, personal communication). The Loyalist Period represents a time of European settlement by American Colonists who were loyal to the British Crown and who were granted lands on San Salvador Island during and after the American Revolution (Gerace, 1982). Most of the land was granted in the eighteenth century and an estimated 2,000 persons occupied the island by the end of the 1780's (Gerace, 1982). A very thick base of a glass bottle was also discovered on the surface of the ridge. The bottle base was from a type of bottle termed a "kick-up" possibly used between the late 1700's and early 1800's (Hoffman, personal communication).

Coastal Ponds

Three shore-parallel ponds are locally contiguous with the landward edge of the gravel ridge. These ponds are formed in the swale developed between the gravel ridge and rising Pleistocene bedrock of the island. The ponds are seasonally flooded with rain water and possibly by marine waters during storms. Salinity of the deepest pond, located at the south-central part of the coast, was 17 parts per thousand as measured during an abnormally wet month of June, 1988. The maximum depth of the pond is 0.8 to 1.0 meters and is filled with water year round. A core collected from the pond 1.5 meters east of the ridge is characterized by a basal peat layer 45 centimeters thick containing abundant gastropods and a cobble size grainstone fragment near the base of the unit (Pace, 1987b). The remainder of the core (153 cm) is characterized by alternating thick beds of ostracode-rich mud (12 to 57 cm), and thinner beds (4 to 9 cm) of fine-to coarse-grained bioclastic sand containing coral, gastropod, and bivalve fragments. A sandy fan-shaped feature, visible from the air, is also present in the southeast corner of this pond. In shape and lateral extent, the sand body resembles a washover fan. Sediment-probing of the south-central pond and another large pond to the north, indicates that sediment in both ponds is 220 centimeters thick near the seaward edge adjacent to the gravel ridge and thins-out to less than 5 centimeters at the mainland edge of the pond. Features associated with the northern pond resemble partially filled tidal channels which provide communication between the pond and the lagoon. These areas are slightly lower in elevation and lined with red mangrove trees that stretch from the pond to the lagoon.

Other low-lying coastal areas are seasonally flooded to form shallow ephemeral ponds. Bottom sediment in these areas forms a thin veneer of fine sand and mud up to 20 centimeters thick. Red and black mangrove trees border the edges of the ponds and adjacent low-lying areas. The landward margin of the ponds and low areas of the coast grades into normal inland portions of the island characterized by lithified Pleistocene aeolian dune ridges and interdune swales.

A stone wall approximately 1 meter in height bisects the ponds; the wall may represent a boundary marker built during the Loyalist Period. The material comprising the wall is different in appearance from the gravel of the ridge and was possibly brought from another source.

DISCUSSION OF THE COASTAL EVOLUTION OF BLACKWOOD BAY

The geomorphic and depositional history of the coastal features along Blackwood Bay may be related to infrequent, severe storm events causing erosional retreat of the bedrock seacliffs, terraces, and shoals that border the coastline, destruction of the fringing reef at the southern edge of Blackwood Bay, and landwand transport of material produced during the storm. Interstorm periods are characterized by low energy tidal conditions and the deposition of sand and mud as a prograding shoreline and general filling of the bay.

Pleistocene Bedrock

Geomorphically, tropical limestone shores differ from siliceous shores because of the soluble nature of the limestone, and the destructive and constructive nature of biogenic processes. The shape, extent, and general nature of the exposed bedrock surface bordering the coast of Blackwood

Bay is controlled by biological erosion followed by reworking by storm currents. Mutler (1969) outlined the various stages of bedrock erosion and cliff line retreat along tropical limestone shores and demonstrated that exposed bedrock is undermined by bioerosion followed by storm waves breaking the remaining weakened rock apart. The erosional modification of the cliffs and terraces at the extreme eastern edge of Blackwood Bay, has created embayments, and collapsed sea caves that follow the pattern expressed by Mutler (1969). The presence of gravel flooring the bottom of the cliffs and sea caves documents the continuous cycle of erosion and breakdown along the exposed coast.

Jointing, and the enhancement of joint patterns by biogenic processes such as mangrove root wedging, and boring by marine organisms have provided the source material for much of the gravel located along the shore. The jointed nature of the bedrock shoal has resulted in production of slab-shaped bedrock fragments that are a common constituent of gravel found higher up on the intertidal portions of the coast. Landwand transport of these slabs, as well as associated smaller clasts of coral, bedrock fragments, and *Strombus gigas* shells are believed to be the result of storm currents either from one storm or a series of storms.

Intertidal Mangroves and Gravel

Mangrove swamps are typical of tropical and subtropical shorelines throughout the world between 25^{0} north and south latitude. The major control of regional mangrove distribution is temperature and more specifically the occurrence of frost. West (1956) suggested that well developed mangrove is found only where average temperatures for the coldest month exceed 20^{0}C. Mangroves are generally found only on low energy accreting shorelines and are largely absent from eroding shorelines, although they may help to stabilize marginal shoreline dynamics (Burns, 1977). The mangrove characterizes sheltered marshes and intertidal mud flats.

Substrate and wave energy impart controls on the local distribution of mangrove. Davies (1977) suggested that mangroves will grow on a variety of sediment substrates including coraline sand, but more commonly on fine grained, soft, organic-rich mud deposited in estuarine and deltaic settings. The mangrove along Blackwood Bay is growing in sandy areas, gravelly areas, and in organic-rich muds. The only relationship that can be made between the trees and substrate is that organic-rich muds are only accumulating in the densest portions of the mangrove forest.

Davies (1977) believed that mangroves are vulnerable to wave action at all stages of their life history, and therefore are absent from exposed stretches of shore and can occur on the open coast in only very low energy environments. The distribution of the mangrove trees along Blackwood Bay's coast may be related to antecedent bedrock topography. The intertidal bedrock shoal acts as a secondary breakwater of wave energy and therefore may create the low energy conditions required by the mangrove. The imbricated gravel and smaller gravels in the upper intertidal zone may also slow incoming water reaching the mangrove. The limiting factor in mangrove colonization of Blackwood Bay is probably energy.

The role that the mangrove trees play in shoreline stabilization or accretion along Blackwood Bay is uncertain. The mangrove trees appear to be taking advantage of areas were sedimentation is occurring. On the other hand, adolescent red mangrove trees have taken root in the lower intertidal areas of the coast and are covered by as much as half a meter of water at high tide. In these cases the limiting factor is time of exposure, which controls the time required for plant respiration. These pioneer plants appear to show that the mangrove fringe is indeed prograding into the bay. The network of prop-roots and entangled debris of the mangrove fringe acts as a sediment trap by causing the deceleration of flowing water and settling out of entrained grains. The land-building role of mangroves has recently been questioned. However, the present consensus would be that mangroves are significant factors of shoreline evolution, but that their significance has sometimes been overstressed (Davies, 1977).

Due to inter-relations between the mangrove fringe and sand bench, time of colonization of Blackwood Bay's shoreline by mangrove vegetation is as yet uncertain. The oldest mangrove tree along the shore may be on the order of tens of years, but the existence of an ancestral coastal mangrove fringe may represent hundreds or thousands of years. Modern coastal mangrove swamps studied by Scholl (1963) suggest that mangrove swamp forests of southwestern Florida began to develop approximately 4,000 years ago as a result of Holocene transgression (post-Wisconsin sea-level rise) across the southwestern margin of the Everglades. Initiation of marine and swamp sedimentation of the Great Bahama Bank is estimated at 3000 to 5000 years ago (Newell and others, 1959). Mangrove colonization along Blackwood Bay could have conceivably begun 3000 to 5000 years ago when the platform was flooded (Pace, 1987b, Andersen and Boardman, 1987).

Sand Bench

Sand beaches to the west of Blackwood Bay, along French Bay, are composed of coarse- to medium-grained bioclastic sediment derived from the reef and other organisms in the shelf environment. The sand bench present inland of the mangroves along Blackwood Bay merges with the beach deposits of French Bay in the absence of mangrove trees. The bench and the most recent beach ridge appears to be inactive at the present time. The dense mass of mangrove fronting the coast is believed to restrict normal wave energy and prohibits net transport of grains along the coast.

If the beach is presently inactive, the beach ridges and swales of the sand bench may represent a series of wave-built storm deposits. Newell and others (1959) described similar alternating ridges and swales along the lagoonal shores of Bimini. They believed that the ridges represented successive storm beaches which were rapidly stabilized by mangrove plants, in particular the red (*Rhizophora mangle*) and black (*Avicennia nitida*). The black mangrove preferentially occupies the most recent shoreward beach ridge along Blackwood Bay as in cases noted by Newell and others (1959). The scenario presented by Newell and others (1959) was characterized by the red mangrove plants forming a protective seaward fringe that effectively anchors the outer ridge, and by trapping sediment they gradually extend the shore until the community advance is temporarily checked by the construction of a new storm beach beyond the mangrove. Due to geomorphic and botanical similarities, the sand bench at Blackwood Bay with its ridge and swale topography may have a similar history.

Gravel Ridge

Fair weather conditions along the coast of Blackwood Bay are not sufficient to account for the large gravel accumulations found in the upper intertidal and backshore areas of the coastline. Hurricanes are known to produce and transport large amounts of gravel debris by momentary high-velocity currents breaking over reef areas. Various authors have described the effects of hurricanes on reefs and coastal areas of south Florida and the Bahamas (Ball and others, 1967, Hurricane Donna; Perkins and Enos, 1968, Hurricane Betsy; Woodley and others, 1981, Hurricane Allen; and Sealey, 1985, Hurricane David). The sedimentary deposits and geomorphic changes in areas such as Blackwood Bay may be related to high-energy currents generated by tropical cyclones. The gravel ridge deposits and the gravel armor in the intertidal-shore zone have characteristics similar to carbonate gravel deposits in south Florida and the Bahamas produced by severe storm events as described by Newell and others (1959); Folk (1966); Ball and others (1967); Perkins and Enos (1968); Adams (1980); Hinman (1983); Scholle, and others (1983); and Scoffin (1987). Hinman (1983) described gravel deposits produced by winter northwesterly storm currents on Singer Beach, along the western part of San Salvador Island, Bahamas. Beaches along other localized parts of San Salvador Island are characterized by similar intertidal and backshore gravel ridge deposits especially on the north and west side of the island, and on cays known as Hinchinbroke Rocks and Middle Cay on the southeast.

Storm currents produce coral rubble that far exceeds the amount produced by normal daily breakdown. An extreme case of reef destruction by hurricane forces was presented by Woodley and others (1981) for the coral reefs of north Jamaica; the planar living areas of the branching *Acropora* were reduced by up to 99 percent along one stretch of reef. In this case, the breaker zone and reef flat were transformed from constructional reefs into a gently sloping rubble rampart. Ball and others (1967) also noted the development of a reef rubble rampart after Hurricane Donna hit the Florida reef tract. The presence of a coral rubble rampart located between the fringing reef and shoreline of Blackwood Bay is evidence for at least one severe storm striking the coast in the Holocene.

The abundance of coral material making up the inland gravel ridge suggests that storm processes are removing material from subtidal portions of the bay and emplacing them along the coast. The 3810 +/- 90 years bp age of a coral clast collected from the surface of one of the gravel lobes suggests that the lobe can be no older than approximately 3900 years (Pace, 1987b, and Pace and others, 1988). However, a younger age for emplacement may be suggested if the coral died prior to deposition. Ball and others (1967) in their study of the effects of Hurricane Donna and Folk (1966) in his morphologic study of the sand cays of Alacran Reef, Yucatan, Mexico suggested that the distance that larger pieces of coral debris move shoreward from the reefs during each storm is measured in feet. The material of the gravel ridge feature may have slowly accreted to the shore. Alternatively, the relatively unbored and unabraded nature of coral on the surface of the ridge would support a rapid emplacement of gravel. Rapid emplacement would limit subaqueous exposure and weathering by boring marine organisms.

The packages of coarsening upward gravel material comprising the inland ridge may represent a series of storm surges during a single storm event, or single storm events over a long period of time. An emergent coral rubble island (Little Molasses, Molasses Reef, Florida Reef Tract) studied by Banks (1959), and later by Ball and others (1967) displayed a similar 0.3 to 0.7m sequence of coarse coral rubble grading downward into fine coral gravel, followed by a thin sand layer and more loose gravel. The coarsening upward nature of the gravel deposit may suggest affinity to fluidized sediment flow or grain flow conditions resulting in inverse grading of beds. Clast-to-clast dispersive pressure created by these conditions act to size-sort grains, pushing larger grains to the top of the flow, the area of least shear-strain rate, and moving smaller grains to the bottom producing reverse grading (Middleton and Hampton, 1973). Little Molasses Island was shown to represent the cumulative product of storm activity which has piled up rubble on the leeward side of Molasses outer reef. The gravel ridge of Blackwood Bay may represent accretion of material from the bay that has accumulated since the bay was initially flooded.

The source of the boulders, cobbles, and fragments of rock that form a large part of the gravel ridge is believed to be the bedrock cliffs, terraces, and shoals that border the coast. The block- or slab-shape of many of the rock fragments of the ridge implies a shape inherited from the jointed bedrock pattern found in the present intertidal zone.

The presence of man-made artifacts such as the box-like structure, pond walls, and glass bottle bottom, possibly related to Loyalist Settlement, on the surface of the ridge suggest that the event which produced the gravel ridge did not occur within the last 200 years. The occurrence of *Strombus gigas* shells (conch) found 40 to 60 centimeters below the surface of the gravel ridge with man-made features suggests that the gravel was not deposited prior to aboriginal settlement.

Coastal Ponds

The ponds occupy a swale area between lithified Pleistocene dunes and the coast and may represent flooded areas resulting from the preexisting bedrock topography. Another possibility is that the emplacement of the gravel ridge and progradation of the shoreline restricted a landward portion of Blackwood Bay to form brackish ponds. Gradual filling and segmentation of the low areas may have followed. The presence of an active tidal channel along the northern-most pond and a feature resembling a washover fan in the large southern pond may support damming and segmentation. The sediment comprising the washover is composed of medium- to coarse-grained bioclastic sand similar to sediment presently occurring in the bay. Within a core taken by Pace (1987b) four thin sand stringers containing marine fauna within a series of mud sequences may represent normal sedimentation in the pond interrupted by sandy washover deposits formed during storms.

CONCLUSIONS

A proposed history for the evolution of the coast is presented based on the areal extent, lateral relations and distribution of the five shore zones and their characteristic deposits, and analogies between published descriptions of similar carbonate deposits and coastlines. Initial conditions of Blackwood Bay prior to Holocene post-glacial transgression and platform flooding approximately 3800 to 3600 years ago (Pace, 1987b; Andersen and Boardman, 1987), are represented by the existence of an exposed Pleistocene bedrock surface covered with a thin soil (Pace, 1987b).

An interpretation of coastal changes occurring along the coast of Blackwood Bay following platform flooding may be as follows: 1. Between approximately 3900 and 200 years before present, constrained by the radiocarbon age of coral material and human artifacts located on the surface of the ridge, a severe storm or series of storms of hurricane magnitude struck the coast resulting in the breakdown of bedrock cliffs, terraces, and shoals, as well as destruction of the offshore reef. Transport and emplacement as ridge deposits may have been synchronous with the storms; or the gravel may have been deposited in a manner whereby each storm that affected the coast created new material and slowly moved the mass shoreward as a sand and gravel slurry and heaped the debris as a discontinuous, poorly sorted, poorly stratified ridge. 2. The sand bench appears to have formed under higher energy conditions than are presently occurring along the coast. This feature may represent sand size material deposited under higher energy conditions either due to the destruction of the protective reef offshore (related to the storms that produced the gravel material) or due to lesser storms producing swash ridges stabilized by mangrove as a series of accretionary ridges. 3. The mangrove vegetation appears to be a relatively new shoreline feature. This is suggested due to lateral relations between the coastal deposits. The gravel ridge and sand bench landward of the mangrove would appear to be older than the mangrove presently occupying upper intertidal regions of the coast. The mangrove and sand bench could also be coeval, representing a series of storm ridges stabilized by the mangrove vegetation. Alternatively the mangrove forest may have been present along the coast since initial bank-top flooding, dependent upon energy conditions within the bay. 4. During later mangrove colonization and stabilization of the shoreline, another storm event occurred. Currents ripped-up molluscan and reef debris from within the bay and joint-separated slabs of bedrock from the intertidal shoals, and deposited the debris as a patchy, locally imbricated gravel layer atop sand, barren bedrock, and a mat of mangrove pneumatophores in the upper intertidal zone. This gravel layer is the youngest, seaward-building unit along the shore of Blackwood Bay and records the latest storm events.

Further analysis of surface sediment and rock samples collected from each of the shore zones, gravel samples from the backshore ridge and intertidal zone, as well as cores collected from the sand bench, ponds and beneath the ridge may provide answers and new theories on the evolution of the coast of Blackwood Bay.

ACKNOWLEDGEMENTS

We are indebted to Don Gerace, director of the College Center of the Finger Lakes, Bahamian Field Station for full logistical and financial support for this project.

The senior author is especially grateful to Matt Wilkerson (Appalachian State University), and Mac and Lee McKee, Bill Balcerzak, and Rice Mellon (Mississippi State University) for their assistance in collecting samples and cores; to John Hebert, Woody Pace, and Dr. John Mylroie for an introduction to Blackwood Bay; to Karen Andersen for her drafting, typing and immeasurable patience; to Dr. James Teeter and Dr. Charles Hoffman for their guidance and inspiration during field work.

REFERENCES CITED

Adams, R.W., 1980, General guide to the geologic features of San Salvador Island, in Gerace, D.T., ed, Field Guide to the Geology of San Salvador (First Edition): Fort Lauderdale, Florida, College Center of the Finger Lakes, Bahamian Field Station, p. 1-66.

Andersen, C.B., and Boardman, M.R., 1987, Sedimentary gradients in a high-energy carbonate lagoon, Snow Bay, San Salvador, Bahamas, in Boardman, M. R.,ed., A field guide to the modern carbonate enironments of San Salvador and Andros Islands: University of Miami, Ohio.

Ball, M.M., Shin, E.A., and Stockman, K.W., 1967, The geologic effects of Hurricane Donna in south Florida: Journal of Geology, v. 75, no. 5, p. 583-597.

Banks, J.E., 1959, Limestone conglomerates (Recent and Cretaceous) in southern Florida: American Association of Petroleum Geologists Bulletin, v. 43, p. 2237-2243.

Burns, L.A., 1977, Mangrove Swamps, in Clark, J.R., ed., Coastal Ecosystem Management, Wiley Interscience Publication, p. 660-665.

Carew, J.L., and Mylroie, J.E., 1985, The Pleistocene and Holocene stratigraphy of San Salvador Island, Bahamas, with reference to marine and terrestrial lithofacies at French Bay, in Curran, H.A., ed., Pleistocene and Holocene carbonate environments on San Salvador Island, Bahamas-Guidebook for Geologic Society of America, Orlando annual meeting field trip: Fort Lauderdale, Florida,

College Center of the Finger Lakes, Bahamian Field Station, p. 11-61.

Davies, J.L., 1977, Geographic variation in coastal development, in Clayton, K.M., ed., Geomorphology Texts, Longman Group Limited, p. 199.

Folk, R.L., 1966, Sand Cays of Alacran Reef, Yucatan, Mexico: Morphology, Journal of Geology, vol. 75, p. 412-437.

Gerace, K., 1982, Three loyalist plantations on San Salvador Island, Bahamas, in Carr, R.S., ed., The Florida Anthropologist: Florida Anthropological Society, vol. 35, no. 4.

Hinman, E., 1983, Recent sedimentary lithotypes and ancient facies analogs, in Gerace, D.T., ed., Field Guide to the Geology of San Salvador (Third Edition): Fort Lauderdale, Florida, College Center of the Finger Lakes, Bahamian Field Station, p. 117-144.

Middleton, G.V., and Hampton, M.A., 1973, Sediment gravity flows: mechanics of flow and deposition, in Middleton, G.V. and Bouma, A.H., co-chairman, Turbidites and deep water sedimentation, Pacific Section S.E.P.M. Short Course, Los Angeles, California, May 1973.

Mutler, H.G., 1969, Field Guide to Some Carbonate Rock Environments: Florida Keys and western Bahamas, Miami Geological Society, Miami, Florida, 159p.

Newell, N.D., Imbrie, J., Purdy, E.G., and Thurber, D.L., 1959, Organism communities and bottom facies, Great Bahama Bank: American Museum of Natural History Bulletin, v. 117, article 4, New York, p. 181-226.

Pace, W., 1987a, Preliminary investigation of a platform margin lagoon-tidal flat complex, Blackwood Bay, San Salvador Island, Bahamas, in Curran, H.A., ed., Proceedings of the 3rd Symposium on the Geology of the Bahamas: Fort Lauderdale, Florida, CCFL Bahamian Field Station, p. 109-115.

Pace, W., 1987b, Sedimentology of a Holocene platform margin carbonate lagoon, San Salvador Island, Bahamas, [M.S. thesis]: Mississippi State, Mississippi, Mississippi State University, 116 p.

Pace, W., Mylroie, J., and Carew, J., 1988, Sedimentology of Holocene, platform margin carbonate lagoon, Blackwood Bay, San Salvador Island, Bahamas [abs.]: Abstracts and Programs of the Fourth Symposium on the Geology of the Bahamas, San Salvador Island, Bahamas, College Center of the Finger Lakes, Bahamian Field Station, p. 28.

Perkins and Enos, P., 1968, Hurricane Betsy in the Florida-Bahama area: Geologic effects and comparison with Hurricane Donna: Journal of Geology, v. 76, p. 710-717.

Scholl, D.W., 1963, Sedimentation in modern coastal swamps, southeastern Florida, American Association of Petroleum Geologists Bulletin, v. 47, no. 8, p. 1581-1604.

Scholle, P.A., Bebout, D.G., and Moore, C.H., eds., 1983, Carbonate Depositional Environments, published by the American Association of Petroleum Geologists, Tulsa, Oklahoma, 708 p.

Scoffin, T.P., 1987, An introduction to carbonate sediments and rocks: Blackie & Son Ltd., London, 267 p.

Sealey, N.E., 1985, Bahamian landscapes: an introduction to the geography of the Bahamas: Collin Caribbean, London, p. 69-70.

Smith, R.R., 1985, Major plant communities on San Salvador Island, the Bahamas, in Smith, R.R., ed., Proceedings of the First Symposium on the Botany of the Bahamas, College Center of the Finger Lakes, Bahamian Field Station, San Salvador, Bahamas, p. 35-51.

West, R.C., 1956, Mangrove swamps of the pacific coast of Columbia, Annals of the Association of American Geographers, volume XLVI, no. 1, p. 98-122.

Woodley, J.D., Chomesky, E.A., Clifford, P.A., Jackson, J.B.C., Kaufman, L.S., Knowlton, N., Lang, J.C., Pearson, M.P., Porter, J.W., Rooney, M.C., Rylaarsdam, K.W., Tunnicliffe, V.J., Wahle, C.M., Wulff, J.L., Curtis, A.S.G., Dallmeyer, M.D., Jupp, B.P., Koehl, M.A.R.,

Neigel, J., and Sides, E.M., 1981, Hurricane Allen's Impact on Jamaican Coral Reefs, Science, Vol. 214, no. 4522, p. 749-755.

GEOELECTRICAL SURVEY OF THE COLUMBUS LANDINGS I REGION OF SAN SALVADOR BAHAMAS

A. W. Gerhard Kunze and Aphrodite K. Sauter,
Department of Geology
The University of Akron
Akron, Ohio 44325

William G. Weir
Standard Oil Production Co.
8404 Esters Blvd.
Irving, Texas 75063

ABSTRACT

An electrical resistivity sounding survey of Columbus Landings I (CL1) and the adjacent northern section of Columbus Landings IV (CL4), San Salvador, Bahamas, was conducted during a ten-day period in December 1987/January 1988, in order to assess the distribution of fresh groundwater of the region. An earlier groundwater hydrology investigation by Greene and Associates (1972) indicated a substantive fresh water lens in the western interior of the region, but little or no fresh water beneath the peripheral hills. Our survey utilized a Soiltest R-50/AKU Stratameter apparatus and employed the Schlumberger electrode configuration. Thirty-two field stations were established, and the resulting VES curves were inverted digitally with the USGS computer program by Zohdy (1974). The computed electrical resistivity layer models were interpreted in terms of aquifer porosities and groundwater salinities. According to World Health Organization Standards, the maximum allowable TDS content of potable water is 1500 ppm. Assuming aquifer porosity of 25%, subsurface temperatures of 25°C, and the formation resistivity factor according to the basic Archie formula, our results also indicate a major fresh water lens in the western interior of the study area, with maximum thickness possibly in excess of 100 feet. In addition, substantial fresh groundwater reserves appear to exist underneath the peripheral hills. Little or no fresh water is evident in the eastern low-lying portion of CL1.

Petrophysical laboratory analysis of twelve bedrock samples from the study area indicate very high near-surface aquifer porosities averaging approximately 30%, and a formation factor significantly different from the Archie formula. Interpretation of subsurface resistivities in terms of this formation factor substantially reduces inferred freshwater thickness at several stations.

In general, it is expected that porosities decrease with depth. We, therefore, surmise that high formation resistivities at depths in excess of 100 feet are the result not of fresh pore water, but of low formation porosities.

INTRODUCTION

This investigation is a continuation of earlier electrical resistivity work in the Columbus Landings region of San Salvador, Bahamas, by Kunze and Burke (1984) and Kunze and Weir (1987), in order to further define the distribution of fresh groundwater of that region. The investigation of the Sandy Point area by Kunze and Burke (1984) revealed a probable major freshwater lens with maximum thickness of approximately 100 feet beneath the southern hills of Sandy Point with two lesser fresh groundwater lobes extending toward the north-east and east beneath the peripheral hills of the Sandy Point peninsula. Little or no fresh groundwater was indicated in the low interior of Sandy Point. A similar peripheral distribution of fresh groundwater was indicated by electrical resistivity measurements on Sandy Hook (Kunze and Weir, 1987), although maximum peripheral fresh groundwater thicknesses there are mostly less than ten feet.

An earlier groundwater survey of the Columbus Landings region by Greene and Associates (1972), utilizing chemical water analysis and electrical conductivity measurements in test wells, revealed a substantial freshwater lens in the low interior of the northern portion of the CL4 area immediately to the north-east of the study area of Kunze and Burke (1984), and extending into CL1, with almost no fresh water beneath the peripheral coastal regions. This situation is nearly opposite to that inferred by Kunze and Burke (1984) and Kunze and Weir (1987), although the failure of Greene and Associates to locate fresh water beneath the peripheral hills may have been largely due to the fact that their test wells could not be drilled in areas with surface elevations greater than 20 feet due to equipment limitations. Thus, the present resistivity survey was undertaken to verify and clarify the distribution of fresh groundwater in the regions to the north and northeast of Sandy Point, specifically in Columbus Landings I and the northern part of Columbus Landings IV.

PROCEDURES

This study was conducted during a ten-day period in December 1987/January 1988 and utilized the Schlumberger electrode configuration with a Soiltest R-50/AKU Stratameter apparatus capable of 500 ft maximum penetration. Vertical electrical sounding (VES) curves were constructed based on current electrode half-spacings (L) ranging from 1.5 ft to 460 ft with six points per decade. Thirty-two field stations, 1000-1500 ft apart, were established along road sides throughout the 1.5 square mile study area. Station locations are shown in Fig. 1. The VES field curves were corrected for apparent resistivity off-sets caused by increases in the potential electrode spacing as described by Kunetz (1966). The corrected VES curves (Fig. 2) were inverted digitally using the USGS computer program by Zohdy (1974). Some model parameters, such as expected half-space resistivities, were specified a priori. The resulting computed resistivity layer models (Fig. 3) were interpreted in terms of groundwater salinities and aquifer porosities based on the following considerations:

Ghyben-Herzberg Lens

Fresh groundwater in the Bahamas occurs largely in the form of irregular, seasonally varying freshwater lenses which tend to float in isostatic equilibrium atop denser saltwater in accordance with the Ghyben-Herzberg principle (Little and others, 1977; Weech, 1982; Tarbox, 1987), such that the ratio of water table elevation to lens thickness is approximately 1:40. The freshwater/saltwater interface at the bottom of the lens can, in principle, be detected by the VES

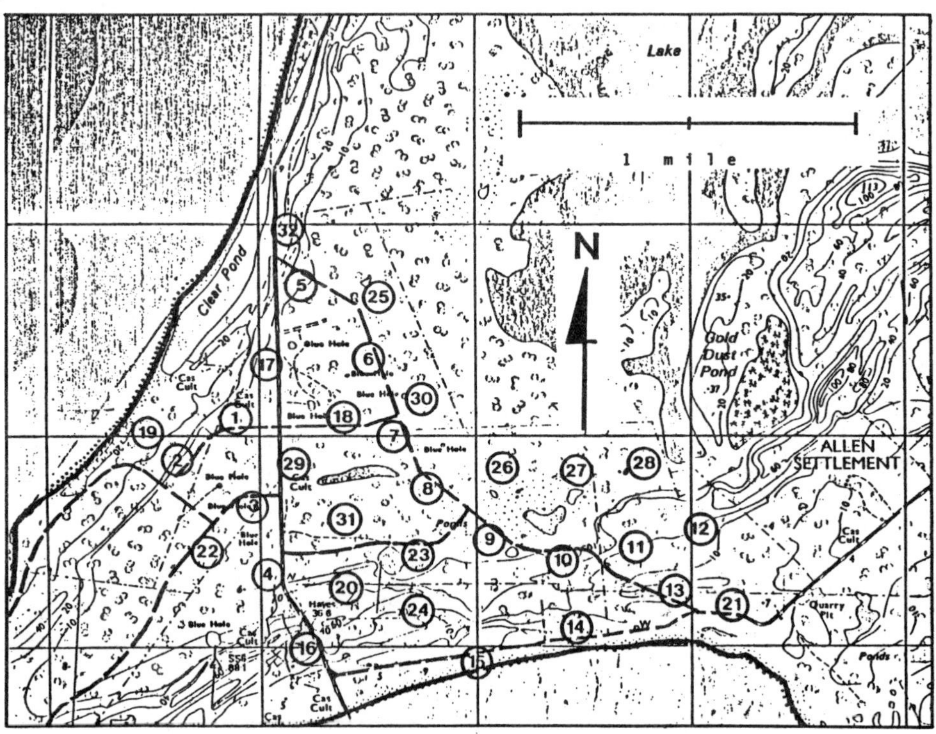

Fig. 1. Station location map. Major roads are shown as dashed lines. Contours in feet.

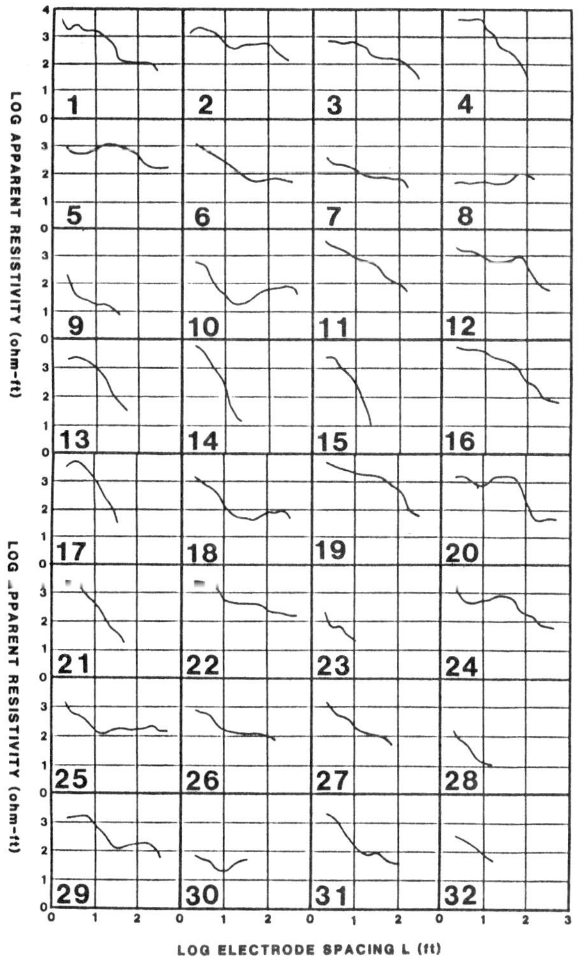

Fig. 2. Corrected VES curves for stations 1 through 32.

method due to the pronounced electrical resistivity contrast between saltwater and freshwater. Knowledge of the depth to that interface permits calculation of the total thickness of the Ghyben-Herzberg lens.

Formation Factor

Formation resistivity (R_o) of saturated rocks is proportional to pore water resistivity (R_w) with the proportionality constant defined as the formation resistivity factor (F):

(1) $$R_o = FR_w$$

F, in turn, is a function of rock porosity (Φ):

(2) $$F = \frac{a}{\Phi^m}$$

where "a" and "m" are characteristic constants. For well-cemented carbonates, F is given by the so-called basic Archie formula:

(3) $$F = \frac{1}{\Phi^2}$$

This relationship should be valid for the deeper aquifers of San Salvador (e.g. marine units 1, 2, and 3 of Little and others, 1977).

A previous petrophysical investigation of the electrical behavior of surface rock samples from San Salvador by Weir and Kunze (1988) yielded a different functional relationship:

(4) $$F = \frac{0.57}{\Phi^{2.4}}$$

The Archie relationship, however, gives comparable results for F in the porosity range of 20%-32%. Hence, the basic Archie formula may also be applicable to porous, near-surface aquifers on San Salvador (e.g. marine unit 4 of Little and others, 1977).

A similar petrophysical analysis of twelve bedrock samples from the present study area (CL1) is described in the appendix. The resulting formation factor is

(5) $$F = \frac{1.07}{\Phi^{2.3}}$$

This equation yields F-values significantly larger than those obtained with the Archie or other commonly used formulas. Therefore, our initial formation resistivity evaluation utilized equation 3, with a subsequent check using equation 5.

Porosity

Helium porosity tests of representative San Salvador lime-stones by Weir and Kunze (1988) showed porosities ranging from 8.2% to 36.1% with an RMS porosity value of 23.5%.

Table 1 shows the results of additional Helium porosity tests performed on twelve bedrock samples from the current study area (CL 1). These porosities range from 8.9% to 39.0% with an RMS value of 30.3%. Little and others (1977) estimate porosities of the shallowest Bahamian aquifers (marine unit 4) to attain values of 40%-50%. Deeper units have reduced porosities due to conversion of aragonite to calcite with its concomitant specific volume increase of 8%, and the introduction of additional cements. For example,

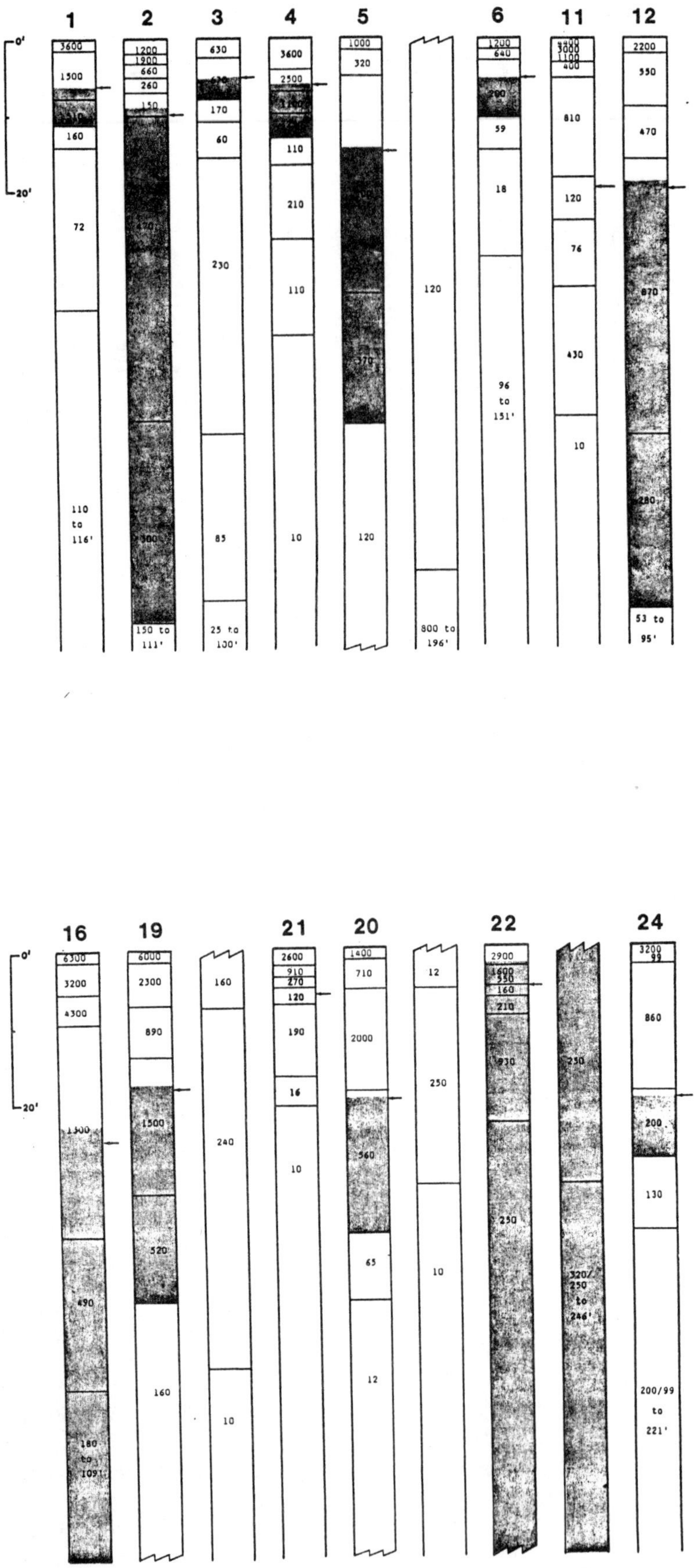

Fig. 3. Computer generated resistivity layer models for selected stations. Sealevel is shown by arrows. Shaded regions indicate possible fresh ground-water. Discussion in text.

Table 1. Rock Sample Data

Site #	Nearest Station	Type	Porosity
1	5	Bioclastic LS	34.5/28.5%
2	10	"	32.5/35.8%
3	25	"	32.6/27.4%
4	30	"	30.5/26.7%
5	16	Eolianite	38.8/39.0%
6	32	Bioclastic LS	11.8/8.9%

the aragonite conversion effect could reduce a 30% porosity by as much as 5.6%. Tightly cemented thin micritic layers or paleosols may have porosities on the order of 5% (Persons, 1975) or less. Nevertheless, most Bahamian subsurface formations appear to be very porous or cavernous to a depth of at least hundreds of feet (e.g. Spencer, 1967), and it is reasonable to assume that the bulk porosity of most Bahamian aquifers is on the order of 25% or more. This porosity value was chosen for initial resistivity model evaluation.

Groundwater Salinity and Resistivity

According to World Health Organization (1963) standards, the maximum allowable total dissolved solids (TDS) content of drinking water is 1500 ppm (Bouwer, 1978), with corresponding resistivity of 11.2 ohm-ft (at 25^oC). The resistivity of seawater (35,000 ppm TDS) at 25^oC is 0.6 ohm-ft. Consequently, according to the Archie formula and assuming formation porosities averaging 25%, the formation resistivity of freshwater aquifers should be at least 180 ohm-ft, and that of seawater saturated aquifers on the order of 10 ohm-ft. Equation 5 predicts 290 ohm-ft and 16 ohm-ft for freshwater and seawater aquifers, respectively. Whatever the formation factor or the formation porosity, the bottom of a GhybenHerzberg freshwater lens should be marked by an 18-fold or greater resistivity contrast.

RESULTS AND INTERPRETATION

Freshwater Resources

Fig. 3 shows fourteen selected resistivity layer models computed with the USGS inversion program by Zohdy (1974). Not shown are models with resistivities too low to be compatible with a fresh groundwater lens (stations 7, 8, 9, 10, 13, 14, 15, 17, 18, 23, 25-32). In Fig. 3, sealevel is indicated by arrows, and potential freshwater bearing layers are shaded. In general, the shaded regions represent minimum resistivities of 180 ohm-ft extending downward from the inferred water table. At station 22, this column includes a narrow layer of resistivity of 160 ohm-ft. This layer, if real, might represent a high-porosity freshwater layer. In accordance with equations 1 and 3:

$$(6) \qquad R_o = R_w/\Phi^2$$

and the porosity Φ corresponding to some formation resistivity R_o is given by:

$$(7) \qquad \Phi = (R_w/R_o)^{0.5}$$

Thus, if R_o and R_w are 160 ohm-ft and 11.2 ohm-ft respectively, Φ is 26.5%, a quite typical value. If equation 5 is used, then

$$(8) \qquad \Phi = (1.07\ R_w/R_o)^{0.435}$$

and Φ is approximately 32.5%. This value is high, but not unreasonable for near-surface layers. Accordingly, we interpret this layer to be part of a freshwater lens. A similar situation exists at station 21, where a 120 ohm-ft layer overlies a potential freshwater bearing layer. The necessary porosity, in this case, is 30.5% (equation 7) or 37% (equation 8). These values are unusually high, hence the Ghyben-Herzberg lens indicated is probably brackish (with lower porosities).

Table 2. Ghyben/Herzberg Lens Parameters

Station	Elevation (feet)	Rel. depth to lens bottom (feet)	Lens thickness (feet)
1	6	11.0	5
2	10	77.2	69
3	5	8.0	3
4	6	12.8	7
5	15	50.6	36
6	5	10.6	6
7	5	-	-
8	5	-	-
9	6	-	-
10	8	-	-
11	20	-	-
12	20	75.3	57
13	8	-	-
14	5	-	-
15	6	-	-
16	25	109.5	87
17	6	7.1	1
18	5	-	-
19	18	45.4	28
20	20	36.8	17
21	6	-	-
22	5	245.5	247
23	6	-	-
24	20	28.3	9
25	4	-	-
26	6	-	-
27	6	7.2	1
28	7	-	-
29	5	7.5	3
30	4	-	-
31	5	5.9	1
32	10	-	-

The Ghyben-Herzberg lens parameters corresponding to Fig. 3 are listed in Table 2. Lens thicknesses are computed by subtracting station elevation from the relative depth to lens bottom and multiplying the result by 1.025 in accordance with the Ghyben-Herzberg relationship. Station elevations are estimated from the 1972 Bahamas Government map of San Salvador and may be in error by ± 5 feet.

More significant errors, no doubt, are introduced by our lack of knowledge of the applicable formation resistivity factor and actual subsurface porosities, as well as by deviations of the actual subsurface from the ideal horizontal layer model. Furthermore, the logarithmic increments of electrode spacings lead to a progressive loss of resolution of subsurface features with increasing depth and electrode spacing. Resolution is also affected by the electrical equivalence of high-resistivity layers of equal transverse resistance (given by the product of layer thickness and layer resistivity) between low-resistivity layers, or of low-resistivity layers of equal longitudinal conductance (given by the ratio of layer thickness to layer resistivity) between high-resistivity layers. Thus, thin low-porosity micritic layers tend to masquerade as much thicker, less resistive, layers in computed resistivity models. The VES curve inversion program used also tends to miss existing or generate non-existing layers of intermediate resistivity between high and low resistivity layers. Consequently, the exact position of the bottom of a freshwater lens may not be identifiable in the computed resistivity models. In fact, only two stations (20 and 21) show the expected diagnostic 18-fold or greater resistivity decrease at depth marking the freshwater/saltwater boundary. Thus the dimensions of the inferred freshwater layers may be greatly in error.

There is significant change in the results if the formation resistivity factor of equation 5 is used. In that case no freshwater is indicated at stations 6 and 24 (unless aquifer porosities are at least 29%), the freshwater column at station 12 is reduced to 33 ft (unless aquifer porosity is at least 25.5%, a quite reasonable value), the freshwater column at station 16 is also reduced to 33 ft (unless aquifer porosity is at least 31%), and at station 22 it is reduced to 19 ft (unless porosity exceeds 27%).

A significant problem is presented by the high resistivity values of the computed layer model of station 22. This station is located near station 31 of Greene and Associates (1972) which indicated a substantial thickness of high-quality fresh groundwater, and a unusually thick freshwater column was therefore expected. However, the initial estimate of 247 ft for the freshwater lens thickness at that site is totally unrealistic. In general, even the most massive Bahamian freshwater lenses are rarely more than 100 ft thick (Cant, 1988). Additional doubts are raised by the fact that station 22 is situated only about 1500 ft from Station 23 of Kunze and Burke (1984) which indicated no freshwater. Finally, according to the Ghyben-Herzberg relationship the water table elevation corresponding to such a lens thickness should be 6 ft above sea level, yet the surface elevation in that area is estimated to be only 5 ft above sea level. Hence, the high subsurface resistivities at station 22 must have another explanation, such as anomalously low subsurface porosity. To explain the model resistivities in terms of seawater saturation requires a porosity of 4.9% for the 250 ohm-ft layer, and 4.3% for the 320 ohm-ft layer (according to equation 7). If equation 8 is utilized, the corresponding porosities are 7.5% and 6.7% respectively. Although these porosity values are not unreasonable, the great thickness of the inferred low porosity layers is unusual. Considering the electrical equivalence of high resistivity layers of equal transverse resistance, the computed 54 ft thick layer of 320 ohm-ft might in reality represent a 4 ft freshwater saturated micritic layer of resistivity 4300 ohm-ft and 5.1% porosity 133 ft below sealevel. Such micritic layers are known to occur at the bottom of at least some freshwater lenses (Persons, 1975; Weech, 1982). Alternately, the bottom of the freshwater lens may be represented by the 930 ohm-ft layer at a depth of 19 feet below sealevel. In either case, the resistivity values at Station 22 remain somewhat enigmatic.

If the deep subsurface porosities are, in fact, significantly lower than 25%, then our interpretation of station 16 is also in error. Specifically, the 180 ohm-ft layer is more likely to represent a saltwater bearing layer with porosity of 5.8% (or 8%) than a freshwater layer with 25% porosity. In that case the bottom of the freshwater lens at that site is 32 feet below sealevel, and the lens is approximately 33 feet thick.

These modified results were used to construct the freshwater isopach map shown as Fig. 4. This map also includes stations 1, 13, 17, 18, 19, 20, 22, 23, 24, 26, 27, and 28 of Burke and Kunze (1984) whose resistivity layer models were reinterpreted according to the slightly different criteria of this paper.

High-Resisitivity Layers at Depth

Table 3 lists pertinent parameters of high-resistivity layers computed by the VES inversion program at twenty stations. As discussed previously, these layers probably represent electrically equivalent low-porosity thin micritic zones or paleosols. They tend to cluster about depths of 20 ft, 30 ft, 60 ft, 90 ft, and 150 ft below sealevel, and may be equivalent to paleosols at 25 ft, 55 ft and 90 ft reported by Little and others (1977). Supko (1977) reports buried paleosols on San Salvador at 51 ft, 81 ft, and 107 ft below sealevel with dolomite below 111 ft (34 m). Rodriquez and others (1988) report a prominent reflector on San Salvador 30-38 m below the surface. This reflector as well as the paleosols of Supko (1977) at 81 ft or 107 ft might correspond to our high-resistivity layer at approximately 90 ft, but, in view of the low resolution of our electrical subsurface models, meaningful correlation is not possible. Alternately, these high-resistivity layers may, of course, represent extended zones of intermediate to low porosities.

Other Correlations

Eight stations were located near blueholes, lakes or wells containing water of known salinity. If the adjacent groundwater has the same salinity, then it is possible to calculate subsurface porosity from the model subsurface resistivity (R_0) using equation 7 or 8. Table 4 lists the pertinent data and the results of such a calculation using equation 7. Inferred near-surface aquifer porosities range from 3.4% to 54.7% and average around 19%. It is, however, unlikely that such surface water has exactly the same salinity as the pore water in the surrounding rock - especially if porosities and permeabilities are low. Hence the low-porosity results are particularly suspect.

SUMMARY AND CONCLUSIONS

Our data indicate a very inhomogeneous subsurface with greatly varying porosities. Near-

Table 3. High Resistivity Layers at Depth

Station	R_o (ohm-ft)	Abs. depth range (ft)	Avg. depth (ft)
1	91	12 - 101	56
3	230	11 - 47	29
4	210	11 - 20	16
5	800	136 - 181	159
6	100	24 - 146	85
8	410	24 - 47	35
10	110	7 - 220	114
11	430	13 - 29	21
16	200	138 - 229	183
18	110	62 - 133	97
19	240	69 - 116	92
20	250	66 - 91	78
22	320	106 - 160	133
25	180	7 - 187	97
27	90	9 - 29	19
29	160	9 - 118	63
30	420	5 - 8	6
31	140	5 - 36	20

Table 4. Porosity Estimates from Salinity and Resistivity Data

Station	R_o (ohm-ft)	Sample Site	Salinity	R_w (ohm-ft)	Porosity
3	667 173 60.4	B.H. #2	23 ppt	0.855	3.4% 7.0% 11.9%
18	93.7 33.5 22.5	B.H. #3	21 ppt	0.92	10.0% 16.6% 20.2%
23	11.2	Pond	38 ppt	0.55	22.2%
25	91.2	B.H. #5	22-25 ppt	0.79-0.89	9.3-9.9%
30	7.5 92.4	B.H. #6	13 ppt	1.48	44.4% 12.7%
31	38.8 25.3	Well	15.5 ppt	1.25	17.9% 22.2%
32	35.8	Clear Pond	35 ppt	0.60	12.9%
15	8.0 1.0	French Pond	70 ppt	0.305	19.5% 54.7%

Salinity data provided by James Teeter, The University of Akron

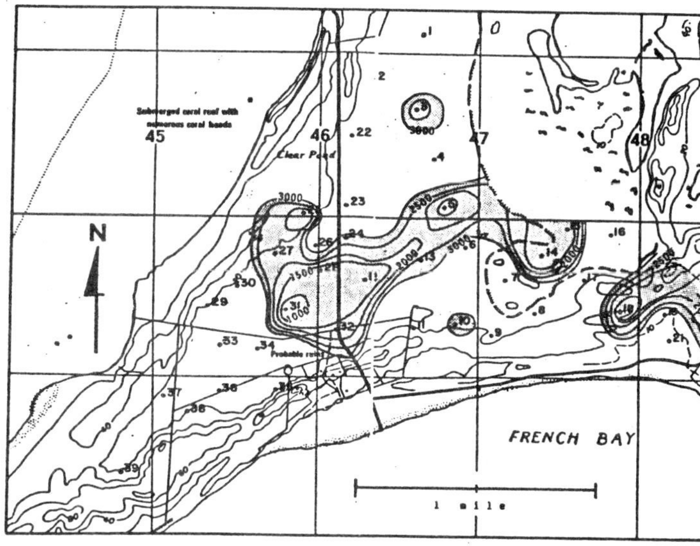

Fig. 5. Freshwater lenses (shaded) determined by Greene and Associates (1972). Freshwater conductivity isolines in micromho/cm. Discussion in text.

surface porosities tend to be very high (perhaps averaging near 30%), but porosities of many of the deeper strata appear to be quite low (5% or less). Such low-porosity regions at depth may represent well cemented limestones or dolomites without caverns. Elsewhere cavernous regions in otherwise low-porosity formations increase the formation bulk porosity affecting resistivity measurements. Because of these porosity variations, estimation of groundwater salinity by the resistivity method is problematic, if not impossible. For example, low-porosity, seawater saturated layers have similar resistivities as high-porosity, freshwater aquifers. This may be the reason for the unrealistic results at station 22. Another major problem hampering the interpretation of resistivity data is the lack of resolution of the deeper layers. As a result of this, the freshwater/saltwater transition zone, which is commonly only a few feet wide (Little and others, 1977), is usually poorly defined or occasionally missed altogether in the computed resistivity layer models. Consequently, the freshwater lens thicknesses depicted in Fig. 4 may be greatly in error - especially the larger values.

In spite of these uncertainties, certain trends in the interpreted fresh groundwater distribution are unmistakable. As in previous investigations on San Salvador (Kunze and Burke, 1984; Kunze and Weir, 1987), there is a definite tendency for fresh groundwater to underly peripheral hills and topographic highs, with generally lesser freshwater accumulations in the low interior. A similar distribution of fresh groundwater on San Salvador was inferred by Davis and Johnson (1988). There is, however, the possibility of an anomalously thick freshwater lens in the western part of the low-lying interior which may reach a maximum thickness on the order of 100 ft near station 22, but the freshwater/saltwater interface at that station is not defined in the corresponding resistivity layer model. We conclude that freshwater thickness at that site must be at least 19 ft. There appears to be almost no fresh groundwater in the eastern part of CL1 except underneath the easternmost hills. The freshwater lens as indicated in Fig. 4 (shaded) bears little resemblance to that of Greene and Associates (1972) as shown in Fig. 5. The isolines represent equal well-water conductivities (in micromho/cm) and are strongly correlated with measured freshwater thicknesses. The maximum conductivity of freshwater as defined by the World Health Organization (1500 ppm TDS), corresponding to a resistivity of 11.2 ohm-ft, is 2930 micromho/cm. The 1500 micromho/cm line corresponds roughly to the 15 ft freshwater isopach line of Greene and Associates (1972). Thus the freshwater lens depicted by Fig. 5 (shaded) is confined largely to the low interior of the study area and attains a maximum thickness of perhaps well over 20 ft

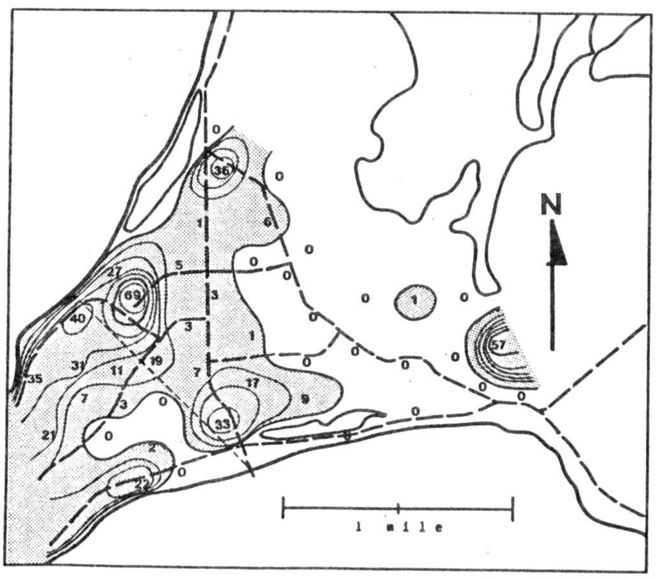

Fig. 4. Inferred fresh groundwater isopach map. Shaded area indicates freshwater lens. Numbers indicate freshwater thicknesses (in feet). Contour interval is ten feet. Major roads are shown as heavy dashed lines. Thin dashed line is study area boundary.

near our station 22. As discussed earlier, no test wells were established in the higher, hilly regions; consequently Fig. 5 does not show most of the freshwater present there. However, the freshwater lens beneath the easternmost hills is evident and confirms our results. On the other hand, our results show virtually none of the fresh groundwater detected by Greene and Associates (1972) in the low-lying eastern interior of CL1. The fact that all surface water recently tested in that region (Table 4) was brackish supports our results. We, therefore, conclude that, on San Salvador, relatively stable major freshwater lenses tend to occur beneath hills and highlands, and that the few freshwater lenses that develop in low-lying areas tend to be unstable and ephemeral. Nevertheless, special conditions - such as the absence of caves - may permit development of a major freshwater lens anywhere on the island.

ACKNOWLEDGEMENTS

We wish to thank the College Center of the Finger Lakes, Bahamian Field Station for providing logistical support. We also wish to express our appreciation to the Standard Oil Production Company, Dallas, Texas, for kindly providing use of their laboratory facilities. Thanks are also due to Jim Teeter for furnishing pertinent water salinity and temperature data, to Tom Quick for analysing water samples, and to Roger Bain for petrological analysis of our rock samples. Financial support for this research was received from the Research (Faculty Projects) Committee of the University of Akron.

APPENDIX

Determination of Formation Resistivity Factor

Twelve bedrock samples from the CL1 area were analysed at the Standard Oil Production Laboratory, Irving, Texas for porosity and rock resistivity (R_0) while saturated with an NaCl solution of known resistivity R_w (0.184 ohm-m). Procedural details are described by Weir and Kunze (1988). Knowledge of R_0 and R_w permits calculation of corresponding formation factors (F):

Table 5. Geoelectrical parameters of rock samples

Sample	R_o (ohm-m)	F	Porosity (%)
1a	2.20	12.0	34.5
1b	3.22	17.5	28.5
2a	3.26	17.7	32.5
2b	2.28	12.4	35.8
3a	2.48	13.5	32.6
3b	3.03	16.5	27.4
4a	3.03	16.5	30.5
4b	3.98	21.6	26.7
5a	1.84	10.0	38.8
5b	1.62	8.8	39.0
6a	25.65	139.4	11.8
6b	54.55	296.5	8.9

(9) $$F = R_o/R_w$$

The twelve measured resistivity (R_o) and porosity (Φ) values as well as corresponding calculated formation factors (F) are listed in table 5.

The formation factors are functionally related to rock porosities in accordance with equation 2.

$$F = \frac{a}{\Phi^m}$$

This relationship may be written in logarithmic form as:

(10) $$\log F = \log a - m \log \Phi$$

A plot of log F versus log Φ thus represents a straight line with slope "-m" and intercept "log a", permitting determination of the parameters "a" and "m" of equation 2.

Fig. 6 is the logarithmic plot of the twelve F and Φ values (actually log F versus -log Φ). The best fitting least squares line has a slope "m" of 2.30 ± 0.08 and intercept "log a" of 0.028 ± 0.046. The parameter "a", therefore, is equal to 1.07 ± 0.11. Hence, based on our samples, the formation factor F for shallow aquifers in the CL1 area is given by:

(11) $$F = \frac{1.07}{\Phi^{2.3}}$$

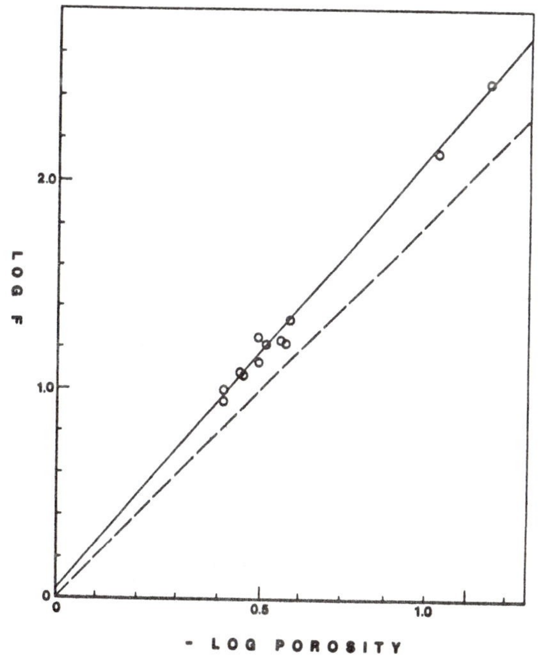

Fig. 6. Scatter plot and least squares trend (solid line) of sample log F versus -log Φ values. The dashed line represents the basic Archie relationship.

REFERENCES CITED

Bouwer, H., 1978, Groundwater Hydrology: McGraw-Hill, New York, p. 360.

Cant, R., 1988, personal communication: Chief Hydrologist, Ministry of Works & Utilities, Nassau, Bahamas.

Davis, L.R., and Johnson, C.R., 1988, A general model of the hydrologic system on San Salvador Island, Bahamas: Abstract and Programs, Fourth Symposium on the Geology of the Bahamas, CCFL Bahamian Field Station, San Salvador, Bahamas, p. 9-10.

Greene, R.W. and Associates, Ltd., 1972, Groundwater hydrology study, Columbus Landings Subdivisions numbers 1, 2, 3, and 4, San Salvador Island, Bahamas: Engineering Report 610-71-53.

Kunetz, G., 1966, Principles of direct-current resistivity prospecting: Gebruder Borntraeger, Berlin, 103p.

Kunze, A.W.G., and Burke, J.C., 1984, A resistivity survey of Sandy Point, San Salvador, Bahamas, in Teeter, J.W., ed., Procedings of the Second Symposium on the Geology of the Bahamas: CCFL Bahamian Field Station, p. 97-112.

Kunze, A.W.G., and Weir, W.G., 1987, Geoelectrical ground-water survey of the Sandy Hook area, San Salvador, Bahamas, in Curran, H.A., ed., Proceedings of the Third Symposium on the Geology of the Bahamas: Fort Lauderdale, Florida, CCFL Bahamian Field Station, p. 81-89.

Little, B.G., Buckley, D.K., Cant, R., Henry, P. W.T., Jefferiss, A., Mather, J.D., Stark, J., and Young, R.N., 1977, Land resources of the Bahamas: a summary: Land Resource Study 27, Land Resources Division, Ministry of Overseas Development, Surrey, England.

Persons, J.L., 1975, The delineation, lithology, and susceptibility to vertical saline communication of a freshwater lens, Cape Eleuthera, Eleuthera, Bahamas: Master's thesis, Wright State University, Dayton, Ohio, 212 p.

Rodriquez, R., Ettenson, F.R., and Goodman,P.T., 1988, Preliminary interpretations of a seismic section from the North of San Salvador, Bahamas: Abstracts and Programs, Fourth Symposium on the Geology of the Bahamas, CCFL Bahamian Field Station, San Salvador, Bahamas, p. 22-23.

Spencer, M., 1967, Bahamas deep test: Bulletin of the American Association of Petroleum Geologists, v. 51, p. 263-268.

Supko, P.R., 1977, Subsurface dolomites, San Salvador, Bahamas: Journal of Sedimentary Petrology, v. 47, p. 1063-1077.

Tarbox, D.L., 1987, Occurrence and development of water resources on the Bahamas Islands, in Curran, H.A., ed., Proceedings of the Third Symposium on the Geology of the Bahamas: Fort Lauderdale, Florida, CCFL Bahamian Field Station, p. 139-144.

Weech, P.S., 1982, A review of groundwater in the Bahamas: Master's thesis, Colorado State University, Fort Collins, Colorado, 112 p.

Weir, W.G., and Kunze, A.W.G., 1988, Geoelectrical properties of selected rock and water samples from San Salvador Island, Bahamas: Bulletin of the Association of Engineering Geologists, v. 25, p. 257-263.

Zohdy, A.A.R., 1974, A computer program for the automatic interpretation of Schlumberger sounding curves over horizontally stratified media: United States Geological Survey Report USGS-GD-74-017, 9 p.

SURFACE TEXTURES OF *NEMASTER RUBIGINOSA* (CRINOIDEA; ECHINODERMATA) SAN SALVADOR, BAHAMAS

Ronald D. Lewis and Mark W. Peebles
Department of Geology
Auburn University
Auburn University, AL 36849

ABSTRACT

The comatulid crinoid *Nemaster rubiginosa* is found off the west coast of the island of San Salvador (Bahamas) in water as shallow as 4 meters. It is abundant enough to contribute ossicles (plates) to the sediment by natural processes.

Arm plates (brachials) and pinnule plates (pinnulars) taken from collected specimens were subjected to laboratory tests of abrasion and dissolution in order to identify the effects of these processes on the echinoderm's stereom. Brachials and pinnulars of *Nemaster* picked from sediment samples were then compared to the test specimens.

Breakage of slender spines is the first recognizable consequence of plate movement. Initial abrasion is indicated by the loss of the large-mesh stereom found on some plates, along with the nearly complete loss of spines. Smoothly rounded plates with polished fulcral ridges result from prolonged abrasion.

Initial dissolution in the laboratory results in a subtle roughening of the stereom's surface. Prolonged dissolution causes enlarged openings in the stereom and sharp-pointed projections. Small solution pits and etched cleavage planes are diagnostic surface details revealed under high magnification.

The plates recovered from the sediment show the effects of abrasion, dissolution, and a range of biologic processes not duplicated in the laboratory experiments. These include localized stereom breakage (bite marks?) and microborings. Plates found in the Pleistocene reef rock exposed on the island demonstrate the potential for future studies of diagenesis using crinoid plates.

INTRODUCTION

In spite of the fact that crinoids were a major component of many shallow marine ecosystems in the geologic past, relatively little research has been done on the taphonomy of modern crinoids. Meyer (1971) and Liddell (1975) documented the rapid rate of skeletal disarticulation in comatulid crinoids at the sediment-water interface, and Meyer (1971) described selective scavenging by the isopod *Cirolana parva.* Within the last five years, research has shown that some fish are active predators of living crinoids (Meyer et al., 1984; Meyer, 1985; Meyer and Meyer, 1986.)

Blyth Cain (1968) laid the foundation for laboratory experiments using crinoid carcasses and skeletal plates. Meyer and Meyer (1986) illustrated separate skeletal plates recovered from sediment samples taken on the Great Barrier Reef and documented their distribution relative to the living crinoids.

The present investigation focuses on the separate plates as sedimentary particles, using laboratory experiments to isolate the effects of dissolution and abrasion on grain shape and surface textures. Our work shows that crinoid plates can be found in Bahamian sediments in sufficient numbers for study even though the living crinoid populations are much smaller than those in the South Pacific. This allows us to compare laboratory observations with the effects of natural taphonomic processes. In addition, we have been able to recover a few crinoid plates from a Pleistocene reef exposed on the island, demonstrating the potential for future studies of crinoid-plate diagenesis. Results to date are summarized below; a larger manuscript, with illustrations, is in preparation.

OBJECTIVES

Our objectives were:

1. To isolate the biostratinomic processes of abrasion and dissolution in the laboratory in order to determine the effects each has on the calcite stereom of a modern crinoid.
2. To examine these effects using the scanning electron microscope and to define diagnostic

surface textures ("signatures") for each process.

3. To apply these diagnostic criteria to Recent and Pleistocene crinoid plates to determine the relative importance of the processes involved in their fossilization.

PART 1. CRINOID PLATES USED AS CONTROL SPECIMENS

Procedures

Whole crinoid specimens were collected by hand using SCUBA in Fernandez Bay at Snapshot Reef and at *Cervicornis* Reef (Telephone Pole Reef), and preserved in buffered formalin. In the laboratory, soft tissue was removed with a solution of chlorox and distilled water.

In order to standardize the elements used in the laboratory experiments, six plate types were defined with a narrow range of morphologic variation and plates were selected which conformed to these norms and which were complete or nearly so.

Results

Each arm of *Nemaster* branches twice to produce a total of 20 arms, and almost every arm plate (brachial) supports one pinnule. Each pinnule contains approximately 23 plates (pinnulars). As a result, one individual crinoid is capable of producing approximately 77,500 separate plates, not counting those formed during arm regeneration.

The six plate types chosen to represent this complex skeleton are listed below:

1. "Small pinnulars"-with 7 or 8 unbroken spines making up a "crown" or "collar" which surrounds a "core" of fine-mesh stereom, and lacking spines on the "shank" of the pinnular.
2. "Large pinnulars"-with 8 or more "collar" spines, commonly fused at their bases, and with small to large spines on the shank.
3. "Small brachials" or "U-shaped brachials"-having a distinctive horseshoe shape, composed mostly of coarse-mesh stereom.
4. "Medium-sized brachials"-with well-developed spines on the aboral margin of facet and with a deep, V-shaped intermuscular groove separating muscle fossae.
5. "Large brachials"-similar to medium-sized brachials, but larger, more elongated, and with smaller spines at margin of facet.
6. Cirrals-both tall and short, with thin rim at periphery of distal facet.

Crinoid stereom (Macurda et al., 1978) is formed by a combination of accretion and resorbtion of previously formed calcite. Accreting ends of the calcite rods which make up the stereom are seen as rounded knobs. The surfaces of the stereom appear smooth at high magnification, except for some irregular pits and microborings at or near the plate margins. Individual plates display repair of injuries by regenerated stereom. Spinose plates commonly have spines with broken tips which are not yet regenerated; some of this breakage undoubtedly took place during collection.

PART 2. ABRASION

Procedures

The primary piece of equipment used in our abrasion experiments was a miniature wave tank consisting of a plastic pan 23cm long, 12.5cm wide, and 4cm deep (Rubbermaid's finest), mounted on roller-skate wheels and driven by a phonograph turntable. A sediment sample taken at Telephone Pole reef was picked to remove all echinoderm grains and 40g of this sample served as the abrasive. The water used was 300ml of buffered sea water (34 parts per thousand). The crinoid plates added to the wave tank consisted of the following carefully examined elements:

Small pinnulars	40
Large pinnulars	40
Small brachials	40
Medium-sized brachials	30
Large brachials	20
Cirrals	40

The wave tank completed 42 cycles per minute, and the grains were observed to move by traction approximately 5 to 15cm on each half cycle, for an estimated transport rate of 4.2 to 12.6 meters per minute.

The wave tank was stopped at intervals and the plates were either sampled (10 specimens of each type were removed and examined), or a complete census was made (all specimens, including fragments, were removed, counted and examined).

Observation Schedule

Interval Time	Elapsed Time	Type of Observation
20 min	20 min	Sample
40 min	1 hr	Census
1 hr	2 hr	Sample
1 hr	3 hr	Census
1 hr	4 hr	Sample
2 hr	6 hr	Census
3 hr	9 hr	Sample
3 hr	12 hr	Census
6 hr	18 hr	Sample
6 hr	24 hr	Census
12 hr	36 hr	Sample
12 hr	48 hr	Census
12 hr	60 hr	Sample
12 hr	72 hr	Census

Results

Diagnostic criteria

Grain morphology:

Some small spines, such as the delicate "collar" spines of small pinnulars, break at the slightest movement, and the amount of spine breakage continues with increasing abrasion. The plates themselves are broken according to definite patterns. For example, after most of the spines of the small pinnulars have been broken off, the remains of the "collar" typically breaks away allowing the fine-mesh "core" to be removed. In some cases, these patterns of breakage allow one to distinguish between breakage caused by abrasion and breakage caused by biologic activity (predators, scavengers, deposit feeders, etc.).

Surface relief decreases and grain outline becomes increasingly rounded. Medium-sized brachials become rounded "buttons" after extreme abrasion. In addition, elements assume their former shapes as the exterior layers of the stereom are worn away, reflecting the importance of the accretionary mode of growth. Nearly cylindrical cirrals become flatter because of preferential wear on lateral surfaces, and extremely abraded pinnulars even reveal former spines (along the shank as well as at the distal end).

Surface textures:

1. Breakage. Spine breakage often occurs during life and with minimal post-mortem transport, but a large number of fractured or chipped rods within the stereom is an indicator of abrasion. Many broken surfaces have rounded or beveled edges.

2. Rounded and polished surfaces. Broken surfaces are blunted, rounded, and polished. Moderate abrasion of brachials results in noticeable polish of the fulcral ridge and surrounding area. Extreme abrasion rounds and polishes other surfaces.

PART 3. DISSOLUTION

Procedures

Slow dissolution

Ten specimens of each of the six plate types were placed in a solution of formic acid with a pH of 3.90. One specimen of each type was removed after ten logarithmic intervals starting at 20 minutes and ending at 170.5 hours.

Rapid dissolution

Specimens were rapidly etched in a solution of hydrochloric acid at a pH of 1.50 for one minute.

Results

Diagnostic criteria

Grain Morphology:

Dissolution affects gross shape less rapidly then abrasion does. For example, brachials retain the V-shaped adoral groove even after the muscle fossae are removed by dissolution.

Surface textures:

1. Pointed terminations of calcite rods. The blunt growing-tips of the calcite rods are sharpened by mild dissolution, and moderate to extreme dissolution destroys continuous rods by severing them at the midsection (the thinnest part). In both cases, pointed projections result.

2. Etched cleavage planes. Visible at high magnification after moderate to extreme dissolution.

3. Dissolution pits. May be difficult to distinguish from microborings, but, unlike microborings, may be found almost anywhere on the plate.

4. Excavated rods. The interior of the calcite rods appears to be more susceptible to dissolution than the outer zone. As a result, extreme dissolution causes hollowing of the rods, forming tubes.

PART 4. RECENT SEDIMENT SAMPLES

Procedures

This part of the study involves sediment samples collected in 1986 at Snapshot Reef and at *Cervicornis* Reef (Telephone Pole Reef) in Fernandez Bay, and a sediment sample taken at a water depth of 90 feet off French Bay. Samples were collected at the uppermost few cm of the sediment with the aid of SCUBA. On average, 5 grams of sediment yielded 1 to 20 crinoid plates depending upon location and grain size sampled.

Results

Crinoid plates recovered from the sediment samples display a wide range of morphologies and surface textures. Some plates are very rounded and polished, duplicating the appearance of the extremely abraded laboratory specimens. Others (even from the same sample) show the diagnostic surface textures of dissolution. However, most plates display evidence for several taphonomic processes, and, for the majority, biologic processes appear to dominate over both abrasion and dissolution. Many plates are broken and worn in ways not predicted by the laboratory abrasion experiments, and some of these show grooves and scratches probably caused by predators, scavengers, and/or deposit feeders. In addition, many plates are partly coated with one or more types of algae(?), whose effects on the stereom are not yet clear. Distinct microborings are present, but inconspicuous on most plates examined to date.

PART 5. PLEISTOCENE ROCK SAMPLES

Procedures

Two Pleistocene reefs on the west side of the island were sampled, Cockburn Town Reef and Sue Point Reef. Rock samples were boiled in Calgon and water, and broken with a rubber mallet. The sediment freed by these methods was washed and screened.

Results

The Cockburn Town reef samples did not yield any crinoid specimens, but some of the samples from the Sue Point Reef contained a few brachials. Diagenetic processes in the exposed reef rock have masked some of the evidence for biostratinomic processes, but at least one plate shows the rounding and polishing associated with extreme abrasion. Most specimens, however, have a very irregular rounded outline possibly caused by diagenetic processes including the activity of the blue-green algae(?) which stain these grains.

GENERAL CONCLUSIONS

Laboratory experiments show that abrasion and chemical dissolution produce distinctive sets of effects on modern crinoid stereom. Aspects of **grain morphology** are most useful in recognizing abrasion--extensive spine breakage and rounding of the plate, as well as distinctive modes of plate destruction. Details of **surface texture** are most diagnostic for dissolution--pointed terminations within the stereom, etched cleavage planes, dissolution pits, and hollowed stereom rods.

These surface textures can be found in crinoid plates recovered from Recent sediment samples, and used to recognize the role played by abrasion and dissolution in the natural habitat. The criteria we define can also be useful in isolating the evidence for biologic activity. Crinoid plates in Pleistocene rocks are rare and their biostratinomic history may be difficult to reconstruct because of the diagenetic overprint.

ACKNOWLEDGEMENTS

We thank Donald T. Gerace, Executive Director of the Bahamian Field Station, and other Field Station personnel for assistance with the logistics of the field work involved in this study. Mr. Clark R. Chambers assisted with sample processing in the laboratory, and Ms. Tracy Jones helped with SEM analysis. We also thank the Auburn University Grant-in-Aid Program for financial assistance.

REFERENCES CITED

Blyth Cain, J.D., 1968, Aspects of the depositional environment and paleoecology of crinoidal limestones: Scottish Journal of Geology, v. 4, p. 191-208.

Liddell, W.D., 1975, Recent crinoid biostratinomy (Abstract): Geological Society of America, Abstracts with Programs, v. 7, p. 1169.

Macurda, D.B., Jr., D. L. Meyer, and M. Roux, 1978, The crinoid stereom, in R.C. Moore and C. Teichert, eds., Treatise on Invertebrate Paleontology, Part T, Echinodermata 2, v. 1, Boulder, Colorado and Lawrence, Kansas, Geological Society of America and University of Kansas, p. T217-T228.

Meyer, D.L., 1971, Post-mortem disarticulation of Recent crinoids and ophiuroids under natural conditions (Abstract): Geological Society of America, Abstracts with Programs, v. 3, p. 645.

Meyer, D.L., 1985, Evolutionary implications of predation on Recent comatulid crinoids from the Great Barrier Reef: Paleobiology, v. 11, p. 154-164.

Meyer, D.L., C.A. LaHaye, N.D. Holland, A.C. Arneson, and J.R. Strickler, 1984, Time-lapse cinematography of feather stars (Echinodermata: Crinoidea) on the Great Barrier Reef, Australia: demonstrations of posture changes, locomotion, spawning and possible predation by fish: Marine Biology, v. 78, p. 179-184.

Meyer, D.L. and K.B. Meyer, 1986, Biostratinomy of Recent crinoids (Echinodermata) at Lizard Island, Great Barrier Reef, Australia: Palaios, v. 1, n. 3, p. 294-302.

HOLOCENE DEPOSITIONAL HISTORY OF CONCEPTION ISLAND, BAHAMAS

Steven W. Mitchell
Department of Geology
California State University
Bakersfield, CA 93311

Nancy Buening
Department of Geology
Sonoma State University
Rohnert Park, CA 94928

John N. Baldwin, Jr.
Earth Sciences Board of Studies
University of California
Santa Cruz, CA 95064

Betsey Westell
Department of Geology
California State University
Los Angeles, CA 90032

ABSTRACT

The surficial rocks of Conception Island can be assigned to four different lithofacies: 1-Holocene multiple beach-dune ridges, 2-late Pleistocene poorly cemented eolianites, 3-late Pleistocene well cemented eolianites, and 4-late Pleistocene subtidal calcarenites. The Holocene depositional history of Conception Island began with an atoll-like island having many inlets to the interior lagoon. Late Holocene accretion of beach-dune ridges enclosed the lagoon to form a tidal creek with a single inlet. Soil erosion, associated with prehistoric and Loyalist plantation clearing, has contributed to the rapid infilling of the tidal creek system over the past 500 years.

INTRODUCTION

Conception Island is the surface expression (about four square kilometers in area) of an isolated platform in the central Bahamas. The platform has an aerial extent of about 65 square kilometers and is generally submerged to a depth of 20 to 30 meters. The island is located 50 kilometers west-southwest of San Salvador island and 20 kilometers northwest of Rum Cay. Conception Island is a Bahamas National Trust Park; it is unoccupied and has no roads or airstrip. The general geography of the island is presented in Figure 1. For reference purposes, the apparently unnamed points have been given the names of the compass directions toward which they point. The coast lines will be referred to in the same way: the coast between North and West Points would be the northwest coast, the coast between West Point and Wedge Point would be the southwest coast, the coast between Wedge Point and East Point would be the southeast coast, and the coast between East Point and North Point would be the northeast coast. Also, the small island northwest of East Point is here referred to as East Point Cay.

Although Conception Island is presently unoccupied, there is evidence that this island was inhabited in prehistoric times (Hoffman, 1988) and a Loyalist plantation, including the remains of buildings, stone fences, and a well still containing fresh water, is located in the area of West Point (Fig. 1).

PURPOSE

This paper summarizes the results of the first geological reconnaissance of Conception Island. There are no previous publications covering any aspect of the geology of the island. A thorough understanding of the Holocene sedimentary development of Conception Island will be extremely useful in interpreting the events associated with the internal flooding of much larger islands in the Bahamas during a very high sea level stand 120,000 - 125,000 years ago (Mitchell, 1987). On a much smaller scale, Conception Island presently probably appears very similar to larger islands (such as Great Inagua Island and Rum Cay) at the peak of this high sea level stand. Also, the small size of the Conception Island Platform should provide a relatively uncomplicated record of the effects of the successive transgressions and regressions affecting the Bahama Archipelago during the late Quaternary. Thus, Conception Island should provide a useful modern analog for interpreting the Quaternary geological history of the Bahamas.

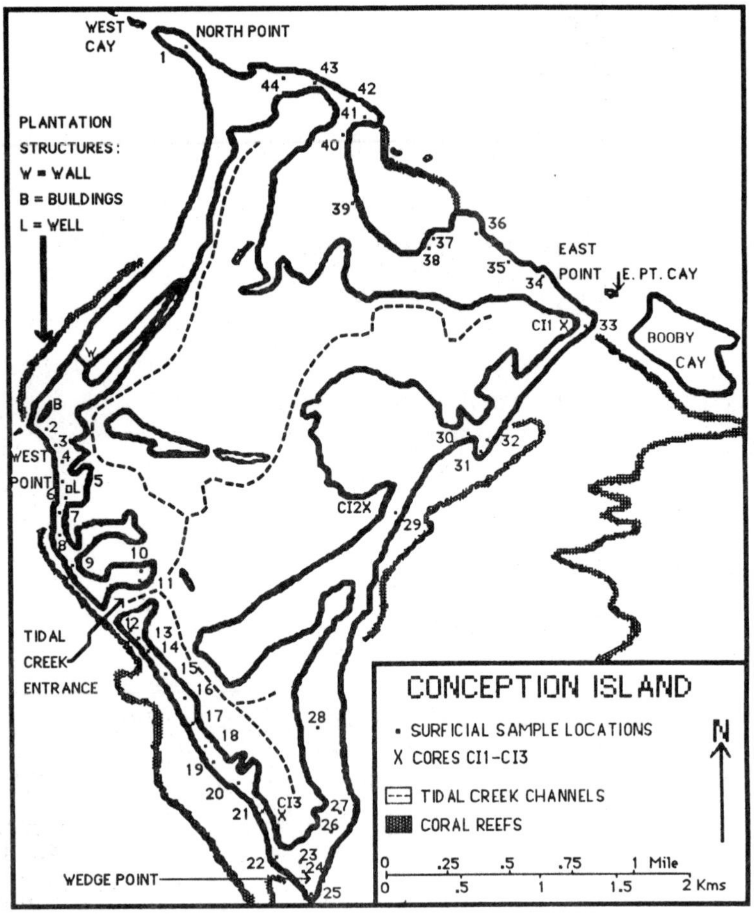

Fig. 1. Map of the geography of Conception Island showing the locations of surficial field samples and shallow cores.

METHODS

A geological reconnaissance of Conception Island was carried out during the summer of 1987 as a part of a Center for Field Research Earthwatch Expedition to the central Bahama Islands. Three days of field work by the authors and 12 volunteer assistants were undertaken in June; the senior author returned to the island in August and conducted one additional day of field survey work with 10 volunteer assistants. Further information on the first geological survey of Conception Island is given by Dumene (1988).

The areas sampled in summer 1987 are shown in Figure 1 and include all coastlines of the island. Samples were collected from all major Holocene and Pleistocene lithofacies encountered. Coastal, lake, and tidal creek sediments were collected. Also, three shallow cores (60-68cms in depth) were obtained from the interior tidal creek system. The locations of lithologic and sediment samples and cores CI1-CI3 are also shown in Figure 1.

Laboratory analyses included thin section petrology, mechanical sediment analysis, and micropaleontologic and grain morphotype analysis.

A Phillips Norelco XRG-3000 X-ray diffractometer was used for the investigation of the bulk mineralogy of sediments and lithologic samples. The silt-clay and sand fractions of the units of cores CI1-CI3, surficial sediments and crusts, and samples of Holocene and Pleistocene lithofacies were analyzed to determine the relative abundances of low and high Mg-calcite, aragonite, and protodolomite. Approximate mineral abundances were determined using a set of mixed standards as described by Blackmon (1962).

MAJOR SURFICIAL LITHOFACIES

Five major surficial lithofacies were mapped and sampled during the field survey of Conception Island. Mineralogic, sedimentologic, and paleontologic analyses of the samples provide evidence for the age and depositional environments of the lithofacies. The approximate distribution of each lithofacies, and the locations of major exposures which were sampled, are presented in Figures 1, 3, and 7. The lithofacies are described below in order of increasing geologic age.

Fig. 2. Aerial photograph of the inlet to the tidal creek system, looking northwest (Booby Cay is in the background).

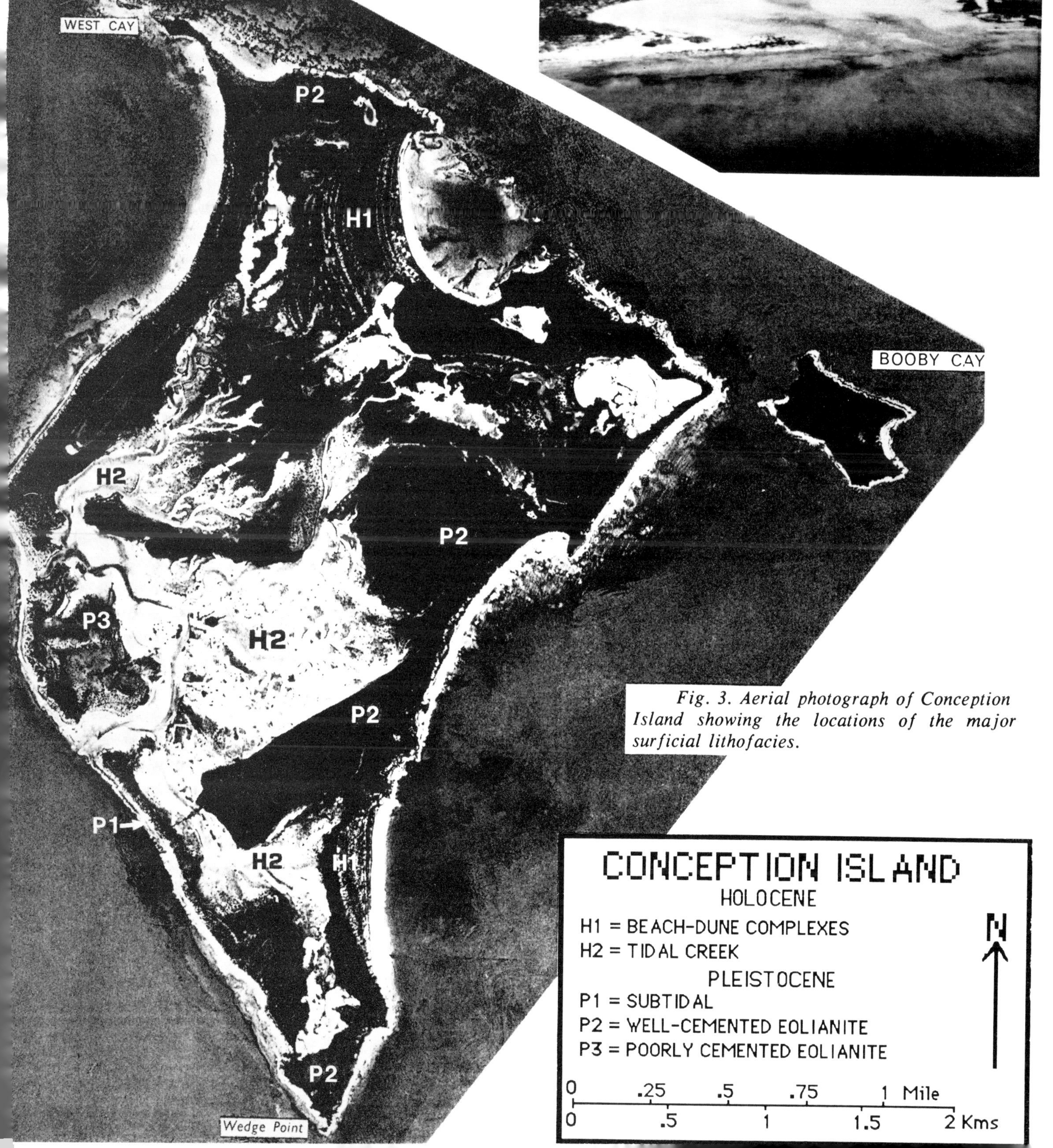

Fig. 3. Aerial photograph of Conception Island showing the locations of the major surficial lithofacies.

Fig. 4. Aerial photograph of the tidal creek system of southern Conception Island, looking southwest (tidal creek inlet is in the left background).

Fig. 5. Ground view of Pleistocene eolianite lithofacies, eastern Conception Island (Loc. 34), looking southeast.

Fig. 6. Ground view of Pleistocene eolianite lithofacies with rhizocretions, eastern Conception Island (Loc. 31), looking northeast.

Holocene Multiple Beach-Dune Complex Lithofacies

Since the last lowest sea level stand about 17,000 years ago, rising sea level has inundated the Conception Island Platform. The island has grown by the accretion of Holocene beach-dune complexes, especially during the past 3,000 years when sea level reached levels near those at present. The sediments of the beach-dune complexes are predominantly skeletal. However, some peloids and ooids do occur in Holocene sediments from the southern coasts of Conception Island (Locs. 19 and 33). The unconsolidated Holocene beach-dune sediments usually contain 65-90% aragonite, 5-30% Mg-calcite, and 3-5% low Mg-calcite. The accretion of Holocene beach-dune ridges has closed off the interior of Conception Island, forming a large, shallow tidal creek system (Figs. 2-4, 7). The creek system would be classified as a single inlet, peloid lithofacies tidal creek using the classification of Mitchell (1987).

Well cemented Holocene beachrock is developed at many locations along the coasts of Conception Island; samples were collected from five of these occurrences: Localities 2, 34, 36, 38, and 42. Aragonite ranged from 50 to 85% in the samples, while Mg-calcite varied between 15 and 35%. Low Mg-calcite was more common (5-15%) than in Holocene beach-dune sediments.

Late Pleistocene Poorly Cemented Eolianite Lithofacies

Following the late Pleistocene highest sea level stand about 120,000 - 125,000 years before present, a series of fairly high sea level stands occurred. Some of these high stands inundated the Conception Island Platform, generating sediments ultimately incorporated into poorly cemented late Pleistocene eolianites composed of fine oopelsparites. This lithofacies seems to be restricted to areas of the northeast and northwest coasts of the island, where it overlies the Well Cemented Eolianite Lithofacies (Fig. 7). A sample from Locality 44 consisted of 60% aragonite, 35% Mg-calcite, and 5% low Mg-calcite -- a composition similar to that of modern coastal dunes. Since the

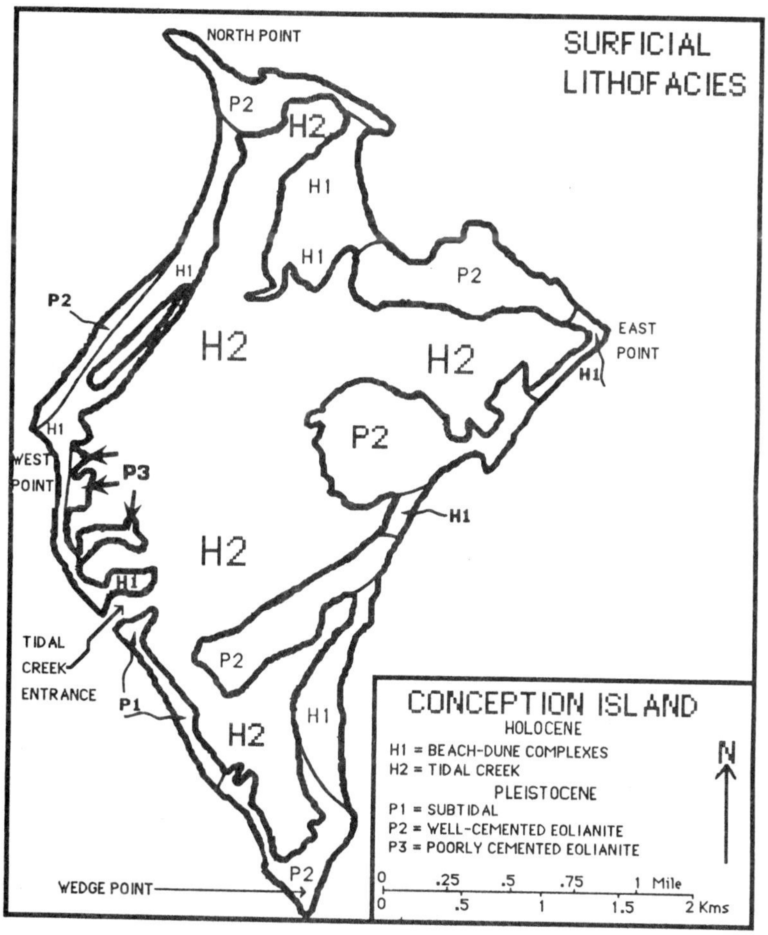

Fig. 7. Preliminary map of the surficial lithofacies of Conception Island.

areal extent of this lithofacies at Locality 44 is not known, its distribution along the northeast coast is not shown in Figure 7. A sample from Loc. 3 (where a thin layer of Holocene dune sand overlies this lithofacies) consisted of 80% aragonite, 5% Mg-calcite, and 15% low Mg-calcite.

Late Pleistocene Well Cemented Eolianite Lithofacies

This lithofacies overlies the Pleistocene Subtidal Lithofacies, which was deposited during a very high sea level stand 120,000 - 125,000 years before present. In areas where the Subtidal Lithofacies is exposed, it grades vertically into the Well Cemented Eolianite Lithofacies, forming a regressive sequence. The Well Cemented Eolianite Lithofacies is composed of coarse oosparites, oopelsparites, and rounded biosparites. Large, individual paleodunes are exposed along the eastern coast of Conception Island; rhizocretions are locally well developed (Figs. 5-6). The land gastropod *Cerion* is a common fossil in this Lithofacies at Loc. 41. The mineralogy of samples of the Well-Cemented Eolianite Lithofacies from ten localities (Locs. 4, 16, 24, 27, 29, 30 31, 41, 42, and 43) was analyzed. Most of the samples contained 75-80% aragonite and 20-25% low Mg-calcite. Two samples (Locs. 41 and 42) were from calcrete zones associated with rhizocretions. These samples had a much lower level of aragonite (25-40%), with low Mg-calcite increasing to 60-75%.

Pleistocene Shallow Subtidal Lithofacies

The oldest surficial lithofacies mapped on Conception Island occurs sporadically along all coasts, where it underlies the younger eolianite units (Figs. 3 and 7). The Pleistocene Shallow Subtidal Lithofacies is best exposed along the coast south of the inlet to the interior tidal creek system (Locs. 12-17). In this area, the Lithofacies consists of a very well cemented zone of coral reef rubble extending for a distance of about 1 kilometer along the coast. The Lithofacies is also exposed below the Pleistocene Well Cemented Lithofacies at locations (for example Locs. 30 and 42) along the eastern coast of the island. Isolated heads of *Diploria strigosa* (Dana) are present at Loc. 30.

QUATERNARY DEPOSITIONAL HISTORY

The late Quaternary surficial lithofacies of Conception Island are suggestive of the following sequence of events:

1. The formation of reef rubble (calcirudites) and other shallow subtidal sediments on bathymetic highs of the platform during a very high sea level stand 120,000 - 125,000 years before present.
2. The deposition of (well cemented) eolianites as a regressive sequence directly overlying the subtidal lithofacies.
3. The later deposition of a series of (poorly cemented) eolianites during high sea level stands in the latest Pleistocene.
4. The flooding of the Conception Island Platform in the mid-Holocene, forming an atoll-like feature (similar to the present Hogsty Reef of the southern Bahamas) with large inlets to the central lagoon occurring on all sides of the island.
5. A rapid late Holocene inflling of the lagoon and closing off of the inlets by the longshore, eolian, and tidal transport of sediments.

LATE HOLOCENE DEPOSITIONAL HISTORY

Interpretation of the Cores

Three shallow cores (60-68cms in depth) were obtained from the edges of the tidal creek system at East and Wedge Points and at about the middle of the southeast coast of the island (Fig. 1). The analysis of cores CI1-CI3 provides detailed information on the last 500 years of the development of the central tidal creek system. The cores were divided into units 3cms or less in thickness. The silt-clay percent, bulk mineralogy, grain morphotypes, and foraminiferal, ostracod, and molluscan biofacies of each unit were analyzed. A summary of the biofacies and silt-clay abundances is given in Figure 8; the levels of low Mg-calcite and protodolomite are plotted in Figure 9.

Biofacies

Mitchell (1984, 1985), Mitchell and Keegan (1987), Mitchell and Sealey (in press), and Mitchell and Sigler (this volume) have developed a classification of foraminiferal, ostracod, and molluscan biofacies for use in evaluating the paleoenvironments of Bahamian tidal creeks and lakes. The Conception Island cores preserve many of these biofacies.

<u>Foraminiferal Biofacies</u>. Three foraminiferal

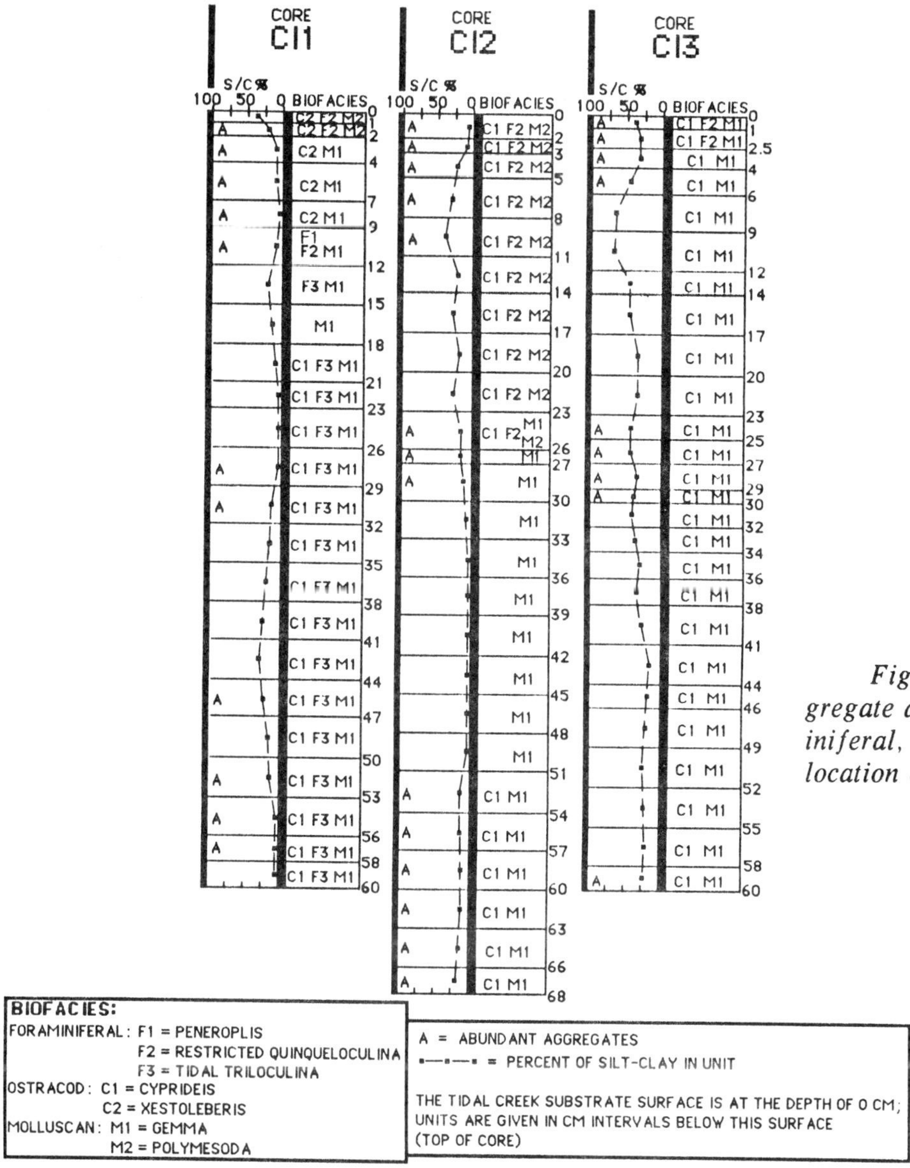

Fig. 8. Summary of silt-clay percent, aggregate abundance, and biofacies (ostracod, foraminiferal, and molluscan) of cores CI1-CI3; the location of the cores is given in Figure 1.

biofacies were encountered in the cores: (F1) *Peneroplis* Biofacies, (F2) Restricted *Quinqueloculina* Biofacies, and (F3) Tidal *Triloculina* Biofacies. The *Peneroplis* Biofacies is dominated by the species *Peneroplis proteus* Orbigny, which is most common in the high salinity portions of the upper reaches of tidal creeks. *Quinqueloculina bosciana* Orbigny and *Q. costata* Orbigny are the characteristic species of the Restricted *Quinqueloculina* Biofacies. This foraminiferal biofacies occurs in hypersaline to brackish lakes and in the most restricted upper reaches of tidal creeks. The Tidal *Triloculina* Biofacies is present in tidal creek channels where salinities and dissolved oxygen levels are intermediate between those of the open coast and the upper reaches of the creek system. *Triloculina bassensis* Parr and *T. bermudezi* Acosta are common species in this biofacies.

Ostracod Biofacies. Two ostracod biofacies were present in the cores: (C1) *Cyprideis* Biofacies, and (C2) *Xestoleberis* Biofacies. The *Cyprideis* Biofacies contains *Cyprideis americana* and *Hemicyprideis setipunctata* (Brady); *Actinocytheris subquadrata* Puri is occasionally present. This biofacies is found in hypersaline lakes and in the restricted upper reaches of tidal creeks. The *Xestoleberis* Biofacies includes the species *Xestoleberis curassavica* Klie and *Dolerocypria inopinata* Klie. These species are most common in slightly hypersaline lacustrine and tidal creek environments with salinities typically ranging between 38 and 42 o/oo.

Molluscan Biofacies. Two molluscan biofacies occur in the cores: (M1) *Gemma* Biofacies, and (M2) *Polymesoda* Biofacies. *Gemma* (Totten) is an abundant bivalve in the mid to upper reaches of

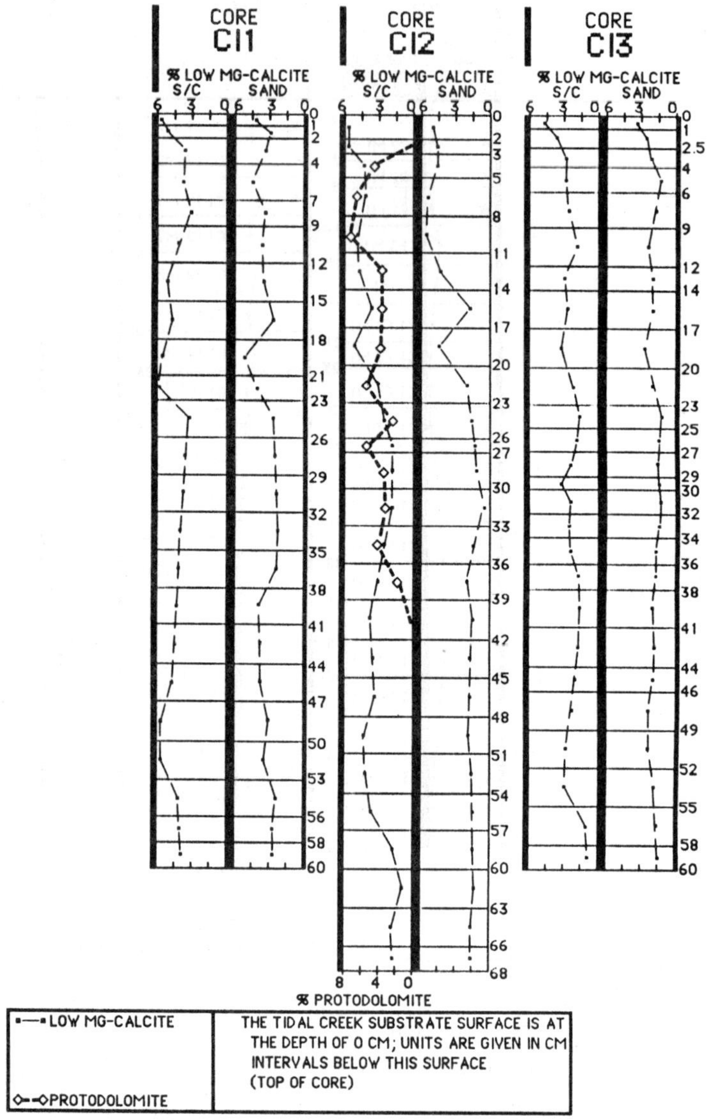

Fig. 9. Mineralogy of the silt-clay and sand fractions of cores CI1-CI3 given in percent of low Mg-calcite and protodolomite; the remainder of each fraction is composed of aragonite and high Mg-calcite.

tidal creeks in the central Bahamas. In the Conception Island cores this species is often found associated with the gastropods *Acteocina canaliculata* (Say), *Cerithidea costata* (da Costa), *Cerithium eburneum* Bruguiere, *Diastoma varium* (Pfeiffer), *Marginella carnea* (Storer), *Olivella* floralia (Duclos), and *Zebina browniana* (Orbigny) and the bivalve *Tellina iris* Say. Bahamian lacustrine environments typically contain the *Polymesoda* Biofacies, which is dominated by bivalves such as *Polymesoda maritima* (Orbigny) and *Anomalocardia brasiliana* (Gmelin). The gastropods *Batillaria minima* (Gmelin) and *Cerithidea costata* (da Costa) are common. The biofacies also occurs in the restricted, uppermost reaches of tidal creeks.

Core CI1

This core was obtained from a small hypersaline lake recently cut off from the tidal creek system at the end of East Point (Fig. 1). The molluscan and foraminiferal biofacies indicate a transition from a tidal creek to a lacustrine environment at a depth of 2cm (Fig. 8). Between 2 and 12cm, the biofacies suggest an environment typical of the upper reaches of the tidal creek system, but only slightly hypersaline. The sand fraction of the sediments from 0 to 9cm is largely composed of the rounded skeletal sand of the nearby beach and low dunes, while fine peloids dominate in the interval from 38 to 60cm. Between 9 and 38cm, the sediment is a variable mix of these two sediment sources. The sediments of core CI1 reflect the infilling of a deep, extensive central tidal creek system to form the present localized lake environment. As the interior creek system shallowed rapidly, changes in the coastline of East Point also seem to have occurred, allowing beach sand to be blown and washed directly into the tidal creek behind East Point. This points to a thinning of the extensive sand flats off the coast of eastern Conception Island and a reduction in the size of the coastal beach-dune ridge making up East Point (Fig. 1).

Core CI2

This core was obtained from a fairly large coastal lake present at about the middle of the southeast coast of the island (Fig. 1). A tidal creek-lake transition at a depth of 23cm is indicated by the biofacies (Fig. 8). Again, the sand fraction of the upper part of the core (0 to 14cm) is dominated by rounded skeletal sand derived from the nearby beach. Fine peloid sand dominates in the depth interval of 51 to 68cm. Between 30 and 51cm the sediment is a mix of fine peloid and coarse rounded beach and contains very little silt-clay. Core CI2 is similar to the East Point core in reflecting a transition from a deeper, better developed interior tidal creek system (depth of 51 to 68cm) to a present lacustrine environment (0 - 23cm). Also, there is similar evidence for coastline modification that increased the likelihood of the transport of beach sand into the interior of the island. The overall sequence of sediments and biofacies in core CI2 suggests that the low coastal beach and dune sediments were breached, forming a local temporary entrance to the tidal creek system. The sediments in the depth interval between 30 and 51cm were deposited during the time this inlet was open. By the time the inlet had begun to close off (23 to 30cm) the infilling of the interior tidal creek had already isolated this eastern arm of the creek system. The final sealing off of the temporary inlet then produced the present lake. The sediments of core CI2 are unusual in that they contain, in only the silt-clay fraction, low levels of protodolomite (Fig. 9). The protodolomite appears after the temporary inlet had breached the coastal beach-dune ridge. The precipitation of protodolomite continues upwards in the core until a maximum is reached in the 8-11cm interval. Its abundance then declines; protodolomite was not found in the upper 3cm of the core. The occurrence of this mineral supports the presence of a locally developed tidal inlet providing a source of marine pore water. The protodolomite appears to have grown around aragonite needles present in the silt-clay fraction of the sediment, similar to occurrences along west-central Andros Island and in Bonefish Pond of New Providence Island (Gebelein and Others, 1980; Mitchell and Sigler, this volume).

Core CI3

Core CI3 was collected from the upper reaches of the creek system about 0.5km north of Wedge Point. The entire length of the core contains the typical tidal creek biofacies. The sand fraction of the sediment includes a variable mix of fine peloids and rounded skeletal grains. The skeletal sand is presumably derived from the adjacent beach and is most common in the core below a depth of 27cm. The core provides evidence for additional coastline modifications along southern Conception Island. The diminution of rounded skeletal grains in the upper 27cm implies that the eastern side of Wedge Point has become more protected with less beach sand being blown and washed into the interior tidal creek system. A likely cause for this change would be an increase in the width and height of the coastal beach-dune ridge running northwest from Wedge Point (Fig. 1).

Tidal Creek Development

Mitchell (1984), Mitchell and Keegan (1987), and Mitchell and Sealey (in press) have proposed that some increases in the silt-clay content of shallow coastal lacustrine cores are the result of soil erosion due to Loyalist plantation clearing in the 1780s. Teeter and Others (1987) correlate the levels of insoluble residues and selected element distributions in a coastal lake core from San Salvador Island with the same event. There was apparently only a single, small Loyalist plantation constructed on Conception Island. Field mapping indicates that the main building structures were limited to the West Point area (Fig. 1). However, the usual farming procedure was to slash and burn large areas for cotton planting. The low coppice vegetation currently characterizing Conception Island is probably due to the fact that much of the northern and eastern sides of the island were cleared by the Loyalists. These are the areas of the best soil, and most likely, the locations of fresh water lenses. Conception Island, then, seems to be an ideal location to test the hypothesis that Loyalist plantation clearing produced significant soil erosion which was introduced into the sedimentologic record of lakes and tidal creeks.

Evidence for soil erosion might be most easily be determined if a clear relationship could be established between soil composition and the sediments occurring in adjacent lakes and tidal creeks. Analysis of the late Pleistocene well-cemented eolianites and subtidal calcarenites demonstrates that soils derived from these sources would have high levels of aragonite (75% or more); the remainder of the sediment would be low Mg-calcite. Soils derived from the later Pleistocene poorly cemented eolianites and unconsolidated Holocene beach-dune complexes would be expected to contain up to 35% Mg-calcite and 15% low Mg-calcite.

The sediments occurring in the tidal creek system, and local lakes, are restricted to peloids and skeletal grains. Aragonite is the dominant mineral present; it is derived from sources such as molluscs, calcareous algae, and scleractinian corals. High Mg-calcite is much less common; sand-sized grains of this mineralogy are derived from miliolinid foraminifera, ostracods, and echinoderms. The sand-sized fraction of the tidal creek and lacustrine sediments does not contain significant levels of any skeletal types known to have a low Mg-calcite mineralogy (for example, planktonic foraminifera, amphineurans, scaphopods, and Ostreid bivalves). Consequently, the predominant source of low Mg-calcite occurring in the sediment of the Conception Island tidal creek system would be the erosion of soils derived from the Pleistocene bedrock.

Figure 9 summarizes variations in the percentages of low Mg-calcite found in the silt-clay and sand fractions of cores CI1-CI3. High Mg-calcite levels were usually very close to those of low Mg-calcite, except in approximately the upper 12cm of all three cores. In this upper portion of the cores, high Mg-calcite increased to several times that of low Mg-calcite. An increase in the abundance of miliolinid foraminifera and ostracods seems to be the reason for the increase. The remainder of each sediment fraction was aragonite, with low levels of protodolomite also occurring in core CI2. The abundances of high and low Mg-calcite are generally quite similar in the complementary silt-clay and sand fractions of each unit (below 12cm) in each core. This indicates that the particles of low and high Mg-calcite are being derived from similar sources with a wide range of particle sizes.

Maximum percentages of low Mg-calcite occur at the depths of about 48-56cm, 18-23cm, and 0-4cm. The 18-23cm range corresponds well with an increase in silt-clay attributed to Loyalist plantation clearing (Mitchell, 1984; Mitchell and Keegan, 1987; Mitchell and Sealey, in press). A similar increase can be observed in cores CI1-CI2 (Fig. 8). If the 18-23cm level sediments are associated with a 1780's plantation clearing event, then the inferred depositional rate suggests that the increase in low Mg-calcite at a depth of 48-56cm would be the result of prehistoric clearing during the 1400s. A similar increase in soil erosion due to prehistoric activity has been reported from southern Samana Cay (Mitchell and Sealey, in press). Cores CI1 and CI2 are nearest to the area of presumed plantation clearing (northwest and southwest coasts) and prehistoric occupation (southwest coast). The increases in the abundance of low Mg-calcite are remarkably similar for the two cores. Core CI3 presents similar, but less pronounced, increases.

Overall, cores CI1-CI3 preserve a history of changes in sediment mineralogy which can be attributed only to different levels of soil erosion. While the erosion may be due in part to changes in precipitation levels, the underlying cause for erosion is believed to be vegetation modification in the form of slash and burn agriculture. Depositional rates for the three cores can be determined by matching the low Mg-calcite maxima with the inferred vegetation modification. Loyalist plantation clearing took place at a depth of about 20-22cm; the end of prehistoric activity (about 1500 AD) occurred at a depth of about 48-50cm. Using these approximate dates, a portion of the late Holocene history of Conception Island can be inferred.

The interior tidal creek system was much deeper and extensively developed prior to about 1600 AD. Accelerated infilling began at this time, caused by coastline modification which increased the influx of coastal beach-dune ridge sediment into the creek system. For example, the beach-dune ridge in the vicinity of core CI2 was breached from about 1500 to 1700 AD. In the 1780s, Loyalist plantation clearing significantly increased the soil erosion marginal to the creek system. The transport of beach-ridge sediments into the creek system reached a maximum along the southeastern side of the island during the past 100 years.

CONCLUSIONS

1. The surficial lithofacies of Conception Island are, in general, very similar to those of other islands in the central Bahamas.

2. The Holocene inundation of the Conception Island Platform produced an atoll-like island similar to the present Hogsty Reef.

3. Variation in the levels of late Holocene low and high Mg-calcite present in the island's central tidal creek sediments document a correlation between prehistoric and Loyalist plantation clearing and levels of soil erosion.

4. The interior tidal creek has undergone rapid infilling during the past 500 years. Coastline modification, allowing greater sediment influx from beach-dune environments, and soil erosion, have been major factors in this reduction in the creek system's extent.

5. The zone of protodolomite occurring in core CI2, along with the extensive distribution of protodolomite in Bonefish Pond on New Providence Island (Mitchell and Sigler, this volume) indicates that the contemporary formation of this mineral may be far more widespread in the Bahamas than previously recognized. New models for the late Holocene occurrence of Bahamian protodolomite may need to be developed.

ACKNOWLEDGEMENTS

The geological survey of Conception Island was carried out as a part of a Center for Field Research Earthwatch Expedition to the central Bahamas in summer 1987. The authors are especially indebted to the volunteers making up Teams 1 and 3 of the expedition and to Mr. Charles King of Daytona Beach, Florida who provided the use of his 42-foot sailboat *Sea Joy* for the Team 3 field research on Conception Island. Mr. Jorge Friese and Mr. Peter Kuska of the Stella Maris Inn on Long Island contributed very important support for the research by making available the 85-foot dive boat *Sol-Mar III* and a plane for aerial photography.

The authors are also indebted to Dr. Neil E. Sealey, Professor of Social Science at The College of the Bahamas, for providing logistical support for Teams 1, 2, and 3 of the Earthwatch Expedition, as well as his loan of aerial photographs of Conception Island. California State University, Bakersfield supported the senior author's research related to this project through a Reid Innovation Grant (Fall 1986), sabbatical leave award (Spring 1987), and a University Research Council release time award (Fall 1988).

REFERENCES CITED

Blackmon, P.D., 1962, Mechanical characteristics and mineralogy of the sediments in Cloud, P.E., ed., Environment of calciumcarbonate deposition west of Andros Island, Bahamas: United States Geological Survey Professional Paper 350, p. 37-64.

Dumene, J E., 1988, Looking for Columbus: Naval History, v. 2, no. 4, p. 33-37.

Gebelein, C.D., Steinen, R.P., Garrett, P., Hoffman, E.J., Queen, J.M., and Plummer, L.N.,1980, Subsurface dolomitization beneath the tidal flats of central west Andros Island, Bahamas in Zenger, D.H., Dunham, J.B., and Ethington, R.L., eds., Concepts and models of dolomitization: Society of Economic Paleontologists and Mineralogists, Special Publication No. 28, p. 31-49.

Hoffman, C.A., 1988, Caribbean in Lynch, T.F., ed., Current research: American Antiquity, v. 53, no. 1, p. 189-191.

Mitchell, S.W., 1984, Late Holocene tidal creek-lake transitions, Long Island Bahamas in Teeter, J.W., ed., Addendum to Proceedings of the Second Symposium on the Geology of the Bahamas, College Center of the Finger Lakes, Bahamian Field Station, San Salvador, Bahamas, p. 1-28.

Mitchell, S.W., 1985, Quaternary lacustrine and tidal creek microbiofacies of the Bahama Archipelago and Florida Keys: Geological Society of America Abstracts with Programs, v. 17, no. 7, p. 666.

Mitchell, S.W., 1987, Surficial geology of Rum Cay, Bahama Islands in Curran, H. A., ed., Proceedings of the Third Symposium on the Geology of the Bahamas, College Center of the Finger Lakes, Bahamian Field Station, San Salvador, Bahamas, p. 231-241.

Mitchell, S.W. and Keegan, W. F., 1987, Reconstruction of the coastlines of the Bahama Islands in 1492: American Archaeology, v. 6, p. 88-96.

Mitchell, S.W. and Sealey, N. E., in press, Late Quaternary depositional history of central Samana Cay, Bahama Islands: Caribbean Journal of Science.

Mitchell, S.W. and Sigler, M. E., this volume, Twentieth Century sedimentological development of Bonefish Pond, New Providence Island, Bahamas.

Teeter, J.W., Beyke, R.J., Bray, T.F., Jr., Brocculeri, T.F., Bruno, P.W., Dremann, J.J., and Kendall, R.L., 1987, Holocene depositional history of Salt Pond, San Salvador, Bahamas in Curran, H.A., ed., Proceedings of the Third Symposium on the Geology of the Bahamas, College Center of the Finger Lakes, Bahamian Field Station, San Salvador, Bahamas, p. 145-150.

TWENTIETH CENTURY SEDIMENTOLOGICAL DEVELOPMENT OF BONEFISH POND, NEW PROVIDENCE ISLAND, BAHAMAS

Steven W. Mitchell and Marcile E. Sigler
Department of Geology
California State University
Bakersfield, California 93311

ABSTRACT

Changes in the sediments and microfaunas of Bonefish Pond during the past 100 years have been investigated by the analysis of surficial sediments, shallow cores, and map evidence. The aerial extent of the creek system has been reduced by about one-half during the past century. During this time there has been a major change in microfaunas and sediment types. Shallow cores indicate that the creek has changed from a predominately carbonate mud environment characterized by the bivalve *Polymesoda maritima*, with locally precipated gypsum zones, to the present carbonate sand environment dominated by macrophyte crusts and the bivalves *Gemma* and *Chione*. The development of macrophyte crusts as a major sediment component seems to have occurred as a result of the increasing industrial and agricultural development of southern New Providence Island. Protodolomite crusts are presently being formed throughout the creek system by the dolomitization of carbonate mud. This investigation presents evidence of a previously unreported marine environment of macrophyte crustal sediments, with associated protodolomite and gypsum, and documents the potential use of molluscan faunas in the interpretation of ancient Bahamian tidal creek environments.

INTRODUCTION

Bonefish Pond is a multi-inlet tidal creek system developed along the south-central coast of New Providence Island about 4 miles southeast of Nassau International Airport (Fig. 1). The creek system covers an area of approximately 2 1/2 square kms and has 3 inlets (here designated A, B, and C) connecting the creek with Bicardi Bay (Fig. 1). The sediments and fauna of this creek have not previously been investigated. In general, little is known about the sedimentology and faunas of the numerous tidal creeks occurring throughout the Bahama Archipelago. The only detailed, published study of a Bahamian tidal creek is one on Pigeon Creek, San Salvador Island (Mitchell, 1987a).

The use of modern tidal creeks as analogs for the interpretation of ancient tidal creek sediments has become increasingly important. The geologic mapping of Rum Cay (Mitchell, 1987b), New Providence Island (Garrett and Gould, 1984), Mayaguana and Great Inagua Islands (Mitchell, 1985a, 1985c; Pierson, 1982), Samana Cay (Mitchell and Sealey, in press), and San Salvador Island (Teeter, 1985; Thalman and Teeter, 1983; Titus, 1987) indicates that a significant portion of the surficial rocks of the Bahamas were formed in Pleistocene tidal creek environments when sea

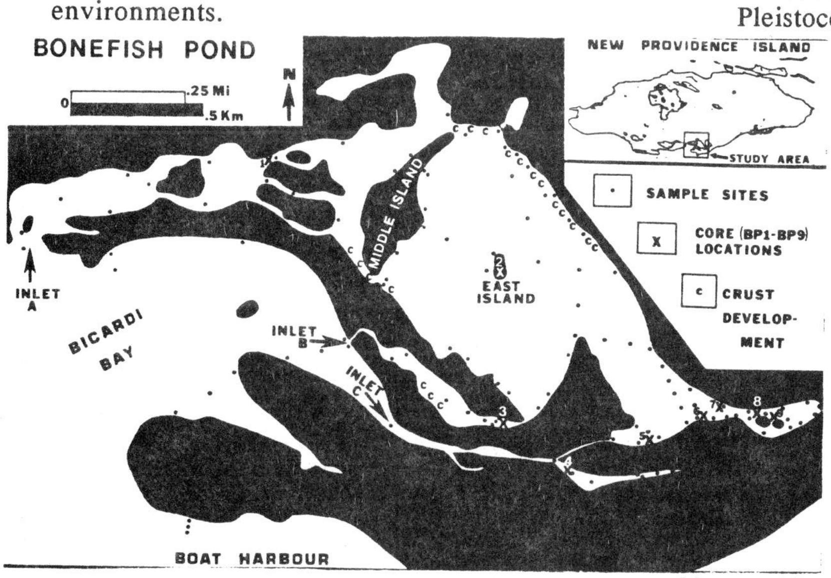

Fig. 1. Map of Bonefish Pond area showing major geographic features, sample locations, core sites, and areas of thin crust development; inset map indicates location of field area on New Providence Island.

level was higher than at present. On some islands (Great Inagua and Mayaguana Islands and Rum and Samana Cays) many interior Pleistocene rock units occurring at elevations up to 5m above present sea level can be assigned to tidal creek lithofacies (Mitchell, 1987b, p. 237; Mitchell and Sealey, in press).

The purpose of this study is to document the sedimentology and microfaunas of Bonefish Pond. The development of a previously unknown sedimentary environment is described and the first in depth investigation of the foraminiferal, ostracod, and molluscan biofacies of a Bahamian tidal creek is presented.

METHODS

A total of 200 surface sediment and alcohol-fixed/rose bengal stained microfaunal samples (upper 0-2cm), 100 water samples, and numerous current measurements were collected from 100 sampling stations in the Bonefish Pond area during 5 sampling periods: June-July 1982, July 1984, October 1985, December 1985, and June 1986 (Fig. 1). Water samples were collected in 10ml vials and analyzed for salinity using a Reichert automatic temperature compensated hand refractometer with an accuracy of $1^{o}/oo$. Current measurements were made using a neutral float suspended several centimeters below the water surface. Distances were determined using a compass and adjusted for the magnetic declination (4.5^{o} west) of the field area.

The sediment samples were collected using cores or small containers. In the laboratory each sample was treated with chlorox until all organic matter had been decomposed. Samples were then wet sieved through a U.S. Standard Series Sieve No. 230 (.06mm or 4ø diameter mesh). The silt-clay and sand subsamples were dried and weighed. To prevent loading on the sieves, the sand was, if necessary, split into 30 to 125 gram subsamples. Selected sand subsamples were then dry sieved through a 1/2ø interval set of U.S. Standard series sieves using a ro-tap for 15 minutes. The fraction retained on each sieve was weighed, and the percent each fraction made up of the total sand sample determined. The weight of the pan fraction was added to the weight of wet sieved silt and clay. Basic sedimentary parameters such as mean grain size, standard deviation, skewness, and kurtosis were then calculated. Visual and thin section identifications of the grain morphotypes occurring in the -2, -1, 0, 1, 2, and 3ø fractions of each sample were also undertaken. The results presented here are based on the silt-clay separation of 100 samples and the 1/2ø interval mechanical sieve analysis of 30 samples. Analyzed samples were selected in order to include all of the major subenvironments of the Bonefish Pond area.

Nine shallow cores were taken from various parts of the creek system. The locations of cores BP1-BP9 are shown in Figure 1. After extrusion, the cores were divided into units 1 to 7cms in thickness. Wherever possible, units were based on textural differences which suggested changes in the depositional environment. Each unit was wet sieved to separate silt-clay and sand, as described above. The dried sand subsamples were examined using a binocular microscope (X7 to X30 magnifications) for the identification of microfaunas and grain types.

A Phillips Norelco XRG-3000 powder X-ray diffractometer was used for the investigation of the bulk mineralogy of sediments. The silt-clay and sand fraction of the units from the 3 longest cores (BP2, BP6, BP8), surficial sediments and crusts, and samples of Pleistocene outcrops were analyzed to determine the relative abundances of low and high Mg-calcite, aragonite, protodolomite, gypsum, and low temperature quartz. Approximate mineral abundances were determined using a set of mixed standards as described by Blackmon (1962, p. 48).

PHYSICAL PARAMETERS

The major physical parameters related to sediment distribution in Bonefish Pond are: 1. salinity, 2. tidal currents, and 3. prevailing winds. The approximate distribution of maximum isohalines for the creek system, based on water samples from 70 stations, is shown in Figure 2. As expected, salinities increase toward the upper reaches of the creek, with a maximum salinity of $60^{o}/oo$ recorded in easternmost areas. However, salinities exceeding $80^{o}/oo$ were found to occur on tidal flats subject to flooding at high tide. During flood tides, the positions of isohalines are shifted a significant distance up creek due to coastal water ($37^{o}/oo$) entering through the 3 inlets; a similar down creek shift in the positions of isohalines occurs during ebb tides. Seasonal variation in salinity was found to occur, but has not been documented in detail.

The positions of the major tidal channels in Bonefish Pond are shown in Figure 3. The figure also presents the maximum recorded tidal current

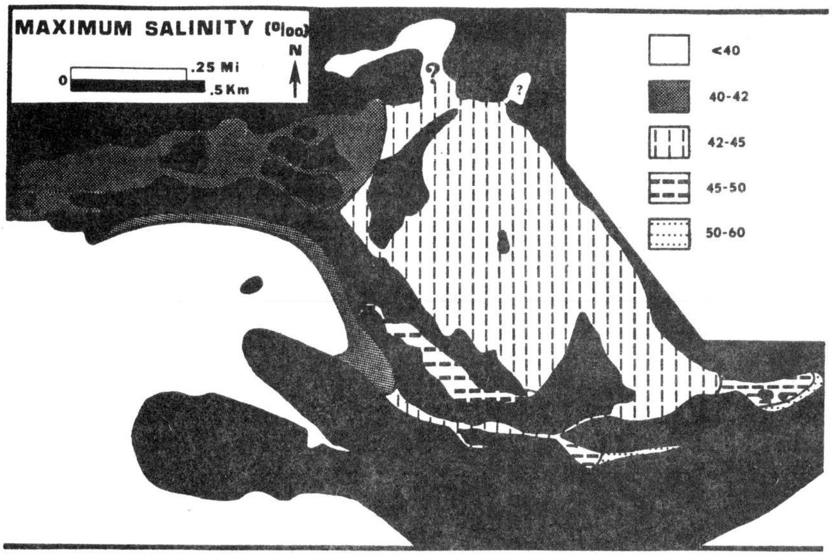

Fig. 2. Approximate distribution of isohalines, Bonefish Pond.

velocities (in meters per minute) for selected parts of the creek system. A maximum velocity of 22 meters/minute was recorded at the narrow channels connecting Inlets A, B, and C to the central and eastern parts of the creek system (Fig. 3).

SEDIMENTOLOGY

Sediment Sources

Five basic grain morphotypes occur in the sediments of Bonefish Pond: 1. aggregates; 2. lithoclasts; 3. skeletal remains; 4. macrophyte crusts; and 5. peloids. Ooids were not found to be present in the sediments of this creek system. Aggregates are broken pieces of thin, cemented sediment crusts containing high percentages of skeletal remains or macrophyte crusts. The locations where thin crusts are presently forming in Bonefish Pond are shown in Figure 1. The thin crusts are composed of protodolomite and varying levels of Mg-calcite and aragonite. Aggregates found in surficial sediments are presumably derived from one of these sources. The peloids are formed by the micritization of fecal pellets. Peloids are occasionally present in areas where sediment crusts are forming. Lithoclasts are eroded pieces of Pleistocene limestones (oosparitic eolianities) and thick Holocene crusts cropping out along the shores of Bonefish Pond or present beneath the unconsolidated tidal creek sediments. Pleistocene exposures are best developed along the southern margins of the creek; this area would be expected to have the highest concentrations of surficial lithoclasts with a low Mg-calcite mineralogy. Lithoclasts derived from thick Holocene crusts can be distinguished from those of Pleistocene origin by bulk mineralogicial analysis. The Holocene crusts are dolomitized carbonate

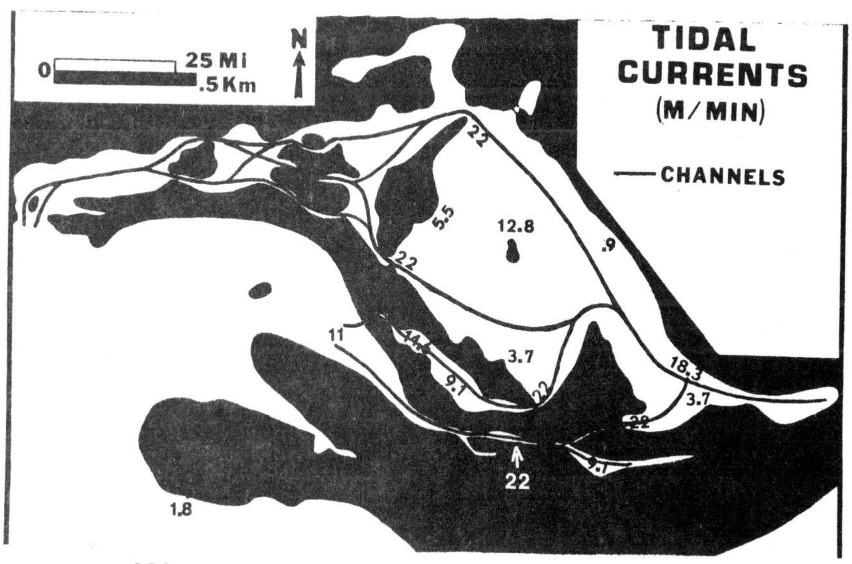

Fig. 3. Distribution of tidal channels and maximum recorded velocities of tidal currents in meters/minute, Bonefish Pond.

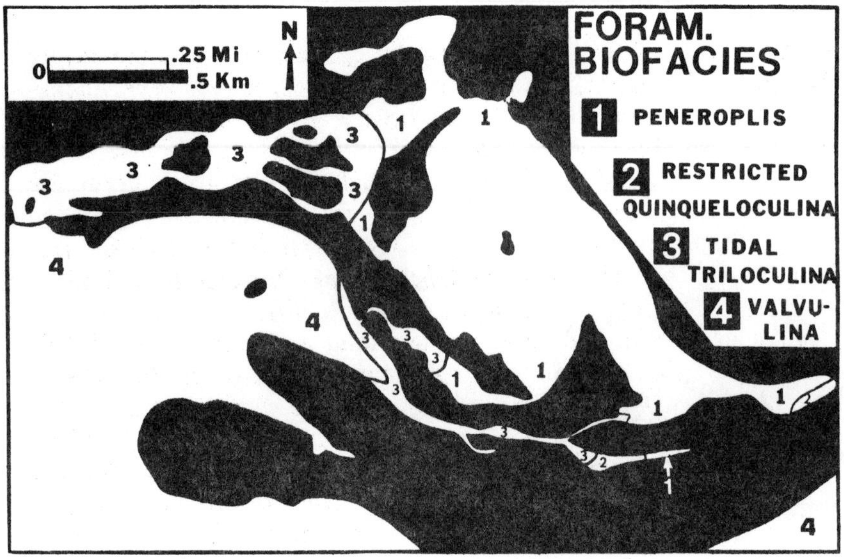

Fig. 4. Approximate distribution of foraminiferal biofacies, Bonefish Pond.

muds which are very finely crystalline and have few skeletal grains. Protodolomite and Mg-calcite are the dominant minerals present; aragonite is less common (under 15%) than it is in associated unconsolidated sediment. Very small amounts of gypsum may also be present. Macrophyte crusts are hollow tubular crusts formed by cyanophytes. Similar crusts, from a freshwater lake in Austria, have been described in detail by Schneider and Others (1983). Macrophyte crusts from Bonefish Pond are generally 0.2-0.4mm in diameter; broken crusts up to 4mm in length occur in the surficial sediments of the extreme southeastern part of the creek system. The crust exterior is usually covered by a network of microphyte crusts similar to those reported by Schneider and Others (1983). This is the first reported occurrence, in the marine environment, of macrophyte crusts as a significant contributor to the local sedimentary record.

A wide variety of skeletal grain types occur in Bonefish Pond. Some types (scleractinian corals, gorgonian spicules, and the branching foraminiferan *Homotrema rubra*) are derived from coastal fringing reefs. Shallow near-shore sand flats have concentrations of other skeletal grain types such as coralline and calcareous algae and infaunal bivalves and gastropods. The coastal sand flats and inlets to Bonefish Pond produce a diversity of skeletal types derived from organisms living as epibiota on the seagrass *Thalassia*. The most abundant types of epibiota are serpulid worms, encrusting melobesoid algae, and the foraminifera *Archaias, Cyclorbiculina, Quinqueloculina*, and *Sorites*. Within the creek the most common skeletal types are generated by the breakdown of calcareous algae (*Acetabularia, Halimeda*, and *Penicillus*) and by epibiota (for example, foraminifera and ostracods) occurring on these algae, *Thalassia*, and on mangrove roots and dead leaves. Infaunal, semi-infaunal, and epifaunal molluscs are other significant contributors to the skeletal sediments of Bonefish Pond. Skeletal grains derived from gastropods and bivalves were present at all sample stations. Shells of the gastropod genus *Caecum (C. cooperi* Smith and *C. pulchellum* Stimpson) are abundant in the sediments of the western and central parts of the creek system.

Transport Mechanisms

The various grain types occurring in Bonefish Pond may form by local current or wave effects on surficial crusts and exposed Pleistocene rocks (producing aggregates or lithoclasts) or by biological processes (producing skeletal grains, macrophyte crusts, and peloids). Once each of these grain types becomes a part of the creek system, there is the opportunity for transportation by tidal currents, waves, and bioturbation. Tidal currents are the chief mechanism of sediment transport within Bonefish Pond.

Sediments are transported up creek (from Inlets A, B, and C) during flood tides and settle out at slack high tide. This results in finer sediments being concentrated in the upper reaches of the creek system and in areas where tidal currents are low due to islands and embayments. However, the hollow tubular macrophyte crusts are preferentially transported down creek by ebb tide currents. Bioturbation mixes the sediment in a local area, moving some fine sediment to the sediment-water interface where it is likely to be transported by tidal currents or waves.

Silt and Clay

The percentages of silt and clay occurring in samples from 100 stations in the Bonefish Pond area are presented in Figure 5. Percents represent the total amount of wet and dry sieved particles less than 0.06mm in diameter. The silt-clay size component of the sediment is greatest in creek embayments and in the uppermost reaches of the creek, and is at its maximum in the extreme southeastern part of the creek system where values reach 55%. Mg-calcite and aragonite are the dominant components of the silt-clay mineralogy. Aragonite makes up 35-50% of the silt-clay sediment fraction in the coastal and western areas of Bonefish Pond. In the upper reaches of the eastern part of the creek, however, aragonite levels drop to 15-25%. The major source of this aragonite silt and clay is the physical and biological reduction of grains of thin-shelled molluscs and the calcified algae *Acetabularia*, *Halimeda*, and *Penicillus* (Neumann and Land, 1975; Stockman and Others, 1967). The sources of the Mg-calcite silt-clay are the macrophyte crusts and miliolinid foraminifera, such as *Peneroplis proteus*, which are abundant throughout the creek system (Gebelein and Others, 1980, p. 43).

One additional source contributes to the silt-clay sediment fraction occurring in Bonefish Pond. The Pleistocene eolianites cropping out along the margins of the creek are undergoing slow chemical and biological decomposition, producing low Mg-calcite grains and concentrations of terrigenous particles (primarily low temperature quartz). The terrigenous particles were incorporated in the eolianites after being transported by trade winds from northern Africa (Mitchell, 1984b). All surficial sediment samples from the interior of the creek system contain small amounts (under 1-2%) of terriginous silt-clay derived from the leaching of Pleistocene eolianites. Coastal sediments do not. The levels of terriginous silt-clay in surficial sediments are highest in the extreme southeastern part of the creek, where thick terriginous clay deposits have been leached from the underlying Pleistocene eolianites. In addition, small amounts of low Mg-calcite are present in the silt-clay fraction of the surficial sediments along the northern and extreme southeastern margins of Bonefish Pond. These are areas where Pleistocene limestones are exposed to surficial weathering.

Lithofacies

Based on the characteristic grain morphotypes present, 12 major lithofacies are recognized in the area of the Bonefish Pond tidal creek system. The 12 lithofacies are listed below along with a brief discussion of the physical and biological process responsible for their occurrence. The distributions of the lithofacies are presented in Figure 6.

1. <u>*Acetabularia-Batillaria* Lithofacies</u>. Dense patches of *Acetabularia crenulata* Lamouroux are developed in the narrow channels connecting Inlets A and B with central and eastern parts of the creek system. The channels have high tidal current velocities which encourage the development of *Acetabularia* and *Penicillus*. *Batillaria*

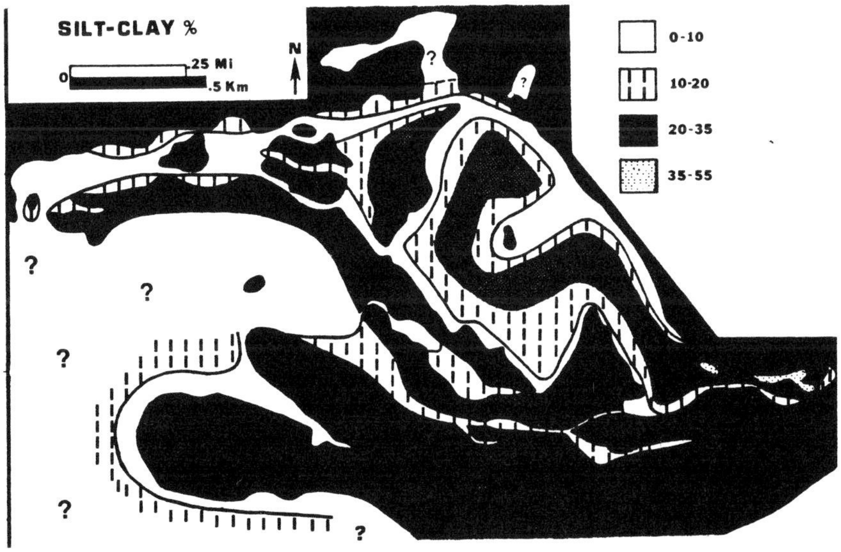

Fig. 5. Percent of silt and clay in surficial sediments, Bonefish Pond.

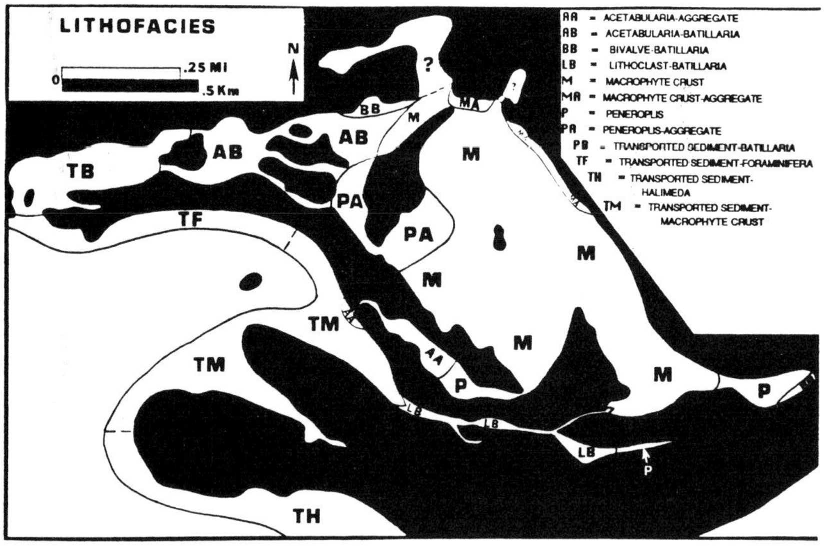

Fig. 6. Distribution of surficial sediment lithofacies, Bonefish Pond.

minima (Gmelin) is a herbivorous gastropod which is abundant in the creek system wherever algal development is significant.

2. <u>*Peneroplis* Lithofacies</u>. Large tests of *Peneroplis proteus* Orbigny are the dominant grain morphotype in shallow areas where salinities are high and a surface floc provides support for detached mangrove leaves, which are colonized by foraminifera (Mitchell, 1987a).

3. <u>Macrophyte Crust Lithofacies</u>. Macrophyte crusts are the dominant grain morphotype in the eastern part of the creek system. The

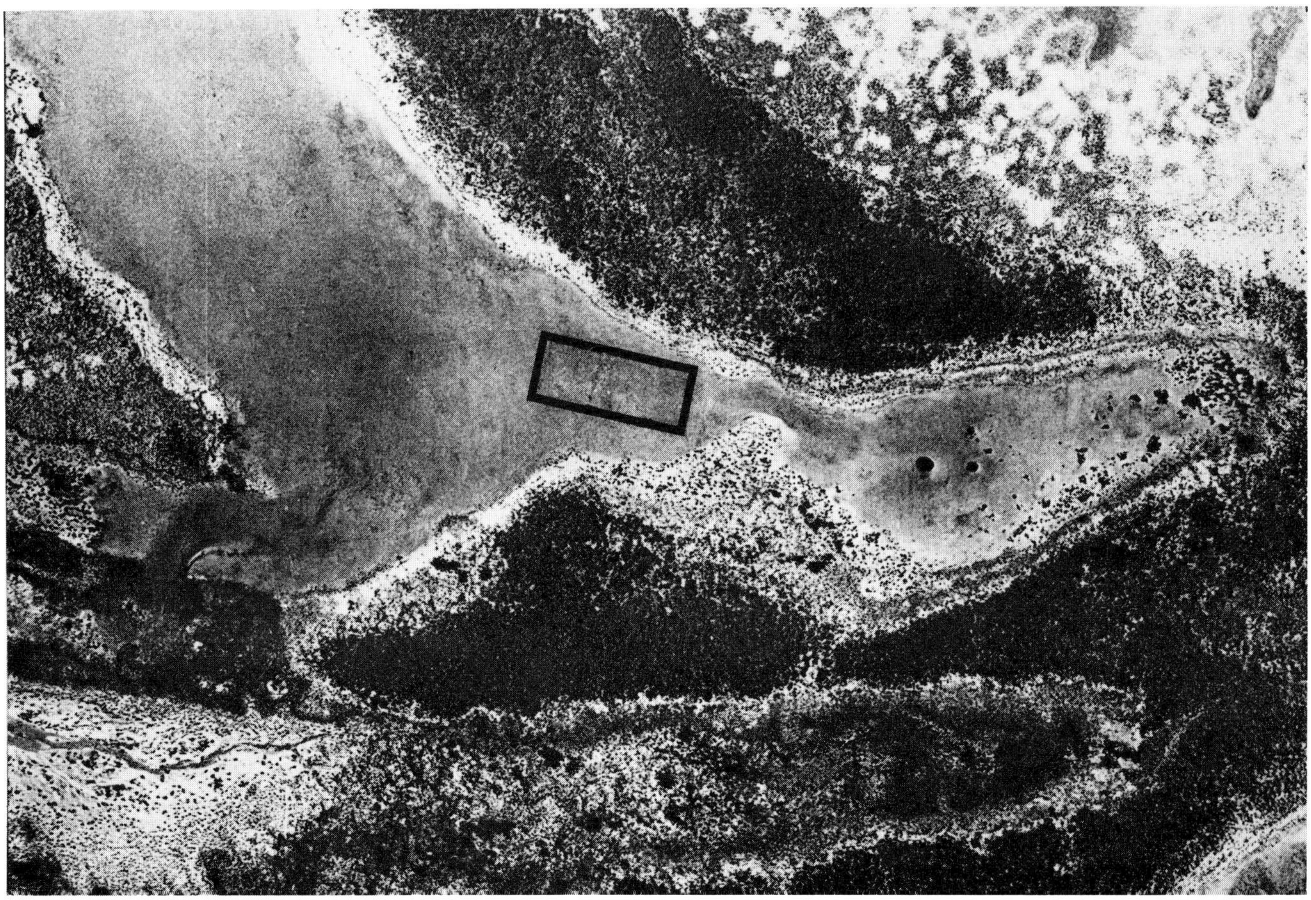

Fig. 7. Aerial photograph of southeastern Bonefish Pond, the occurrence of cyanophytes producing macrophyte crusts is indicated by the rectangle.

Fig. 8. Aerial photograph of eastern Bonefish Pond; arrows point to beaches, sand bars, and islands formed by the deposition of macrophyte crustal sediments.

field staining (with rose bengal) of sediment samples, and an associated organic floc, indicates that the cyanophytes producing the macrophyte crusts are living in an area of about 100 by 250 meters in the extreme southeastern part of the creek (Fig. 7). The hollow, cylindrical crusts are transported down creek by ebb tides, forming sand beaches, bars, and East Island (Fig. 8). The macrophyte crustal sediments become progressively finer northwest from the production area. They do not occur as significant grain types in sediments west of western Middle Island. However, macrophyte crusts have been transported through Inlet C to form an important component of surficial sediments in eastern Bicardi Bay.

4. *Acetabularia*-Aggregate Lithofacies.

5. *Peneroplis*-Aggregate Lithofacies.

6. Macrophyte Crust-Aggregate Lithofacies. A thin carbonate crust is formed below organic flocs up creek from Inlets A and B, along the edges of the creek channels north and south of Middle Island, and below an intertidal organic mat along the creek shore northwest of East Island (Fig. 1). The protodolomite crusts provide a source of aggregates for the three lithofacies, which are distinguished by the differences in types of grain morphotypes cemented into aggregates.

7. Bivalve-*Batillaria* Lithofacies. The sediments of two areas of the creek are characterized by a very fine sand fraction. These environments appear to be optimum for burrowing infaunal bivalves; most grain morphotypes are derived from the shells of the bivalves *Polymesoda* and *Anomalocardia*, as well as the associated gastropods *Batillaria minima* and *Cerithidea costata*.

8. Lithoclast-*Batillaria* Lithofacies. This lithofacies is limited to the narrow tidal channels connecting Inlet C with the southeastern part of the creek system. Due to high current velocities, a thick Holocene crust and Pleistocene bedrock is frequently exposed in the channels. The thin sediment covering consists of eroded lithoclasts and *Batillaria minima* shells; worn macrophyte crusts are also present in abundance.

9. Transported Sediment-*Batillaria* Lithofacies.

10. Transported Sediment-Foraminifera Lithofacies.

11. Transported Sediment-Macrophyte Crust Lithofacies.

12. Transported Sediment-*Halimeda* Lithofacies. The characteristic sediment of coastal areas adjacent to Bonefish Pond is a light to dark gray, reworked, and much transported skeletal sand. This transported sediment is combined with recently generated skeletal grains to produce four lithofacies. The Transported Sediment-*Batillaria* Lithofacies occurs in the lower reaches of the Inlet A entrance into the creek system, where *Batillaria minima* is abundant. This indicates that there is a net up creek movement of coastal transported sediments through Inlet A. Conversely, broken and abraded macrophyte crusts are a significant grain type along the coast adjacent to Inlet C. The resulting Transported Sediment-Macrophyte Crust Lithofacies occurs because there is an extensive net down-creek movement of sediment through Inlet C. The Transported Sediment-*Halimeda* Lithofacies occurs where *Halimeda* is abundant along the open coast of Boat Harbour. Finally, a Transported Sediment-Foraminifera Lithofacies is present along the coast between Inlets A and B west of the coastal occurrence of macrophyte crusts. The tests of abundant coastal foraminifera such as *Quinqueloculina* and *Triloculina* are combined with the gray transported sediment to form this lithofacies.

BIOFACIES

Mitchell (1984a, 1985b), Mitchell and Keegan (1987), and Mitchell and Sealey (In Press) have developed a classification of foraminiferal and ostracod biofacies based on the detailed analyses of the surface sediments of tidal creeks of Florida and the Bahamas. The foraminiferal and ostracod biofacies of Bonefish Pond are generally similar to those encountered in other Bahamian tidal creeks. The molluscan biofacies of Bahamian tidal creeks previously have not been investigated in detail. The ostracod and molluscan biofacies classification presented here for Bonefish Pond is based on the distribution of living individuals determined by rose bengal staining. The foraminiferal biofacies classification is based primarily on test preservation and abundance, since only dried microfaunal samples were analyzed; usually these samples did not provide adequate preservation of living, stained foraminifera. Consequently, the ostracod and molluscan biofacies distributions most accurately reflect the relationships between physical parameters and biofacies distributions in Bonefish Pond.

Foraminiferal Biofacies

The distribution of foraminiferal biofacies is presented in Figure 4. Four major foraminiferal biofacies are recognized in the Bonefish Pond area:

1. *Peneroplis* Biofacies. The sediments of most of the central and eastern portions of the creek system contain an abundance of large, very well preserved tests of *Peneroplis proteus* Orbigny. The fine sand fraction of the sediment contains numerous tests of *Miliolinella suborbicularis* (Orbigny). The optimum occurrence of these species appears to be limited to the areas of Bonefish Pond with the highest salinities and lowest dissolved oxygen levels and current velocities. The *Peneroplis* Biofacies is associated with lithofacies dominated by macrophyte crusts or aggregates.

2. Restricted *Quinqueloculina* Biofacies. The most extreme upper reaches of the southeastern part of the creek system are characterized by the foraminifera *Quinqueloculina bosciana* Orbigny and *Q. costata* Orbigny. These species are typically abundant in hypersaline to brackish lakes in the Bahamas (Mitchell, 1984 and 1985b). This suggests that tidal effects are minimal in these two restricted areas.

3. Tidal *Triloculina* Biofacies. The tidal channels connecting Inlets A, B, and C to the central and eastern parts of Bonefish Pond contain a limited foraminiferal fauna dominated by living specimens of *Triloculina bassensis* Parr and *T. bermudezi* Acosta. Tests of *Quinqueloculina bosciana* Orbigny, *Q. costata* Orbigny, *Q. tenagos* Parker, and *Q. poeyana* Orbigny are also abundant. The substrates in the three channels are largely composed of transported sediment (A), aggregates (B), or lithoclasts (C). Salinities vary considerably every several hours depending on whether the flow is an ebb or flood current.

4. *Valvulina* Biofacies. This biofacies is present in the coastal areas adjacent to Bonefish Pond. The most abundant species are the agglutinated foraminifera *Clavulina tricarinata* Orbigny, *Textularia agglutinans* Orbigny, and *Valvulina oviedoiana* Orbigny. Other common species include *Articulina mucronata* (Orbigny), *Peneroplis pertusus* (Forskal), *Pyrgo subsphaerica* (Orbigny), *Quinqueloculina lamarckiana* Orbigny, *Q.*

subpoeyana Cushman, *Sorites marginalis* (Lamarck), *Spiroloculina antillarum* Orbigny, *Triloculina carinata* Orbigny, *T. linneiana* Orbigny, and *T. rotunda* Orbigny. Abraded representatives of this biofacies are common throughout the western and central parts of the creek system, where they are transported by flood tide currents.

Ostracod Biofacies

Two major ostracod biofacies are recognized in the Bonefish Pond area: 1. an assemblage dominated by *Cyprideis americana*; and 2. an assemblage characterized by *Bairdia harpago* Kornicker. The distribution of these ostracod biofacies is presented in Figure 9. Virtually all sediment samples contained several living representatives of one or the other of these two taxa. In addition, living specimens of *Cytherella arostrata* Kornicker were often found associated with *Bairdia harpago*. The dead shells of other species were common throughout most of the creek system: *Actinocytheris subquadrata* Puri, *Aurila floridana* Benson and Coleman, *Dolerocypria inopinata* Klie, *Hemicyprideis setipunctata* (Brady), *Loxoconcha postdorsolata* Puri, *L. purisubrhomboidea* (Edwards), *Perissocytheridea bicelliforma* Swain, *Puriana floridana* Puri, *Reticulocytheris multicarinata* (Swain), and *Xestoleberis curassavica* Klie. Since none of these 10 species was found living in sediment samples, it is inferred that they occur on algal or rocky substrates. The *Cyprideis* Biofacies occurs in restricted areas where salinities are the highest. The *Bairdia* Biofacies occurs in coastal areas and where salinities are lowered by flood tide currents entering Inlets A, B, and C.

Molluscan Biofacies

The distribution of molluscan biofacies is presented in Figure 10. Four major molluscan biofacies are present:

1. Gemma Biofacies. This biofacies is characterized by the bivalves *Gemma gemma* (Totten) and *Laevicardium laevigatum* (Linnaeus). Well preserved dead shells of *Acteocina canaliculata* (Say) were usually found with the living *Gemma*, suggesting that the two species have a similar distribution. These species are abundant in the creek system southeast of Middle island. The *Gemma* Biofacies occurs where salinities are high, dissolved oxygen levels low, and the substrate is a fine to coarse sand composed of macrophyte crusts.

2. Polymesoda Biofacies. Sediments in two areas of Bonefish Pond are unusually fine-grained: (a) Along the shore north of Middle Island; and (b) in the extreme southeastern arm of the creek system. The two areas seem to provide an optimum substrate for the infaunal bivalves *Polymesoda maritima* (Orbigny) and *Anomalocardia brasiliana* (Gmelin) and the gastropod *Cerithidea costata* (Da Costa).

3. Chione Biofacies. A very small species of the bivalve genus *Chione* is very abundant throughout the western part of the creek system. The species occurring in Bonefish Pond seems to be most similar to *Chione pubera* (Bory Saint-Vincent). However, *C. pubera* is much larger and has different coloration. The *Chione* Biofacies is apparently dominated by an undescribed species of *Chione*. Living representatives of *Chione* and *Caecum nitidum* Stimpson were found together in most sediment samples from this area. The *Chione*

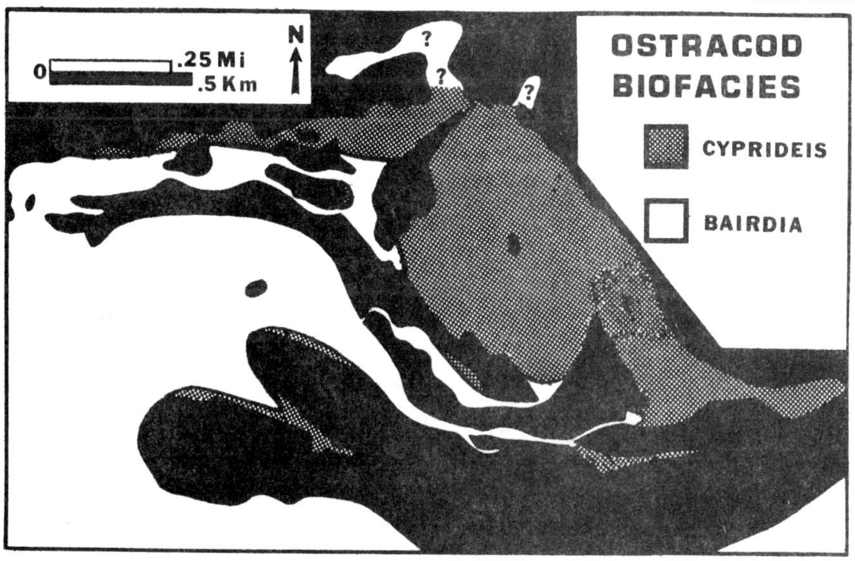

Fig. 9. Distribution of ostracod biofacies, Bonefish Pond.

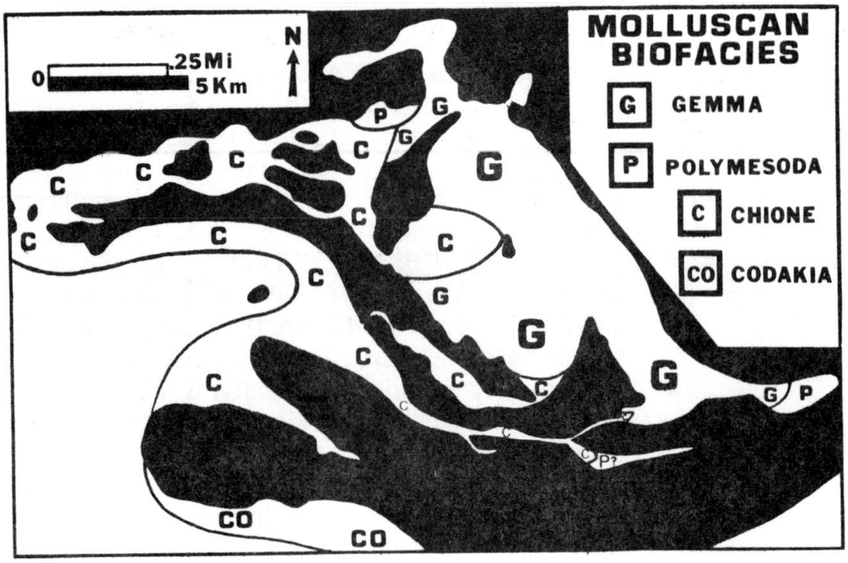

Fig. 10. Distribution of molluscan biofacies, Bonefish Pond.

Biofacies occurs where maximum salinities are about 45^{0}/oo and dissolved oxygen levels higher than those in the upper creek system. *Chione* sp. cf. *C. pubera* is most abundant in substrates consisting of the gray transported sediment and/or fine, broken aggregates or macrophyte crusts.

4. Codakia Biofacies. The bivalves *Codakia orbicularis* (Linnaeus) and *Codakia costata* (Orbigny) are typical of the diverse bivalve fauna occurring along the open coastal sand flats of Boat Harbour. Other living molluscs recovered from sediment samples include *Caecum plicatum* Carpenter, *C. nitidum* Stimpson, *Littorina angulifera* Lamarck, and *Brachiodontes exustus* (Linnaeus).

TWENTIETH CENTURY SEDIMENTOLOGICAL DEVELOPMENT

Core Analysis

Recent changes in the sedimentology and microfaunas of central and eastern Bonefish Pond have been investigated through the analysis of nine shallow cores (the locations of cores BP1-BP9 are given in Figure 1). All of the cores reached a thick Holocene crust or the Pleistocene limestones underlying the uncemented tidal creek sediments. Variations in the silt-clay percent, the abundance of selected grain types (aggregates, gypsum crystals, lithoclasts, and macrophyte crusts), and the biofacies present (foraminiferal, molluscan, and ostracod) in each core are summarized in Figure 11. All cores document considerable increases in silt-clay with increasing depth. During the sedimentary intervals representing the highest silt-clay deposition, gypsum crystals (disks) occur in cores BP1, BP2, BP5, and BP6. Lithoclasts are most abundant in the lowest and/or uppermost sediments of the cores; in most cases the lithoclasts contain significant levels of protodolomite. Macrophyte crusts are limited to the upper 8 1/2 to 13cms of the cores (Fig. 11).

The mineralogy of the silt-clay and sand fractions of each unit in cores BP2, BP6, and BP8 were determined by X-ray diffraction analysis. In general, dolomitization increases with depth, and there is a corresponding decrease in the amount of aragonite present. Low levels (less than 2%) of terrigenous quartz occur in all silt-clay fractions except those from the upper 5cms of core BP2. These sediments were deposited in a small hypersaline lake formed on East Island. It is unlikely that the terriginous silt-clay, leached from Pleistocene eolianites and transported through the creek by tidal currents, would be found in this local lacustrine environment.

Cores BP2-BP9 contain the *Peneroplis* foraminiferal biofacies and the *Cyprideis* ostracod biofacies at all levels. All nine cores contain two molluscan biofacies: 1. an upper (surface to a depth of approximately 10-12cm) *Gemma* Biofacies; and 2. a lower (below a depth of 10-12) *Polymesoda* Biofacies. In the cores, the *Gemma* Biofacies commonly contains the species *Gemma gemma* (Totten), *Acteocina canaliculata* (Say), and *Chione* sp. cf. *C. pubera*. A similar assemblage of dead shells is present today throughout most of the central and eastern parts of the creek system due to the tidal current mixing of the living *Gemma* and *Chione* Biofacies. These species are not present, or are rare, in the *Polymesoda* Biofacies of the cores. *Polymesoda maritima* is

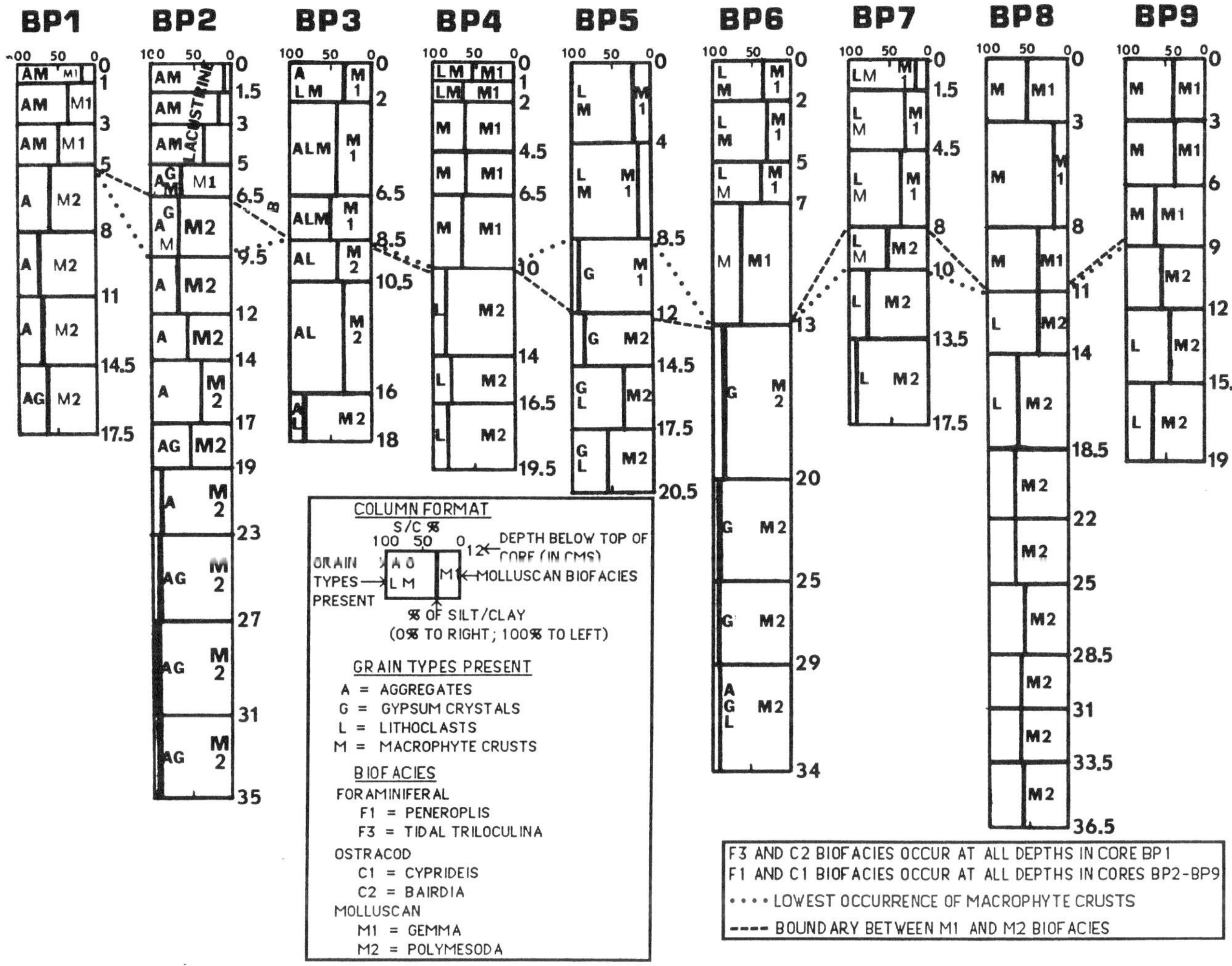

Fig. 11. Silt-clay variation, biofacies, and occurrence of selected grain morphotypes in shallow cores from Bonefish Pond. Units are given in cm intervals below the top of the core, which is the creek or lake sediment surface. Core locations are shown in Figure 1.

abundant in the lower parts of the cores; it is absent, or rare, in the upper 10-12cm (Fig. 11). The boundary between these biofacies coincides with a marked decrease in silt-clay percent in most cores. The change in molluscan biofacies seems attributable to a significant change in water turbidity, which would significantly affect these infaunal suspension feeding bivalves. At present, the *Polymesoda* Biofacies is limited to two small areas of the tidal creek system where the sand fraction of the sediment is very fine (Fig. 10). An increase in lithoclasts in the upper part of several cores (BP3-BP6) seems to reflect the heightened erosion and transport of Pleistocene limestones or Holocene crusts exposed along the margins of the creek system. This implies an increase in current velocities and is consistent with the corresponding decrease in silt-clay percents. The sand and silt-clay fractions at the base of two cores (BP2, BP6) contain significant levels (10-20%) of low Mg-calcite. This suggests that, in these areas, the cored sediment sequence directly overlies eroding Pleistocene limestones. Aggregates occur only in core BP3 which is located in an area where thin protodolomite crusts are presently forming. Since aggregates are common to abundant throughout core BP3, it appears that the conditions necessary for producing thin sediment crusts have continued in this area over a relatively long period of time.

Map Evidence

Map evidence of changes in the geography of the Bonefish Pond creek system during the past 100 years exists and indicates that much of the sedimentary sequence recorded in the upper portion of the cores was deposited in the Twentieth Century. An enlarged portion of an 1881 British survey map of southern New Providence

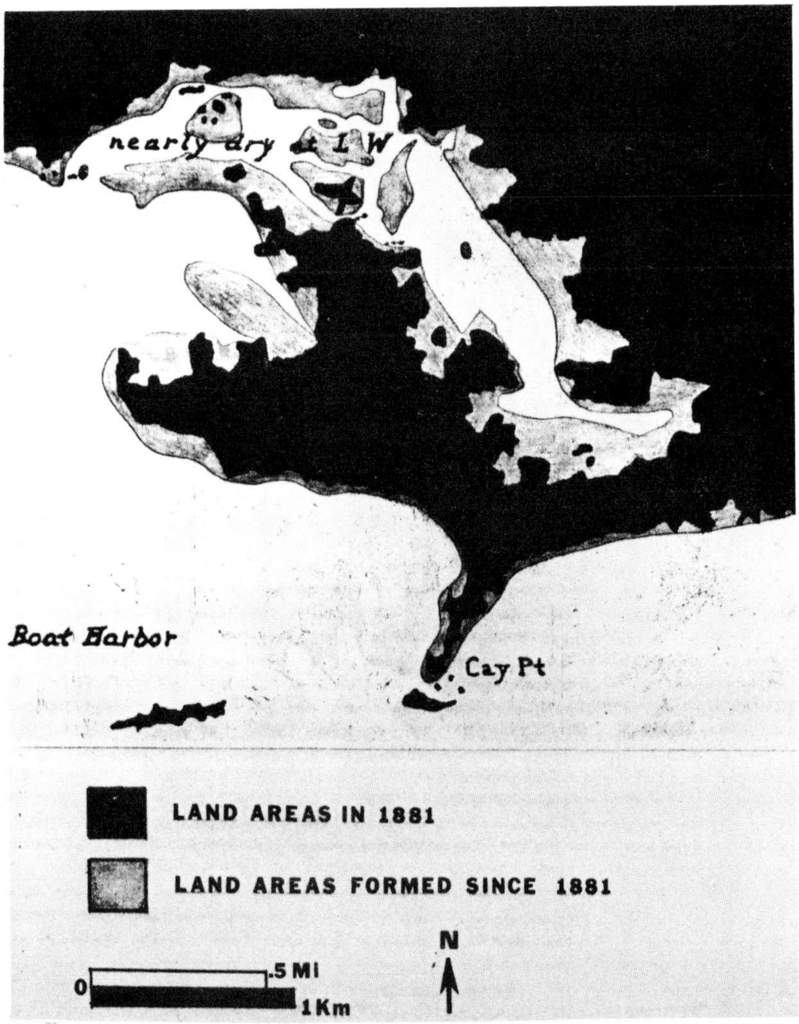

Fig. 12. Map of Bonefish Pond showing land areas in 1881 (black) and land areas formed within past 100 years (gray). Base map is from United States Hydrographic Office (1936).

island is presented in Figure 12. Land areas in 1881 are shown in black; land areas formed since 1881 (as determined from detailed aerial photographs and field surveys) are shown in gray. The map, aerial photograph, and field evidence documents that there have been very significant changes in the geography of Bonefish Pond over the past 100 years. The area covered by the tidal creek system has been reduced by about one-half. Coastline accretion has narrowed Inlet A to about one-sixth of its width in 1881. In addition, the islands restricting the western half of the creek have been developed during the past 100 years. Also, the shorelines of the eastern part of Bonefish Pond have been straightened and East Island has been formed by the deposition of sediments (Fig. 12).

DISCUSSION AND CONCLUSIONS

The evidence provided by present surficial sediment and biofacies distributions, shallow core analyses, and the evaluation of maps and aerial photographs supports a preliminary scenario for the development of Bonefish Pond over the past 100 years. In the Nineteenth Century, the creek system was a relatively extensive carbonate mud flat dominated by hypersaline foraminiferal, ostracod, and molluscan biofacies. The shallow mudflat ("nearly dry at low water" according to the 1881 British survey map) was geochemically suited for the widespread formation of protodolomite and gypsum crystals (disks). The protodolomite seems to have been precipitated in an unlithified carbonate mud, with the dissolution of aragonite taking place. A similar origin for protodolomite, occurring along western Andros Island, has been proposed by Gebelein and Others (1980). The gypsum disks are similar to ones discovered in shallow lakes on Rum Cay, Great Exuma Island, and northern Long Island (Mitchell, 1987b). The offshore area of Barcardi Bay seems to have been shallower than at present. The 1881 British survey map indicates that the Bay is "clear white sand [with] depth from 1 to 2 feet" (Fig. 12). Through the past 100 years, longshore transport and net up creek sediment transport have reconfigured the major entrance to the creek system, Inlet A, by narrowing the inlet and by developing islands in the inlet channel.

In the past few decades, the industrial development of south-central New Providence Island has modified Bonefish Pond in another way. Macrophyte crusts typically occur in the fresh water sublittoral environments of hardwater lakes rich in dissolved calcium carbonate. These environments may be subject to low levels of industrial pollution (Schneider and Others, 1983). It is hypothesized that the development and high level of productivity of the macrophyte crust-producing cyanophytes in Bonefish Pond are due to water chemistry changes caused by the increasing agricultural and industrial development of the area of southern New Providence Island adjacent to the creek system. The upper 10-12cms of all of the 9 shallow cores record the advent of major changes in the sedimentation and biofacies of Bonefish Pond, due apparently, to the explosive development of cyanophytes in the extreme southeastern part of the creek (Fig. 7). At present, thin protodolomite crusts are forming in several areas of central Bonefish Pond (Fig. 1). Much thicker Holocene protodolomite crusts occur along the margins of the central and eastern parts of the creek system. These thick crusts are largely composed of particles in the silt-clay size range. It is tentatively suggested that the thicker protodolomite crusts formed prior to the onset of macrophyte crust production. At present, 12 lithofacies, 4 foraminiferal biofacies, 2 ostracod biofacies, and 4 molluscan biofacies are recognized in the Bonefish Pond area. They have developed over the past few decades in response to shifting environments: one characterized by the widespread precipitation of protodolomite and gypsum has given way to one dominated by macrophyte crustal sediments.

ACKNOWLEDGEMENTS

The authors are indebted to Dr. Neil E. Sealey, Professor of Social Science at The College of the Bahamas, for providing logistical support for the field research described here. In addition, he provided maps and aerial photographs essential to dating the geographic development of Bonefish Pond. We also gratefully acknowledge the support of the faculty and students of The College of the Bahamas, who participated in the field research during a COB Visiting Professorship held by the senior author in Fall 1985. California State University, Bakersfield supported the senior author's research related to this project through a Reid Innovation Grant (Fall 1986), sabbatical leave award (Spring 1987), and University Research Council release time award (Fall 1988).

REFERENCES CITED

Blackman, P.D., 1962, Mechanical characteristics and mineralogy of the sediments in Cloud, P.E., ed., Environment of calcium carbonate deposition west of Andros Island, Bahamas: United States Geological Survey Professional Paper 350, p. 37-64.

Garrett, P., and Gould, S.J., 1984, Geology of New Providence Island, Bahamas: Geological Society of America Bulletin, v. 95, p. 209-220.

Gebelein, C.D., Steinen, R.P., Garrett, P., Hoffman, E.J., Queen, J.M., and Plummer, L.N., 1980, Subsurface dolomitization beneath the tidal flats of central west Andros Island, Bahamas in Zenger, D.H., Dunham, J.B., and Ethington, R.L., eds., Concepts and models of dolomitization: Society of Economic Paleontologists and Mineralogists, Special Publication No. 28, p. 31-49.

Mitchell, S.W., 1984a, Late Holocene tidal creek-lake transitions, Long Island, Bahamas in Teeter, J.W., ed., Addendum to Proceedings of the Second Symposium on the Geology of the Bahamas, College Center of the Finger Lakes, Bahamian Field Station, San Salvador, Bahamas, p. 1-28.

Mitchell, S.W., 1984b, Geology of Great Exuma Island: Field Guide, Second Symposium on the Geology of the Bahamas, College Center of the Finger Lakes, Bahamian Field Station, San Salvador, Bahamas, 45 p.

Mitchell, S.W., 1985a, Surficial geology of the southernmost Bahama Islands: Geological Society of America Abstracts with Programs, v. 17, no. 2, p. 125.

Mitchell, S.W., 1985b, Quaternary lacustrine and tidal creek microbiofacies of the Bahama

Archipelago and Florida Keys: Geological Society of America Abstracts with Programs, v. 17, no. 7, p. 666.

Mitchell, S.W., 1985c, Quaternary eustatic sedimentary accretion of southern Bahamas Archipelago: American Association of Petroleum Geologists Bulletin, v. 69, p. 289.

Mitchell, S.W., 1987a, Sedimentology of Pigeon Creek, San Salvador Island, Bahamas in Curran, H.A., ed., Proceedings of the Third Symposium on the Geology of the Bahamas, College Center of the Finger Lakes, Bahamian Field Station, San Salvador, Bahamas, p. 215-230.

Mitchell, S.W., 1987b, Surficial geology of Rum Cay, Bahama Islands in Curran, H.A., ed., Proceedings of the Third Symposium on the Geology of the Bahamas, College Center of the Finger Lakes, Bahamian Field Station, San Salvador, Bahamas, p. 231-241.

Mitchell, S.W., Baldwin, J.N., Jr., Buening, N., and Westell, B., this volume, Holocene depositional history of Conception Island, Bahamas.

Mitchell, S.W. and Keegan, W.F., 1987, Reconstruction of the coastlines of the Bahama Islands in 1492: American Archaeology, v. 6, p. 88-96.

Mitchell, S.W. and Sealey, N.E., in press, Late Quaternary depositional history of central Samana Cay, Bahama Islands: Caribbean Journal of Science.

Neumann, A.C., and Land, L.S., 1975, Lime mud deposition and calcareous algae in the Bight of Abaco, Bahamas: a budget: Journal of Sedimentary Petrology, v. 45, p. 763-786.

Pierson, B.J., 1982, Cyclic sedimentation, limestone diagenesis and dolomitization in Upper Cenozoic carbonates of the southeastern Bahamas: Unpublished Ph.D.Dissertation, University of Miami, 343 p.

Schneider, J., Schroder, H.G., and Le Campion-Alsumard, Th., 1983, Algal micro-reefs -- coated grains from freshwater environments in Peryt, T.M., ed., Coated grains: Berlin, Germany, Springer-Verlag, p. 284-298.

Stockman, K.W., Ginsburg, R.N., and Shinn, E.A., 1967, The production of lime mud by algae in south Florida: Journal of Sedimentary Petrology, v. 37, p. 633-648.

Teeter, J.W., 1985, Pigeon Creek Lagoon, a modern analogue of the Pleistocene Granny Lake Basin in Curran, H.A., ed., Pleistocene and Holocene carbonate environments on San Salvador Island, Bahamas: Geological Society of America, 1985 Annual Meeting, Guidebook for Field Trip no. 2, p. 147-160.

Thalman, K.L., and Teeter, J.W., 1983, A Pleistocene estuary and its modern analogue, San Salvador, Bahamas in Gerace, D.T., ed., Proceedings of the First Symposium on the Geology of the Bahamas, College Center of the Finger Lakes, Bahamian Field Station, San Salvador, Bahamas, p. 18-21.

Titus, R., 1987, Geomorphology, stratigraphy, and the Quaternary history of San Salvador in Curran, H.A., ed., Proceedings of the Third Symposium on the Geology of the Bahamas, College Center of the Finger Lakes, Bahamian Field Station, San Salvador, Bahamas, p. 155-164.

United States Navy Hydrographic Office, 1936, West Indies - Bahama Islands - New Providence Island - original British surveys to 1881: 21st edition, scale 1:36,350, H.O.1377.

MODERN STROMATOLITES AND ASSOCIATED MATS: SAN SALVADOR, BAHAMAS

Conrad A. Neumann, Brad M. Bebout, Linda R. McNeese,
Charles K. Paull, Hans A. Paerl
Marine Sciences Program
University of North Carolina
Chapel Hill, North Carolina 27599

ABSTRACT

Stromatolitic growths of Mg-calcite are forming subtidally in several hypersaline lagoons on San Salvador, Bahamas. A study in Storrs Lake reveals that they range from locally thickened laminated crusts to club-shaped heads over 1 m in diameter and up to 1 m tall in water depths up to 2 m deep. The surface of the sediment and the heads is everywhere draped by a dark green, cohesive, microbial mat. Heads in water depth less than 1 m are larger, have a deeply pitted top, and protrude 10 cm or more above the sediment surface, while those in deeper water protrude only a centimeter or more above the mat level and have a smooth, crenulate surface. The exterior of the heads is densely cemented while the interiors often crumble into loose stacks of millimeter thick laminae and thicker intervals of thrombolitic, non-layered growth. The top exterior few centimeters of all sampled heads exhibit a more massive structure while walls, bases and inner portions can be finely laminated. The enclosing muddy sediment is composed of carbonate, gypsum and organic matter. Measured salinities range from 60 to over 100 parts per thousand and may vary throughout the seasons. The ubiquitous microbial mat is made up of filamentous and unicellar cyanobacteria as well as anaerobic photosynthetic bacteria. Gradients of oxygen were measured by microelectrodes to be over 4 times atmospheric concentration during daylight in the upper 5 mm below which anoxic conditions abruptly set in. Corresponding pH in the high oxygen zone went over 9 and fell to close to 7 in the anoxic zone beneath. SEM photos reveal euhedral crystal faces of apparently freshly precipitated Mg-calcite on the small hard nodes that form along the mat/head contact. Commonly, however the surfaces reveal evidence of episodes of dissolution indicating a complex alteration of deposition and dissolution on and within the calcified heads. It appears at present that these calcified heads undergo both thrombolitic and stromatolitic intervals, and that they accrete by chemical precipitation at the mat/head contact as well as internal cementation. Foreign material incorporated into the stromatolites indicates some accretion by trapping but it appears to be a very secondary process.

INTRODUCTION

The major macroscopic evidence that life existed on earth as early as 3.5 billion years ago is in the form of fossil stromatolites, "organo-sedimentary structures produced by the sediment trapping, binding, and/or precipitation as a result of the growth and metabolic activity of micro-organisms, principally cyanophytes" (Awramik and Margulis, 1976). These ancient bulbous upgrowths exhibit a wide variety of sizes and shapes. They are often finely or coarsely laminated (i.e., Garrett, 1970), but can also be massive with a clotted microfabric yielding a form for which Aitken (1967) reserves the term, "thrombolite." Pre-Cambrian and early Phanerozoic forms are very fine-grained carbonate, sometimes cherty, and are usually associated with marine facies ranging from hypersaline bays to open shelves (Griffin and

---

A complex peripheral terminology has grown associated with stromatolites and thrombolites in an attempt to clarify description, elucidate origin or point out process. Typical is a recent review by Burne and Moore (1987) which suggests a series of terms centering upon "microbialite." The result is a good review of the complexities of the associated phenomena and a scheme of terms that mirrors the complexity rather then reducing it. "Lithoherm," for instance is proposed in relation to some stromatolitic forms well beyond its original definition as a "subphotic" feature (Neumann, et al., 1977).

Awramik, this vol.). It is often not clear if the ancient forms were mostly precipitated or if they accumulated by trapping and subsequent cementation of particles from the water column. Questions of the calcification of these earliest macrofossils, the organisms responsible and the associated ecological conditions are of great interest to the evolution of life and the conditions on earth under which life pioneered.

Mats of blue-green algae (now "cyanobacteria") were suspected as the causative agent for stromatolites, and many studies of living mat communities ensued. Living mats, however, are commonly soft and rarely build up into layered carbonate accumulations of the proportions, complexity and abundance of ancient stromatolites. Rarely, they are calcified such as some *Scytonema* mats at Andros, Bahamas (Monty, 1967). Commonly, however, they are soft and composed of a single, organic mat that rapidly recycles, usually due to grazing by higher organisms (Garrett, 1970), or by physical destruction (Neumann, et al, 1970).

The ancient stromatolites remained mute on the subject of their exact origin, or of the environmental parameters responsible for their existence, distribution or form, until the early 1960's when lithified stromatolitic growths of the size, shape and internal structure very similar to ancient forms were discovered in hypersaline Shark Bay on the remote west coast of Australia (Logan, 1961). Although recent work has shown that they are complex and enigmatic, the inference was then that they were modern forms in the process of formation, and thus a living analog for the ancient. Since then there have been other modern, currently accreting, lithified, stromatolitic bodies found in settings ranging from high energy to low energy supratidal to subtidal marine, as well as fresh, brackish and hypersaline water chemistries (i.e. Logan, 1961; Eggleston and Dean, 1976; Halley, 1976; Osborne, et al, 1982; Dill, et al, 1986; Cohen and Thouin, 1987; Braithwaite, et al, 1989). These include precipitated ("skeletal") forms as well as those created mostly by trapping and binding of sediment (i.e., the Shark Bay forms as well as the recently described Exuma examples of Dill, et al, 1988). The mineralogy of the precipitated forms is usually aragonite or low-Mg calcite. They can be layered (stromatolitic) or massive (thrombolitic). McNeese (1988) and Griffin (1988) have recently tabulated the properties of recent stromatolites. The few generalizations that do emerge are that the modern fresh water forms are usually fine grained, and thus possibly precipitated, while the marine forms are composed mostly of trapped and subsequently cemented sediment (Griffin and Awramik, this volume). As we might have suspected, there is no single analog to a 3.5 billion year past. Good modern candidates for the fine-grained, laminated, Pre-Cambrian and early Paleozoic precursors are still needed. The fine-grained, laminated stromatolites forming in Storrs Lake may help provide them.

The occurrence of stromatolites in Storrs Lake, San Salvador Island, Bahamas (Fig. 1) has been known for several years (J. Teeter, pers. comm.) and was first reported by Hattin (1982). Mann and Hoffman (1984) described soft, flat, algal structures at the nearshore periphery which were uncalcified and unlaminated. Cowin (1985) mentioned stromatolitic layers when he described the ostracode stratigraphy of the lake. The occurrence of large calcified heads offshore was reported to us by J. Teeter and D. Gerace (1985, pers. comm.) The occurrence and distribution of Storrs Lake stromatolites have recently been described by Mann and Nelson (in press). Pentecost (this vol.) describes small, bulbous, radially calcified heads in the shallow, nearshore periphery of Storrs Lake and associates them with *Scytonema* growth. It wasn't until 1987 when we waded out into waist-deep water, loosened a one-meter-large specimen from its crust-like base beneath the organic mud with a crowbar, and wrestled the calcified mass ashore that we realized that here indeed were large, layered, fine-grained, mushroom-shaped stromatolites that appeared to be good candidates for Pre-Cambrian and early Phanerozoic counterparts.

McNeese (1988) described the calcified elements of the crusts, bulbous upgrowths and stromatolitic heads of Storr's Lake as part of a Master of Science thesis. Work is continuing on the Holocene history of lake deposition, the isotopic geochemistry of the hypersaline system as well as the production, metabolism and calcification of the cyanobacteria associated with the stromatolite and crust formation. Reported here are the occurrence and description of the stromatolitic structures found within the lake and some properties of the cyanobacterial mats associated with them.

STORRS LAKE

Late Holocene sea level rise resulted in the interior flooding of an enclosed, karstified, Pleis-

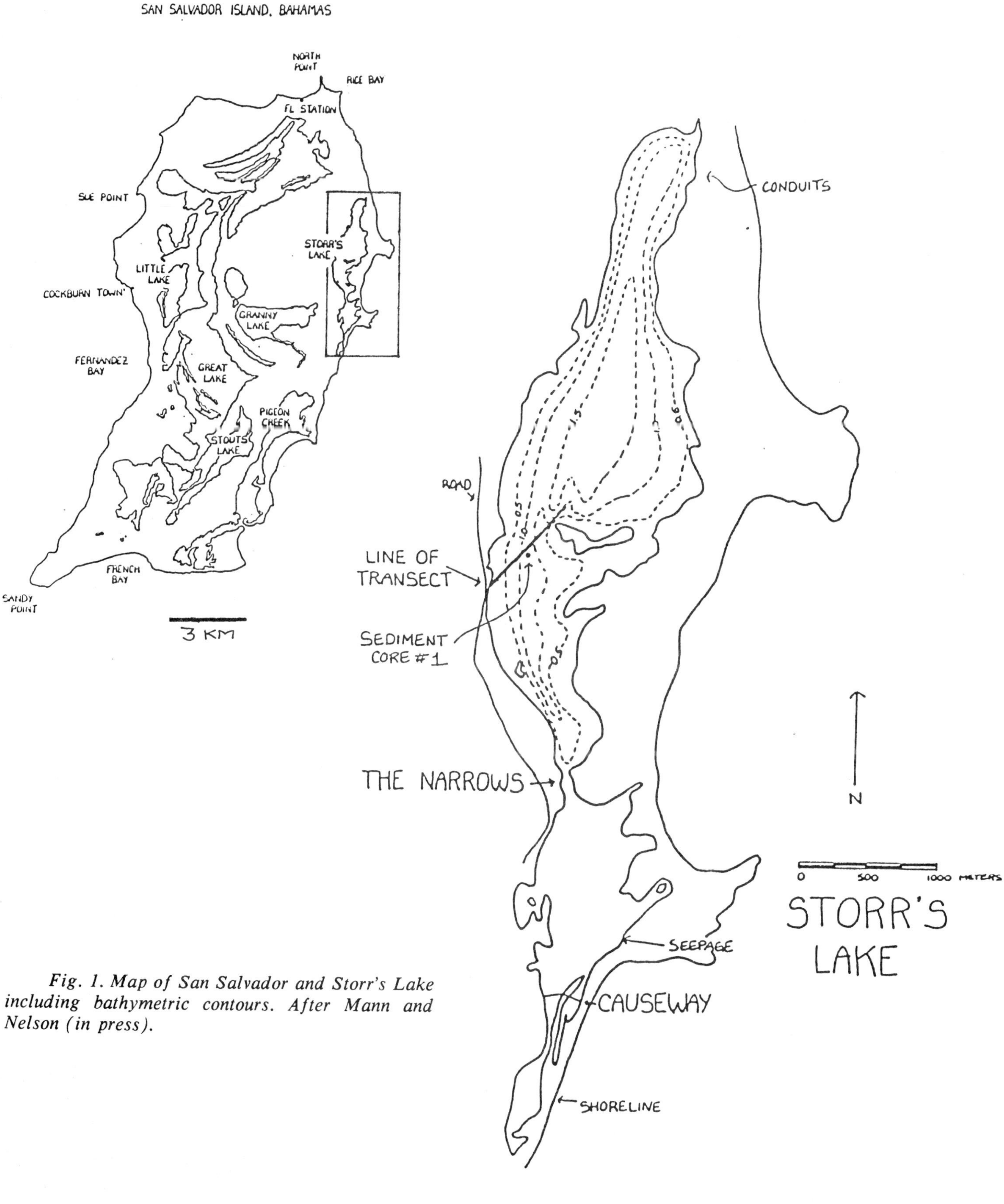

Fig. 1. Map of San Salvador and Storr's Lake including bathymetric contours. After Mann and Nelson (in press).

tocene, interdune depression on the east coast of San Salvador Island and created Storrs Lake (Fig. 2) which is now 7.3 km long, 1.4 km wide and 1 to 2 m deep. Mangroves rim the periphery. Conduits in the eastern bedrock ridge and seepage through Holocene sand afford limited exchange with the ocean. Evaporation exceeds rainfall and ground water influx so that the observed salinities within the lake are elevated. Average rainfall varies from lows of less than 2" in March to over 10" in October (Bahamas Meteorological office, compiled data). Limited data suggest that lake salinity varies inversely with rainfall (McNeese, 1988). March-April maximums of over 100 ppt. have been recorded. West season minimum salinities have not been determined. A six month, Dec. to May, record fluctuates from 70 to 100 ppt. around a 85 ppt. mean (McNeese, 1988). Ostracode assemblages indicate that the lake has been hypersaline since its formation with only local, sporadic incursions of open marine conditions in the area called, "The Narrows" (Corwin, 1985). The water color in the lake varies from brown to brownish green and has been reported by locals to "go white" on occasions (D. Gerace, pers. comm.). Turbidity measured by secchi disc in Dec. 1987 was only 46 cm.

SEDIMENTS

Lake floor sediments are up to 2 m thick and are covered by a thick, tough, leathery, microbial mat which becomes less cohesive and more flocculent as water depth increases to 1.5 to 2 m. The upper 10 cm. of mat-bound sediment is an organic-rich carbonate which is 44 to 74 percent organic matter as determined both by digestion of organics and dissolution of carbonate (McNeese, 1988). The carbonate fraction of the surface sediment is mostly very fine to fine-grained, sand-sized, sub-spherical particles of micritic Mg-calcite averaging 10 mole per cent Mg (McNeese, 1988). The XRD peaks of the sediment Mg-calcite from individual layers are sharper than those of the calcified heads, suggesting that the heads are composed of a broader spectrum of Mg-calcite phases deposited over a greater time span. There are only trace amounts of forams and disarticulated ostracodes in the surface sediment. The inference is that the bulk of the mat-bound, disseminated, sedimentary Mg-calcite is nondetrital and precipitated somehow within or beneath the cyanobacterial mat.

MICROBIAL MATS AND ASSOCIATED CRUSTS

Mats

Microbial mats dominate the biota of Storrs Lake. Their consistency characterizes the sediment surface and varies from thick, uncalcified gelatinous accumulations nearshore to cohesive leathery mats with disseminated calcite on and between the lithified, stromatolitic heads offshore. In the deepest parts of the lake at 1.5 to 2 m the mats lose their strong cohesiveness and are not apparent by macroscopic observation. Mann and Nelson (in press) report that the leathery mats on the soft sediment surface between the stromatolites can at times withstand the weight of a person. The mats are dark green on the surface with a pink zone of anaerobic, photosynthesizing bacteria a few mm beneath.

Nearest the shore is a periodically exposed gelatinous mat sometimes covered by a white, organic crust of desiccated cyanobacteria. Beyond this the subaqueous mat becomes thick, coagulated and characterized by large diameter (20 cm) hemispheroids of clear, gelatinous, extracellular ectoplasm. These are the largely noncalcareous "thrombolitic pie mounds" of Mann and Nelson (in press) and the *Phormidium* mounds of Mann and Hoffman (1984) as well. Except for disseminated carbonate grains, these inshore mats are uncalcified and extend to water depths of about 30 cm. The gelatinous mounds are often buoyed to the surface during the afternoon as a result of gas accumulation within them. The gas is presumably mostly but not necessarily entirely photosynthetically produced oxygen. The organic mats quickly become anaerobic with depth. Disturbance effuses hydrogen sulfide which is immediately obvious by smell.

All described mats from Storr's Lake are very complex assemblages of filamentous and coccoid cyanobacteria, bacteria, and diatoms. Some of the micro-organism which are present in the mat are listed below:

CYANOBACTERIA	*BACTERIA*	*DIATOM*
Phormidium	*Beggiatoa* (colorless sulfur)	*Navicula*
Microcoleus	*Chromatium* (purple sulfur)	
Calothrix		
Spirulina		
Scytonema		

Figure 2

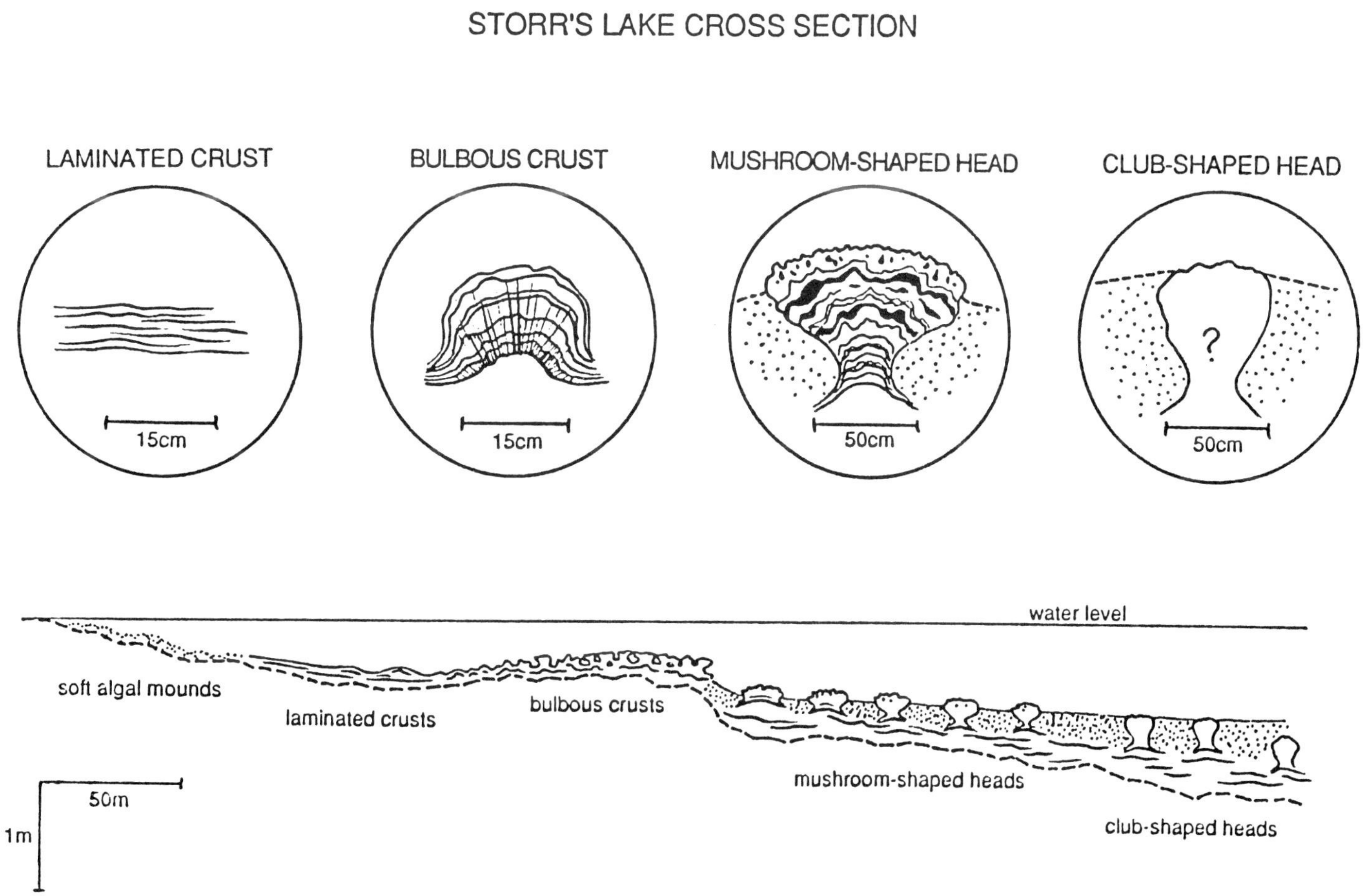

Figure 3

Crusts

Proceeding offshore, the onset of intensive calcification within and beneath the mats is obvious by the feel of the mats underfoot. Laminated, calcified crusts begin to appear within 50 m from shore in water depths that were about 20 cm at the time of observation in Dec., a particularly wet time. All depths are approximate and will vary with seasons. Mann and Nelson (in press) report seasonal variations of 16 cm in water level and infer that changes of up to 26 cm are possible. Near shore the crusts are horizontal, friable and foliate, but 150 to 200 m from shore in water depths of only 30 to 50 cm the laminated crusts begin to swell into locally mounded masses of 2 to 5 cm relief. These grow thicker and more calcified and coalesce into an anastomosing flattened surface of friable, bulbous crusts and heads. Head level stays at or near 10 cm water depth. An extensive terrace that can bear the weight of passers by is formed by the coalesced heads. This is the "hard calcareous table" of Mann and Nelson (in press). It is within this zone that low relief (10 cm), surficially blackened heads of radiating calcified elements occur. Pentecost (this vol.) has identified these radial calcified elements as those associated with the cyanobacterium, *Scytonema*. The surface mat here is also bright green with a deep pink zone a few millimeters beneath.

At about 250 m from shore along the study transect the terrace formed by the coalesced bulbous crusts and heads suddenly becomes more open and individual heads appear beyond a distinct drop off created by the sudden termination of the bulbous crust bench. Figure 3 is an offshore transect that depicts the distribution and general character of the calcified stromatolitic elements of the lake. Lakewide reconnaissance carried out in March 1989 confirms the general offshore distribution which characterizes the study transect.

Mat Chemistry

Oxygen microelectrode profiles were made on two of the inshore mats; (1) the innermost mat with the clear, ectoplasm hemispheroids, and (2) the next adjacent offshore mat which overlies the calcareous bulbous crusts that construct the inshore bench. A third, offshore mat was also profiled. This is the "leathery" mat that extends lakeward from the end of the bulbous crust shelf and is found on and between the calcified stromatolitic heads (Fig. 3).

Where the mat extends over the upper surface of the heads, it is broken only by centimeter-size, white, densely calcified nodes which grow upward as small pinnacles and ridges from the surface of the stromatolite head. Microelectrode profiles of dissolved oxygen made on the three mat types are shown in figure 4. Profiles made in the afternoon, when photosynthesis has been active for some time, show remarkably high oxygen concentrations in the upper 5 to 10 mm of the mats. The most extreme case is the four fold increase of oxygen above atmospheric concentration in the leathery mat. This is the same mat abutting and covering the offshore calcified heads. The oxygen concentration peaks at 400 percent of saturation at only 5 mm into the mat whereupon it rapidly goes to 0 and the mat becomes completely anaerobic. Post-respiration intervals reveal the reverse. They show no oxygen maxima, just a rapid decrease from saturation to anaerobiosis.

Daytime pH profiles were obtained by a standard combination electrode (Markson 985B) mounted on a micromanipulator which incrementally penetrated the mats (Fig. 5). "Stromatolite mat" refers to a mat in the "leathery mat" zone of the large heads that is actually on top of and covers the calcified stromatolitic head. The "leathery mat" is the cohesive mat between the heads. In both cases the pH rises to values over 9 coincident with the oxygen maxima at 5 mm depth in the mat.

The extreme oxygen maxima over such short mat intervals reveals high levels of photosynthetic activity with the subsequent uptake of comparable concentrations of CO_2. This loss of carbon dioxide produces high pH levels during the day and should achieve the reverse at night. The suggestion is that the intensely photosynthesizing mats create the potential to precipitate calcium carbonate in daylight and dissolve it at night. The crusts and heads then may only reflect a small <u>net</u> accumulation of an alternating process that fluxes large concentrations of calcium carbonate into and out of the system as a result of the photosynthetic process alone. This process may in turn be superimposed upon an overall seasonal effect, so that net accretion may be contained within seasonal brackets in water chemistry parameters. That statement must remain as vague as it is until the lake is "lived with" long enough to understand its seasonal chemistry and the response of the mats

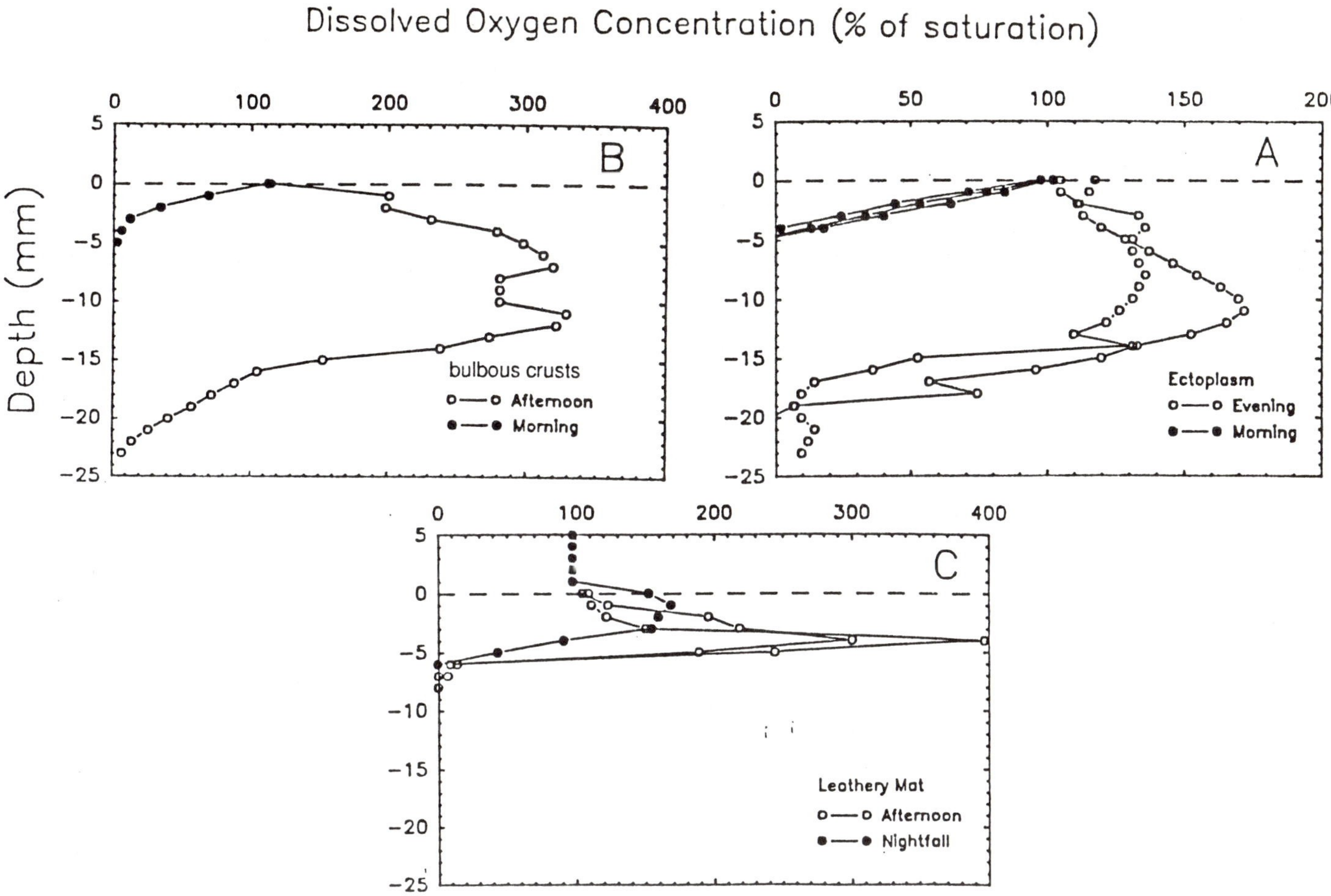

Fig. 4. Dissolved oxygen concentrations within three microbial communities at Storr's Lake, measured with oxygen microelectrodes. Diel variability in the oxygen concentration can be seen.

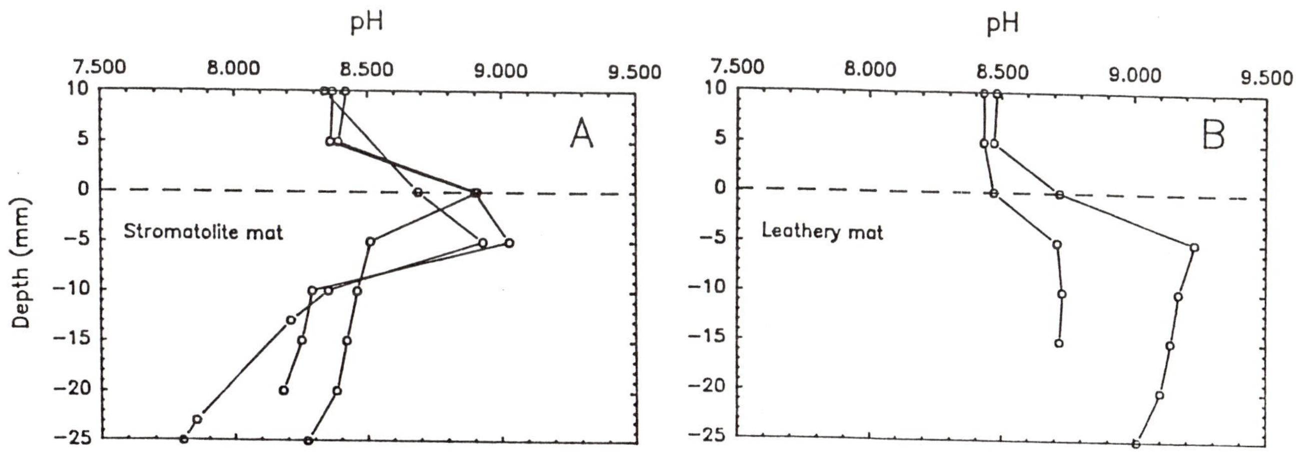

Fig. 5. Profiles of pH within the stromatolitic and leathery mat communities at Storr's Lake. Replicate profiles are plotted on the same axes.

and the calcified heads to such climatic forcing effects as evaporation, temperature, wind, etc.

The ability of the mats to precipitate calcitic mineral matter directly to the lake sediment is graphically demonstrated by the SEM image prepared from fresh mat material of euhedral mineral material in intimate association with a cyanobacterial filament (Fig. 6). The crystals on the mat filaments are apparently the same Mg-calcite that XRD reveals to make up both the sediment and stromatolite carbonate. If the mats precipitate Mg-calcite, as based on present knowledge they appear to do, the question arises, "why don't the mats precipitate more widely and form extensive crusts rather than discrete heads of localized calcification?" Episodes of widespread surficial crust formation has apparently gone on in the past because there are at least two intervals of buried, laminated crusts within the sediment column. The shallowest crust locally drapes buried heads that must date back to a former episode of stromatolite formation. How stromatolitic heads are initiated and why they maintain specific morphologies is a concern of ongoing interdisciplinary research.

THE CALCIFIED STROMATOLITIC HEADS

Morphology

Beginning in water depths of 60 to 80 cm discrete, lithified, stromatolitic heads are encountered along the study transect beyond the shelf of bulbous crust (Figs. 7A, B). The heads are spaced two to several meters apart. These are the "hard pinnacle mounds" of Mann and Nelson (in press). The largest heads are over a meter in diameter, up to a meter tall, and exhibit either a club or mushroom shape. Heads are found along both ends of the study transect and diminish in frequency with depth and distance into the lake. The belt of stromatolite development is wider on the western or landward side of the lake. There are two basic head morphologies separated by depth and distance from shore.

The inshore heads are larger, more spread out at the top and exhibit more relief above the sediment surface (about 10 cm) than the heads found in water deeper than one meter (Fig. 3). The inshore heads hereafter referred to as "mushroom-shaped heads," have flat tops indented by many pits 5 to 10 cm deep. The green leathery mat extends onto the head surface and covers the pits. Pinnacles (nodes) and sharp walls of dense calcite separate the pits and protrude above the mat surface. The nodes are often pinkish white upon retrieval and appear to be covered by clear mucilage. The bottom of the pits are in the anaerobic zone below the algal mat and are often soft and friable in contrast to the intervening indurated nodes and ridges.

The heads are of an irregularly circular outline. The sides, or walls, are ringed by irregular folds or annular ridges which encompass the lower periphery of the crown of the mushroom (Fig. 7A, B). The larger of these appear to be continuous around the circumference of the head. Some of the heads exhibit larger folds that actually overhang and give the head a true mushroom shape. The head narrows downward abruptly to a distinct, smooth, hard neck which can be as small as one third the size of the head, commonly 20 to 30 cm in diameter. This neck is also smooth walled and surficially indurated (Fig. 7A, B). The neck is often based upon a bulbous thickening of a finely laminated crust buried in the sediment. On one occasion where an excavation was made below the basal crust, an older antecedent head was found upon which the surface head had frown subsequent to the interruption represented by the interval of laminated crust. The crusts descent deeper into the sediment with distance from shore and disappear toward the lake axis. They begin at or a few centimeters beneath the sediment surface at the lakeward terminus of the bulbous crust terrace and descent to depths of 30 to 40 cm beneath the sediment surface.

The offshore heads, beginning at depths of 1.2 to 1.5 m are more club-shaped and are relatively taller and narrower than the inshore wider, mushroom-shaped heads. They also differ by exhibiting very little relief above the mat/sediment surface, only a few centimeters and are most easily distinguished from the mushroom-shaped forms because of the surface pattern of the top of the head. The club-shaped heads exhibit a tightly anastomosing series of low, rounded, indurated ridges and rarely show the elevated nodes. The surface pattern is more crenelated and reminiscent of a tightly convoluted brain (Fig. 8).

A cross section of one of the smaller, whole, mushroom-shaped heads was prepared by vacuum embedding the head in resin and sectioning it (Fig. 9). The internal structure is characterized by laminar horizontal and massive vertical elements which are cut by large voids. The voids appear to be empty (water filled) except for scant geopetal

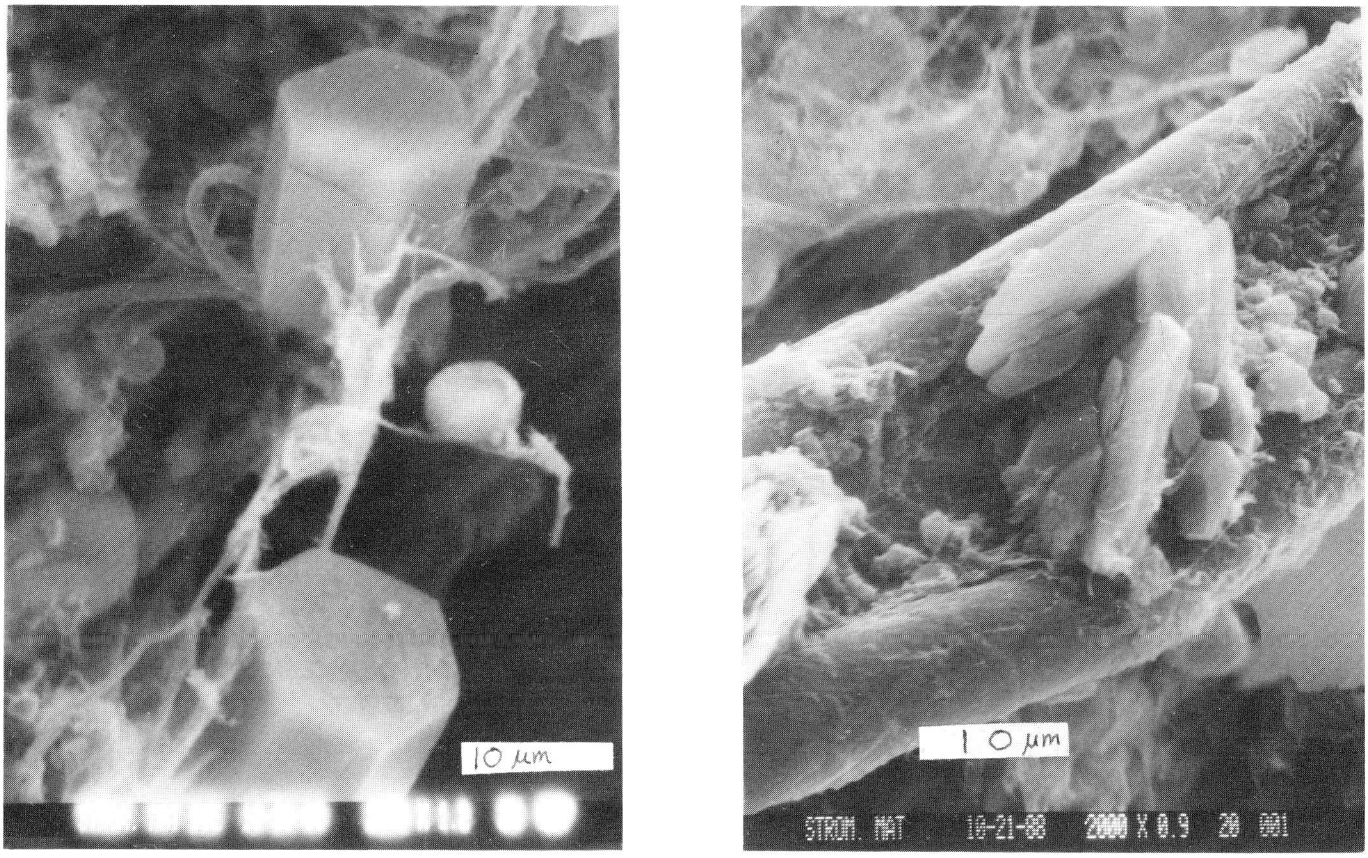

Fig. 6. Scanning electron micrographs of two examples of crystal formation in close association with cyanobacteria from a stromatolitic head.

Fig. 7a and b. Photo of inshore, mushroom-shaped stromatolite head.

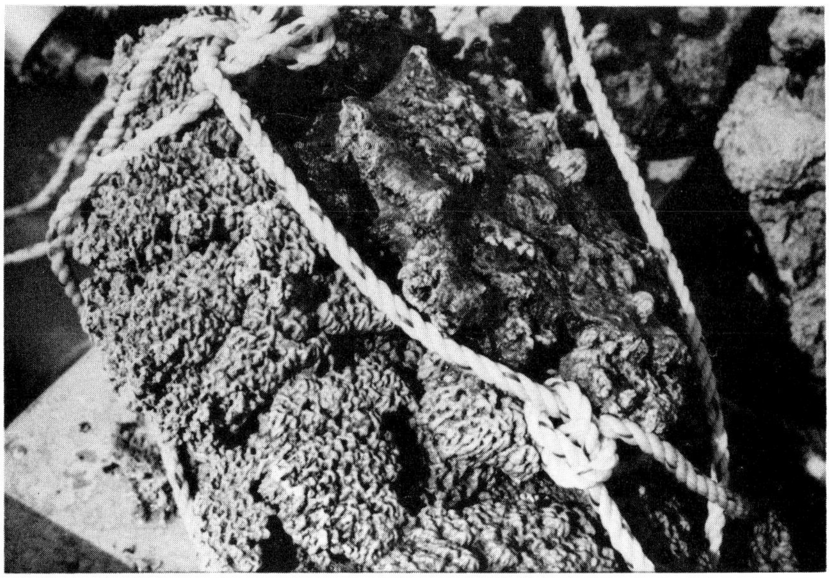

Fig. 8a. Surface pattern and overlying mat of offshore club-shaped stromatolite head.

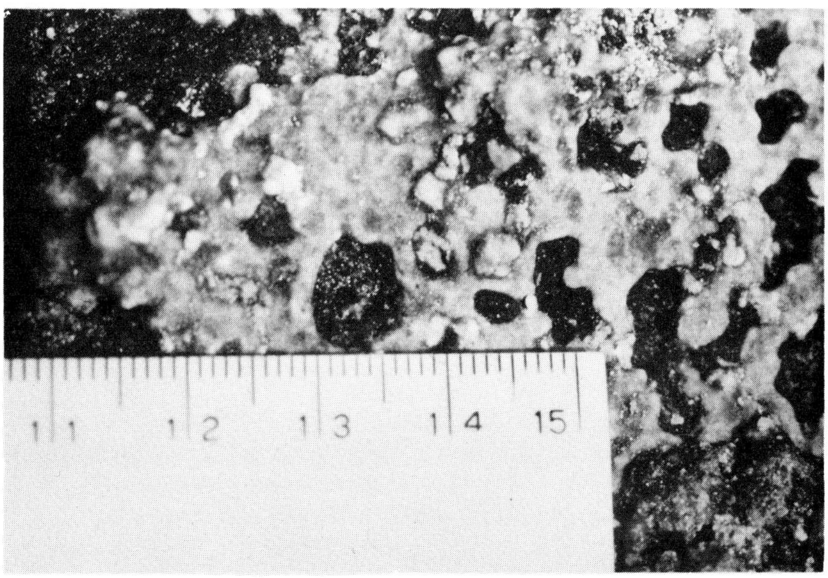

Fig. 8b. Detail of surface carbonate crust forming on top of club-shaped head and appearing to coalesce to form a thin lamina between the nodes on top of the mat. (1.25X)

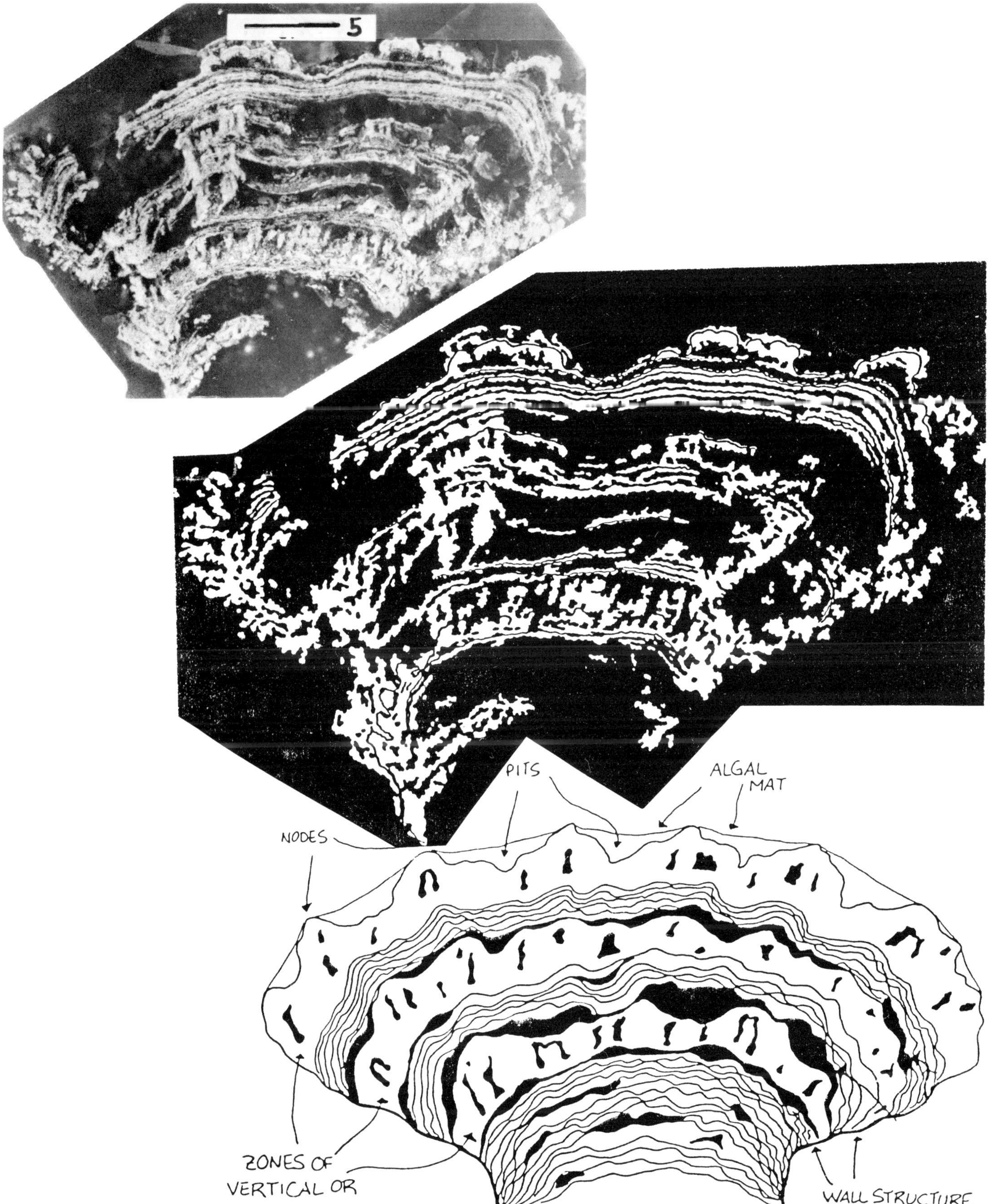

Fig. 9. Schematic cross-section of a "mushroom-shaped" stromatolite.

infill of internal organic and carbonate debris at the base. Repeating sequences of thin continuous laminae (0.5 to 2 mm thick) are broken only vertical, finger like, roughly radiating masses that are truncated below the present surface by a laminated interval. The nodes on the surface appear to be a new generation of re-established radial finger. The nodes and radial fingers appear to be macroscopically massive relative to the fine laminar crusts. They could be considered as "thrombolitic" elements within an overall stromatolitic structure except that microscopically they are made up of discontinuous laminar elements. Some intervals are dominated by these dense, irregularly radial, macroscopically massive, calcified structures while others are obviously, thinly laminated and friable. An interior wall on the right side appears to have been occluded by later lateral growth. The outwardly smooth walls of the neck and sides of the head are built up of thin (1 mm), discontinuous, crenelated laminae (Fig. 10B).

Mineralogy and Petrography

Mineralogical analysis of 31 samples of calcified components of the heads by XRD (walls, nodes, laminae) reveals a spectrum of Mg-calcites ranging from 12 to 17 mole percent. No aragonite or lo-Mg calcite was detected.

Petrographic analysis of embedded thin sections was performed on the various components of the lithified crusts and heads. The following forms of Mg-calcite were identified, and their frequency of occurrence estimated by 100 point counts (McNeese, 1988).

Form	Frequency
Spar/Microspar (5-20 microns)	
blunt needles or flattened spheres - crudely radiating, or random, "rice-shaped" needles	19%
Calcite microspar - lining and/or filling pore space	8%
bladed/platy crystal clusters - intergrown, variable, blunt-edged shapes	11%
rhombohedral crystals	trace
Micrite (<5 microns)	
peloidal-clusters and larger, indistinct, spherical masses	21%
clotted - dark colored, irregular patches	27%
Skeletal - includes foraminifera, ostracod, and gastropod	trace
Pore space	14%

The head interiors are characterized by a large amount of void space, almost no skeletal detrital components, and the bulk of the material is made up of several forms of micrite or microspar. Micrite occurs as clotted fabric, peloidal clusters and indistinct subcircular masses. Its color and opacity suggest that it is not only fine grained but also contains organic material, whereas the spar and microspar exhibit more clarity and crystalline form. The cloudy, dark micrite may be an early, rapid and original precipitate while the spar and microspar are remobilization products of multiphase dissolution and redeposition by internal crystal growth, i.e. cements. Dark cross cutting surfaces suggest organic concentrations at surfaces of dissolution akin to microdisconformities.

The thin sections and the micro-oxygen/pH profiles both suggest that photosynthetically driven, mat-mediated, micro-environments are subject to short intervals alternately favoring precipitation and then dissolution. This might provide an explanation of net accretion at the photosynthetically active, mat-covered stromatolite head, but it doesn't explain how the head-like forms are produced nor what modifications might follow in the anaerobic interior of the heads.

The surficial nodes in thin section (Fig. 10 A) are not massive but are thinly and discontinuously laminated with upwardly fining elements that also suggest rapid surficial precipitation followed by a slower, coarser, more crystalline infill and/or reorganization below. The outermost layers of the nodes are the thickest and least laminated peloidal and clotted micrite (Fig. 10 B, C). The nodes then appear to be sites of episodic accretion and perhaps dissolution as well. The smooth outer wall of the neck at the base of the stromatolite reveals crenulate, discontinuous laminae in thin section (Fig. 10 D). The downward

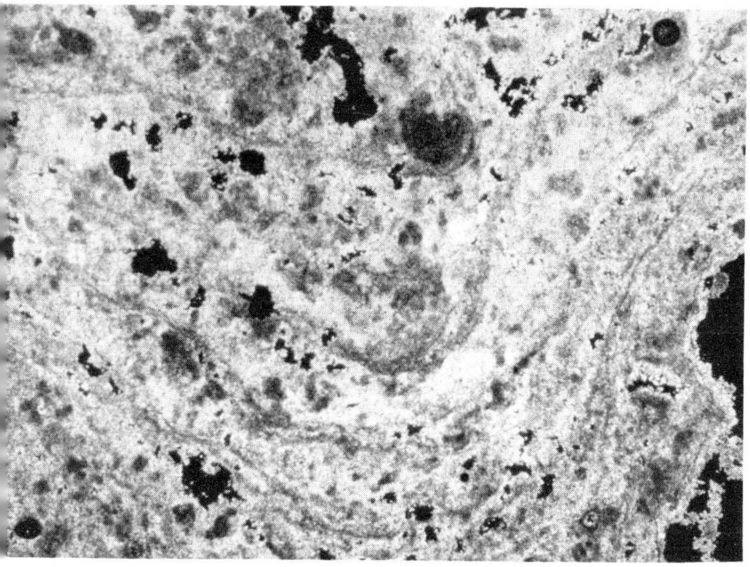

(a) Node with laminae.

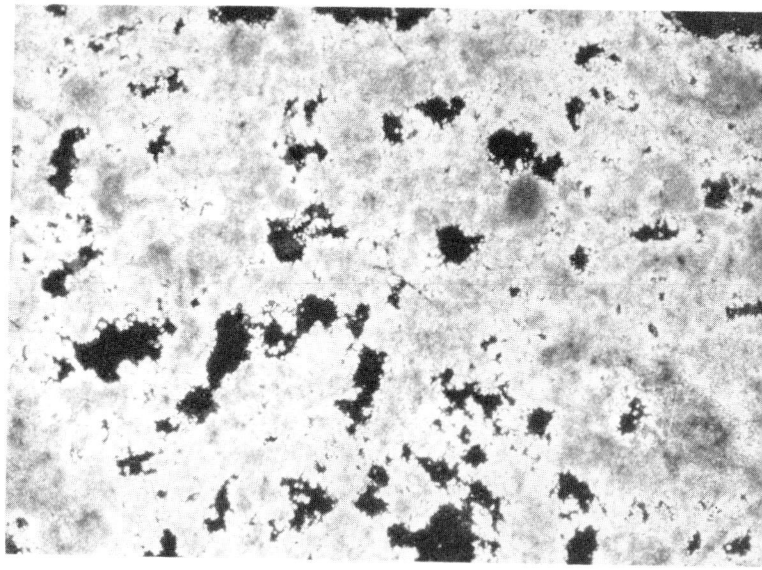

(b) Peloidal and clotted micrite rimmed with microspar.

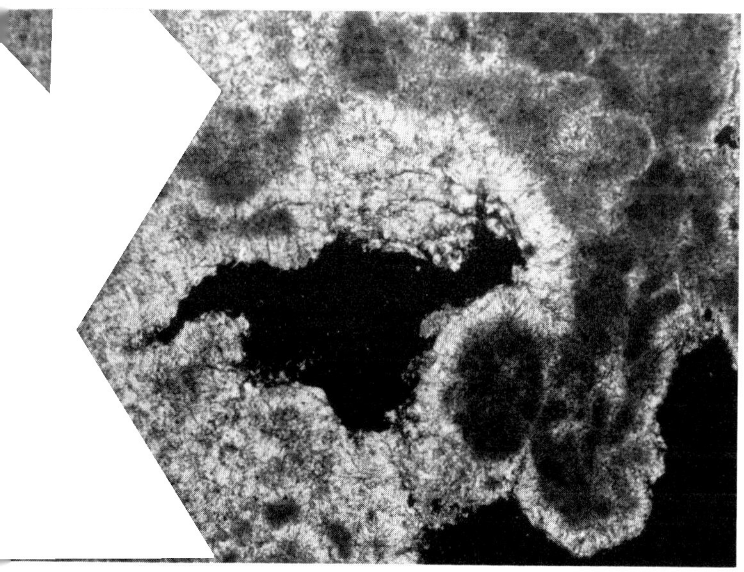

(c) Microspar lined cavity in clotted and peloidal micrite.

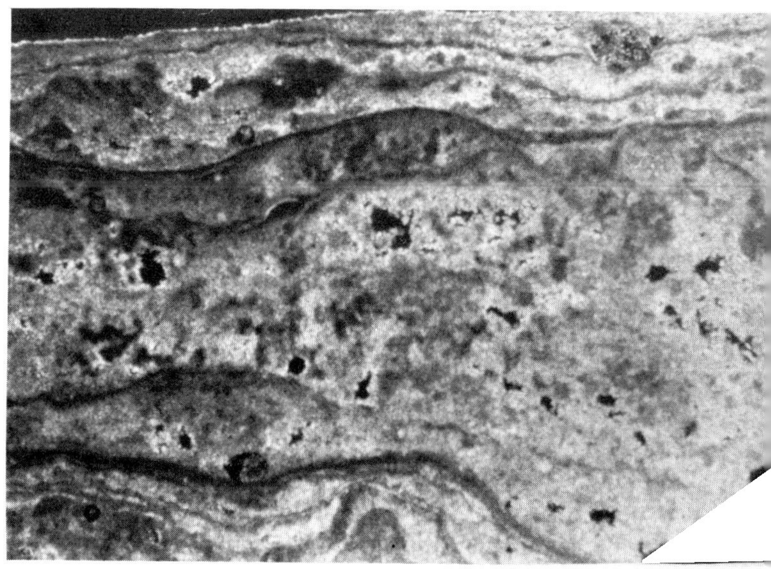

(d) Crenulated laminae on wall of stromatolitic neck.

Fig. 10. Thin section view of stromatolite head. Scale; longitudinal field of view is: (a) 3.5mm (b) 3.5mm (c) 1.4mm (d) 3.5mm.

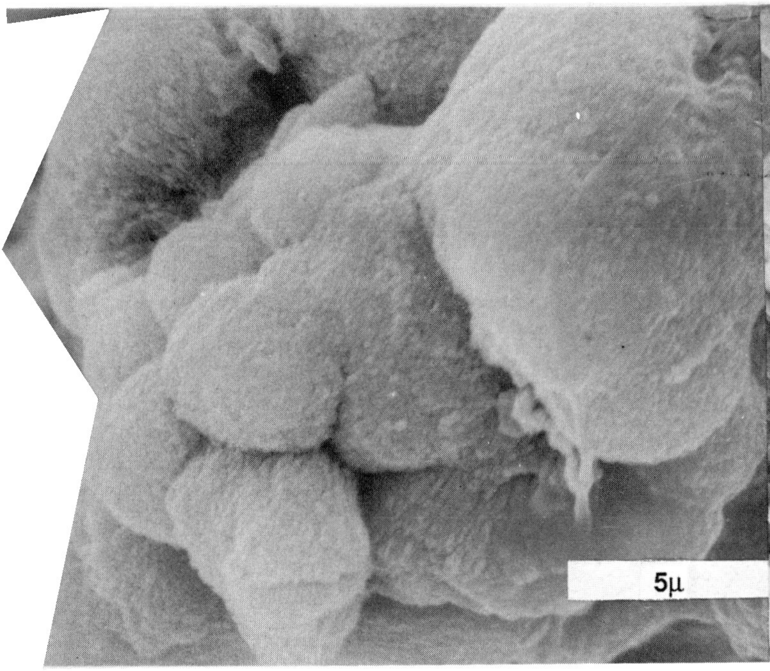

(a) Spherical clusters.

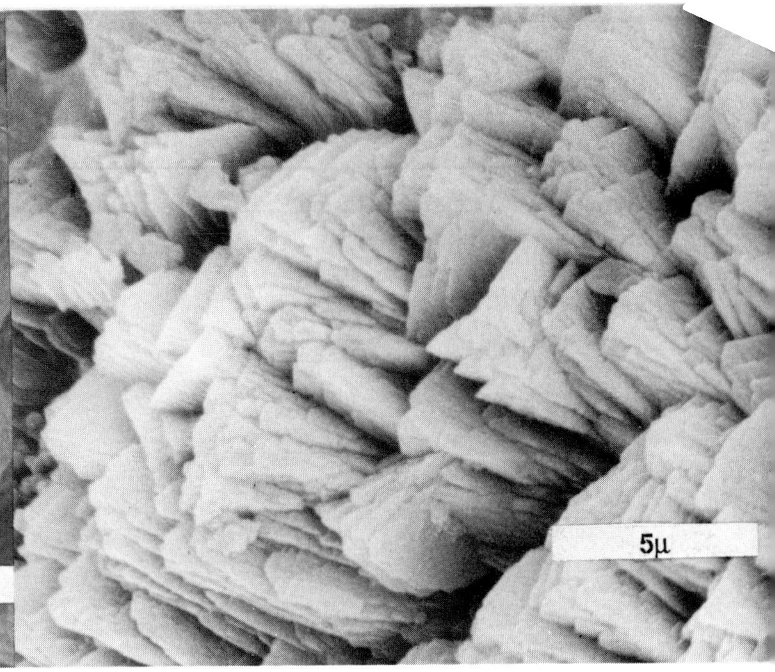

(b) Bladed crystals.

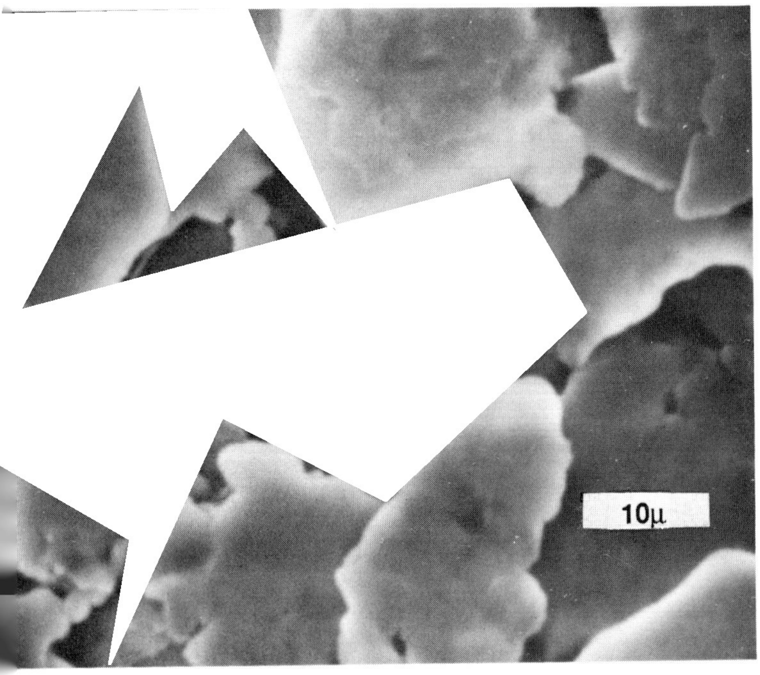

(c) "Jigsaw" fabric from laminae.

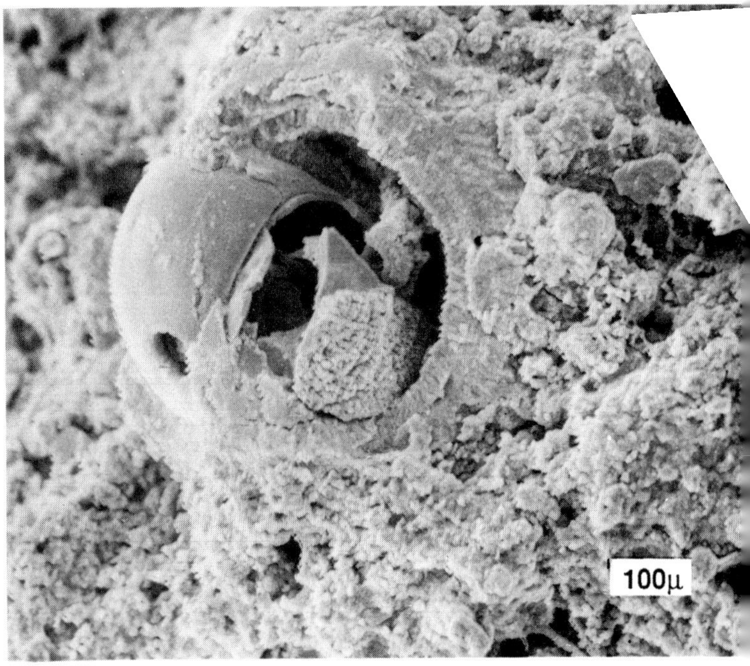

(d) Skeletal fragment, gastropod.

Fig. 11. SEM photos of stromatolite heads.

progression from peloidal and clotted micrite to microspar is even more distinct here than in the nodes on the head surface.

One limited but interesting observation provides a clue for one possible process of formation of the fine laminae seen in the heads. Figure 8B is a macroscopic view of the surface of a small mushroom-shaped head. Thin calcareous crusts are apparently overgrowing and coalescing on top of the algal mat that fills the larger pits. This will result in thin "ice-like" overgrowth that will join to form a continuous lamination which will abut the intervening nodes. Calcareous accretion then appears to occur both on and extending between the nodes and ridges of the stromatolite head surface.

Scanning Electron Microscopy

Some of the various forms of Mg-calcite observed by SEM are:

1. spherical clusters (Fig. 11 A),
2. stacks of sharp or blunt, bladed crystals (Fig. 11 B),
3. stacks of "jigsaw" crystals-irregularly shaped, individual flakes (Fig. 11 C), and
4. skeletal - gastropod, foraminifera (Fig. 11 D).

The preliminary interpretation of the SEM views is that the calcareous elements of the heads are dominated by a very patchy distribution of euhedral, sharp-bladed, crystal clusters as well as very finely crystalline spherical structures and also finely laminated plate-like structures. Partially dissolved counterparts of each of these fabrics are also common.

The SEM views supports the contention, drawn from both the microelectrode data and the thin sections, that the mineralized elements of the stromatolitic heads are Mg-calcite that appears to be undergoing alternate phases of fine-grained precipitation followed by phases of partial dissolution, and then subsequent regrowth of void-filling microspar cements. This is not at all incompatible with a system that is undergoing daily fluctuations due to photosynthetic processes. it does raise questions, however, of what limits or controls this mineral growth, and how fast is its net accumulation, and how is the characteristic mushroom- or club-shape achieved?

Radiocarbon Measurements

The radiocarbon values (from Beta Analytic, Coral Gables, FL) are given in Table 1. Radiocarbon measurements of the organic carbon as well as the carbonate (mineral) carbon contained in surface layers of a mushroom-shaped stromatolite (S3-4) both contained "modern" values and thus suggest recent mineralization. Values of nearly two thousand years, however, are obtained from the other three samples, one from the bulbous crusts closer to shore, one from a buried stromatolite encrusting fossil wood (mangrove?), and the third from the encrusted wood itself. Due to the "hardwater effect", radiocarbon measurements may tell more of the flux of ground water, and thus "dead" carbonate carbon, than of the true temporal age of the precipitated materials. The modern dates, on the other hand, indicate

Radiocarbon Measurements

The radiocarbon values (from Beta Analytic, Coral Cables, FL) are given below:

Lab. No.	Sample Description	C-14 Age	$\delta^{13}C$	^{13}C adj.age
Beta-26843	SSLM-1: Bulbous, Mg-calcite crust, ~100m from shore, 50 cm. water depth	1440±70BP	-1.7°/oo	1830±70BP
Beta-26844	SSLM-2: Wood (root) from SSLM-3	2360±70BP	-23.9°/oo	2380±70BP
Beta-26895	SSLM-3: Encrusting Mg-calcite stromatolite buried 30cm below sed. surface in club-shaped head zone, ~500m from shore in 1.2m depth	2310±70BP	-0.7°/oo	2710±70BP
Beta-28725 Beta-28726	S 3-4 surficial mat (organic) and underlying carbonate crust (inorganic) from upper surface of mushroom-shaped Stromatolite head 300 m from shore. Depth to top of head about 1 m.		-19.2°/oo -0.8°/oo	105.6±1%BP 107.7±1%BP (modern)

TABLE 1

that some precipitation is independent of the hard water effect. Either the flux of dead carbon into the lake is not constant and/or the formation of the stromatolites is episodic over periods of thousands of years. The fresh appearance of crystal faces on stromatolite surfaces refutes long periods of quiescence as does the intimate association of mat, head surface and calcified nodes. The isotope problem begs an answer and calls for a more rigorous program of sampling and measurements. Our exploratory values suggest a deeper complexity that may be due to wide seasonal variation, fractionation and/or recent diagenetic reorganization, or to yet unknown factors. It is apparent that the next chapter of our research should concentrate upon the isotopic character of the water, the organisms, the stromatolitic heads and crusts, as well as the sediments. Perhaps the cycling between these various pools will become assessable so that ages can be calculated.

At this point it is exciting to consider that we appear to have in Storrs Lake a calcifying microcosm of what may well have occurred over much of the earth over most of the geologic past. We have only scratched the surface and look forward to these curious stromatolites to shed more light on the dark past; a past they share with the earliest of living things.

ACKNOWLEDGEMENTS

We are particularly indebted to D. Gerace, Executive Director of the Bahamian Field Station and J. Teeter, University of Akron, for sharing with us their knowledge of the stromatolitic heads in Storrs lake. We thank the staff of the Bahamian Field Station for assisting us in our work. John mann, University of Illinois, was kind enough to share with us a preprint of his important reconnaissance of the stromatolites of Storrs Lake. Miles Crenshaw has been an extremely valuable colleague throughout field and laboratory phases. Karen Griffin, T. Palmer and R. Dill joined us in the field and we benefitted by their stimulating discussion and suggestions. Shirley Kilpatrick and Victor Zabielski gave generously of their time and patience in preparing this manuscript and the figures.

BIBLIOGRAPHY

Aitken, J.D., 1967, Classification and environmental significance of cryptalgal limestones and dolomites with illustrations from the Cambrian and Ordivician of southwestern Alberta: Jour. of Sed. Petrology, vol. 37, no. 4, p. 1163-1178.

Awramik, S.M., 1984, Ancient stromatolites and microbial mats: in Cohen, Castenholz, Halvorson (eds.) Microbial Mats: Stromatolites.

Braithwaite, C.J.R., Casanova, J., Frevert, T., and Whitton, B.A., 1989, Recent stromatolites in landlocked pools on Aldabra, western Indian Ocean: Palaeogr., Palaeocl., Palaeoecol., vol. 69, p. 145-165.

Burne, R.V., and Moore, L.S., 1987, Microbialites: organosedimentary deposits of benthic microbial communities. Palaios, vol. 2, p. 241-254.

Cohen, A.S. and Thouin, C., 1987, Nearshore carbonate deposits in Lake Tanganyika: Geology, vol. 15, p. 414-418.

Corwin, B.N., 1985, Paleoenvironments, using Holocene Ostracoda, in Storr's Lake, San Salvador, Bahamas: Masters Thesis, Dept. of Geology, University of Akron.

Dill, R.F., Shinn, E.A., Jones, A.T., Kelly, K., and Steinen, R.P., 1986, Giant subtidal stromatolites forming in normal salinity water: Nature, vol. 234, p. 55-58.

Eggleston, J.R. and Dean, W.E., 1976, Freshwater stromatolitic bioherms in Green Lake, New York: in Walter, M.R. (ed.) Stromatolites: Developments in Sedimentology: Elsevier, Amsterdam, Oxford, New York. p. 479-488.

Garrett, P., 1970, Phanerozoic stromatolites: Noncompetitive ecologic restriction by grazing and burrowing animals: Science, vol. 169, p. 171-173.

Griffin, Karen M., 1988, Sedimentology and Paleontology of Thrombolites and Stromatolites of the Upper Cambrian Nopal Formation and Their Modern Analog on Lee Stocking Island, Bahamas. University of California of Santa Barbara, Santa Barbara, CA, Unpublished M.S. thesis, 131 p.

Halley, R.B., 1976, Textural variations within Great Salt Lake algal mounds in Stromatolites: Developments in Sedimentology 20, M.R. Walter (ed.), Elsevier, Amsterdam, Oxford, New York, pp. 435-445.

Hattin, D.E., 1982, Holocene lithification of carbonate sediments San Salvador Island, Bahamas: Proc. First Symp. on Geology of Bahamas, San Salvador, Bahamas, Bahamian Field Station, p. 45-62.

Logan, B.W., 1961, Cryptozoan and associated stromatolites from the recent, Shark Bay, Western Australia: Jour. of Geol., vol. 69, p. 517-533.

Mann, C.J. and Hoffman, L.R., 1984, Algal mounds in Storr's Lake, San Salvador, Bahamas: Proc. 2nd Symp. on Geology of the Bahamas, CCFL Bahamian Field Station, p. 41-51.

Mann, C.J. and Nelson, W.M., (in press), Stromatolitic structures in Storr's Lakes, San Salvador Island, Bahama Islands: Palaios.

McNeese, L.R., 1988, Modern Stromatolites in Hypersaline Storrs Lake, San Salvador, Bahamas, unpublished Masters Thesis, Dept. of Geology, University of North Carolina, Chapel Hill, NC.

Monty, C.L.V., 1967, Distribution and structure of recent stromatolites algal mats, eastern Andros Island, Bahamas: Soc. Geol. Belgique Annales,v. 90, Bull. 3, p. B55-B100.

Neumann, A.C., C.D.Gebelein and T.P. Scoffin, 1970, The composition, structure and erodability of subtidal mats, Abaco, Bahamas: Jour. Sed. Petrol., 40, p. 274-297.

Neumann, A.C., J.W. Kofoed, and G.H. Keller, 1977, Lithoherms in the Straits of Florida. Geology, vol. 5, no. 1, p. 4-10.

Osborne, R.H., Licari, G.R. and Link, M.H., 1982, Modern lacustrine stromatolites, Walker Lake, Nevada: Sedimentary Geology, vol. 27.

Walter, M.R. (ed.), 1976, Stromatolites, Developments in Sedimentology 20, Elsevier, New York, 790 pp.

SEDIMENTOLOGY OF A HOLOCENE PLATFORM-MARGIN CARBONATE LAGOON: BLACKWOOD BAY SAN SALVADOR ISLAND, BAHAMAS

Woody Pace
Marathon Oil Company
5555 San Felipe
Houston, Texas 77056

John E. Mylroie
Department of Geology and Geography
Mississippi State University
Mississippi State, MS 39762

James L. Carew
Department of Geology
The College of Charleston
Charleston, South Carolina 29424

ABSTRACT

Blackwood Bay, at the southern end of San Salvador Island Bahamas, is a quiet, shallow water lagoon. It is bounded to the north and east by the island, to the west by French Bay, and to the south by a barrier reef-rubble rampart along the platform margin. The lagoon floor consists of elevated mounds carpeted by *Thalassia* and other sea grasses and algae with intervening deeper sandy areas. The sandy areas contain occasional patch reefs, coral rubble, and chunks of *Thalassia*-bound sediment that have caved-off undercut *Thalassia* mounds.

Detailed analysis of surface samples and cores show the lagoon to have four main facies: 1) barren mobile sand facies; 2) sparse *Thalassia* facies; 3) moderate *Thalassia* facies; and 4) dense *Thalassia* facies. Among the four facies, the mobile sand facies contains the highest percentage of sand and the lowest percentages of mud, high-magnesium calcite, and living *Thalassia*. With increasing amounts of *Thalassia* (sparse to dense *Thalassia* facies) progressively lesser amounts of sand and greater amounts of mud and high-magnesium calcite occur. The cores exhibited a coarse sand and gravel base that typically graded upward into fine-grained deposits as a set of fining-upward sequences.

The increase in mud and high-magnesium calcite with increasing *Thalassia* abundance reflects the baffling effect of the grasses and the high-magnesium calcite secreting epibionts encrusting the grass blades. The overall fining upward core sequences represent the Holocene marine transgression across the lagoon, followed by progressive energy decrease as the barrier reef developed on the ocean side of the lagoon. The repeated fining upward sequences in the cores, scattered coral rubble debris, undercut banks of the *Thalassia* mounds with intervening sand scour pits, and the rubble rampart nature of the current barrier reef suggest repeated storm modification of Blackwood Bay.

INTRODUCTION

Blackwood Bay is located at the southern end of San Salvador Island, in the eastern portion of the large embayment known as French Bay (Fig. 1). The bay is bounded to the north and east by beaches and Pleistocene outcrops, to the south by a shallow-water reef-rubble rampart that trends east-west along the platform margin, and to the west it is open into French Bay. Immediately seaward of the reef, the upper fore--reef slope has a gentle gradient with well developed spur and groove topography. This area is barren of coral except for a few small isolated colonies. Further seaward, the fore-reef slope plunges nearly vertically to abyssal depths. Wilson (1975) classified this type of platform margin setting as a framebuilt reef rim.

The platform edge is an area of very high energy, as waves approaching from the south shoal rapidly and produce very large breakers.

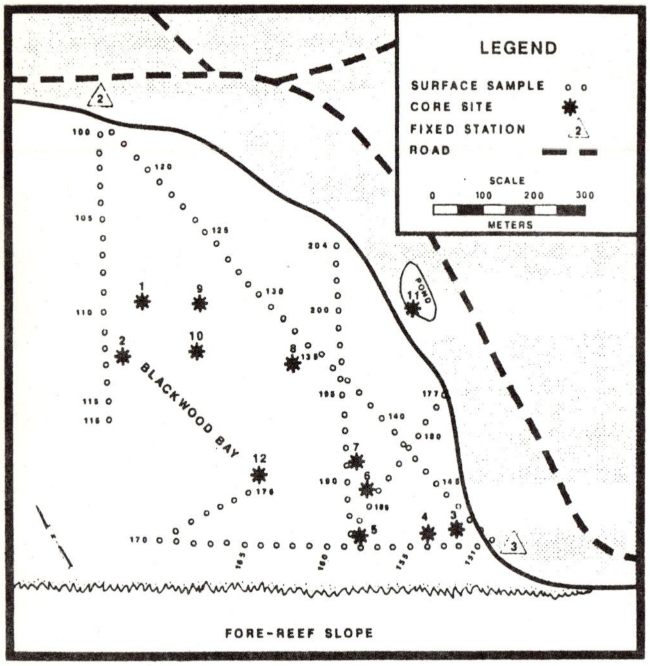

Fig. 1. Map of the study area, Blackwood Bay, showing location of cores and surface samples.

Active reef growth is minimal here except for small encrusting forms such as the hydrozoan *Millepora*. The majority of the platform edge consists of cobble to boulder size coral skeletal material sometimes bound together by *Millepora*. The barrier effect of this boulder rampart is chiefly responsible for the quiet water, low energy conditions in Blackwood Bay.

The back-reef contains many features characteristic of shallow water lagoonal settings. Water depth in the study area ranges from 0-2 meters at mean low tide. Three major features result in topographic irregularities on the lagoon floor: (1) *Thalassia* mounds and meadows; (2) intermound areas that are bowl-shaped depressions containing mostly sand and gravel; and (3) small patches of coral.

The boundaries between the sandy depressions and the *Thalassia* meadows are usually very steep with roots exposed along the steep face. Most of the steepest grass "cliffs" are oriented east-west and face south toward the platform edge and prevailing current. In most cases the current has undercut the lower roots leaving the upper *Thalassia* mat hanging over into the adjacent sand pit.

Several workers have referred to these sandy depressions as "blow-outs" or "deflation pits: (Ball, 1967; Perkins and Enos, 1968; Wanless, 1974; Patriquin, 1975; and Hine and Neumann, 1977). This has led to some confusion because these terms were originally applied to eolian environments (McKee and Ward, 1983). Herein, such features are referred to as subaqueous scour pits.

The subaqueous scour pits range in size from several meters to tens of meters across (Fig. 2). Most of the pits contain large sand-wave bedforms. These mega-ripples have the coarsest material in their troughs and the finest material on the crest, which is typical of subaqueous bedforms. Scour pits that are situated near the platform edge tend to contain large boulders.

Along the northern and eastern portions of the bay a prominent intertidal zone is present. During low tide, paleosol-covered Pleistocene bedrock and *Thalassia*-covered mudflats with abundant *Neogoniolithon* become subaerially exposed. Black mangrove (*Avicennia*) trees inhabit the landward portions of the intertidal zone and are presently overgrowing relict beach deposits. The relict beach deposits are interspersed with outcrops of paleosol-covered Pleistocene bedrock, and these grade westward into active sand beaches.

Inland of the mangroves, many large lobes of coral boulder debris are present. These lobes are best observed where they encroach on a small pond in the northeast portion of the study area. The individual lobes are one to two meters high and two to five meters across, and are composed of cobble to boulder-sized coral rubble. These individual lobes coalesce to form a mega-lobe. A large (15x10cm) piece of *Montastrea annularis* was collected from one of these lobes. XRD analysis showed the sample to be 100% aragonite, and radiocarbon analysis yielded an age of 3810±90 yBP, implying that the rubble lobes are less than 3900 years old (see Caputo and Kramer, this volume).

BOTTOM FACIES

Based upon field observations and analyses of more than 50 surface sediment samples, four facies or bottom types have been recognized in Blackwood Bay. These are:

1. Barren Mobile Sand Facies
2. Sparse *Thalassia* Facies
3. Moderate *Thalassia* Facies
4. Dense *Thalassia* Facies

Fig. 2. Subaqueous scour pit. Note the boulder-sized debris. Relief on the Thalassia *cliff face is 1 meter. Water depth: 2 meters.*

Fig. 3. Barren mobile sand facies containing asymmetrical wave ripples. Grass covered mound at left is 1 meter across. Water depth: 3 meters.

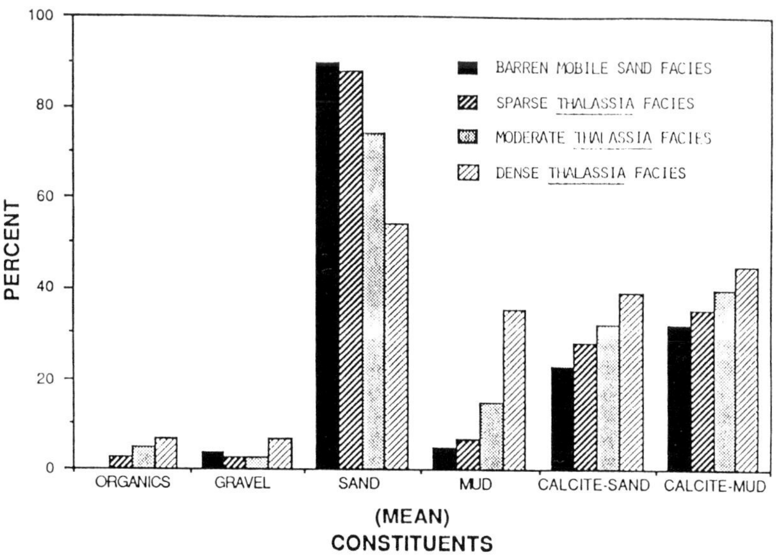

Fig. 4. Graph showing percent of constituents for each of the four facies from Blackwood Bay. Grain sizes and organics are weight percents; calcite values determined from XRD peak height ratios.

Fig. 5. The sparse Thalassia *facies. Blades are heavily encrusted with epibionts. Note the short, stunted nature of the blades. Field of view is approximately 20cm. Water depth: 2 meters.*

Fig. 6. The moderate Thalassia *facies. Blades are long and healthy compared to those of the sparse* Thalassia *facies. Spotted Trunkfish is approximately 5cm. wide. Water depth: 1.6 meters.*

Fig. 7. The dense Thalassia *facies. Note the abundance of rhizomes. Water depth: 1.8 meters.*

Barren Mobile Sand Facies

The barren mobile sand facies that occupy the subaqueous scour pits derives its name from a similar facies found in Snow Bay (to the east of the study area) and described by Andersen and Boardman (1987). Coarse to fine sand with little mud, no sea grass, and small amounts of algae are characteristics of the barren mobile sand facies. Asymmetrical wave ripples are a common feature (Fig. 3).

Fourteen surface samples were collected and analyzed from the barren mobile sand facies. On average, sand particles (2.0-0.065mm) comprise 91% (Fig. 4), and mud content comprises an average of only 5.3% of the total sample. (All grain size and organic content percentages reported in this paper are weight percents, mineralogic percents were calculated from XRD peak height ratios). The average mineralogy for the sand is 24.2% high-- magnesium calcite and 75.8% aragonite. The average mineralogy for the mud is 33% high- magnesium calcite and 67% aragonite. Megascopic observations indicate that the sediment of the barren mobile sand facies is a coarse to fine grained, moderately to well sorted bioclastic sand. The major constituents include abraded coral, mollusc, and foraminifera debris, with minor amounts of *Halimeda*, echinoderm, and ostracod fragments.

Thalassia Facies

Thalassia testudinum (Turtle grass) and other sea grasses such as *Syringodium* and *Halodule* are present in varying concentrations throughout the study area. Based upon visual observations of *Thalassia* density, three facies termed sparse, moderate, and dense *Thalassia* facies were defined.

Sparse *Thalassia* Facies. The sparse *Thalassia* facies supports a very limited *Thalassia* growth with somewhat greater amounts of *Syringodium* (Fig. 5). The sand is moderately to well sorted and is well rounded. Foraminifera, algae and coral fragments dominate most of the samples analyzed. Sand size particles average 88.1% for the 12 samples collected (Fig. 4). Gravel and mud average 3.0% and 6.3% respectively. High-magnesium calcite in the sand and mud fractions averaged 27% and 35% respectively, and aragonite comprised 73% and 65% respectively.

Moderate *Thalassia* Facies. Criteria for recognition of the moderate *Thalassia* facies was abundant *Thalassia* growth, but with the substrate still clearly visible when viewed from above (Fig. 6). Sediment in the moderate *Thalassia* facies is moderately to poorly sorted. Foraminifera, mollusc, and echinoderm fragments are the most abundant constituents. The gravel percentage is essentially the same as in the sparse *Thalassia* facies, but the greater mud content averages 15.2% while sand comprises 74% of the 8 samples collected (Fig. 4). High-magnesium calcite in the sand and mud fractions averaged 32% and 40% respectively, and aragonite comprises 68% and 60% respectively.

Dense *Thalassia* Facies. Dense grass growth that completely shields the substrate from view, characterizes the dense *Thalassia* facies (Fig. 7). Gravel is slightly more abundant than in the other facies, and comprised 5.5% of the 5 samples analyzed (Fig. 4). The amount of sand was markedly decreased to an average of only 53.7%, while the mud content (averaging 35%) was the highest among all the facies. High-magnesium calcite in the sand and mud fractions averaged 38% and 44% respectively, and aragonite averaged 62% and 56% respectively.

Megascopic observations indicate that the sand fraction of the dense *Thalassia* facies is composed of abundant foraminifera and mollusc fragments, echinoderm spines and pieces, ostracods, and *Halimeda* chips. Gravel-sized particles are almost exclusively whole mollusc shells and coral fragments.

Correlations Among the Surface Sediment Data

Along any of the transects in Blackwood Bay, the habitats or bottom facies of adjacent sample localities may differ drastically. Comparing parameters along transects would be meaningless because of the sudden changes in the facies. Therefore, the surface sediment samples were split into the four facies based on the relative abundance of *Thalassia* grass, and statistical correlation tests were made on several parameters in each of the facies.

The following parameters were correlated within each of the facies: percent organics, percent gravel, percent sand, percent mud, percent calcite in the sand fraction, percent calcite in the mud fraction, and water depth. The results of the correlation study are shown in Table 1. The bold numbers represent significant correlations.

In all four facies the percent sand versus

	Barren Mobile Sand Facies	Sparse Thalass. Fac.	Moderate Thalass. Fac.	Dense Thalass. Fac.
Organics vs. Gravel	0.21	0.04	**0.79**	**0.82**
Organics vs. Sand	-0.39	**-0.78**	**-0.97**	**-0.85**
Organics vs. Silt/Clay	0.08	**0.8**	**0.9**	**0.88**
Organics vs. Calcite-Sand	-0.12	0.08	**0.89**	**0.91**
Organics vs. Calcite-Silt/Clay	-0.15	-0.05	**0.82**	**0.87**
Organics vs. Water Depth	-0.05	-0.16	-0.2	-0.39
Gravel vs. Sand	-0.24	-0.52	-0.24	-0.42
Gravel vs. Silt/Clay	-0.21	0.11	0.1	-0.03
Gravel vs. Calcite-Sand	-0.28	0.14	0.32	0.04
Gravel vs. Calcite-Slit/Clay	-0.07	-0.001	0.32	0.8
Gravel vs. Water Depth	0.35	-0.21	0.18	-0.62
Sand vs. Silt/Clay	**-0.86**	**-0.89**	**-0.97**	**-0.87**
Sand vs. Calcite-Sand	0.21	-0.33	**-0.93**	**-0.9**
Sand vs. Calcite-Silt/Clay	-0.35	-0.28	-0.41	0.02
Sand vs. Water Depth	0.48	0.45	0.18	0.03
Silt/Clay vs. Calcite-Sand	-0.15	0.4	0.22	-0.03
Silt/Clay vs. Calcite-Silt/Clay	0.44	0.47	0.52	-0.49
Slit/Clay vs. Water Depth	-0.23	-0.18	-0.22	0.25
Calcite-Sand vs. Calcite-Silt/Clay	-0.19	0.51	0.61	-0.28
Calcite-Sand vs. Water Depth	0.34	-0.13	-0.03	0.14
Calcite-Silt/Clay vs. Water Depth	-0.65	-0.09	0.24	-0.68
Number of Samples	n=14	n=12	n=8	n=5
Critical Value	C.V.=.514	C.V.=.553	C.V.=.666	C.V.=.811

Table 1. Correlation of sedimentary parameters among the four facies. Boldface indicates a significant correlation.

the percent mud showed a strong inverse relationship. No other parameters exhibited significant correlations among all four of the facies. Parameters that revealed correlations among two or more of the facies included: organics versus mud (direct relationship), organics versus sand (inverse relationship), organics versus calcite in the sand fraction (direct relationship), organics versus calcite in the mud fraction (direct relationship), and sand versus calcite in the sand fraction (inverse relationship). The remaining combinations of parameters showed no significant correlations.

DISCUSSION OF MODERN FACIES

In Blackwood Bay, currents and bottom topography play an important role in the distribution and texture of the sediments. The shallow water conditions in the lagoon enable the currents to affect bottom topography which in turn exhibits controls on the currents. Such a feed-back mechanism was documented by Ball (1967). Each of the facies studied has its own characteristic bottom topographic setting that results in a predictable suite of sediments.

Barren Mobile Sand Facies

The barren mobile sand facies is characteristic of the highest energy areas of the bay. During calm conditions in the lagoon, subaqueous asymmetrical wave ripples form on the sand sheets indicating that they are probably being reworked by normal currents. Correlations show that the amount of sand (averaging 91%) increases as mud content decreases suggesting that winnowing by currents is very effective.

Patch reefs along the platform edge appear to contribute significantly to the sediments of the adjacent barren mobile sand facies. Sands in this area contain large amounts of coral debris. In contrast, sands collected farther shoreward are predominantly composed of foraminifera and algal debris. Almost all of the sediments within the barren mobile sand facies are rounded to well rounded and polished, indicating that mechanical breakdown is dominant over biological breakdown.

Thalassia Facies

The influence of *Thalassia* grass on bottom

topography and currents has been well documented (Ginsburg and Lowenstam, 1958; Swinchatt, 1965; Scoffin, 1970). As currents pass over the *Thalassia* meadows, the velocity is reduced and sediment in suspension is deposited. Through time, carbonate mud accumulates and produces a local buildup supported by *Thalassia* grass. Within the study area, small isolated mounds and extensive meadows of sparse to dense growth of *Thalassia* occur.

Several studies have indicated that *Thalassia* mounds form in topographic lows (Howard and others, 1970; Scoffin, 1970; Zieman, 1972; Ebanks and Bubb, 1975; Turmel and Swanson, 1976; Boscence and others, 1985; Andersen and Boardman, 1987). In Blackwood Bay, the living *Thalassia* mounds may have originally formed in topographic lows, but their present position shows no relationship to underlying bedrock topography. Sediment probing across many of the mounds revealed no depressions in which the mounds could have originated. Fig. 8 is an interpretive illustration of the origin of *Thalassia* mounds in Blackwood Bay.

Associated with the *Thalassia* buildups are significant amounts of carbonate mud. Particular attention has focused on the sources of this mud. Cloud (1962) in his study of the Great Bahama Bank concluded that much of the carbonate mud was produced by direct precipitation. Lowenstam (1955) discovered that calcareous green algae disintegrate upon death into needles and plates small enough to form carbonate mud. Additional studies have shown that the breakdown of calcareous algae is a major contributor of carbonate mud (Neumann, 1965; Stockman and others, 1967; Neumann and Land, 1975; Bosence and others, 1985). Physical and chemical breakdown can also contribute to the mud fraction (Folk and Robles, 1964; Neumann, 1965; Matthews, 1966; Stieglitz and Wise, 1969; Hay and others, 1970; Howard and others, 1970; Stieglitz, 1970). Carbonate production by epibionts encrusting *Thalassia* blades is also an important contributor of lime mud (Land, 1970; Patriquin, 1972; Nelson and Ginsburg, 1986). In Blackwood Bay, skeletal breakdown (by mechanical and algal processes) and epibiont growth are the most important contributors of carbonate mud.

<u>Sparse</u> <u>*Thalassia*</u> <u>Facies</u>. The sparse *Thalassia* facies is similar in many ways to the barren mobile sand facies. Sand content is high (averaging 88%) and mud content is low (averaging 6.3%). Grains, especially foraminifera, are rounded and polished indicating reworking and mechanical

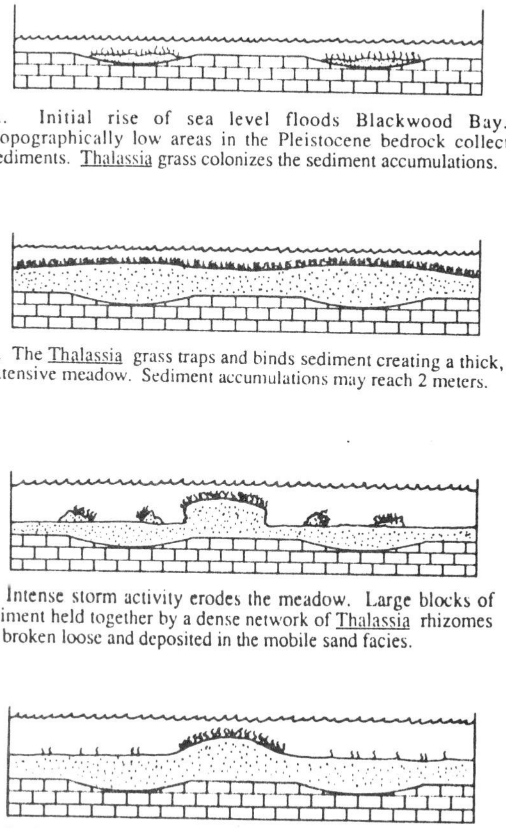

A. Initial rise of sea level floods Blackwood Bay. Topographically low areas in the Pleistocene bedrock collect sediments. <u>Thalassia</u> grass colonizes the sediment accumulations.

B. The <u>Thalassia</u> grass traps and binds sediment creating a thick, extensive meadow. Sediment accumulations may reach 2 meters.

C. Intense storm activity erodes the meadow. Large blocks of sediment held together by a dense network of <u>Thalassia</u> rhizomes are broken loose and deposited in the mobile sand facies.

D. Sediments of the mobile sand facies entombs the breakdown blocks of <u>Thalassia</u>. The original <u>Thalassia</u> mound remnant stabilizes. Mound position is no longer related to underlying Pleistocene bedrock topography.

Fig. 8. Interpretive illustration of the origin of Thalassia *mounds in Blackwood Bay.*

breakdown are dominant processes.

Correlation data reveals that the amount of organics (primarily *Thalassia* blades and rhizomes), sand, and mud are interrelated (Table 1). Organic content shows a direct relationship with mud content and an inverse relationship with sand. The slight increase in mud content with increased organics is attributed to a very limited baffling effect of the sparse *Thalassia* grass.

<u>Moderate</u> <u>*Thalassia*</u> <u>Facies</u>. The trapping and binding of sediment in the moderate *Thalassia* facies is fairly efficient. Sand content averages 74% and mud content averages 15%. The decrease in sand and the increase in mud with respect to the mobile sand facies is a direct result of the baffling effect of the grass.

Epibionts of red algae and foraminifera composed of high-magnesium calcite, and aragonitic serpulid worms encrust the broad, thin blades of *Thalassia* (Land, 1970; Patriquin, 1972; Nelson and Ginsburg, 1986). High-magnesium calcite averages 32% in the sand fraction and 40% in the mud fraction. In contrast, the barren mobile sand facies contains only 24% high-magnesium calcite in the sand fraction and 33% in the mud fraction. The increased amount of high-magnesium calcite is attributed to the presence of epibionts, and its high percentage in the mud fraction (40%) indicates that epibionts are a significant contributor of carbonate mud. The majority of the sand grains are angular, indicating that biological breakdown dominates over mechanical breakdown (Swinchatt, 1965; Scoffin, 1970).

Dense *Thalassia* Facies. Sediments of the dense *Thalassia* facies are similar to those of the moderate *Thalassia* facies but mud content is very high (averaging 35%) and sand content is very low (averaging 54%). The increased mud content is a result of the more efficient baffling of the dense *Thalassia*.

High-magnesium calcite is more abundant in the dense *Thalassia* facies than any other facies. The sand fraction is 38% high-magnesium calcite and the mud fraction contains 44% high-magnesium calcite. The large amounts of *Thalassia* grass provide an extensive habitat for epibionts. The increased levels of high-magnesium calcite reflects the abundance of epibionts encrusting the dense grass.

DEPOSITIONAL EVOLUTION OF BLACKWOOD BAY

Blackwood Bay, situated on the southern edge of the San Salvador platform, provides an unique setting in which to study modern carbonate sedimentation. To the south, the lagoon is bounded by the platform edge and steep fore-reef slope. To the north and east, Pleistocene outcrops border Blackwood Bay. The closeness of the lagoon to the platform edge and the ramp-like nature of the platform above the fore-reef slope indicate that as sea level encroached upon the platform, the "carbonate factory" was initiated early and *in situ* sedimentation began immediately. As no low-magnesium calcite was detected in any of the sediments of the bay, reworked Pleistocene material has not been a significant contributor to the sediment.

Based on analyses of cores, surface sediment samples, and study of the present facies distribution, the depositional evolution of Blackwood Bay can be presented as follows. Prior to flooding, Blackwood Bay existed as an exposed Pleistocene bedrock surface covered with a thin soil. The remnants of this soil has been preserved in the form of a dark red paleosol exposed in many parts of the study area. The presence of black micrite pebbles in the base of core 9 (Fig. 9) is also indicative of subaerial exposure (Strasser, 1984). These well rounded pebbles are interpreted as relics of paleosol reworked by waves, currents, and storm activity, and represent the only detected Pleistocene sediments in Blackwood Bay.

The timing of the marine inundation of Blackwood Bay can be surmised from data obtained from two cores collected in Snow Bay to the east. The basal sands yielded radiocarbon dates of 3850±190 yBP and 3630±120 yBP (Anderson and Boardman, 1987). These dates are also consistent with the Holocene chronology of Carew and Mylroie (1987) and the age (3810±90 yBP) of the coral from the landward coral rubble lobes previously mentioned. Although it cannot be proven that the sediments were formed where found today, it appears that the gently sloping Pleistocene floors of Snow Bay and Blackwood Bay were flooded around 3800 years ago.

As sea level inundated the ramp-like platform, high energy conditions existed in the partially flooded lagoon. The absence of any barrier reef framework or consolidated coral rubble along the platform edge resulted in the high energy conditions. Cores 3, 4, 9 and 12 have very coarse coral fragments which are believed to be derived from the fore-reef slope or from local patch reefs colonizing hardgrounds within the early lagoon.

Point count analysis of the sediments in core 3 indicates that molluscs, echinoderms, *Lithophyllum*, *Halimeda*, and *Neogoniolithon* algae were all very abundant during the flooding. Extensive hardgrounds, depressions in the Pleistocene bedrock surface, and initial *Thalassia* sediments would have provided excellent habitats for the those organisms (Howard and others, 1970; Scoffin, 1970; Zieman, 1972; Ebanks and Bubb, 1975; Turmel and Swanson, 1976; Bosence and others, 1985; Andersen and Boardman, 1987). While algae and other organisms were flourishing, foraminifera and ostracod abundance was low.

The continued transgression of the sea resulted in the deposition of overall fining-upward sequences present in all eleven of the lagoon cores. Boardman and others (1987) in a study of three Bahamian lagoons found that the depth and type of seaward sill, along with sea level rise, control the deposition of sediments. In Blackwood Bay, stabilization of the coral-rubble rampart along the platform edge provided a sill that produced a gradual decrease in physical energy conditions that produced the fining upward sequences. Localized buildups of patch reefs and colonization of current-baffling *Thalassia* grass also resulted in decreased energy conditions.

The present day sediment accumulation patterns observed in Blackwood Bay are a result of several processes. Sediments of the *Thalassia* facies are angular, poorly sorted, and contain elevated amounts of mud that reflect the low energy conditions brought on by the baffling effect of the grass. Higher velocity currents active in the barren mobile sand facies are evidenced by the clean, well rounded sediments found there. However, the process that exerts long term control over sediment distribution appears to be storm activity.

EVIDENCE FOR STORMS

The importance of storms within carbonate settings has been stressed by many workers (Ball, 1967; Hayes, 1967; Perkins and Enos, 1968; Patriquin, 1975; Hine and Neumann, 1977; Morton, 1981). In Blackwood Bay, cores 3, 4, 8, 9, 10, and 12 are characterized by small packages of fining--upward sequences, the basal portions of which are characteristically medium to coarse sands of the barren mobile sand facies deposited during peak storm activity. The remaining portion of the fining-upward sequences within the cores are fine sands, silts, and muds of the *Thalassia* facies deposited during waning storm activity and calm, more normal conditions.

Core 11, collected in a nearby pond, also shows evidence of storm activity. The base consists of a dark brown peat. Above the peat is a thick sequence of dark gray muds. Laden with ostracods, this deposit is typical of San Salvador pond environments (Teeter and Thalman 1984). Encased within the mud section are several thin layers of sand. Three of these layers are composed of marine sediment identical to those currently forming in the lagoon. The layers are thought to be the result of storm events that transported lagoonal sediments landward and deposited them in the pond.

During the latter days of December 1986, San Salvador Island was hit by an intense storm that lasted for two days. In Blackwood Bay, entire blocks of sediment held together by *Thalassia* rhizomes were torn from their mounds, thus producing subaqueous scour pits (Fig. 10). Sediments of the barren mobile sand facies immediately began entombing the breakdown blocks (Fig. 11). Evidence for a past similar event can be found in core 6 (see Fig. 12). That core contains one large (~120 centimeters) fining-upward sequence. The base of the core is a medium to fine moderately sorted bioclastic sand. This grades upward into a well sorted very fine bioclastic sand with abundant *Thalassia* rhizomes. The center of the sequence (40 to 80 centimeter level) shows abundant *Thalassia* rhizomes encased above and below by clean, well sorted sands. This anamolous package of *Thalassia* rhizomes is interpreted to be a breakdown block similar to the ones described above.

The presence of the cobble to boulder size coral rubble in many parts of the lagoon and back-beach area suggests that storms are important to the morphology of the lagoon. Although a coral-rubble boulder rampart presently forms the barrier to the south, this rubble has been stabilized on the platform edge by encrusting algae and *Millepora* and many of the primary internal coral structures have been obliterated, making taxonomic identification difficult. In contrast, the coral rubble deposited as lobes in the back-beach area shows no signs of having been submerged on the lagoon floor for very long. The coral rubble there is not bored or encrusted and preservation of primary structures is excellent. Most of the coral can be identified to the species level.

As mentioned above, the age (3810 +/- 90 yBP) of a coral sample from one of these lobes seems to coincide with the flooding of the bank top. The age of the coral and the excellent preservation of primary structures suggest that the coral was uprooted, transported across Blackwood Bay and deposited at its current location during one severe storm event.

Today, there exists a dynamic equilibrium between *Thalassia* grass colonization of the barren mobile sand facies and mound erosion primarily by storm activity. The fining-upward packages capped with *Thalassia* rhizomes in core 9 suggest that stabilization of the barren mobile sand facies by *Thalassia* is a continually occurring process. *Thalassia* breakdown blocks observed today and in core 6 suggest that erosion of *Thalassia*-stabilized sediment has also been occurring in the lagoon. Mound stabilization and subsequent erosion of those mounds results in an equilibrium reflected in the present day mound and scour pit topography in Blackwood Bay.

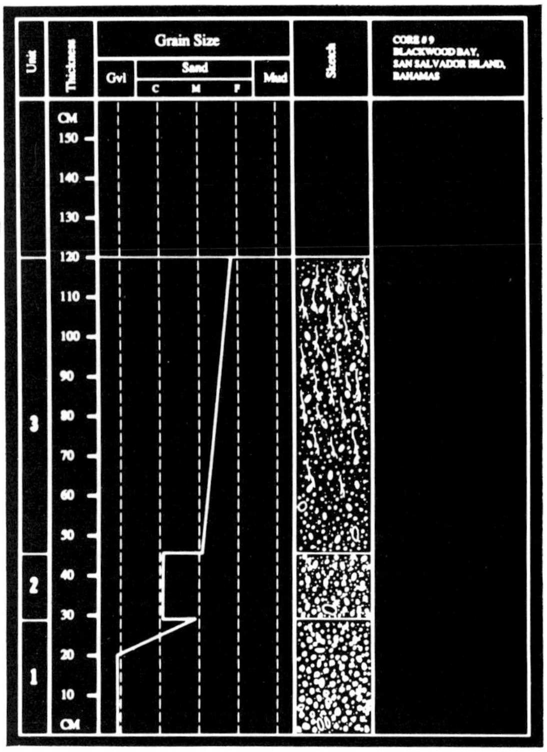

Fig. 9. Graphic display of core #9, collected in the center of the Blackwood Bay lagoon. Moderate Thalassia *growth on the bottom. Water depth: 54cm.*

Fig. 10. A breakdown block of Thalassia *produced during a severe storm that struck Blackwood Bay in December, 1986. Water depth: 1.7 meters.*

Fig. 11. Photo of Thalassia *breakdown blocks entombed by sediments of the barren mobile sand facies. Water depth: 1.7 meters.*

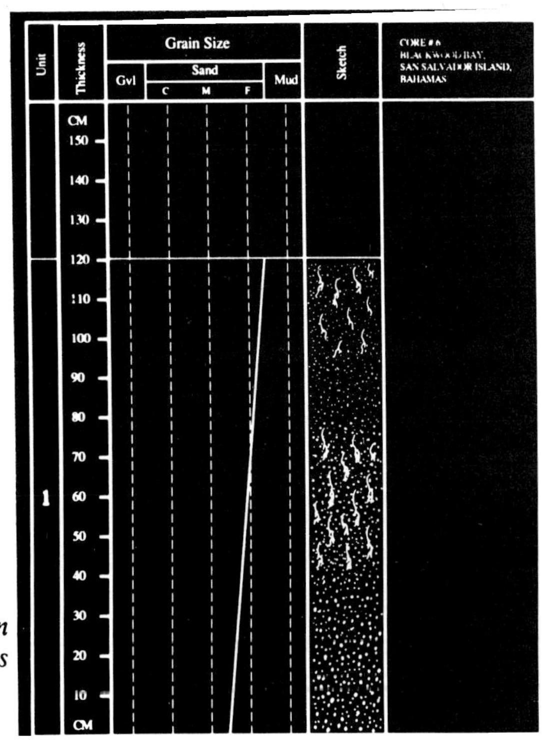

Fig. 12. Graphic display of core #6, collected from an elevated moderate Thalassia *area surrounded by numerous subaqueous scour pits. Water depth: 76cm.*

SUMMARY AND CONCLUSIONS

Blackwood Bay is a medium to low energy lagoon shielded from full wave energy by a boulder-rampart platform edge to the south. The fore-reef slope is very steep, plunging nearly vertically to abyssal depths.

Four sedimentary facies are recognized based on field observations and laboratory analyses of more than 50 surface sediment samples and twelve cores. The barren mobile sand facies consists of a clean, medium to well sorted, well rounded bioclastic sand which is largely deposited and reworked during sporadic high energy conditions produced by storms. Sediment composition varies depending on the location within the lagoon. Foraminifera, coral debris and algal fragments are most common, with coral debris very abundant along the platform edge. Mineralogically, the barren mobile sand facies contains the highest proportion of aragonite.

The sparse *Thalassia* facies is very similar to the barren mobile sand facies. Sediments are moderately to well sorted and do not contain appreciable mud. Sediment composition and mineralogy are almost identical to that of the barren mobile sand facies.

The moderate *Thalassia* facies exhibits signs of the baffling effect of the grass. Mud averages 15% of the samples analyzed. The sands are unabraided and angular, indicating that biological breakdown is dominant. Foraminifera, mollusc fragments, echinoid spines and pieces, algae, and ostracods dominate this facies.

The dense *Thalassia* facies shows the greatest effect of baffling as indicated by the very high average mud content (35%). Sand content is the least of any facies, averaging only 54%. Mineralogically, this facies contains the highest amount of high-magnesium calcite (38% of the sand and 44% of the mud). The steady increase in the amount of high-magnesium calcite with increasing amounts of grass is interpreted to be a function of epibionts inhabiting the surface of the grass blades. The increased level of magnesium calcite in the muds also suggests that epibionts are a significant contributor of lime mud in Blackwood Bay.

Summary of the Depositional History of Blackwood Bay

1. Holocene sea level rise flooded the majority of the San Salvador platform between 3600 and 4000 years ago.

2. Approximately 3800 years ago a major storm deposited gravels, sands and coral rubble throughout the lagoon. Much of the coral boulder rubble was concentrated into lobes deposited in the back-beach area.

3. As sea level continued to rise, stabilization of coral rubble by encrusting algae and *Millepora* along the platform margin reduced incoming wave energy and created quiet water conditions in the lagoon. The colonization of *Thalassia* on early lagoonal sediments reduced current energy. The decreasing energy conditions resulted in overall fining-upward sequences found in all eleven cores collected in the lagoon.

4. Storm activity reworked sediments producing small fining-upward sequences within an overall fining-upward package. The lower portion of each sequence represents deposition within the barren mobile sand facies while the upper portions show evidence of *Thalassia* grass colonization. The baffling effect of the grass trapped and bound sediment thus producing the *Thalassia* facies at the top of each sequence.

5. Today, a dynamic equilibrium exists between barren mobile sand facies stabilization by *Thalassia*, and *Thalassia* mound erosion by storm activity. This results in the present mound and subaqueous scour pit topography.

ACKNOWLEDGEMENTS

We are indebted to the Bahamian Field Station and its staff on San Salvador Island, Bahamas for logistical and financial support for this research. Special thanks to Dr. Donald Gerace and his wife, Kathy, for their support of and interest in this research.

Radiocarbon analysis was conducted at the Radiocarbon Lab of the University of Texas Balcones Research Center.

Marathon Oil Company, Houston, Texas, was responsible for photographic reproduction of the figures for this paper.

Many people contributed valuable assistance in the field. Thanks to Eric Babillas, Agnes Davis, Donna Devery, Cindy Guess, Mark Haverstein, Ben Jones, Joel Kostka, Elaine Kotter, Tom McClure, Robert Olive, Ruby Olive, Gena Pace, Tracy Sanders, Dave Shapiro, and Marc Wildly. We are especially grateful to John Hebert for his assistance in collecting the surface sediment samples and the cores.

REFERENCES CITED

Anderson, C.B., and Boardman, M.R., 1987, Sedimentary gradients in a high-energy carbonate lagoon, Snow Bay, San Salvador, Bahamas: CCFL Bahamian Field Station Occasional Paper 1987 No. 2, 30 p.

Ball, M.M., 1967, Carbonate sand bodies of Florida and the Bahamas: Journal og Sedimentary Petrology, v. 37, p. 556-591.

Bosence, D.W., Rowland, R.J., and Quine, M.L., 1985, Sedimentology and budget of a Recent carbonate mound, Florida Keys: Sedimentology, v. 32, p. 317-343.

Carew, J.L., and Mylroie, J.E., 1987, Refined geochronology of San Salvador Island, Bahamas, Proceedings of the third symposium on the geology of the Bahamas: Fort Lauderdale, Florida, CCFL Bahamian Field Station, p. 35-44.

Cloud, P.E., 1962, Environment of calcium carbonate deposition west of Andros Island, Bahamas: United States Geological Survey Professional Paper 350, 138p.

Ebanks, W.J., and Bubb, J.N., 1975, Holocene carbonate sedimentation Matecumbe Keys tidal bank, South Florida: Journal of Sedimentary Petrology, v. 45, p. 422-439.

Folk, R.L., 1974, Petrology of sedimentary rocks. Austin, Texas, Hemphill Publishing Company, 159 p.

Ginsburg, R.N., and Lowenstam, H.A., 1958, The influence of marine bottom communities on the depositional environment of sediments: Journal of Geology, v. 42, p. 687-689.

Hay, W.W., Wise, S.W., and Stieglitz, R.D., 1970 Scanning electron microscopy of fine-grain size biogenic carbonate particles: Transactions of the Gulf Coast Association Geological Society, v. 20, p. 287-302.

Hayes, M.O., 1967, Hurricanes as geological agents: case studies of hurricanes Carla, 1961, and Cindy, 1963: University of Texas, Bureau of Economic Geology, Report of Investigations, v. 61, 54 p.

Hine, A.C., and Neumann, A.C., 1977, Shallow carbonate bank-margin growth and structure, Little Bahamas Bank, Bahamas: American Association of Petroleum Geologists Bulletin, v. 61, p. 376-406.

Howard, J.F., Kissing, D.L., and Lineback, J.A., 1970, Sedimentary facies and distribution of biota in Coupon Bight, lower Florida Keys: Geological Society of American Bulletin, v. 81, p. 1929-1945.

Land, L.S., 1970, Carbonate mud: Production by epibiont growth on *Thalassia testudium*: Journal of Sedimentary Petrology, v. 40, p. 1361-1363.

Lowenstam, H.A., 1955, On the origin of sedimentary aragonite needles of the Great Bahama Bank: Journal of Geology, v. 65, p. 364-375.

Matthews, R.K., 1966, Genesis of Recent lime mud in southern British Honduras: Journal of Sedimentary Petrology, v. 36, p. 428-454.

McKee, E.D., and Ward, W.C., 1983, Eolian environment, in Scholle, P.A., Bebout, D.G., and Moore, C.H., eds., Carbonate depositional environments: American Association of Petroleum Geologists Memoir 33, p. 131-170.

Morton, R.A., 1981, Formation of storm deposits by wind forced currents in the Gulf of Mexico and the North Sea, in Nio, S.O., Shottenheim, R.T.E., and Van Weering, T.J.C.E., eds., Holocene marine sedimentation in the North Sea Basin: International Association Sedimentology Special Publication, v. 5, p. 123-138.

Nelson, J.E., and Ginsburg, R.N., 1986, Calcium carbonate production by epibionts on *Thalassia* in Florida Bay: Journal of Sedimentary Petrology, v. 56, p. 622-628.

Neumann, A.C., and Land, L.S., 1975, Lime mud deposition and calcareous algae in the Bight of Abaco, Bahamas: a budget: Journal of Sedimentary Petrology, v. 42, p. 195-204.

Patriquin, D.G., 1972, Carbonate mud production by ebibionts on *Thalassia*: an estimate based on leaf growth data: Journal of Sedimentary Petrology, v. 42, p. 687-689.

Patriquin, D.G., 1975, Migration of blowouts in seagrass beds at Barbados and Capriacove, West Indies, and its ecological and geological implications: Aquatic Botany, v. 1, p. 163-189.

Perkins, R.D., and Enos, P., 1968, Hurricane Betsey in the Florida-Bahamas area: geologic effects and comparison with Hurricane Donna: Journal of Geology, v. 76, p. 710-717.

Scoffin, T.P., 1970, The trapping and binding of subtidal carbonate sediments by marine vegetation in Bimini lagoon, Bahamas: Journal of Sedimentary Petrology, v. 40, p. 249-273.

Stieglitz, R.D., 1970, Scanning electron microscopy of the fine fraction of recent carbonate sediments from Bimini, Bahamas: (unpublished Ph.D. dissert.), University of Illinois, Urbana, 130 p.

Stockman, K.W., Ginsburg, R.N., and Shinn, E.A., 1967, The production of lime mud by algae in south Florida: Journal of Sedimentary Petrology, v. 37, p. 633-648.

Strasser, A., 1984, Black-pebble occurrence and genesis in Holocene carbonate sediments (Florida Keys, Bahamas, and Tunisia): Journal of Sedimentary Petrology, v. 54, p. 1097-1109.

Swinchatt, J.P., 1965, Significance of constituent composition, texture, and skeletal breakdown in some recent carbonate sediments: Journal of Sedimentary Petrology, v. 35, p. 71-90.

Teeter, J.W., and Thalman, K.L., 1984, Second symposium on the geology of the Bahamas field trip to Pigeon Creek, in Teeter, J.W., ed., Proceedings of the second symposium on the geology of the Bahamas: Fort Lauderdale, Florida, CCFL Bahamian Field Station, p. 177-186.

Turmel, R.J., and Swanson, R.G., 1976, The development of Rodriguez Bank, a Holocene mudbank in the Florida reef tract: Journal of Sedimentary Petrology, v. 46, p. 497-518.

Wanless, H.R., 1974, Mangrove sedimentation in geologic perspective: Miami Geological Society Memoir 2, p. 190-200.

Wilson, J.L., 1975, Carbonate facies in geologic history: Berlin, Springer-Verlag, 471 p.

Zieman, J.C., 1972, Origin of circular beds of *Thalassia* (Spermatophyta: Hydrocharitaceae) in south Biscayne Bay, Florida and their relationship to mangrove hammocks: Bulletin of Marine Science, v. 22, p. 559-574.

NOTES ON THE TAPHONOMY OF COMMON SHALLOW-WATER BENTHIC FORAMINIFERA FROM SAN SALVADOR, BAHAMAS

Mark W. Peebles* and Ronald D. Lewis
Department of Geology
Auburn University
Auburn University, AL 36849

ABSTRACT

Four different processes were investigated with regard to their effects on the preservation potential of benthic foraminifera common in the reefs and lagoons around San Salvador. Twenty-two different species, comprising approximately eight percent of the total specimens studied, showed signs of selective predation as evidenced by large borings that were perpendicular to test surfaces and that had beveled edges. While there was selective predation of foraminifera, this predation was not species selective. The most abundant species were the most heavily preyed upon in any given area. It was also discovered that a foraminiferan, *Rosalina* sp., used the tests of larger, commonly dead, foraminifera as a substrate.

Microborings are found in virtually any exposed carbonate substrate in the tropical Bahamian waters. Though most studies of microborings have found no differences between substrates, this study demonstrates that there are differences among the host taxa with respect to the extent of infestation and the location of infestation within the test. The primary differences are at the suborder level; members of the Suborder Rotaliina are not as bored as are specimens of the Suborder Miliolina taken from the same sample. Although the causes of these differences are not clear, it is suspected that differences in test wall composition may be a factor.

Abrasion and dissolution leave recognizable signatures on the test surface. Abrasion can break off large areas of the test, but it works relatively slowly and more frequently polishes the surface without causing extensive damage. The effects of dissolution are harder to detect at low magnifications than are those of abrasion. However, even small amounts of dissolution can affect the integrity of the test by removing surface layers and by weakening the test structure. Dissolution, including that caused by microboring organisms, is the primary process affecting preservation in this reefal environment.

INTRODUCTION

The taphonomy of shallow-water benthic foraminifera has received relatively little attention in contrast to deeper water species (studies of dissolution) and in contrast to other shallow-water organisms. Taphonomic processes affecting shallow benthic foraminifera have been suggested by comparing living and dead assemblages (Murray, 1973; Douglas et al., 1980; and Smith, 1987), but relatively few studies have isolated taphonomic processes and examined them in detail. Notable exceptions are: Murray and Wright, 1970 (dissolution), Butcher and Steinker, 1979 (fungal and bacterial degradation in *Archaias angulatus*), Collen and Burgess, 1979 (the diagenetic processes of dissolution, overgrowth, and recrystallization in *Notorotalia*), and Cottey and Hallock, 1988 (abrasion, dissolution and breakage for *Archaias angulatus*).

Predation also plays a significant role in the preservation of foraminifera. Indiscriminate deposit feeders may ingest living and dead foraminifera as part of the sediment, or smaller predators may selectively prey on foraminifera, leaving behind bored and otherwize damaged tests. Tests of foraminifera have been recovered from the digestive tracts of gastropods (Walker, 1971; Buzas and Carle, 1979), polychaetes (Lipps and Ronan, 1974; Buzas and Carle, 1979), decapods (Buzas and Carle, 1979), and fish (Daniels and Lipps, 1978; Buzas and Carle, 1979). Bored tests of foraminifera have been described by Sliter (1971), Hickman and Lipps (1983), and Arnold et al. (1985). These borings were inferred to have been the result of gastropod and nematode predation.

*Present Address: Department of Marine Science, University of South Florida, St. Petersburg, FL 33701

Our purpose is to review the senior author's research on the taphonomy of the common foraminifera found at San Salvador. An important theme of the several investigations described herein is the comparison of effects among a number of taxa to ascertain potential preservational biases. This research was done as part of the requirements for the Master of Science degree at Auburn University, and with substantial assistance from the personnel of the Bahamian Field Station on San Salvador. A complete history of previous research is beyond the scope of this summary, but can be found in the references cited and in Peebles, 1988; Peebles and Lewis, 1988; and Peebles and Lewis, in press.

METHODOLOGY

All samples used in this study were collected during the summer of 1986 from five areas around the island of San Salvador. The samples were collected directly by using SCUBA equipment and small hand-held coring tubes. Environments sampled include: (1) a shallow *Thallassia* (turtle grass) meadow near Cut Cay (water depth 1m), (2) a protected patch reef in French Bay dominated by *Montastrea annularis* and *Acropora palmata* (water depth 6.5m), (3) a shallow back-reef near Gaulin's Reef (water depth 3m), (4) an unprotected shallow-water patch reef dominated by *Montastrea annularis* (Snapshot Reef, water depth 4m), and (5) a similar reef dominated by *Acropora cervicornis* and *M. annularis* (Telephone Pole Reef, water depth 5m).

The samples were stained with rose bengal at the Bahamian Field Station in order to identify specimens living at the time of collection (Walton, 1952). As noted by other workers (Benda and Puri, 1962; Atkinson, 1971; Le Calvez and Cesana, 1972; Martin and Steinker, 1973) this technique often fails to stain living specimens and sometimes stains dead tests. Therefore, no attempts were made to count living specimens. A large proportion of one species, *Discorbis rosea*, appeared to be vividly stained in addition to its characteristic red coloration. The majority of these were apparently living as shown by the occurrence of protoplasm in many randomly sampled tests. No other species studied was as consistently stained nor had as many specimens with visible protoplasm. Specimens showing abrasion, breakage, and dissolution textures were assumed to be dead at the time of collection.

Representative specimens from each of the studies were coated with a gold-palladium alloy using the Auto Conductovac IV sputter coater, examined using an ISI40 scanning electron microscope, and photographed with Kodak TMAX 400 ASA film. Illustrations will be found in the references cited above.

PREDATION

Approximately three hundred foraminifera were counted in two samples from each locality, and the bored foraminifera were picked and mounted for further study. Specimen totals and the number of Milolina, Rotaliina, and Textulariina were noted from each sample site. Measurements were taken of the bore-hole sizes using a stage micrometer. The borings were classified into large and small predationary borings. Specimens were counted if they contained one or more borings. Total specimens bored were recorded, along with the number of Miliolina, Rotaliina, and Textulariina bored by predators, and the number of borings made by *Rosalina* sp.. The total number of species at each site, the number of the two most common species, and the number of species preyed upon at each site were also recorded. Two distinct size ranges of borings are found in this study. Small borings range in size from 10μm to 30μm and have a mean diameter of 20μm. Large borings range in size from 40μm to 100μm and have a mean diameter of 59μm. Borings in both size ranges are circular to oval in shape, commonly have beveled edges, and are perpendicular or nearly perpendicular to the test wall. The specimens with this type of boring comprise 8.3% of the 2974 specimens. Several of the borings in the small size range resembled those classified as nematode borings (Sliter, 1971), but we consider it unlikely that nematodes produced them. We tentatively identify both the large and the small borings as naticid gastropod borings.

Ninety percent of the bored tests had the small-size borings, suggesting that juvenile gastropods are the primary hole-producing predator on the benthic foraminifera of San Salvador. Our data indicates that predation is opportunistic rather than selective; the predators feed on the most abundant foraminiferal taxa, whatever that happens to be.

A small foraminiferan, *Rosalina* sp., was found in its boring in the test of 43 specimens. These borings differ from those described above by being strongly beveled on one side, failing to penetrate through the test wall, and containing a specimen of *Rosalina* sp. The tests bored by *Rosalina* appeared to have been dead at the time of collection, thus ruling out predation. Specimens of *Rosalina* sp. were also noted on other highly weathered skeletal types and on unidentifiable sediment grains. Of the 43 tests bearing *Rosalina*, 32 were *A. angulatus*, with the remaining 11 comprising assorted miliolines and textulariines. This also suggests a nonselective pattern.

MICROBORINGS

Only one sample was used in this investigation in order to compare the effects of microboring in different genera. The sample used in this study was taken at a depth of three meters from a patch reef dominated by *Montastrea annularis*, located in French Bay, off the southern tip of the island.

Six genera were selected for this study (*Discorbis, Amphistegina, Quinqueloculina, Peneroplis, Cyclorbiculina*, and *Archaias*) because of their high abundance in all samples. Representative specimens were embedded in Spurr's low viscosity embedding medium in the manner described by Golubic et al., 1970, and the calcite tests were dissolved in acid. Resin casts of the micoborings were then examined using SEM (Peebles and Lewis, 1988).

Microboring intensity was characterized by the use of four categories: not infested, sparsely infested, moderately infested, and heavily infested. The areas of the tests infested with microborers were designated by three categories: proloculus (if the majority of casts occurred either directly on the proloculus or in the group of chambers immediately adjacent to the proloculus), outer margin (if the majority of casts were noted on the outer wall or near the group of chambers directly adjacent to the outer wall and far away from the proloculus), and scattered (if the casts were found in all areas of the test except the proloculus).

The rotaliines *Discorbis* and *Amphistegina* are the only taxa that consistently showed specimens with no visible endoliths. In the bored specimens, the degree of infestation was slight, and was most often noted in the outer margin.

Among the Miliolina, *Quinqueloculina* and *Peneroplis* commonly contained some microboring casts, although the degree of infestation is often slight. *Peneroplis* is the only genus that consistently contained casts on or near the proloculus. In specimens which were heavily bored, the majority of casts were scattered throughout the test, with only a few cases of heavy infestation along the outer margin.

In *Archaias* and *Cyclorbiculina* the majority of specimens were moderately to heavily infested. Microboring casts were most often scattered indiscriminately throughout the test in *Cyclorbiculina.* In *Archaias* many of the casts were located along the outer margin. Though a few casts were noted in the proloculus area in both taxa, they never constituted the majority of casts found in any specimen.

SURFACE TEXTURES

Ten species were chosen for this study: the miliolines *Archaias angulatus, Borelis pulchra, Cyclorbiculina compressa*(?), *Peneroplis proteus, Pyrgo comata*, and *Quinqueloculina tricarinata*; the rotaliines *Amphistegina* sp., and *Discorbis rosea*; and the textulariines *Clavulina tricarinata*, and *Valvulina* sp. (classification of Loeblich and Tappan, 1964). For each species eighty specimens were selected which best exhibited the unaltered characteristics of each taxon, and one test per species was chosen as a control. Ten specimens from each taxon were used for the dissolution experiment and 60 per taxon were used in the abrasion experiment. The remaining 100 specimens were used in preliminary experiments.

The six hundred specimens used for the abrasion experiment were weighed to the nearest one hundredth of a gram and were then placed in a one dram bottle 12mm in diameter containing buffered seawater with a pH of 7.45. This bottle was then placed into a standard tumbling barrel. The tumbling barrel rotated at 40 revolutions per minute, while the smaller bottle inside rotated approximately 280 revolutions per minute. In order to approximate the relative distance the specimens moved, it is estimated they tumbled from 1-3mm with every revolution. Weights were recorded and specimens were taken periodically for 376 hours. One specimen of each taxon was extracted at selected intervals for examination with a scanning electron microscope. After the 376th hour the tumbling barrel was allowed to run continuously for 1624 hours bringing the total time to 2000 hours, or the equivalent of 34-100km of movement. Susceptibility to abrasion was estimated by the number of specimens remaining after 2000 hours of tumbling. If all specimens of a species were completely destroyed, the degree of damage at 376 hours was used for ranking. Types of damage taken into account were the amounts of outer wall remaining, the amount of breakage, and the percent weight lost by each taxon.

Dissolution was studied in two simple experiments designed to indicate changes in the various taxa over time. The first experiment began by placing ten specimens of each species into a petri dish containing formic acid with a pH of 3.90. One specimen of each taxon was then removed for examination at logarithmic intervals starting at twenty minutes and ending at 10,240 minutes (170.5 hours). In the second experiment, the effects of extreme dissolution were examined by rapidly etching tests in hydrochloric acid with a pH of approximately 1.5 for one minute. Susceptibility to dissolution was estimated from the degree of damage visible after 10,240 minutes in the formic acid solution. Types of damage noted were the approximate amounts of material removed in each species and the approximate time for complete dissolution in the hydrochloric acid.

Foraminifera extracted from two Pleistocene Cockburn Town Reef rock samples were examined for evidence of abrasion and dissolution textures. The 1-kilogram limestone samples were disaggregated by boiling them in a solution of approximately 2.0 gm/l sodium hexametaphosphate (calgon) and water for approximately five minutes. The samples were then crushed by hand and again boiled in the calgon and water solution for approximately five minutes. The resulting sediment was placed in an ultrasonic bath for eight hours, and was washed in tap water to remove the clay-sized particles. A total of 38 specimens were found in the two samples. These consisted of 36 specimens of *Archaias* sp., one quinqueloculine test, and one unidentified milioline.

Both of the rotaliine genera used in the laboratory experiments are radial-walled (after Loeblich and Tappan, 1964) with the surface appearing smooth at low SEM magnifications, while at higher magnifications the borders of the oriented crystallites can faintly be seen. The majority of the miliolines studied exhibit three layers. An outer, unordered layer of very fine particles is evident on most tests and has been described previously for other miliolines (e.g., Walker, 1971). Underneath this layer lies a thin veneer of oriented calcite laths that vary slightly in shape and size with different species. The lowermost layer is composed of unoriented crystallites in an organic matrix.

After 64 hours of tumbling (1-3km), 48% of the specimens exhibited a slight degree of damage. Most of this is evident in small breaks and polishing around the perimeter or aperture of all of the specimens. Several specimens of *Pyrgo*, *Clavulina*, and *Valvulina* are altered by the appearance of small holes or are missing large portions of a chamber. However, other than the missing holes or chambers, the degree of damage is very slight and would be difficult to recognize without undamaged control specimens for comparison.

Damage is much easier to document after 376 hours (6-18km), with 72% of the tests exhibiting damage visible at a magnification of 50X. Portions of the tests that are elevated above other test surfaces show the most damage. Evidence of this is illustrated by missing chamberlet walls in the soritids, flattened ornamentation in *Quinqueloculina*, broken apertures and faint scratches in the rotaliines, and a loss of detail in the sutures of the textulariines. The thin veneer of agglutinated(?) particles is removed from most of the miliolines, and, especially in *Borelis*, the thin veneer of oriented laths is beginning to be removed. All of the specimens exhibit some degree of polishing, but the rotaliines are the only suborder that are visibly scratched.

After 2000 hours (34-100km), all of the small specimens, including all of the *Borelis* and *Discorbis* specimens, were destroyed. The percentage of other species that were destroyed varies; 88% of the *Pyrgo*, 81% of the *Amphistegina*, 71% of both *Peneroplis* and *Clavulina*, 68% of the *Quinqueloculina*, 59% of the *Valvulina*, 34% of the *Archaias*, and 4% of the *Cyclorbiculina* specimens. All of the remaining specimens were damaged in the same ways as discussed for 376 hours, except to a greater degree. The veneer of oriented crystallites was not completely removed from all of the miliolines. The rotaliines exhibit more scratches and pits than do the miliolines and textulariines. A textulariine surface texture discernible at high magnifications includes ridges of cement surrounding areas from which the agglutinated particles have been removed.

Dissolution textures are coarser than abrasion textures. After 20 minutes in the formic acid, all of the fine agglutinated particles on the miliolines were removed, and, after 80 minutes, the thin veneer of oriented calcite laths was completely removed. Once this outer layer was removed, the surface texture of the miliolines consisted of only the unoriented calcite crystallites, which appeared the same regardless of the degree of dissolution, or the pH of the acid used, until the tests were completely dissolved. At lower magnifications (less than 500X) the crystallites could not be seen, and the tests were characterized by a very coarse, uneven appearance.

The rotaliines were less susceptible to dissolution than were the miliolines or the textulariines, as has been reported by other workers (Murray and Wright, 1970; Walker, 1971). Dissolution textures in the rotaliines were not readily apparent, using magnifications of less than 100X.

The two Pleistocene samples investigated were difficult to interpret due to the high degree of diagenetic cementation and crystal overgrowth common in carbonate environments (Collen and Burgess, 1979). All of the specimens were coated with small, acicular aragonite(?) crystals. Approximately half of the specimens had visible cement or particles cemented to a portion of the test. However, a majority (30 out of the 38 specimens) exhibited probable dissolution textures, a rough surface from which much of the outer wall had been removed. The remaining 8 specimens had probable abrasion textures as indicated by the removal of surface detail.

DISCUSSION

Taphonomy is an important subdiscipline of paleontology that has not yet fully been exploited in the study of foraminifera. Predation, as evidenced by borings, is a topic that illustrates the intimate relationship between taphonomy and paleoecology. It is an ecologic study in that it is a record of the interactions between groups of organisms during life. It is a taphonomic study by virtue of its preservation in skeletal parts after predation causes the death of the foraminiferan, and because the test's preservation potential could be adversely affected by the borings. The predators are inferred to be juvenile naticid gastropods feeding in an opportunistic pattern. Recognition of such predation patterns in the fossil record could possibly aid in delineating generalized paleoenvironments. For example, opportunistic patterns could indicate an abundant and diverse prey in an unrestricted environment, while species-selective predation could imply fewer and less diverse prey in a restricted environment. Predational borings have been implicated in some environments as the cause for preferential dissolution of heavily preyed species by opening more surface area to dissolution (Douglas, 1973). Although this has been suggested, and probably occurs, borings have also commonly been found preserved in fossils (e.g., Sliter, 1975).

In addition to predation, it was found that a small foraminiferan, *Rosalina* sp., commonly bores into the tests of larger, usually dead foraminifera. The larger foraminifera tests apparently were used as a substrate. Only miliolid and textulariid species were bored in every location where *Rosalina* sp. occurred, with no observed borings on rotaliid species. In the second investigation of this thesis it was found that miliolids were infested by microboring organisms to a greater degree than were rotaliids. Thus, there is also the possibility that the larger foraminiferan tests provided a food source for *Rosalina* sp..

Unlike previous studies of microborers, the second study indicates some consistent differences in infestation among the host organisms. These differences are evident at the suborder level, particularly between the Rotaliina and the Miliolina in the degree and in the test location of infestation. Less obviously, the type of microboring casts seem to represent different assemblages in the two suborders studied. Fundamental differences in test composition and microstructure could be the major factor in these microboring distributions. By simply exposing more surface area to dissolution they could negatively bias the preservation potential of the affected taxa. There is also the possibility that in certain situations they could positively bias the preservation potential by the formation of a micritic crust.

Abrasion and dissolution textures, treated in the third study, may be used as environmental indicators because each leaves a recognizable signature on the surface of the test. One slightly surprising conclusion of this study is that dissolution appears to be the predominant destructive force acting on the tests found at the island. This supports a recent geochemical investigation that indicated a greater degree of dissolution in shallow carbonate environments and on the continental shelves than in the deep sea (Walters and Burton, 1987). This is caused primarily by the high degree of infaunal activity and by the large amounts of unstable carbonates (aragonite and high-magnesium calcite) in shallow environments. Dissolution is a pervasive process that happens very quickly over geologic time. After only moderate amounts of dissolution, many surface features are destroyed, and further dissolution may be difficult to recognize. However, abrasion does not destroy surface features as rapidly or as completely, and may prove to be more useful in determining relative transport distance or sediment residence time.

Taphonomy has grown from being simply a description of the preservational processes and how they bias the fossil record into a subdiscipline in which the goal is the retrieval of all possible information. Each of the processes described in this summary plays a role in test destruction, but at the same time adds ecologic and environmental information that would not be apparent in a study of the foraminifera *per se*. Predation, because it focuses on the interactions between species, adds more ecologic information to an environmental investigation than would a simple cataloging of species distributions. Microboring studies hold great promise for environmental interpretations. With further research from San Salvador and a variety of other locations it may be possible to classify microborings into easily recognizable sub-environments. A combination of foraminifera and the microborings preserved in their tests may permit finer resolution of paleoenvironments.

With the identifications of abrasion and dissolution textures it is possible to make estimates of the amount of transportation and/or sediment residence time. After further study, abrasion and dissolution textures could also aid in delineating environments, especially if abrasion and dissolution indices are made from many areas.

ACKNOWLEDGEMENTS

We thank Donald T. Gerace, Executive Director of the Bahamian Field Station, for his assistance with the logistics of the field work involved in this study, and acknowledge the following for financial support of field operations: (MWP) The College Center of the Finger Lakes, The Alabama Academy of Science, and The Geological Society of America; (RDL) the Auburn University Grant-in-Aid Program. We also thank Roland R. Dute and Ann E. Rushing, Department of Botany-Microbiology, Auburn University, for their assistance with the SEM photography; Charles E. Savrda, Department of Geology, Auburn University, for his many valuable discussions concerning methodology and results; and Ira F. Holston, Department of Geology, Auburn University, for his valuable input concerning carbonate dissolution; and Robert A. Gastaldo, Department of Geology, Auburn University, for his critical reading of the manuscript.

REFERENCES CITED

Arnold, A.J., F. D'Escrivan and W.C. Parker, 1985, Predation and avoidance responses in the foraminifera of the Galapagos hydrothermal mounds: Journal of Foraminiferal Research, v.15, p.38-42.

Atkinson, K., 1971, The relationship of Recent foraminifera to the sedimentary facies in the turbulent zone, Cardigan Bay: Journal of Natural History, v. 5, p. 385-489.

Benda, W.K., and H.S. Puri, 1962, The distribution of foraminifera and ostracoda off the Gulf coast of the Cape Romano area, Florida: Gulf Coast Association of Geological Societies Transactions, v. 12, p. 303-341.

Butcher, W.A., and D.C. Steinker, 1979, Fungal and bacterial degradation of the test of *Archaias angulatus* (Foraminifera): Micron, v. 10, p. 35-36.

Buzas, M.A., and K.J. Carle, 1979, Predators of foraminifera in the Indian River, Florida: Journal of Foraminiferal Research, v. 9, p. 336-340.

Collen, J.D., and C.J. Burgess, 1979, Calcite dissolution, overgrowth and recrystallization in the benthic foraminiferal genus *Notorotalia*: Journal of Paleontology, v. 53, p. 1343-1353.

Cottey, T.L., and P. Hallock, 1988, Test surface degradation in *Archaias angulatus*: Journal of Foraminiferal Research, v. 18, n. 3, p. 187-202.

Daniels, R.A., and J.H. Lipps, 1978, Predation on foraminifera by Antarctic fish: Journal of Foraminiferal Research, v. 8, p. 110-113.

Douglas, R.G., J. Liestman, C. Walch, G. Blake, and M.L. Cotton, 1980, The transition of live to sediment assemblage in benthic foraminifera from the Southern California Borderland, in M.E. Field, A.H. Bouma, I.P. Colburn, R.G. Douglas and J.C. Ingle, eds., Quaternary Depositional Environments of the Pacific Coast, Pacific Coast Paleogeography Symposium 4, p. 257-280.

Golubic, S., G. Brent, and T. LeCampion, 1970, Scanning electron microscopy of endolithic algae and fungi using a multipurpose casting-embedding medium: Lethaia, v. 3, p. 203-209.

Hickman, C.S., and J.H. Lipps, 1983, Foraminifer-ivory: selective ingestion of foraminifera and test alterations produced by the neogastropod *Olivella*: Journal of Foraminiferal Research, v.3, p.108-114.

Le Calvez, Y., and D. Cesana, 1972, Detection de l'etat die vie chez les foraminiferes: Annales de Paleontologie (Invertebrae), v. 58, p. 129-134.

Lipps, J.H., and T.E. Ronan, 1974, Predation on foraminifera by the polychaete worm *Dioptera*: Journal of Foraminiferal Research, v. 4, p. 139-143.

Loeblich, A.R., Jr., and H. Tappan, 1964, Sarcodina, chiefly "Thecamoebians" and Foraminiferida, in R.C. Moore, ed., Treatise on Invertebrate Paleontology, Protista 2, pt. C, University of Kansas Press, 900 p.

Martin, R.E., and D.C. Steinker, 1973, Evaluation of techniques for recognition of living foraminifera: Compass, v. 50, p. 26-30.

Murray, J.W., 1973, Distribution and Ecology of Living Benthic Foraminiferids: Crane Russak and Company, New York, 274 p..

Murray, J.W., and C.A. Wright, 1970, Surface textures of calcareous foraminiferids: Palaeontology, v. 13, p. 184-187.

Peebles, M.W., 1988, Taphonomy of common shallow-water benthic foraminifera from San Salvador, the Bahamas: Unpublished M.S. Thesis, Auburn University, Auburn, Alabama, 120 p.

Peebles, M.W., and R. D. Lewis, 1988, Differential infestation of shallow-water benthic foraminifera by microboring organisms: possible biases in preservation potential: Palaios, v. 3, p. 345-351.

Peebles, M.W., and R. D. Lewis, in press, Surface textures of benthic foraminifera from San Salvador, Bahamas: Journal of Foraminiferal Research.

Sliter, W.V., 1971, Predation on benthic foraminifera: Journal of Foraminiferal Research, v. 1, p. 20-29.

Smith, R.K., 1987, Fossilization potential in modern shallow-water benthic foraminiferal assemblages: Journal of Foraminiferal Research, v. 17, p. 117-122.

Walker, D.A., 1971, Etching of the test surface of benthonic foraminifera due to ingestion by the gastropod *Littorina littorea* Linne: Canadian Journal of Earth Science, v.8, p. 1487-1491.

Walters, L.M. and E.A. Burton, 1987, Active dissolution in modern marine carbonate sediments: Global implications [abstr.]: Geological Society of America, Abstracts with Programs, v. 19, n. 7, p. 880.

Walton, W.R., 1952, Techniques for the recognition of living foraminifera: Cushman Laboratory of Foraminiferal Research Contributions, v. 3, p. 56-60.

STRATIGRAPHIC DISTRIBUTION AND ASSOCIATIONS OF TRACE ELEMENTS IN VADOSE-ALTERED MULTICOMPONENT CARBONATE ASSEMBLAGES

Ronald C. Pelle and Mark R. Boardman
Geology Department
Miami University
Oxford, Ohio 45056

ABSTRACT

Trace element geochemistry is being increasingly used to determine diagenetic trends in carbonates (Pingitore, 1978; 1982; Veizer, 1983; Brand and Veizer, 1980). Recent studies have developed theories on the mechanisms involved in the redistribution of trace elements in carbonates and laid the groundwork for further research. However, the effects of vadose diagenesis on the distribution of trace elements in carbonate assemblages remains poorly understood.

Analysis of multicomponent carbonates (i.e., stable and metastable carbonate minerals, oxides, organic matter, and non-carbonate minerals) from Andros Island and San Salvador Island, Bahamas provides information on trace element associations among various components within carbonate units and the vertical and lateral stratigraphic distribution of trace elements, particularly in relation to paleosol horizons. Three geochemically distinct zones may be observed within a single vadose-altered carbonate dune: 1. paleosols; 2. "zone of influence"; and 3. host rocks. Each of these zones can be recognized by the trace element signatures associated with the exchangeable carbonate and oxide fractions of multicomponent carbonates.

Multivariate statistical analysis reveals the presence of five geochemically-distinct groups within the Pleistocene and Holocene dunes of the Bahamas: 1. paleosols; 2. early diagenetic dune sands; 3. recrystallized dune sands; 4. soil influenced type I; and 5. soil influenced type II.

A statistical model (polar ordination) is developed to account for the diagenetic trends observed within the Bahamian carbonate dune sands. Variability among the units studied is accounted for by progressive aragonite to diagenetic low-Mg calcite transformation and by varying degrees of pedogenesis. Although time is a factor in the development of the trends recognized, other factors related to vadose diagenesis are more important. The trace element signals associated with vadose diagenesis result from the degree of chemical diagenesis, the type and timing of paleosol development, and the extent of bioturbation and fracturing of the host unit. These mechanisms play the most important role in the geochemical signature present in vadose-altered carbonate sands.

INTRODUCTION

Carbonate rocks and sediments are comprised of a variety of constituents, including: the carbonate minerals aragonite, high-Mg calcite, and low-Mg calcite; oxides; organic matter; and non--carbonate minerals. Such "multicomponent" carbonates have diagenetically-mobile and immobile trace elements associated with various constituents of the rock or sediment.

In recent years, studies concerning the occurrence and mobility of trace elements in carbonate rocks and sediments have provided both practical and theoretical models for chemical controls and pathways during carbonate diagenesis (Land, 1967; Kinsman, 1969; Friedman, 1969; Pingitore, 1978; 1982; Brand and Veizer, 1980; and Veizer, 1983). The precipitation or dissolution of calcium carbonate is a wet chemical reaction, and trace elements from solution such as Sr^{++}, Mg^{++}, Fe^{++}, Mn^{++}, and Na^{+} can substitute for Ca^{++} in the $CaCO_3$ lattice (Bathurst, 1975; Brand and Veizer, 1980). Trace elements can be incorporated into carbonate minerals in several ways: 1. substitution for calcium in the $CaCO_3$ lattice; 2. interstitially between lattice planes; 3. in lattice positions that are open due to defects in the structure; and 4. in non-carbonate inclusions, such as silicate impurities and fluid inclusions (Veizer, 1983).

The original carbonate minerals, aragonite (A) and high-Mg calcite (HMC), will possess a trace element signature derived from the ambient sea-water in which they formed. Some modification by biogenic processes may occur. This is particularly important for low-strontium molluscan aragonite. Upon exposure to meteoric waters,

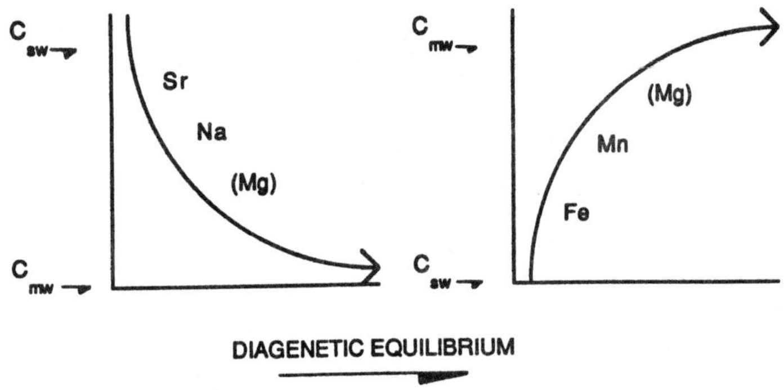

Fig. 1. Elemental changes expected during increasing diagenetic equilibrium with meteoric water. C_{mw} and C_{sw} are calcium carbonate in equilibrium with meteoric water and sea-water, respectively. Sr and Na, present in higher concentrations in sea water, show a decrease in concentration as diagenetic equilibrium with meteoric water is approached. Fe and Mn, present in higher concentrations in meteoric water, should show a concomitant increase. Mg can either increase or decrease with progressive equilibrium with meteoric water, depending on original mineralogy; high-Mg calcite, low-Mg calcite, or aragonite (after Brand and Veizer, 1980).

these minerals partly or completely dissolve, exchange, and mix their trace elements with those of the meteoric water, and subsequently reprecipitate as diagenetic low-Mg calcite (dLMC) (Pingitore, 1978; 1982; Brand and Veizer, 1980). This dLMC will have a trace element signature shifted in the direction of equilibrium with the meteoric water from which it precipitated (Brand and Veizer, 1980). As fresh water diagenesis proceeds, a progressive loss of strontium and sodium, along with a concomitant increase in iron and manganese should occur (Fig. 1). Such elemental behavior is identifiable after extensive diagenetic equilibration, as in the Silurian Read Bay Formation and the Mississippian Burlington Limestone (Brand and Veizer, 1980); but what role does early vadose diagenesis play in the establishment of these trends, and are they identifiable in younger subaerially-exposed carbonates?

Previous workers have collected samples from wide geographic areas and/or concentrated on examining specific allochems (Brand and Veizer, 1980; Pingitore, 1978). In doing so, 1. the stratigraphic variability in trace element composition has been neglected and remains unknown, and 2. the relative importance of the non-carbonate fractions in determining trace element compositions of a multicomponent carbonate system has been ignored or brushed aside. In addition, previous workers have utilized single-dissolution procedures for the extraction of trace elements from carbonate rocks and sediments (Veizer and Demovic, 1974; Pingitore, 1978; Brand and Veizer, 1980; and others). These treatments commonly consist of dilute HCl solutions aimed at dissolving the calcium carbonate component of limestones. However, experimental studies show that while treatment of rocks and sediments with dilute HCl is effective at dissolving calcium carbonate, trace elements can be leached from oxides, organic matter, and the lattices of some clay minerals (Heath and Dymond, 1977; Tessier et al., 1979; Pelle, 1989). As a result of these shortcomings, the modes of occurrence (associations), sources and stratigraphic variability of trace elements in subaerially-exposed carbonates, and the physical controls responsible for the redistribution of trace elements through a carbonate section are not well documented.

The objective of this study is to examine carbonate dune sands and the exposure surfaces/paleosols associated with them to determine the effects of early vadose diagenesis, paleosol formation, and mineral insoluble residue on the distribution of trace elements in eolian carbonate sands. In particular, the associations of trace elements with various fractions of multicomponent carbonate rocks, and the vertical and lateral stratigraphic variability of trace elements as related to diagenetic changes produced in the vadose environment are examined. This study attempts to develop a model for vadose diagenesis in the Pleistocene and Holocene vadose-altered carbonate sands of the Bahamas. It is hoped that the diagenetic trends and processes identified in this study can be useful to the interpretation of ancient carbonate sequences.

METHODOLOGY

Field Methods

Sites on San Salvador Island and Andros Island, Bahamas (Fig. 2) were chosen for this study because of the geographic isolation from most sources of non-carbonate minerals and

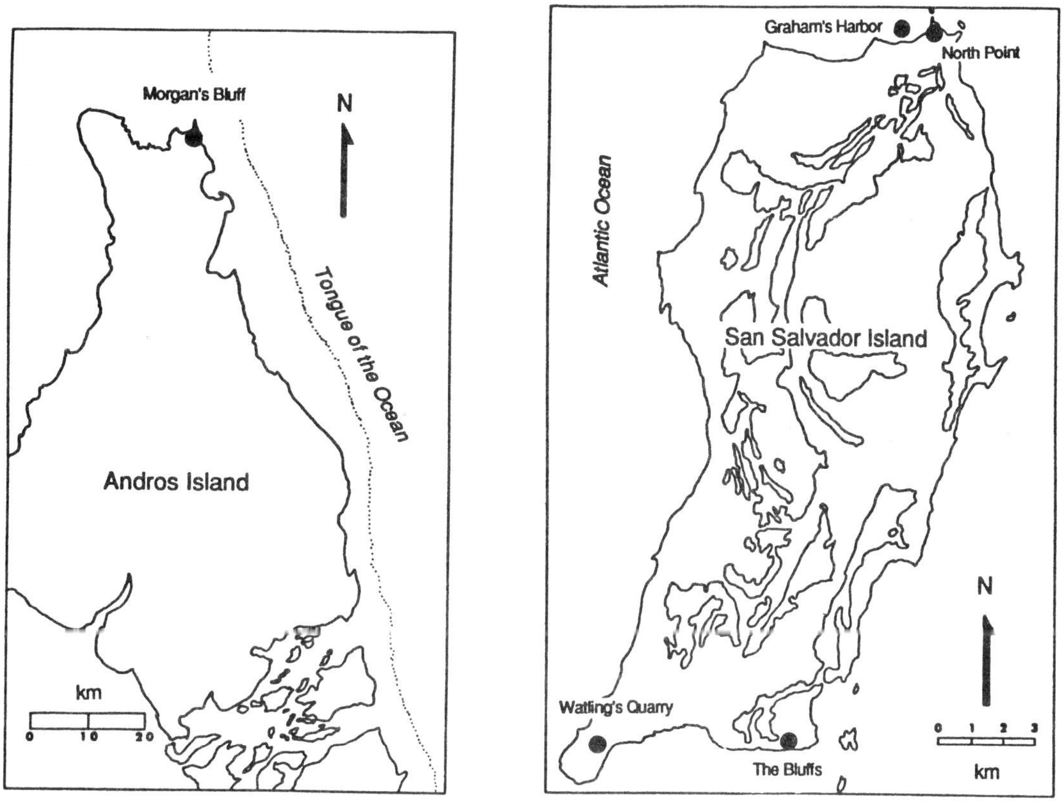

Fig. 2. Sample sites on Andros Island (A) and San Salvador Island (B), Bahamas.

because of the well-developed paleosols present on both islands. A total of 122 limestone and paleosol samples were collected from eolian dune localities on San Salvador Island and Andros Island. In addition, ten sediment samples were obtained from a core taken from Graham's Harbor, a lagoon north of San Salvador Island, to provide a baseline or starting point for measuring diagenetic alterations in subaerially-exposed, vadose-altered rocks.

Outcrops were sampled at approximately 50cm intervals along vertical transects to provide information on the vertical stratigraphic distribution of trace elements within a single outcrop. Parallel transects were sampled on two outcrops (Morgan's Bluff, Andros Island and Watling's Quarry, San Salvador) to examine lateral variability in trace element distributions on the outcrop scale.

Lab Methods

A selective sequential dissolution procedure, modeled after the method developed by Tessier and others (1979) was utilized in this study for the extraction of trace elements from rock and sediment samples. This procedure is capable of chemically separating trace elements into distinct fractions (Tessier et al., 1979), including: 1. exchangeable (adsorbed trace elements); 2. bound to carbonates; 3. bound to Fe-Mn oxides; and 4. bound to organic matter (see Pelle, 1989, for detailed methodology). A fifth fraction, composed of trace elements in lattices of stable minerals, is not considered to be diagenetically mobile (Tessier et al., 1979), and this "residual fraction" is considered to be beyond the scope of this study.

Calcium, magnesium, strontium, sodium, potassium, manganese, iron, aluminum, and silicon concentrations were determined on a Beckman SpectraSpan V argon plasma spectrometer (DCP). All discussion concerning elemental concentrations in the carbonate fraction is based on the total carbonate (insoluble residue-free) basis.

Percent insoluble residue was determine by dissolving the calcium carbonate from 1 g (dry weight) of each sample in 10% (v/v) HCl. The insoluble residue was collected on pre-weighed 0.45μm Millipore filter paper, dried, and reweighted.

Mineralogy was determined by x-ray diffraction on a Phillips XRG-3000 x-ray diffractometer using Cu-Kα a radiation. Percent aragonite was calculated using ratios of peak areas of aragonite and calcite and comparison to mixtures of biogenic carbonates (Neumann, 1965; Boardman, 1976).

Multivariate Statistics

Multivariate statistical analyses were performed to aid in the evaluation of stratigraphic trends and trace element associations. Two-way cluster analysis and polar ordination techniques (Sepkoski and Sharry, 1976a; 1976b) were applied to the mineralogical and geochemical data set to provide information on the geochemical similarities, relationships, and trends among samples within the study (see Pelle, 1989, for detailed statistical methodology).

RESULTS

Site Descriptions

Graham's Harbor. Graham's Harbor is a silled, high-energy lagoon located at the northern tip of San Salvador Island. The 3m sediment core extracted from Graham's Harbor is characterized by three distinct facies. The base of the core is a grapestone/bioclast "wacke-packstone" facies. This facies grades upward into an abraded grain "grapestone" and is capped by a *Halimeda*, foram--rich "packstone" facies (Colby and Boardman, 1988).

North Point. North Point is a Holocene eolian dune complex comprised of peloidal sands less than 6500 years old, based on C-14 dates of bulk sand (Boardman et al., 1987). This dune is characterized by fresh exposures adjacent to Graham's Harbor. North Point does not contain a paleosol, but is characterized by a poorly-developed soil and minor development of rhizocretions.

Watling's Quarry. Located in the southwest portion of San Salvador Island, Watling's Quarry is an 8.5m Pleistocene dune ridge approximately 120,000 years old (Carew and Mylroie, 1985). This outcrop is characterized by a lower paleosol that separates an underlying eolian skeletal grainstone from the overlying oolitic eolianite. The outcrop is capped by a thin laminar calcrete that extends down the flank of the dune. This calcrete has numerous micritic stringers (Braithwaite, 1983) associated with it that extend well below the exposure surface into the main body of the rock along fractures and more permeable laminae. Two sampling transects were taken at this location: one through the middle of the exposed dune, and a second along a cut near the dune flank.

The Gulf. The Gulf was formed from the coastal erosion of a Pleistocene dune somewhat younger than Watling's Quarry (Carew and Mylroie, 1985). This section is characterized by a well-developed paleosol profile similar to the generalized calcrete profile proposed by Read (1974). The section is composed of peloidal grainstone host rock that shows signs of pedogenesis upsection. Overlying the host rock is a mottled horizon which grades upward into a thoroughly bioturbated, massive calcrete with extensive, complex, well-preserved rhizocretions. The upper portion of the outcrop is a brecciated zone capped by a well-developed laminar calcrete.

Morgan's Bluff. This well-cemented Pleistocene dune deposit, located on the northern end of Andros Island, is approximately 250,000 years old and contains two paleosol units, separated by a thin eolian section located at and near the top of the dune. The lower paleosol is preserved across the length of the exposure, but the upper paleosol has been eroded away from the northern end of the section. The two paleosols merge in a karst collapse structure near the southern end of the outcrop. Two sample transects were taken at this location, one near the middle of the exposure and another to the south, through a zone where the two paleosols mix.

Geochemical Results

In agreement with Walls and others (1977) and Brand and Veizer (1980), analyses of element concentrations in this study indeed show that carbonate rocks and sediments are a "multicomponent" geochemical system. Each component or fraction within the system contains diagenetically-mobile elements that contribute to the geochemical signature of the whole rock. Although the carbonate fraction dominates, other fractions (exchangeable and oxides) are important in understanding the diagenetic processes operating within the system as a whole.

Exchangeable Fraction. Elements associated with the exchangeable fraction include calcium, sodium, potassium, and strontium. High concentrations of calcium and strontium in marine sediments from Graham's Harbor indicate that treatment of samples with 1M $MgCl_2$ to strip away the adsorbed cations results in the dissolution of a small percentage of the total calcium carbonate available. Calculations and experimental results indicate the dissolution of no more than 0.6% of the total calcium carbonate available (Pelle, 1989). While this analytic problem increases the apparent concentrations of exchangeable cations, the over-

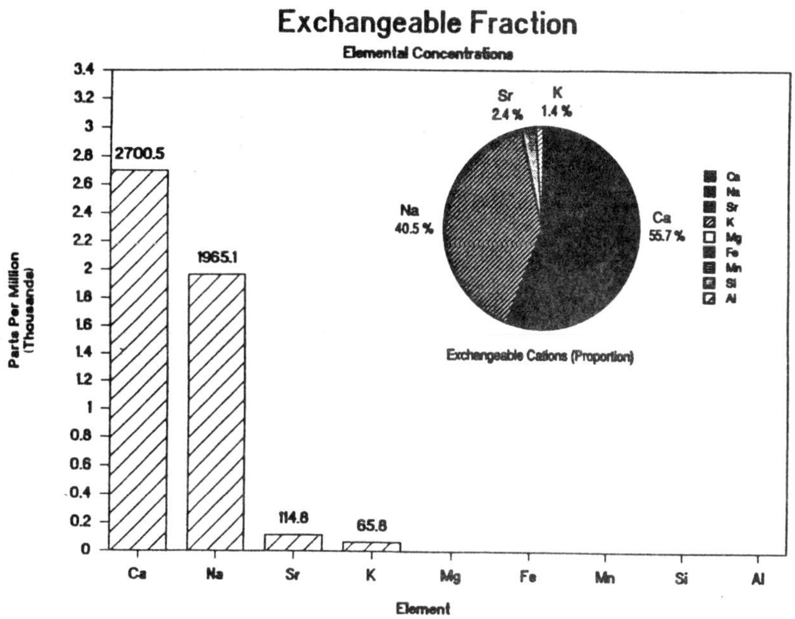

Fig. 3A.

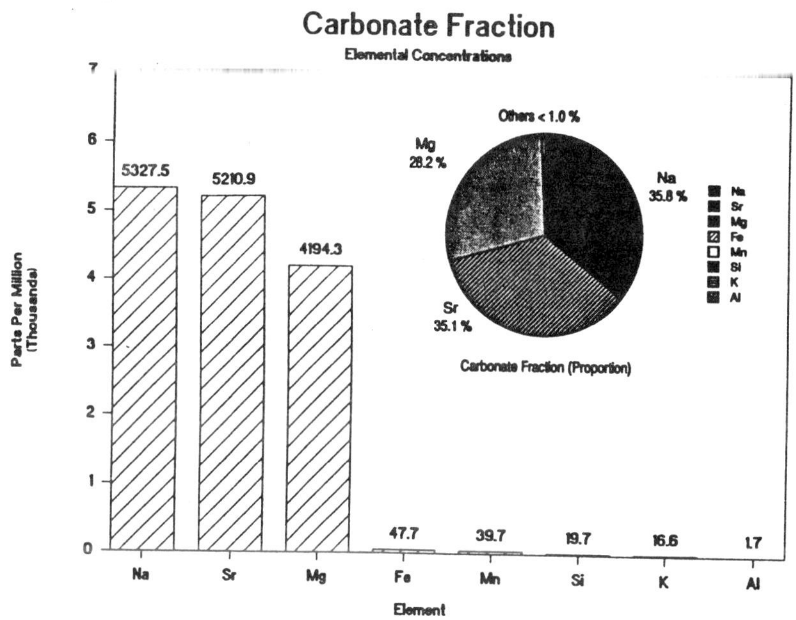

Fig. 3B.

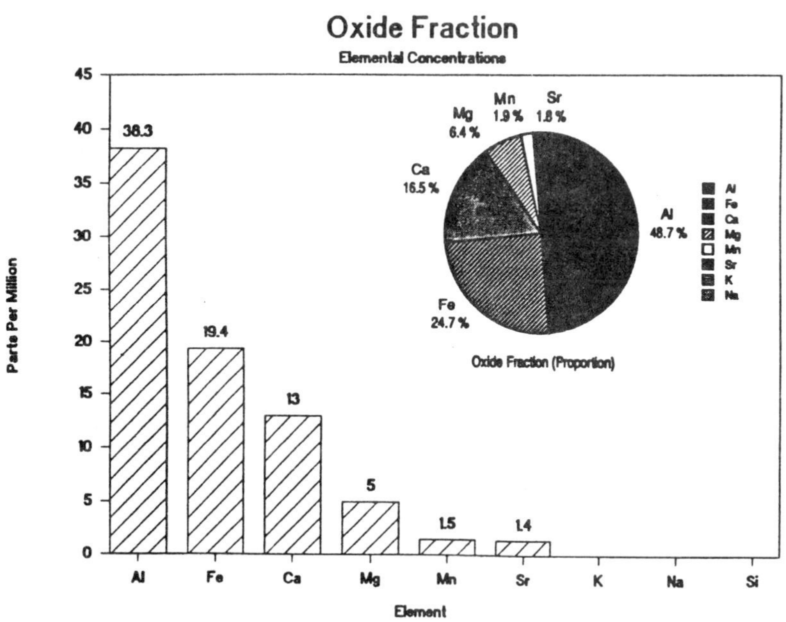

Fig. 3C.

Fig. 3. Average element concentrations associated with the (A) exchangeable fraction, (B) carbonate fraction, and (C) oxide fraction of samples studied. Pie charts illustrate relative proportions of cations present in each fraction.

all elemental trends and relationships are little affected. Fig. 3-A illustrates uncorrected element concentrations and proportions in the exchangeable fraction.

Carbonate Fraction. Elements associated with the carbonate fraction in this study include calcium, magnesium, strontium, sodium, potassium, iron, manganese, aluminum, and silicon. Elemental concentrations observed are in general agreement with previous studies (Kinsman, 1969; Pingitore, 1978; Dulin, 1984; Span and Gaillard, 1986).

The presence of aluminum and silicon associated with the carbonate fraction may seem surprising at first glance. However, Brand and Veizer (1980) attribute aluminum concentrations in the Read Bay Formation (Arctic Canada) and the Burlington Limestone (Iowa and Missouri) to acid leaching of insoluble residues in their samples. Walls and others (1975) discuss the presence and mobility of silica in paleosols of the Newman Limestone (Slade Formation) of northeastern Kentucky. In light of these studies, the presence of aluminum and silicon in some Bahamian rocks, particularly paleosols, should not be unexpected. The significance of these elements will be discussed in a later section (on paleosols). Fig. 3-B illustrates element concentrations and proportions associated with the carbonate fraction (excluding Ca).

Oxide Fraction. The oxide fraction is dominated by aluminum, iron, calcium, magnesium, manganese, and strontium. Sodium is present in high concentrations as well, but this is most likely due to insufficient rinsing of the sodium acetate-acetic acid leach used in the dissolution of the carbonate fraction (Pelle, 1989). Fig. 3-C illustrates elemental concentrations and proportions associated with the oxide fraction. High concentrations of aluminum, iron, and manganese in the oxide fraction are associated with samples having a high insoluble residue content. The linear correlation between iron and insoluble residue (Fig. 4) indicates that the distribution of trace metals in the oxide fraction is strongly influenced by the presence of insoluble residue, an observation also noted by Dulin (1984).

Oxide coatings on carbonate grains protect them from dissolution during the leaching procedure for the carbonate fraction. Calcium, magnesium, and strontium concentrations in the oxide fraction may result from the dissolution of carbonate grains after their protective oxide coatings are dissolved.

Organic Matter Fraction. Elemental analysis reveals no significant concentrations of trace elements associated with residual organic matter. Although organic matter is thought to be an excellent scavenger for trace elements, particularly in soil horizons, the well-drained, sandy soils in this semi-arid environment apparently permit the rapid oxidation of organic matter, leaving behind a paraffin-like material and resistant structural (non-humified) organic matter residues (Tessier et al., 1979). Oxidation effectively releases any scavenged trace elements into the system.

Geochemical/Mineralogical Profiles. Geochemical analyses performed on samples collected along vertical transects at each of the outcrops reveal distinct vertical trends in the geochemistry of the vadose-altered dune rocks. Figs. 5

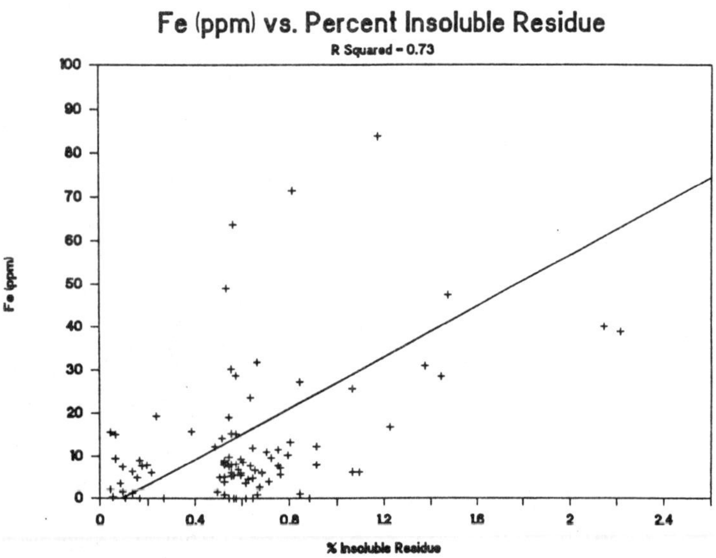

Fig. 4. Plot of Fe (ppm) versus percent insoluble residue for the oxide fraction of vadose- altered sands. The linear correlation (r = 0.85) explains 73% (r^2) of the variance in iron concentration as a result of the presence of insoluble residue. This suggests a strong relationship between trace metals in the oxide fraction and percent insoluble residue. Higher concentrations of iron and % insoluble residue are accounted for in the correlation coefficient and linear regression pictured; however, the scales were adjusted down to better illustrate the majority of the data points.

Fig. 5A.

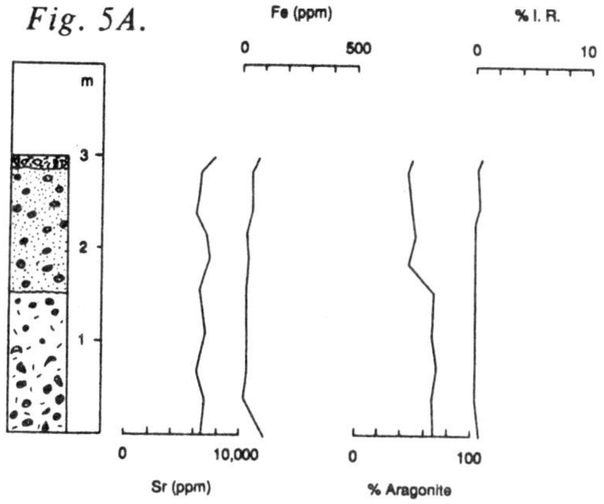

Fig. 5B.

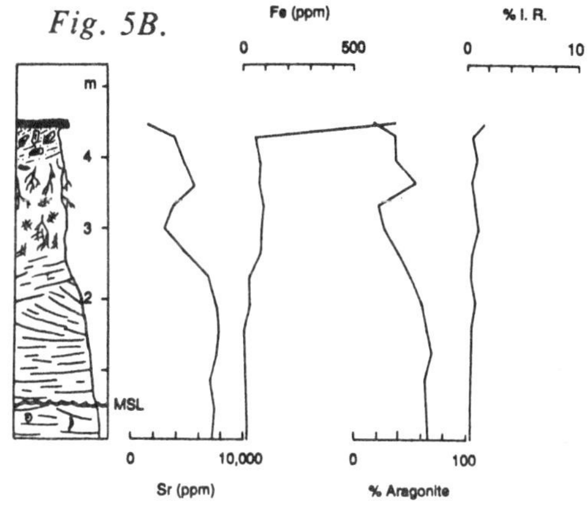

Fig. 5C.

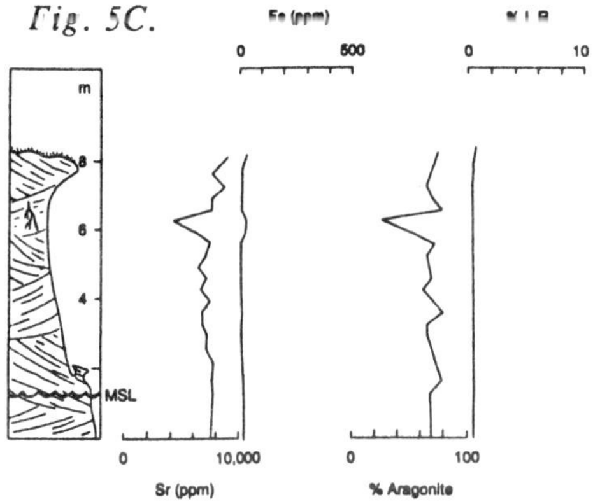

Fig. 5. Geochemical/mineralogical profiles for the carbonate fractions of (A) Graham's Harbor, (B) the Gulf, and (C) North Point.

and 6 illustrate geochemical/mineralogical profiles for the carbonate fraction along each of the outcrops examined in this study. The carbonate fraction was chosen to illustrate the results of these analyses because it contains the highest concentrations of diagenetically mobile elements. Strontium and iron concentrations, percent insoluble residue, and percent aragonite are illustrated in the profiles because they are useful indicators of diagenetic trends that will be discussed in later sections.

Cluster Analysis. Two-way cluster analysis was used to compare 132 samples, each with 11 variables. Results of the two-way cluster analysis (Fig. 7) performed on geochemical and mineralogical data indicates the presence of five geochemically-distinct rock types within Bahamian dune units. A sixth cluster represents samples from the sediment core extracted from Graham's Harbor. The diagenetic significance of these clusters and names assigned to them will be discussed in a later section.

DISCUSSION

The stratigraphic variability of trace elements in vadose-altered carbonates can be discussed as vertical or lateral stratigraphic variability within a single outcrop (intra-outcrop) and lateral variability between outcrops (inter-outcrop). Diagenetic trends and processes responsible for variability in the stratigraphic distribution of trace elements are identified and discussed.

Intra-Outcrop Variability

Vertical Trace Element Distribution. Three geochemically-distinct zones are recognized within a single vadose-altered outcrop: 1. paleosols; 2. zone of influence; and 3. diagenetically-altered host rock. Development of the three zones results from the progressive meteoric diagenesis of the exposed multicomponent carbonate sands and the geochemical interaction between the exchangeable,

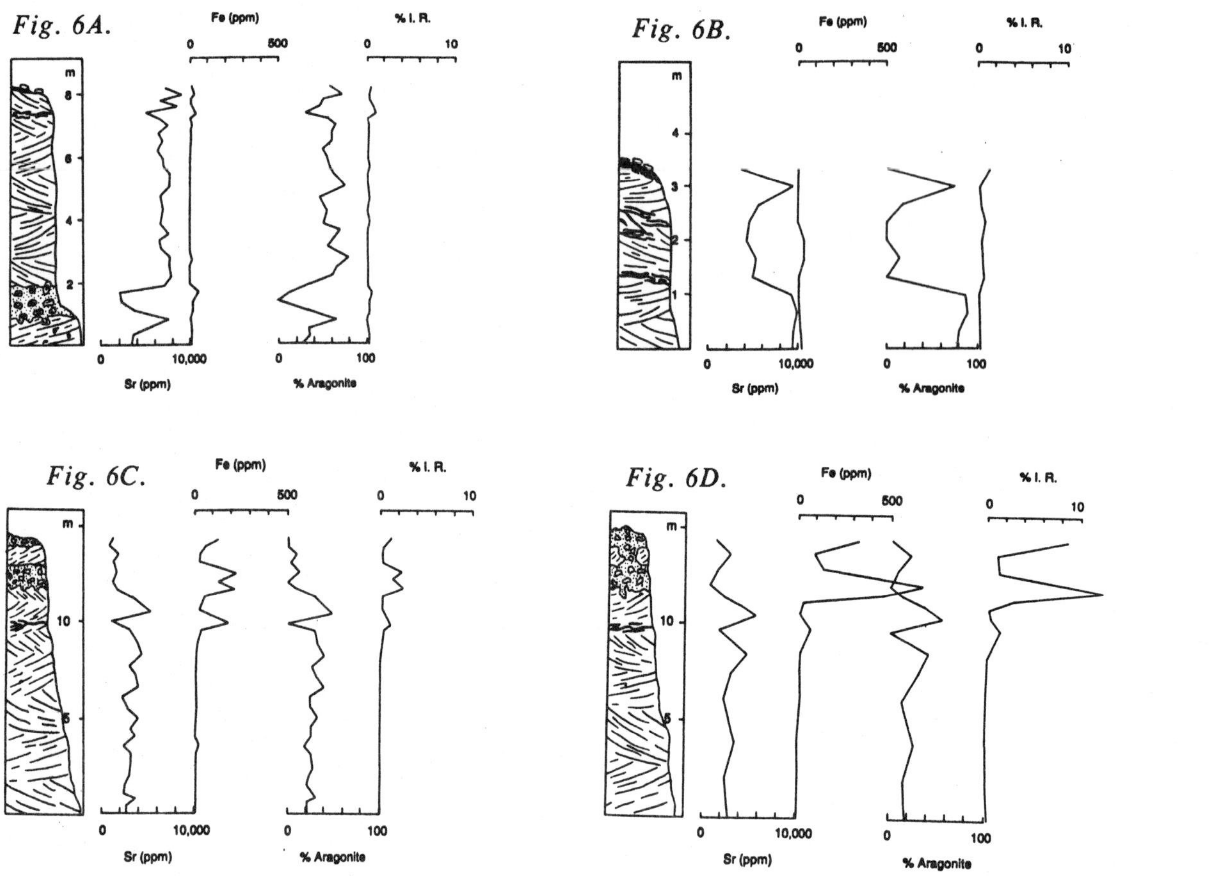

Fig. 6. Geochemical/mineralogical profiles for the carbonate fractions of Watling's Quarry transects 1 and 2 (A and B, respectively) and Morgan's Bluff transects 1 and 2 (C and D, respectively).

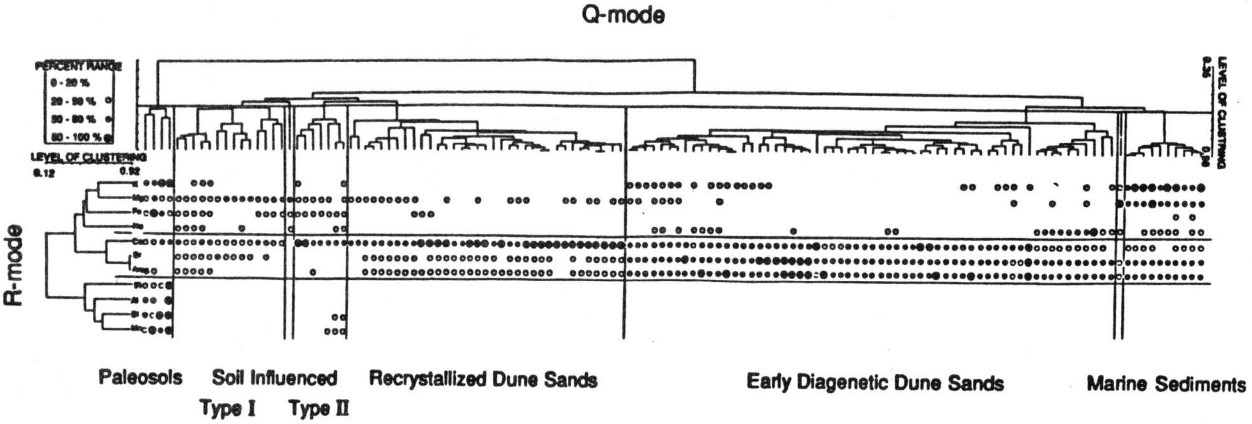

TWO-WAY CLUSTER ANALYSIS: VADOSE-ALTERED EOLIAN SANDS

Fig. 7. Symbols shown in the two-way cluster diagram refer to percent-range transformed data. That is, the sample with the highest value for a given variable is assigned a value of 100%. The sample with the lowest value for the same variable is assigned 0%. All remaining samples are expressed as a percentage of the maximum value. This data transformation is repeated for each variable, thus eliminating problems in comparing samples with variables having different orders of magnitudes (e.g., percent insoluble residue and ppm Sr).

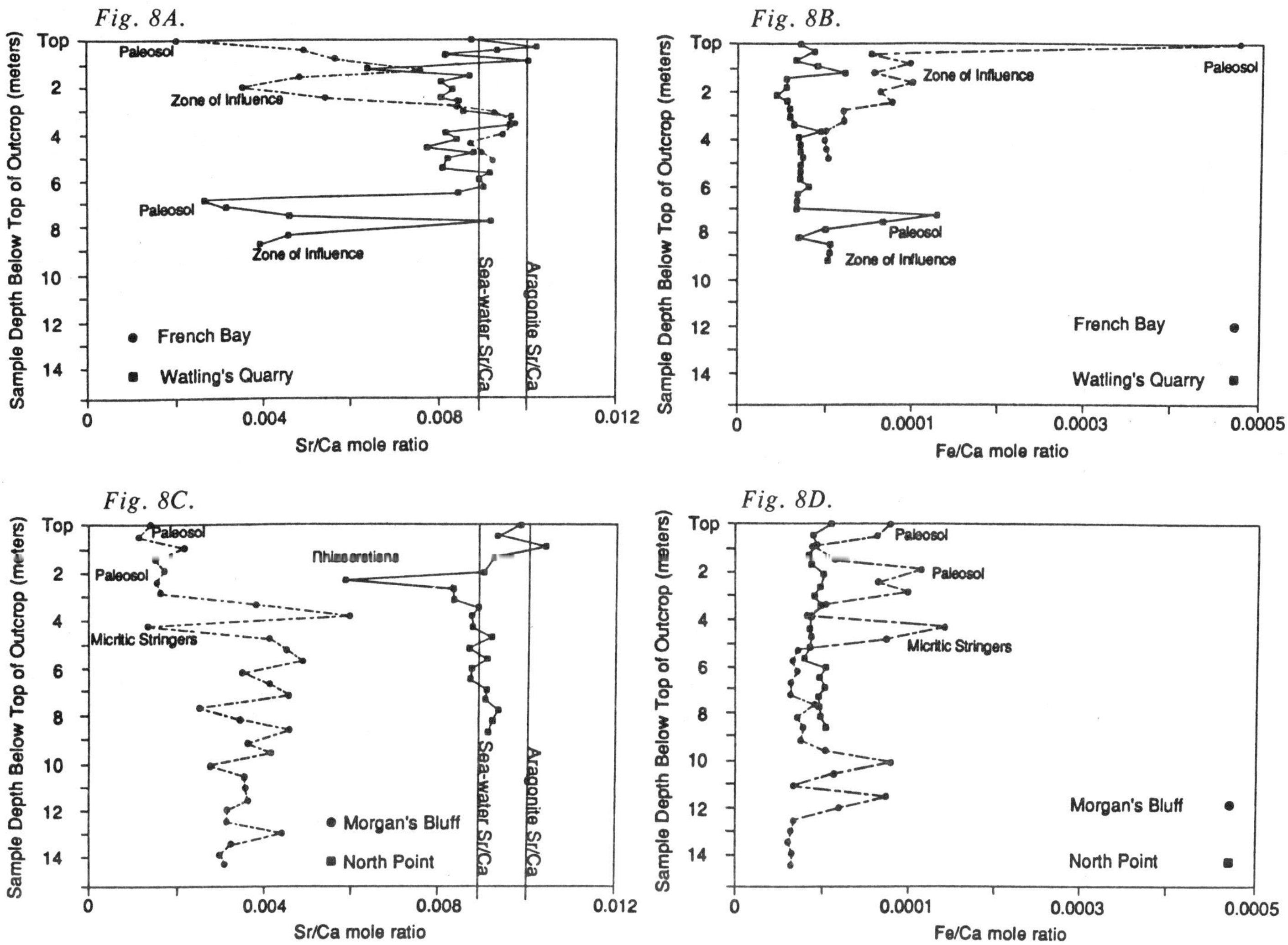

Fig. 8. $^{Sr}/_{Ca}$ ratios for Watling's Quarry and French Bay (A), $^{Fe}/_{Ca}$ ratios for Watling's Quarry and French Bay (B); $^{Sr}/_{Ca}$ ratios for North Point and Morgan's Bluff (C); $^{Fe}/_{Ca}$ ratios for North Point and Morgan's Bluff (D).

carbonate, and oxide fractions. The three diagenetic zones are recognizable from the trace element geochemistry associated with each fraction of the multicomponent system, but the carbonate fraction has the highest trace element concentrations and provides the most information about the diagenetic alterations occurring within the carbonate sands. Fig. 8 illustrates the vertical stratigraphic variability of strontium and iron in terms of $^{Me}/_{Ca}$ ratios in the carbonate fraction of outcrops examined in this study.

Paleosols: Paleosols (caliche, calcrete, etc.) are deposition and diagenetic products of soil-forming processes acting on limestones exposed to meteoric waters (Braithwaite, 1983). In terms of trace element composition, the most important processes operating in the development of paleosols on exposed carbonate sands are the accumulation of insoluble residues and the dissolution of metastable carbonate minerals and reprecipitation of diagenetic low-Mg calcite.

The major source for mineral insoluble residues in the Bahamas is aerosol dust carried from the Sahara Desert on trade winds (Dulin, 1984; Eaton, 1986; McCartney, 1987). Saharan dust has also been shown to be a major contributor of insoluble residues in Barbados, West Indies, and Miami, Florida (Savoie and Prospero, 1977). This dust can be identified in deep-sea sediment cores (Eaton, 1986) and in paleosols of the Bahamas (McCartney, 1987). Dulin (1984) identified oxide coatings on Saharan-derived insoluble residues as a major source for the trace elements iron and manganese in Bahamian sediments. Iron and manganese, along with aluminum and silicon, have been shown to be closely associated with insoluble residues in this study, as well (Fig. 4).

Significant concentrations of aluminum and silicon in vadose-altered carbonate sands are restricted primarily to paleosols. Aluminum dominates the oxide fraction of the multicomponent system, and is present in high concentrations in paleosols. In subaerially-exposed carbonate sands, silicon is found in measurable concentrations exclusively in paleosol profiles. These findings agree with the results of Brand and Veizer (1980) and Walls and others (1975), in that both elements are associated with insoluble residues and/or paleosols. The surprising aspect of these two elements is their association with the carbonate fraction in any appreciable concentration.

Aluminum may occupy crystal sites similar to those occupied by magnesium and iron (Hem, 1978), and the aluminum ion is small enough (0.45Å) to fit in interstices in the calcite lattice. However, aluminum rarely occurs in natural water in concentrations greater than a few tenths of a ppm, and the cation Al^{3+} dominates only in solutions with a pH of less than 4.0. At pH between 4.5 and 6.5, polymerization occurs which results in units of various sizes (some small enough to qualify as ions) with structural patterns of gibbsite (Hem, 1978). It is likely, therefore, that the concentrations of aluminum associated with the carbonate fraction are present due to this polymerization process or leaching of insoluble residues, in agreement with Brand and Veizer (1980). Although aluminum is probably not directly associated with the carbonate fraction, its presence is significant because of its relationship to insoluble residues and, in the case of Bahamian carbonate sands, paleosols. A pH below 6.5 is highly unlikely in carbonate rocks. A thick soil with elevated partial pressure of CO_2 could locally create low pH values.

Silica has a low solubility at pH values below 9.0. Localized solutions with a pH above 9 could be attained in soil profiles by the combined effect of calcite precipitation and cation exchange and, under these conditions, silica could be released from clays. Once in solution, silica is present as silicic acid (H_4SiO_4) or as a colloid (Walls et al., 1975). Under some conditions, it is possible for minor concentrations of silica to be mobilized and included in the precipitating carbonate phase, possibly as fluid inclusions, in Bahamian paleosols. This process is believed to be responsible for the formation of chert nodules in paleosols of the Slade Formation (Walls et al., 1975), and the presence of silica in Bahamian paleosols may be the first stages of chert formation.

Strontium and sodium have low concentrations in paleosols because of the A to dLMC transformation. Under the influence of meteoric waters, metastable carbonate grains (A) dissolve and release strontium, sodium, and calcium into the system. As dLMC precipitates, strontium and sodium are preferentially excluded from the calcite lattice, while iron and manganese are preferentially included (Brand and Veizer, 1980; Pingitore, 1982). Oxide coatings on insoluble residues, concentrated in paleosols, provide the source for iron and manganese, which can substitute for Ca^{++} in the precipitating calcite phase, thereby increasing the concentrations of these two metals. This dissolution and reprecipitation process leads to the progressive depletion of strontium and sodium and an increase in iron and manganese in the carbonate fraction of paleosols, as predicted by Brand and Veizer (1980). Fig. 9 shows the abrupt increase in iron concentrations associated with the carbonate fraction of the paleosol at the Gulf, San Salvador Island.

Zone of Influence: The zone of influence can be thought of as a geochemical gradation between the overlying paleosol and the underlying host rock. The main processes affecting the geochem-

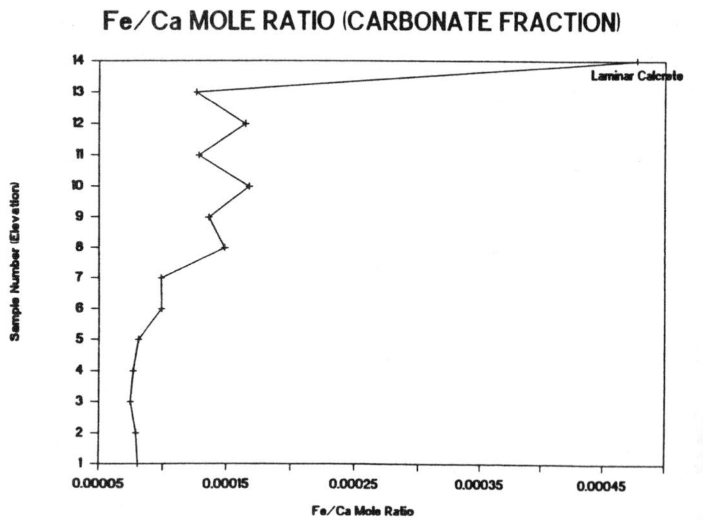

Fig. 9. Iron to calcium mole ratios in the carbonate fraction at the Gulf, San Salvador Island. Note the dramatic enrichment in iron at the laminar calcrete (paleosol). The gradual rise through the middle of the section is attributed to the zone of influence.

ical signature of the zone of influence are the A to dLMC transformation and physical alterations to the host rock associated with pedogenesis. The dissolution/reprecipitation of $CaCO_3$ in the zone of influence operates in much the same fashion as described in the paleosols. However, the geochemical signature within the zone of influence is intermediate between paleosols and the remainder of the host rock.

Important indicators of ongoing diagenesis resulting from the A to dLMC transformation is apparent in the increased $^{Sr}/_{Ca}$ ratios in the exchangeable fraction, high iron and manganese concentrations in the carbonate fraction, and low strontium concentrations in the carbonate fraction. $^{Sr}/_{Ca}$ ratios associated with the exchangeable fraction in the zone of influence are elevated well above those expected for either seawater or aragonite (Fig. 10). Such high ratios can be produced only by the dissolution of A and subsequent reprecipitation of dLMC. As calcium is used in the precipitation of calcite, and strontium is preferentially excluded from the calcite lattice, the ratio of $^{Sr}/_{Ca}$ in the diagenetic solution must increase unless the strontium is removed with the bulk aquifer solution. In Bahamian vadose-altered carbonates, the strontium is retained as exchangeable cations perhaps due to the evaporation of pore waters, thereby increasing the $^{Sr}/_{Ca}$ ratio.

Diagenetic alterations in the zone of influence are enhanced by pedogenesis. Brecciation, the development of micritic stringers that extend downward along fractures and permeable laminae, bioturbation by land plants aid in the downward movement of meteoric waters, insoluble residues, and trace elements and insoluble residue into the host rock. As such, these features are important physical mechanisms for the redistribution of trace elements throughout a vadose-altered carbonate unit. Roots, preserved in carbonate sands as micritized rhizocretions, enhance the movement of water through the body of rock. As such, roots play a major role in the alteration of aragonite to calcite and the transfer of iron and manganese from the paleosol profile into the underlying zone of influence. While not reaching the elevated values observed in paleosols themselves, micritic calcite within the rhizocretions and in micritic stringers can be much higher in iron, manganese, and insoluble residue than the surrounding host rock (Fig. 8), attesting to their direct connection within the paleosol. In addition, grains in and near these features show varying degrees of alteration, ranging from partial to complete micritization.

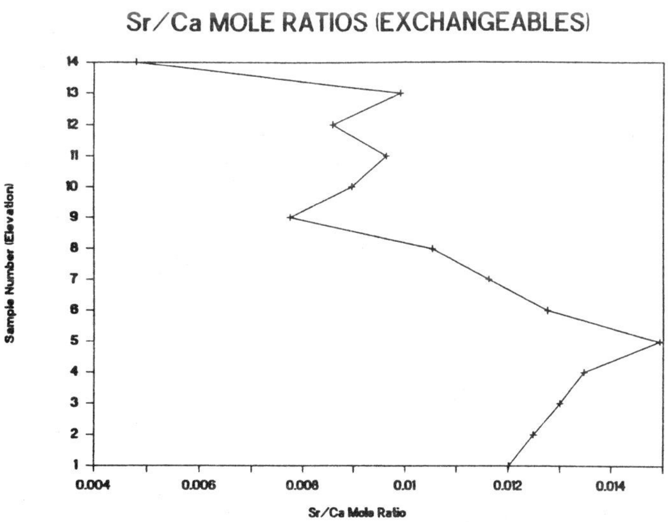

Fig. 10. Strontium to calcium ratios in the exchangeable fraction at the Gulf. Increase in the ratio above that expected for aragonite can only result from the dissolution of aragonite and reprecipitation of diagenetic low-Mg calcite, thereby increasing the $^{Sr}/_{Ca}$ ratio.

Host Rock: The third geochemical zone observed in vadose-altered carbonate sands is characterized by the least diagenetic alteration of the three zones. Much of the original depositional trace element signature is preserved in younger dunes (North Point). Being removed from the effects of paleosol formation, diagenetically-altered host rocks are characterized by low concentrations of iron and manganese and low percentages of insoluble residue, while aragonite and higher strontium concentrations are typically preserved, as illustrated by two-way cluster analysis (Fig. 7).

Known to be the most soluble of the common carbonate minerals (Plummer and McKenzie, 1974; James and Choquette, 1984), high-Mg calcite is virtually non-existent in the Bahamian sand dunes of this study. X-ray diffraction of all samples reveals the presence of HMC in only two dune samples -- both taken from the subtidal portion of outcrops (one from North Point, the other from the Gulf). North Point sands have been dated (C-14) at between 6190 (near the base of the exposure) and 5670 years B.P. (near the top). The presence of HMC in the two subtidal samples and in Graham's Harbor sediments, coupled with the lack of HMC in all North Point sands strongly suggests that the dissolution of HMC is a very rapid process in the vadose en-

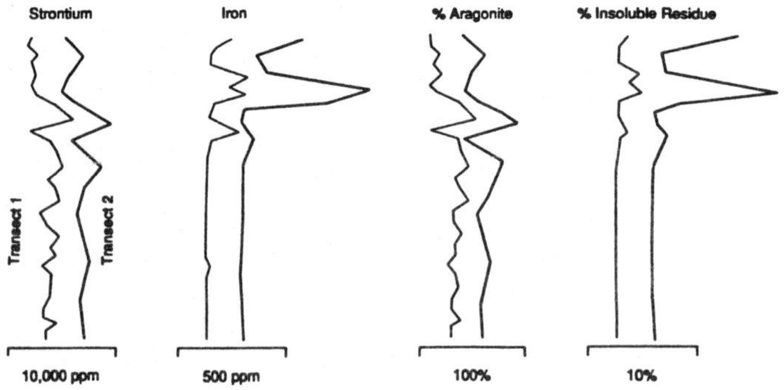

Fig. 11. Geochemical/mineralogical profiles of parallel transects sampled at Morgan's Bluff, Andros Island. Despite the erosion of the upper paleosol in places, the two profiles are remarkably similar.

vironment (near complete dissolution in less than 6000 years).

The host rocks can be altered by more than the loss of HMC. Evidence pointing to the A to dLMC transformation is not recognized in younger rocks. Minor cements in most host rocks may originate from dissolution of HMC or from an external source. Even the oldest, most altered host rocks observed in this study have considerable aragonite preserved. Host rocks at Morgan's Bluff, Andros Island underlay two paleosols and are thus believed to be about 250,000 years old, are very well cemented, and some grains show evidence of micritization and/or dissolution. However, x-ray diffraction of these samples indicates the presence of about 25% aragonite -- significantly more than is found in paleosols or zones of influence in younger rocks on San Salvador. Temporal considerations will be further discussed in a later section.

Lateral Trace Element Distribution: The presence or absence of features associated with paleosols may impart lateral trace element variability with a single outcrop. Parallel transects were sampled on two outcrops (Watling's Quarry and Morgan's Bluff) to identify the lateral variability in trace element distribution. Lateral variability is due primarily to the discontinuous nature of paleosols. Paleosols are typically not laterally continuous over a large area to begin with and, in addition, may be eroded away.

Morgan's Bluff does not display a great amount of lateral variability over the length of the outcrop. The outcrop is characterized by two paleosols across much of the outcrop, but the upper paleosol has been eroded away at some locations. In other locations, the two paleosols merge due to the development of karst collapse structures. This may allow for minor geochemical variability within the outcrop; however, geochemical profiles of the two transects are remarkably similar (Fig. 11). The similarity of the two profiles indicates that the eroded section of paleosol played a role in the establishment of the geochemical signature, and that this signature has been preserved.

On the other hand, Watling's Quarry shows a great deal of lateral variability over a very short distance. The upper paleosol at Watling's Quarry is a laminated calcrete that extends down the flank of the dune. This calcrete is characterized by numerous micritic stringers that extend well below the surface along more permeable laminae and fractures. Samples taken near the flank of the dune are considerably different from those collected through the exposed center of the dune (Fig. 6-C,D Results), where micritic stringers are less abundant. The presence of the micritic stringers imparts a much more pronounced zone of influence along the dune flank, with higher iron, manganese, and aluminum concentrations and lower strontium and sodium concentrations than samples collected from the middle of the dune.

In short, the major controlling factor for lateral variability within a single outcrop is the presence or absence of features associated with paleosols. However, a section of paleosol missing due to erosion may not cause an increase in variability, as seen at Morgan's Bluff.

Inter-Outcrop Variability

Geographic Variability. The only variability in trace element concentrations of Bahamian carbonate sands solely attributable to geographic location occurs in the exchangeable fraction. Concentrations of exchangeable sodium and potassium are controlled by the proximity of the outcrop to salt water spray. Table 1 compares the average exchangeable concentrations from the four outcrops sampled in this study along with Graham's Harbor marine sediments.

The Gulf and North Point, San Salvador Island are both subjected to almost continual contact with salt water spray. These outcrops are located along the coast and rise directly

		Morgan's Bluff	Watling's Quarry	The Bluffs	North Point	Graham's Harbor
Na	x	546.41	1526.12	3344.95	4093.15	2550.54
	σ	1127.92	1413.83	1432.95	2220.47	996.46
K	x	14.57	13.07	156.95	207.79	106.55
	σ	29.49	18.99	61.68	115.23	44.43
Sr	x	48.35	221.87	68.98	94.24	72.19
	σ	24.99	42.90	16.68	22.56	10.43
Ca	x	2579.00	2635.65	3064.73	2739.32	2928.87
	σ	557.22	687.74	1057.72	362.61	436.89

Table 1. Mean and standard deviation of exchangeable elements in outcrops and sediment samples. North Point and the Gulf have significantly higher concentrations of sodium and potassium than Morgan's Bluff and Watling's Quarry due to their association with a salt water spray source for the cations.

from sea-level. They also lack vegetation that might baffle the influx of aerosol salts derived from the sea spray. On the other hand, Watling's Quarry and Morgan's Bluff outcrop at higher elevations and are surrounded by vegetation. As such, these outcrops are much less subject to direct contact with salt water.

Samples collected from North Point and the Gulf contain significantly higher concentrations of sodium and potassium than samples from Watling's Quarry and Morgan's Bluff (based on student T-tests), where salt water spray is less of a factor. Fig. 12 illustrates the $^{K}/_{Na}$ ratios of the four outcrops and the salt water dilution line. North Point and the Gulf contain significantly higher concentrations of these two elements and

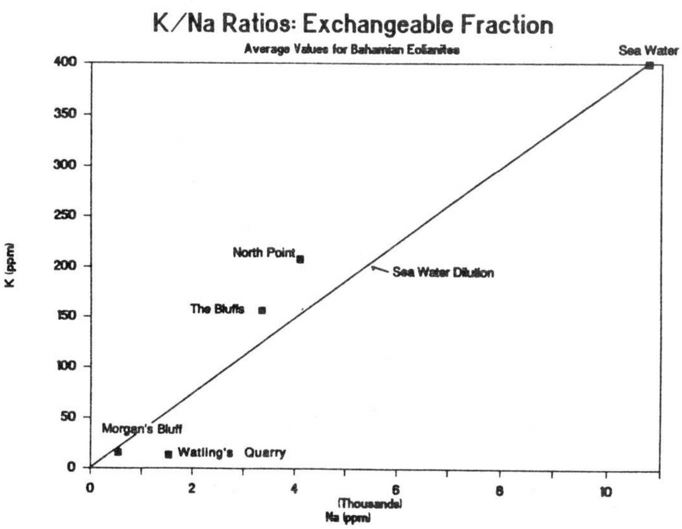

Fig. 12. Potassium to sodium ratios in the exchangeable fraction of Bahamian carbonate dunes.

are slightly enriched in potassium relative to the ratio expected for salt water. This potassium enrichment (sodium depletion) can be accounted for by the relative mobilities of the two cations. Sodium is more mobile than potassium in the exchangeable fraction (Fetter, 1984) and the selective removal of sodium causes enrichment of potassium in the exchangeable.

Watling's Quarry and Morgan's Bluff, on the other hand, are depleted in potassium relative to sodium (Fig. 12). This indicates that the source for these elements in the exchangeable fraction may be something other than sea water. Naturally-occurring carbonates have considerably higher concentrations of sodium than potassium (Fig. 3-B). This leads to the interpretation that the source for sodium and potassium at outcrops not influenced by direct contact with salt water is most likely the dissolution of calcium carbonate.

Statistical Variability. Multivariate statistical analyses, performed on the carbonate fraction of all samples in this study, were used to describe the variability in geochemistry between outcrops of different ages and geographic locations. Cluster analysis identifies five geochemically-distinct clusters, or groups, of diagenetically-altered rocks in the Bahamian dunes studied. A sixth cluster, unaltered marine sediments from Graham's Harbor, was shown to be statistically different from any of the vadose-altered rocks. The two-way cluster diagram illustrated in Fig. 7 depicts the variables (R-mode) that are responsible for clusters of geochemically-different samples (Q-mode; see Results). Table 2 presents averages and standard deviations of values for each variable in the groups identified by cluster analysis. Student T-tests performed on this data indicate that the clusters are indeed statistically significantly different from one another. Table 3 is a dissimilarity matrix showing the percentage of variables that are significantly different between two clusters.

Examination of the rock types present in each of the clusters and the variables responsible for the clusters leads to the interpretation that the six clusters represent varying degrees of progressive meteoric diagenesis related to paleosol formation. Clusters are defined in terms of increasing diagenetic alteration (towards equilibrium with meteoric waters) as follows:

Marine Sediments: All samples in this cluster were taken from the sediment core in Graham's Harbor. These samples have never been subjected

	% IR	% Aragonite	[Mg]	[Ca]	[Sr]	[Si]	[Fe]	[Mn]	[Al]	[K]	[Na]
Paleosols	4.4	10.3	5737.6	376035.4	1225.1	392.0	530.2	148.0	45.6	59.8	2766.9
	5.2	8.2	2305.9	12213.7	826.6	131.2	172.4	68.0	29.7	19.2	2306.6
Soil Influenced Type II	1.6	8.4	4093.3	383345.0	1499.0	59.9	151.5	83.0	2.6	14.0	4485.9
	0.4	8.6	717.0	7460.1	406.5	50.3	58.2	13.8	3.6	4.1	3321.6
Soil Influenced Type I	0.8	14.6	6806.2	363550.0	3825.8	3.0	39.4	65.9	0.2	8.3	5323.2
	0.1	14.8	1631.3	9191.3	1311.3	7.6	34.6	19.0	0.3	10.9	2530.9
Recrystallized Dune Sands	0.6	27.1	3828.3	368709.9	3140.4	0.0	13.4	40.6	0.2	4.4	1488.9
	0.1	7.8	986.0	7508.3	818.8	0.1	13.3	15.1	0.9	5.6	975.9
Early Diagenetic Dune Sands	0.6	62.6	2554.7	373117.2	7326.3	2.5	5.5	35.3	0.0	14.6	7571.2
	0.2	11.0	1083.1	13243.2	976.1	4.9	7.9	6.6	0.1	11.9	6087.0
Marine Sediments	0.9	57.4	9599.6	354059.2	6492.3	42.8	52.4	49.3	2.1	69.5	7289.2
	0.1	11.3	917.6	9838.4	256.7	5.7	24.3	18.2	1.3	8.8	786.4

Table 2. Mean and standard deviation of geochemical/mineralogical variables for clusters delineated by cluster analysis.

to meteoric waters and represent a baseline or point of reference against which diagenetic alterations in the subaerially-exposed carbonates can be measured.

Early Diagenetic Dune Sand: These samples are located within the "diagenetically-altered host rock zone" within the outcrops sampled. Samples belonging to this cluster are characterized by lower magnesium and potassium concentrations than are found in marine sediments. These rocks have been subjected to relatively little diagenesis as shown by the presence of high strontium concentrations and percent aragonite.

	Paleosols	Soil Influenced Type I	Soil Influenced Type II	Recrystallized Dune Sands	Early Diagenetic Dune Sands	Marine Sediments
Paleosols						
Soil Influenced Type I	73%					
Soil Influenced Type II	54%	73%				
Recrystallized Dune Sands	100%	82%	91%			
Early Diagenetic Dune Sands	91%	73%	91%	91%		
Marine Sediments	91%	82%	82%	91%	82%	

Table 3. Dissimilarity matrix calculated from the results of Student T-tests performed on variables between clusters. Values in matrix are the percentage of variables that are significantly different (at 95% confidence interval) between two groups (i.e., 91% of the variables are significantly different between palesols and marine sediments).

Recrystallized Dune Sands: Recrystallized dune sands are characterized by the continued depletion of strontium and the decrease of aragonite preserved in the rock. Sodium is also present in very low concentrations. As such, this cluster represents the next step in the progressive diagenesis of the vadose-altered carbonate sands.

Soil Influenced Type I: This cluster consists of rocks that have been subjected to some degree of pedogenesis. They are altered from the early diagenetic dune sands by soil-forming processes including brecciation and intense bioturbation by land plants. Rocks belonging to this cluster display a further decrease in strontium and aragonite and the incorporation of iron in most samples.

Soil Influenced Type II: Rocks altered by soil-forming processes acting on recrystallized dune sands belong to this cluster. These rocks are characterized by the near total depletion of strontium and aragonite and the incorporation of significant amounts of iron in all samples. Silicon and manganese may also be present in significantly higher concentrations.

Paleosols: This cluster consists of rocks owing their origin to soil-forming processes and meteoric waters. Paleosols are characterized by very low strontium, sodium, and percent aragonite, with significant enrichment in iron, manganese, aluminum, silicon, and percent insoluble residue.

Polar ordination analysis was used to illustrate gradational relationships between the samples in the data set (Fig. 13). This analysis aids in the recognition of the environmental factors responsible for the geochemical variation.

In ordination, samples are plotted along axes whose endpoints represent samples exhibiting maximum statistical dissimilarity (Springer and Bambach, 1985; Pelle, 1989). The spatial arrangement of each sample along the axes de-

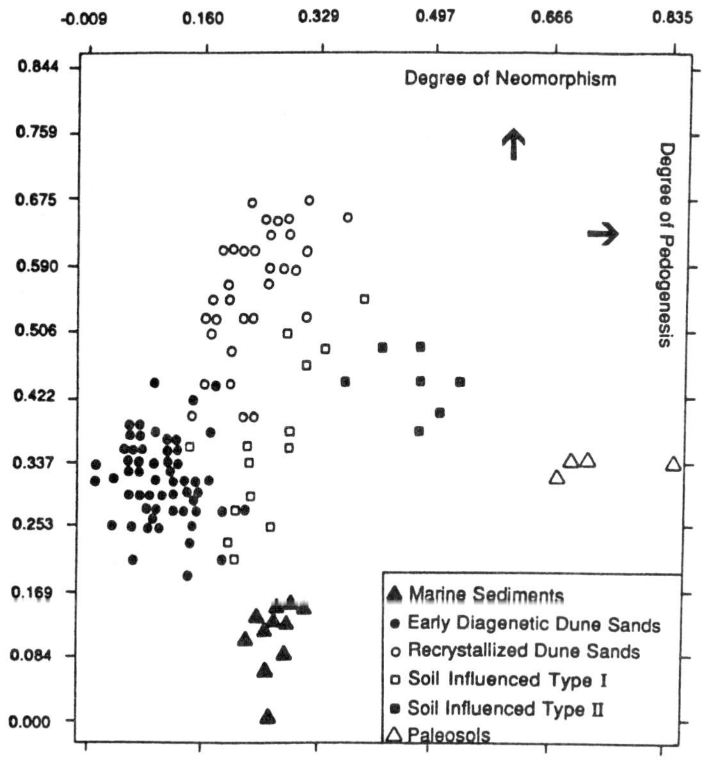

Fig. 13. Polar ordination illustrates gradational relationships (between samples) caused by diagenesis.

pends on the relationships to the endpoints of the axes. That is, the closer a given sample lies to an endpoint, the more similar it is to that endpoint. This allows for the recognition of trends among the samples that can be explained in terms of processes acting on the samples. To aid in the identification of processes responsible for sample locations along the axes, it is useful to examine the nature of the endpoints.

Axis One: Cophenetic Correlation Coefficient = 0.6883

Endpoint 1: WQ-18 (Watling's Quarry). This sample is a very poorly cemented, friable grainstone. Grains consist of well-rounded peloids and/or ooids, with very few recognizable skeletal grains; there is less than 0.6% insoluble residue. X-ray diffraction reveals the presence of 73% aragonite. Geochemically, it contains high concentrations of strontium (7672ppm) and low manganese (28ppm) and iron (below detection limits).
Endpoint 2: MBL-11 (Morgan's Bluff). This sample is a red, well-indurated brecciated matrix paleosol characterized by a red matrix and partial to complete micritization of grains and intraclasts. Insoluble residue accounts for over 12% of the rock's mass. X-ray diffraction indicates an absence of aragonite. Geochemically, the sample is characterized by very low strontium (763ppm) and high concentrations of iron (664ppm), manganese (106ppm), silicon (508ppm), and aluminum (75ppm).

Axis One is interpreted as resulting from diagenetic changes associated with soil-forming processes. Samples along this axis grade from being unaffected by pedogenesis to being the product of pedogenetic processes.

Axis Two: Total Cophenetic Correlation Coefficient (for two axes) = .8601

Endpoint 1: GH-5 (Graham's Harbor). Graham's Harbor sediments have not been subjected to meteoric diagenesis. The sample is composed of 69% aragonite and 31% HMC. Geochemically, the sediment has high concentrations of strontium (6408ppm), sodium (7582ppm), and potassium (70ppm), and low concentrations of iron (60ppm) and aluminum (<1ppm).

Endpoint 2: MB-15 (Morgan's Bluff. This sample is a very well-cemented oolitic grainstone. Grains show evidence of dissolution and are chalky in appearance. Aragonite has been reduced to 25% of the rock by progressive neomorphism. Geochemically, the sample contains low concentrations of strontium (2094ppm), sodium (779ppm), and potassium (below detection limits). Low concentrations of iron (16ppm), manganese (43ppm), and aluminum (below detection limits) attest to the disassociation of this sample with soil-forming processes.

Axis Two can be interpreted as resulting from the progressive alteration of sediments exposed to vadose diagenesis. Samples along this axis display a gradation from unaltered marine sediments to highly neomorphosed, well-cemented rocks that have not been influenced by paleosol formation.

Taken as a whole, these samples represent a geochemical continuum in the progressive diagenetic alteration of subaerially-exposed carbonate rocks. The progressive alteration observed in dune sands of widely-different ages and geographic locations is remarkably similar to the theoretical trends envisioned by Brand and Veizer (1980) (Fig. 14).

With increasing diagenetic equilibration with meteoric waters, there is a progressive decrease in strontium and sodium and a concomitant increase in iron and manganese. Aluminum and silicon are observed primarily in association with insoluble residues, and therefore show an increase in concentration with alterations due to soil-forming processes. Magnesium does not exhibit a clear pattern of either increase or decrease, and this is believed to result from varying amounts of magnesium in the original mineralogy (Brand and Veizer, 1980).

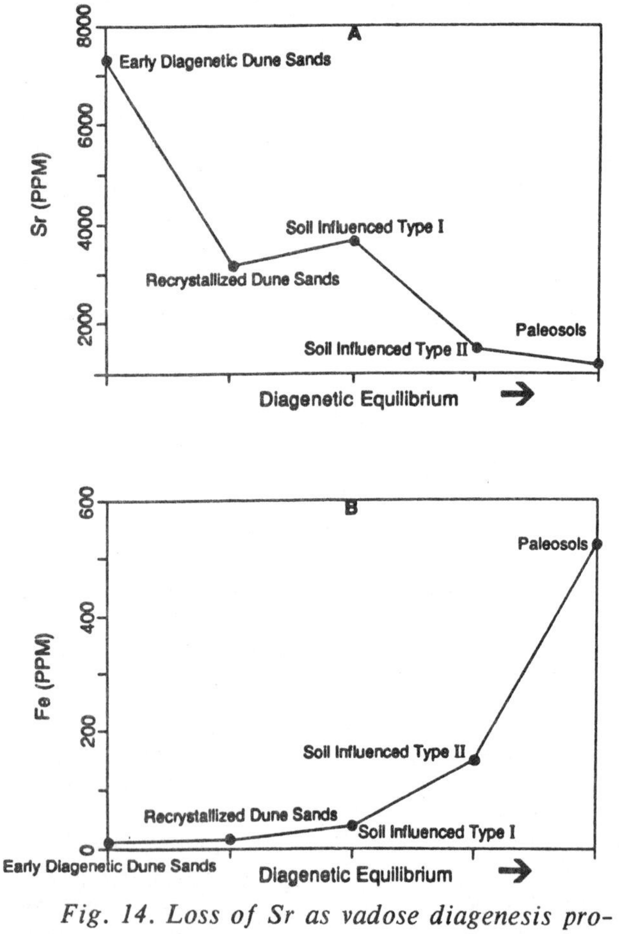

Fig. 14. Loss of Sr as vadose diagenesis progresses toward equilibrium with meteoric waters (A); increase in Fe as vadose diagenesis progresses toward equilibrium with meteoric waters (B). Values reported are average concentrations for each of the diagenetic groups delineated by cluster analysis.

Temporal Considerations

The progressive diagenesis toward equilibrium with meteoric waters in vadose-altered rocks should not be confused with time. Although time is an important factor in diagenetic alterations, other factors are more important and tend to override the effects of time (Fig. 15). The lack of a definite pattern such as the one present in Fig. 14 demonstrates the lack of control of age alone on the formation of the various diagenetic groups.

Early diagenetic dune sands and paleosols are shown to vary widely in age, yet samples within each group are geochemically similar -- regardless of age. The geochemical similarity of paleosols of differing ages is easily explained by the constant nature of pedogenic processes and products through time. In addition, paleosols may be thought of as an end member product of meteoric diagenesis. However, the similarity between the diagenetically-altered host rocks of widely different ages is less easily explained if trends predicted by Brand and Veizer (1980) are primarily time controlled. Examination of geochemical and lithologic characteristics of the outcrops indicates that the primary controls for progressive diagenesis in exposed dune sands are the timing of paleosol development on the exposed sands and the type of paleosol formed.

Samples classified as early diagenetic dune sands range in age from less than 6500 years at North Point (Boardman et al., 1987) to about 125,000 years at Watling's Quarry (Carew and Mylroie, 1984). The geochemical similarity between Watling's Quarry and North Point would be surprising if time were the major consideration. Watling's Quarry would be expected to show a higher degree of vadose alterations including lower strontium concentrations, reduced percentages of aragonite, and increased cementation.

The geochemical similarity between these two outcrops can be explained by the formation

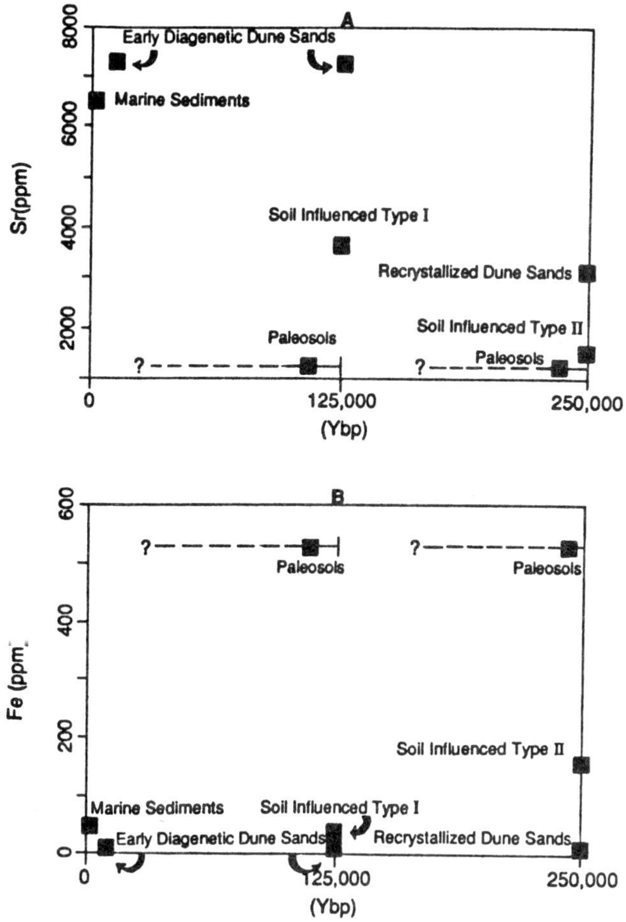

Fig. 15. Sr concentrations of vadose-altered carbonates as related to age (A); Fe concentrations as related to age (B). Values reported are average concentrations for each of the diagenetic groups delineated by cluster analysis.

of the impermeable laminar calcrete cap on Watling's Quarry soon after sand deposition ceased. Such a cap is an effective barrier to the downward movement of meteoric waters, producing a geochemically "closed" dune system. Systems such as this are characterized by low water-rock ratios in which the water, because of its constant interaction with solids, loses its chemical identity and its composition is controlled (buffered) by the rock itself (Veizer, 1983). This closed diagenetic system has effectively preserved the original geochemistry of the rocks at Watling's Quarry.

On the other hand, dune sands that have a permeable, brecciated matrix type paleosol or no paleosol associated with them at all are much more susceptible to meteoric diagenesis. Fresh water is able to move through the permeable soil (if present) and into the rock body. This type of dune system is characterized by high water-rock ratios, in which the dissolving $CaCO_3$ does not effectively alter the bulk water chemistry (Veizer, 1983). Morgan's Bluff and the Gulf are examples of outcrops characterized by a diagenetically "open" system. These outcrops show much more diagenetic alterations to the dune sands -- particularly the strong development of zones of influence beneath the paleosols.

SUMMARY AND CONCLUSIONS

The study of trace and minor element concentrations in vadose-altered eolian sands of San Salvador Island and Andros Island, Bahamas, has provided information on trace element associations, stratigraphic variability, and mechanisms for redistribution in carbonate sequences.

Carbonate rocks and sediments are defined as multicomponent systems with trace elements associated with several different fractions in addition to $CaCO_3$. The use of a sequential extraction procedure allows for the chemical separation of trace elements into fractions that are useful in identifying trace element sources and trends. This has led to several conclusions about elemental associations in vadose-altered carbonate sands.

Elements associated with the exchangeable fraction include potassium, sodium, calcium, and strontium. Concentrations of K and Na are highly variable and are dependent primarily on geographic location (proximity to the coast). $^{Sr}/_{Ca}$ ratios in the exchangeable fraction are shown to be useful indicators of on-going diagenesis. $^{Sr}/_{Ca}$ ratios elevated above ratios expected in sea-water and aragonite are caused by dissolution of aragonite and subsequent precipitation of diagenetic low-Mg calcite with preferential exclusion of Sr from the precipitating phase.

Elements associated with the $CaCO_3$ fraction behave in a predictable manner as diagenetic equilibrium with meteoric waters is approached. Na and Sr show a systematic decrease; Fe, Mn, and Al display a concomitant increase in concentration with progressive equilibrium. Sources for trace elements in the carbonate fraction include sea-water, the original carbonate phase, and mineral insoluble residues concentrated in paleosol profiles.

Metals associated with the Fe-Mn oxide fraction include Fe, Mn, and Al. These metals are present in appreciable quantities only in and near (within the zone of influence) paleosol horizons. The abundance of these metals in paleosol horizons and the strong positive correlation with the percentage of insoluble residue suggests that the Fe, Mn, and Al in the carbonate fraction is derived from oxide coatings of insoluble residue within the paleosols.

Organic matter is of minor importance in the distribution of trace elements in vadose-altered carbonate sands. Most organic matter in this environment is readily oxidized, releasing trace elements to the exchangeable fraction or to be carried away in the bulk aquifer solution.

Variability in trace element distribution is best described in terms of vertical and lateral variability within a single outcrop (intra-outcrop) and lateral variability among outcrops of different age and/or geographic location (inter-outcrop).

Three geochemically-distinct zones, related to the presence of a paleosol, may be recognizable within a single vadose-altered outcrop: 1. metal-rich, Sr-depleted paleosols; 2. zone of influence associated with paleosols; and 3. early diagenetically-altered zone that retains much of the original elemental signature. Lateral variability within a single outcrop may be accounted for by the discontinuous nature of paleosols accounts for increase variability at and near exposure surfaces.

Development of the zone of influence and the redistribution of trace elements into the vadose-altered carbonate sands is dependent on several physical mechanisms: 1. simple downward percolation of meteoric waters through metal-rich soil/paleosol horizons; 2. bioturbation, particularly by roots of land plants; and 3. fracturing of the host rock.

Multivariate statistical analyses (cluster analysis and polar ordination) identify five geochemically-distinct groups of rocks present within vadose-altered dunes of differing ages and degree of diagenesis in the Bahamas: 1. paleosols; 2. soil influenced Type I; 3. soil influenced Type II; 4. recrystallized dune sands; and 5. early diagenetic dune sands. A sixth group, marine sediments, is also shown to be geochemically distinct from the vadose-altered sands.

While trace elements associated with the carbonate fraction behave in a predictable manner as diagenetic equilibrium with meteoric waters is approached, it should be clear that diagenetic equilibrium does not necessarily imply increasing age. Although age is certainly a factor in carbonate diagenesis, the timing and type of paleosol that develops on exposed carbonate rock has a direct bearing on the effectiveness of elemental redistribution through a section.

Early-formed, impermeable laminated calcretes can be an effective seal against the movement of meteoric waters through a carbonate unit, producing a geochemically "closed" system. Permeable, brecciated-matrix soils and/or later-formed soils do not effectively "close" a carbonate section, allowing for the progressive neomorphism of metastable carbonate minerals to diagenetic low-Mg calcite with the accompanying trace element signature. This allows older rocks with impermeable paleosol caps to be preserved with little geochemical alteration (presumably with intact primary porosity), relative to rocks of the same age within the impermeable cap.

The associations, trends, and relationships of trace elements recognized in this study should prove valuable to the interpretation of ancient vadose-altered carbonate sands.

ACKNOWLEDGEMENTS

We would like to thank the many people who helped with this study. Field assistance was provided by Paul Bergstrand and Norm Colby and the participants of the 1986 and 1987 field course, Modern Carbonate Deposition. Tom Musselman provided essential help with the laboratory analysis, and without him, the analytic results would have been impossible. We also thank the Bahamian Field Station (formerly the College Center of the Finger Lakes) and the very helpful staff, directed by Dr. Don Gerace and Kathy Gerace. We are also grateful to Dr. John Mylroie and Dr. James Carew for providing continuously stimulating commentary and insights on the geology of San Salvador, the Bahamas, and on the nature of the world. On Andros Island, the staff of the Forfar Field Station (International Field Studies), led by Dr. Ben Bohl, provided logistical support and were very helpful. Financial support was provided by a grant from Sigma Xi (to Pelle) and by Miami University through the workshop, Modern Carbonate Deposition (directed by Boardman). Without all this help, it just wouldn't have happened.

REFERENCES CITED

Bathurst, R.G.C., 1975, Carbonate Sediments and their Diagenesis: Elsevier Scientific Publishing Company, Amsterdam, 658 p.

Boardman, M.R., 1976, Lime mud deposition in a tropical island lagoon, Bight of Abaco,

Bahamas: Unpublished Master of Science Thesis, Curriculum of Marine Sciences, University of North Carolina, Chapel Hill, North Carolina, 121 p.

Boardman, M.R., Carew, J.L., and Mylroie, J.E., 1987, Holocene deposition of transgressive sand on San Salvador, Bahamas: Geological Society of America Abstracts with Programs, v. 19, p. 593.

Braithwaite, C.J.R., 1983, Calcrete and other soils in Quaternary limestones: structures, processes and other applications: Journal of the Geological Society of London, v. 140, p. 351-363.

Brand, U. and Veizer, J., 1980, Chemical diagenesis of a multicomponent carbonate system - 1: Trace elements: Journal of Sedimentary Petrology, v. 50, p. 1219-1236.

Carew, J.L. and Mylroie, J.E., 1985, The Pleistocene and Holocene stratigraphy of San Salvador Island, Bahamas, with reference to marine and terrestrial lithofacies at French Bay, in Curran, H.A., ed., Pleistocene and Holocene Carbonate Environments on San Salvador Island, Bahamas: CCFL Bahamian Field Station.

Colby, N.D. and Boardman, M.R., 1988, Vertical sequence of a high-energy carbonate lagoon; Graham's Harbor, San Salvador, Bahamas: Society of Economic Paleontologists and Mineralogists Annual Midyear Meeting Abstracts, v. 5, p. 11.

Dulin, L.A., 1984, Manganese and iron in a sediment core from Northwest Providence Channel, Bahamas: a comparison of paleoclimates: Unpublished Master of Science Thesis, Geology Department, Miami University, Oxford, Ohio, 87 p.

Eaton, M.R., 1986, Origin of insoluble residue in a deep-sea sediment core from Northwest Providence Channel, Bahamas: Unpublished Master of Science Thesis, Geology Department, Miami University, Oxford, Ohio, 91 p.

Fetter, C.W. Jr., 1980, Applied Hydrogeology: Charles E. Merrill Publishing Company, Columbus, Ohio, 488 p.

Friedman, G.M., 1969, Trace elements as possible environmental indicators in carbonate sediments, in Friedman, G.M., ed., Depositional Environments in Carbonate Rocks: Society of Economic Paleontologists and Mineralogists, Special Publication Number 14, p. 193-200.

Heath, G.R. and Dymond, J., 1977, Genesis and transformation of metalliferous sediments from the East Pacific Rise, Bauer Deep, and Central Basin, northwest Nazca Plate: Geological Society of America Bulletin, v. 88, p. 723-733.

Hem, J.D., 1978, Study and interpretation of the Chemical Characteristics of Natural Water, Second Edition: Geological Survey Water-Supply Paper 1473: United States Government Printing Office, Washington, D.C., 363 p.

James, N.P. and Choquette, P.W., 1984, Limestones - the meteoric diagenetic environment: Geoscience Canada, v. 11, p. 161-194.

Kinsman, D.J.J., 1969, Interpretation of Sr^{2+} concentrations in carbonate minerals and rocks: Journal of Sedimentary Petrology, v. 39, p. 486-508.

Land, L.S., 1967, Diagenesis of skeletal carbonates: Journal of Sedimentary Petrology, v. 39, p. 980-1006.

McCartney, R.F., 1987, Origin and diagenesis of paleosols in the Bahamas: Unpublished Master of Science Thesis, Geology Department, Miami University, Oxford, Ohio, 88 p.

Neumann, A.C., 1965, Processes of recent carbonate sedimentation in Harrington Sound, Bermuda: Bulletin of Marine Science, v. 15, p. 987-1035.

Pelle, R.C., 1989, Stratigraphic distribution and associations of trace elements in vadose-altered multicomponent carbonate sands: Unpublished Master of Science Thesis, Geology Department, Miami University, Oxford, Ohio, 110 p.

Pingitore, N.E. Jr., 1978, The behavior of Zn^{2+}

and Mn^{2+} during carbonate diagenesis: theory and applications: Journal of Sedimentary Petrology, v. 48, p. 799-814.

Pingitore, N.E. Jr., 1982, The role of diffusion during carbonate diagenesis: Journal of Sedimentary Petrology, v. 52, p. 27-39.

Plummer, L.N. and Mackenzie, F.T., 1974, Predicting mineral solubility from rate data: application to the dissolution of magnesian calcites: American Journal of Science, v. 274, p. 61-83.

Savoie, D.L. and Prospero, J.M., 1977, Aerosol concentration statistics for the Northern Tropical Atlantic: Journal of Geophysical Research, v. 82, p. 5954-5964.

Span, D. and Gaillard, J-F., 1986, An investigation of a procedure for determining carbonate-bound trace metals: Chemical Geology, v. 56, p. 135-141.

Springer, D.A. and Bambach, R.K., 1985, Gradient versus cluster analysis of fossil assemblages: a comparison from the Ordovician of southwest Virginia: Lethaia, v. 18, p. 181-197.

Tessier, A., Campbell, P.G.C., and Bisson, M., 1979, Sequential extraction procedure for the speciation of particulate trace metals: Analytical Chemistry, v. 51, p. 844-850.

Veizer, J., 1983, Chemical diagenesis of carbonates: theory and applications of trace-element technique in Arthur, M.A., Anderson, T.F., Kaplan, I.R., Veizer, J., and Land, L.S., eds., Stable Isotopes in Sedimentary Geology: Society of Economic Paleontologists and Mineralogists, Short Course Notes 10, Ch. 3, 100 p.

Walls, R.A., Harris, W.B. and Nunan, W.E., 1975, Calcareous crust (caliche) profiles and early subaerial exposure of northeastern Kentucky: Sedimentology, v. 22, p. 417-440.

Walls, R.A., Ragland, P.C. and Crisp, E.L., 1977, Experimental and natural early diagenetic mobility of Sr and Mg in biogenic carbonates: Geochemica et Cosmochemica Acta, v. 41, p. 1731-1737.

OBSERVATIONS ON THE *SCYTONEMA* MATS OF SAN SALVADOR, BAHAMAS

Allan Pentecost
Division of Biosphere Sciences
King's College
Kensington Campus
Camden Hill, London W8 7AH

ABSTRACT

Mats of the cyanobacterium *Scytonema myochrous* are described from a range of supralittoral and littoral lacustrine sites on San Salvador, Bahamas. *Scytonema* occurred as flat, polygonal mats, as low rounded colonies and as irregular, partially lithified heads and crusts. Little difference was found in the species composition or biomass per unit area but biomass was extremely low (cell volume excluding sheath) at all sites.

Just below the mat surface, low magnesian calcite accreted, probably through precipitation *in situ* as a result of evaporation and/or photosynthesis. Lithification was infrequent and the deposits were mostly poorly consolidated with high porosity (62-18%). Growth of stromatolitic algal heads in Storr's Lake was undetectable during July using carborundum marking and the growth and calcification of *Scytonema* in relation to salinity is briefly discussed.

INTRODUCTION

Cyanobacterium mats are a characteristic feature of many low-lying tropical landscapes, where higher plant growth is restricted by high salinity or sediment instability. It has been realized for several decades that these mats are of more than passing interest to geologists, for under certain conditions the mats may calcify, or trap and bind suspended particles, then become preserved as stromatolites.

Many types of cyanobacterium mats have been described and among the most important are those formed by *Scytonema*. These were first described from the Bahamas by Black (1933). Investigations in the Florida everglades by Dachnowski-Stokes & Allison (1928) resulted in further discoveries and it soon became apparent that *Scytonema* was a major primary producer and sediment stabilizer in the region. In a series of papers, Monty (1972; 1976) and Monty & Hardie (1976) have further described these mats in some detail and Gleason (1972) made further investigations in the Everglades. Although Andros Island, with its extensive inland marshes remains the *locus classicus* for *Scytonema*, the mats are well developed on other islands. On San Salvador, they are commonly encountered around the lakes and in places they become partially lithified. This paper describes some of these mats and their significance as precipitators of calcium carbonate.

METHODS

Mats were sampled during July, 1987 from 7 sites on San Salvador (Fig. 1) and also from Richmond Hill, Great Exuma. After careful removal of the mat, vertical columns measuring c. 2 x 2 mm were cut through the water-saturated material with a razor and the columns divided into a series of segments 1-4 mm deep to investigate the depth-distribution of cyanobacteria. Each block was decalcified with EDTA, briefly homogenized with a shearing piston and aliquots of suspension counted under an inverted microscope. By calculating the average cell volumes of each species present, the abundance of each species could be expressed on a unit volume basis. Details of the method are described elsewhere (Pentecost, 1982).

The associated mat sediments were carefully detached, washed to remove soluble salts, dried at 105 C and weighed. They were then soaked in air-free water under partial vacuum, weighed, decalcified and reweighed. Finally the organic matter content was calculated by weighing after ignition at 500 C. Selected samples were critical-point dried with carbon dioxide, Au-Pd coated and examined with a Philips EM 501B scanning electron microscope. Other samples were embedded with epoxy resin and sectioned for petrological examination.

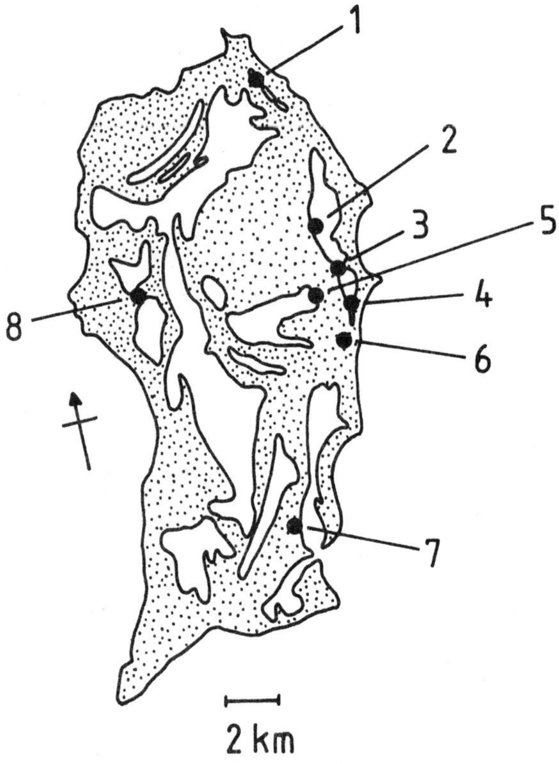

Fig. 1. Sampling sites for Scytonema *on San Salvador. **1.** Upper Fresh Lake; **2.** small inlet on Storr's Lake west shore; **3.** Storr's Lake narrows, west shore; **4.** Storr's Lake causeway; **5.** Granny Lake ponds; **6.** Fortune Hill Settlement pond; **7.** pond at Trail Farm Settlement; **8.** algal marsh complex north of Little Lake.*

RESULTS

General Structure

Scytonema myochrous (Dillw.) Ag. is widely distributed around the San Salvador lakes, occurring in areas of low-lying water-saturated ground. The colonies were rarely submerged and normally confined to the littoral fringe. Several mat morphologies were apparent. Thin black to olive-brown turfs, often with abundant desiccation cracks (Fig. 2) were common. The small polygons bore spiky tufts of vertically-arranged filaments resting upon soft, laminated carbonate sediment. These mats were well developed in many of the small ponds to the east of Granny Lake, the marshes north of Little Lake and in small patches along the shore of Storr's Lake. Many of the circular ponds forming in the limestone depressions east of Granny Lake showed well marked concentric zonation of cyanobacteria probably related to the degree of exposure to air. The centres of the circular depressions contained shallow hypersaline water with flocculent organic debris often fringed with *Batophora* (Fig. 3). This was surrounded by a narrow (30-100 cm) slightly raised *Scytonema* mat saturated but not submerged in water which merged into a darker zone, slightly lower in elevation composed of blistered *Scytonema - Gloeocapsa* mat associated with a flaky carbonate crust. Finally, there was an outer smooth black zone colonized by *Schizothrix calcicola* s. lato, *Gloeocapsa* and occasional tufts of *Lyngbya aestuarii*.

Algal heads composed predominantly of *Scytonema* were also seen in some of the ponds. The heads were varied in shape and consisted of laminate concretionary structures 5-20 cm in diameter and 2-30 cm tall (Fig. 4), with their upper surfaces usually emergent but saturated through capillary action. The heads were poorly lithified and generally easy to cut with a knife. Similar heads were found just offshore at Storr's lake narrows.

At the margin of Trail Farm Settlement, pond mats were less well-developed and consisted of discrete dull black, slight raised mounds 5-10 cm across and up to 1 cm deep on soft, unlaminated sediment overlying limestone. Smooth, thin black mats of *Scytonema* mixed with *Gloeocapsa* occurred on soft mud at the edge of Fortune Hill Settlements pond and also below the causeway at Storr's Lake. Halite was present at both sites. The mats of upper Fresh Lake resembled the polygonal mat described above but were attached to emergent lignum forming a causeway halfway along the lake. These mats had undergone lithification just below the *Scytonema* layer.

Species Distribution

The abundance of cyanobacteria within selected mat profiles is shown in Fig. 5. *Scytonema* was the dominant cyanobacterium in all profiles with greatest biomass in the upper 4 mm of mat. The biomass (cell volume excluding the sheath) ranged from 3ppt at Storr's Lake narrows to a maximum of 24 ppt at Fresh Lake. Apparently health cells were encountered at a depth of 13 mm at Storr's Lake narrows (Fig. 5A0 where light intensity must have been negligible. Empty sheaths of *Scytonema* were abundant throughout all sections and far outnumbered sheaths with trichomes.

The remaining cyanobacteria contributed little to total biomass. Most frequent were aggregates of coccoid *Gloeocapsa* cells with hyaline,

Fig. 2. Scytonema *mat, Granny Lake ponds showing desiccation polygons. Bar 5 cm.*

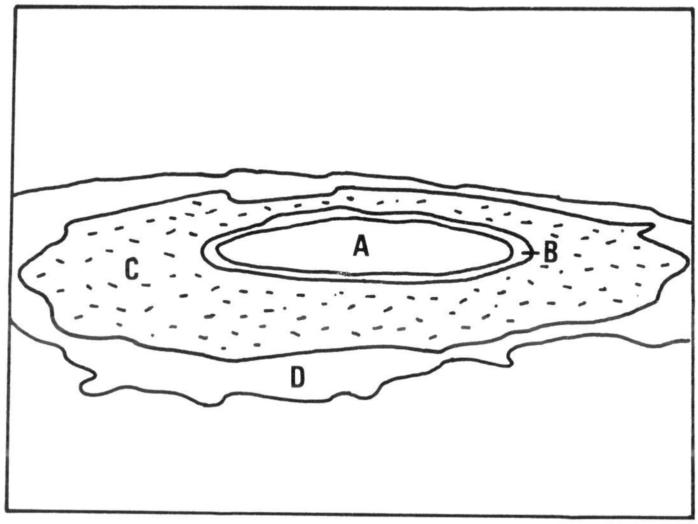

*Fig. 3. Zonation of cyanobacteria in a Granny Lake pond. **A**. shallow open water; **B**.* Scytonema *fringe; **C**. mixed* Scytonema - Gloeocapsa *carbonate crust; **D**. smooth* Schizothrix - Gloeocapsa *mat. Central pool 2.5 m diameter.*

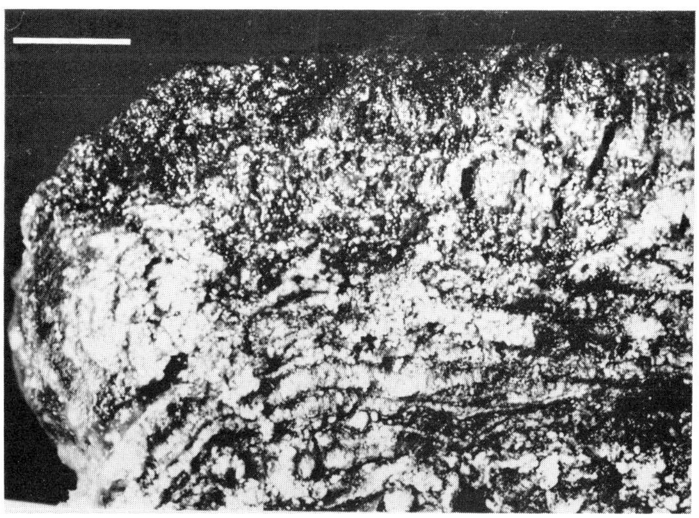

Fig. 4. Section through an algal head from Granny Lake ponds. Dark upper surface shows vertically-oriented filaments of Scytonema. *Note compressed laminae of* Scytonema - Gloeocapsa *alternating with carbonate-rich layers below. Bar 5 mm.*

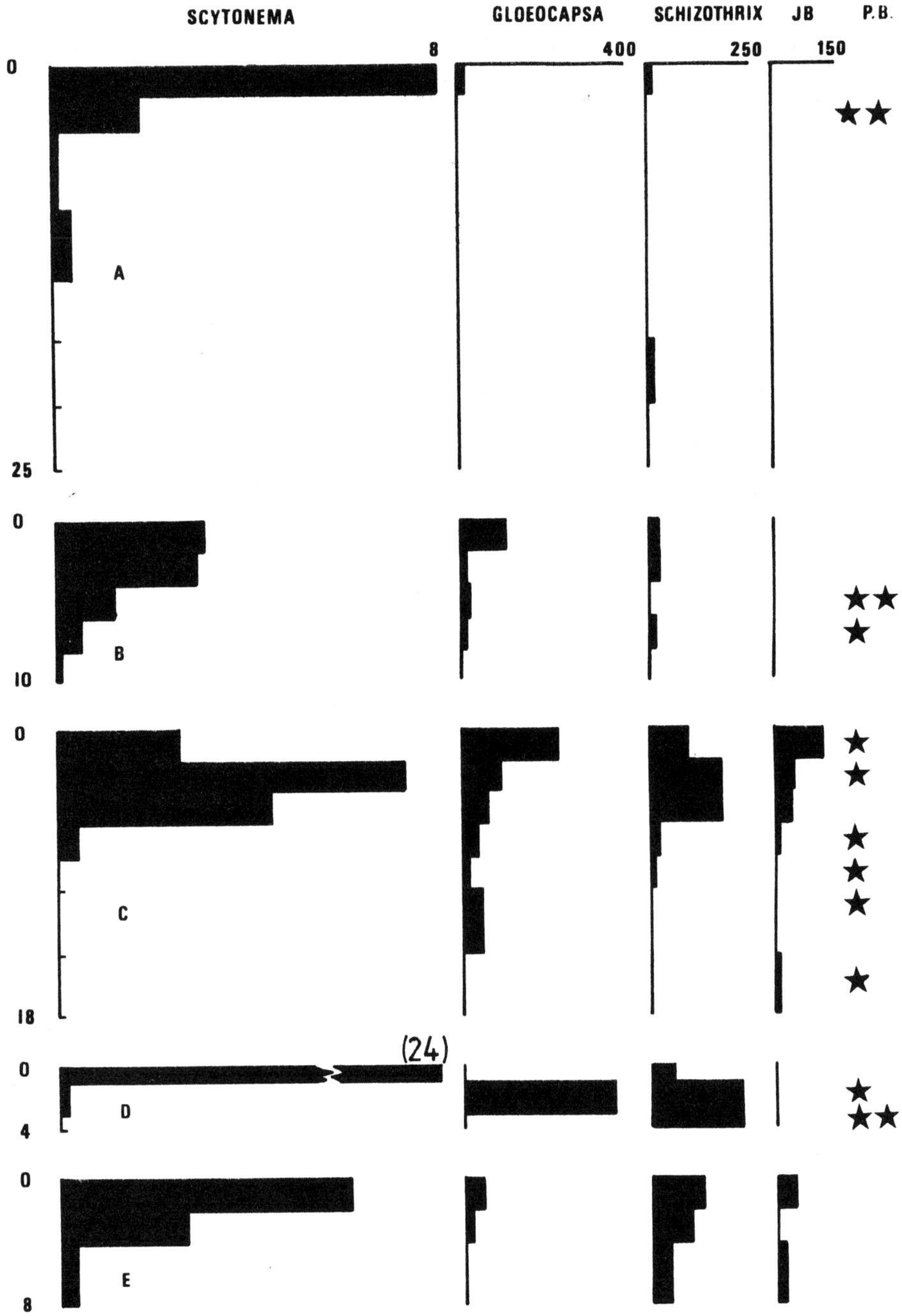

Fig. 5. Species composition of Scytonema *mats expressed as the volume occupied by cells. Abscissa:* Scytonema *volumes in parts per thousand, remaining volumes in parts per million. Ordinate: depth from mat surface (mm).*

***A.** partially lithified algal head, Storr's Lake narrows;* ***B.** unlithified mat, narrows;* ***C.** soft algal head, Granny Lake ponds;* ***D.** lithified crust on submerged wood, upper Fresh Lake;* ***E.** unlithified mat from the supralittoral algal marsh, Richmond Hill, Great Exuma.* ***JB** =* Johannesbaptista *sp.* ***PB** = photosynthetic bacteria (presence indicated by stars).*

brown and occasionally violet-pigmented sheaths. The cells ranged from 1-8 μm in diameter. A distinctive form with exceptionally large cells, resembling *G. gigas* was found in small numbers at Storr's and Granny Lakes and small numbers of *Aphanocapsa* were also detected. *Gloeocapsa* is difficult, if not impossible to separate from poorly developed *Entophysalis*. Well-developed mats of *Entophysalis* were often associated locally with the *Scytonema* mats and much of the coccoid material sampled may in fact belong to this taxon.

Fine filaments of *Schizothrix calcicola* s. lato were encountered in all mats but the species rarely formed discrete layers. It was frequently seen in the algal heads of Granny Lake and at Fresh Lake, the filaments were associated with the lithified underlayer (Fig. 5D). A species of *Johannesbaptista* with filaments 3-6 μm wide was found in small numbers, particularly at Fresh Lake. Although none of these cyanobacteria matched the biomass level of *Scytonema*, their distribution with depth followed a similar pattern. A few other cyanobacteria were occasionally seen and included a *Calothrix* at Fresh Lake and a

Phormidium with trichomes 7 μm wide at Granny Lake. Diatoms were seen at all sites, but were generally rare. Only at Storr's Lake was a thin slime containing *Nitzschia* cells occasionally observed. Purple photosynthetic bacteria were commonly encountered 2-10 mm below the mat surfaces (Fig. 5), but no attempt was made to identify or enumerate them.

The algal distribution patterns in all mats was essentially the same, despite considerable differences in mat morphology. The polygonal mats sampled at Storr's Lake (Fig. 5B) closely resemble the mats of the supralittoral marine marsh at Richmond Hill, Great Exuma (Fig. 5C). The algal heads (Figs 5A and C) were sampled to a greater depth, but floristically closely resembled the polygonal mats.

All mats were associated with fine sediment which was generally unconsolidated. Polygonal mats at the edge of Storr's Lake were associated with uncemented micrite, but the top 2 mm of mat was virtually free of sediment (Fig. 6). Petrological sections revealed irregular or tubular clumps of micrite attached to the *Scytonema* sheaths. Deeper within the mat the tubes coalesced, but there was no evidence of recrystallization immediately below the mat. The partially lithified heads from the same lake were similar consisting of fine, mostly anhedral calcite crystals 1-6 μm in diameter. In all of these mats, the porosity was high and consequently the calcium carbonate content correspondingly low (Table 1). Low magnesian calcite was identified in all samples using X-ray diffraction and atomic absorption spectrophotometry. Traces of dolomite were found in the Granny Lake samples. All sites were brackish to hypersaline (Table 1) and in many isolated pools, salt crystallized out on the mat surface. Carborundum marking of algal heads at Storr's Lake narrows and the inlet over a period of 7 days, failed to reveal any algal growth.

Fig. 6. Tuft of Scytonema *filaments from the surface of a polygonal mat at Storr's Lake narrows. Note filament aggregation, false-branching of filaments and sparse patches of adherent micrite. SEM micrograph. Bar 100 μm.*

DISCUSSION

Before sediment accumulation rates can be estimated beneath these mats, more data on the growth of *Scytonema* are urgently required. Black (1933) made no measurements, but notes that *Scytonema* growth was extremely slow on Andros Island. Monty (1976) noted a 2-3 cm increment in mat thickness in 8 weeks, equivalent to 350-500 μm per day which is certainly much faster than the growth of the algal heads in Storr's Lake measured in July. In cool-temperate climates, Thunmark (1926) and Pentecost (unpublished) observed rates of 1-3 mm per annum but these are unlikely to reflect the growth potential in lower latitudes. Salinity may be responsible for these differences. On Andros Island, *Scytonema* mats are best developed on the freshwater marshes and when the water becomes brackish, the growth of *Scytonema* is replaced by *Schizothrix* (Monty, 1976) suggesting that high salinity might be detrimental to growth. *Scytonema* is typically a freshwater genus, occurring abundantly in the Everglades and calcareous marshes worldwide, but it also occurs adjacent to many saline lakes where it may be subject to seepages of fresh groundwater. Thus, bursts of growth may occur when salinity falls, possibly explaining the wide range of growth rate encountered. Many of the San Salvador mats appear to fall into this category and on the west shore of Storr's Lake, some evidence for fresh water seepage was obtained from the salinity measurements (Table 1). A similar phenomenon could be responsible for the lithified *Scytonema* heads of Lake Clifton, Western Australia (L. S. Moore, pers. com.). A few species of *Scytonema* are truly marine and must, therefore, grow at high salinity. One form is

Table 1

Chemical and physical data of Scytonema mats 15-20.7.1987

Location	t °C	o/oo salinity	% vol $CaCO_3$ *	% weight $CaCO_3$	% organic	% residue	Mg mol %
Storr's Lake narrows							
polygonal mat	36	24	28 (0-4)	93.3	6.2	0.5	2.1
algal heads	32	83	22 (5-10)	92.6	7.2	0.3	2.0
Granny Lake pond							
Gloeocapsa/Scytonema crust	32	47	19 (0-4)	74.8	18.9	6.4	0.7
algal heads	32	148	-	-	-	-	-
Fresh Lake							
crust on lignum	31	-	38 (1-2)	87.3	11.5	1.2	2.1

* depth sampled below surface (mm) in parentheses.

associated with the stromatolites of Hamelin Pool, Western Australia (Logan et al., 1974) where the salinity ranges from 55-70 o/oo.

Scytonema remains moist by capillarity and is rarely seen submerged by water. The large surface area exposed by the upward-growing filament bundles, combined with the dark coloration caused by a sheath pigment must ensure rapid evaporation of water and the concentration of salts. Such conditions probably promote the precipitation of calcium carbonate and this must be considered a likely origin for the associated sediment. Photosynthesis removes carbon dioxide from the water and can also precipitate carbonate and there is some evidence that this occurs in the freshwater marshes of Andros Island and the Everglades (Gleason, 1972; Monty, 1976; Monty & Hardie, 1976). It may also occur in the San Salvador mats, where despite the low cell biomass, a rapid turnover could give high rates of photosynthesis. Since chemolithotrophic (photosynthetic) bacteria were also abundant just below the mat surface, subsurface, photosynthetic precipitation of calcite is also a possibility.

The scarcity of obvious detrital material and the occurrence of micrite tubes surrounding the sheaths of *Scytonema* strongly suggest *in situ* precipitation. Calcification of *Scytonema* in freshwater environments has been observed widely (Gleason, 1972; Gebelein, 1976; Lowenstam, 1986; Pentecost & Riding, 1986) and calcite is often nucleated directly upon the sheath surface. However, although the carbonate is closely associated with the San Salvador *Scytonema*, no evidence of direct nucleation on the sheath was found, suggesting a more passive role for the organism in these sites. Lithification of *Scytonema* heads has been previously described from Andros Island (Gebelein, 1976; Monty, 1976) but appears to be the exception rather than the rule. Shinn et al (1969) reported *Scytonema* cementation from Andros Island with small amounts of associated fine-grained dolomite. Field evidence also suggests that calcification is inhibited at high salinities (Monty, 1972; 1976) though it is apparent that on San Salvador at least, this may not be the case. Stromatolite domes composed of *Scytonema* were reported by Monty (1972) from flats bordering lakes on Andros Island with calcite laminae, and are figured in Monty & Hardie (1976).

Modern stromatolites have been described from San Salvador from two areas. Dakoski & Bain (1984) found a range of mats and stromatolites in Granny Lake and the adjacent ponds. Lithified columnar stromatolites occurred in the latter, but the associated microorganisms were not described. Similar forms were observed in 1987 by this author and colonized by *Gloeocapsa*. Laminae of high magnesian calcite were reported associated with many of these mats. In Storr's Lake embayment, Mann & Hoffman (1984) found unlaminated algal heads in shallow water associated with two *Phormidium* species. In shape and size they resemble the *Scytonema* heads described here, but none of the former was collected in 1987.

The *Scytonema* heads of San Salvador, in contrast to the polygonal mats, had a laminate structure, particularly those at Granny Lake. The laminae appear compressed below, probably as a result of the high initial porosity and organic matter content (Table 1).

ACKNOWLEDGEMENTS

Gratitude is extended to Dr. D. T. Gerace for assistance in the collecting of material and the suggestion of sampling sites. Professor M. Audley-Charles for the use of X-ray diffraction equipment. Research grants from the Central Research Fund, University of London and the Royal Society (IGCP Fund) are gratefully acknowledged.

REFERENCES CITED

Black, M., 1933, The algal sediments of Andros Island, Bahamas, Phil. Trans. R. Soc. B, 272: p. 165-192.

Dachnowski-Stokes, A.P. and Allison, R.V., 1928, A preliminary note on the blue-green algal marl in Southern Florida in relation to the problem of coastal subsidence, J. Wash. Acad.Sci. 18: p. 476-480.

Dakoski, A.M. and Bain, R.J., 1984, Recent microbial mats and stromatolites of Granny Lake, San Salvador, Bahamas, in Proceedings of the 2nd Symposium on the Geology of the Bahamas, ed. J.W. Teeter, p. 17-26, CCFL Bahamian Field Station.

Gebelein, C.D., 1976, Open marine subtidal and intertidal stromatolites, Florida, Bahamas, and Bermuda, in Stromatolites, ed., M.R. Walter, p. 381-388, Elsevier, Amsterdam.

Gleason, P.J., 1972, The origin, sedimentation and stratigraphy of a calcitic mud located in the southern freshwater Everglades, Thesis, Department of Geosciences, Pennsylvania University, University Park, Pennsylvania.

Logan, B.W., Hoffman, P., and Gebelein, C.D., 1974, Algal mats, cryptalgal fabrics and structures, Hamelin Pool, Western Australia, in Evolution and diagenesis of Quaternary carbonate sequences, Shark Bay, Western Australia Ed. Logan, B.W., p. 140-194, American Association of Petroleum Geologists Memoir 22.

Lowenstam, H.A., 1986, Mineralization processes in monerans and protoctists, in Biomineralization of Lower Plants and Animals, Eds. B.S.C. Leadbeater & R. Riding, p. 1-17, Oxford: Clarendon Press.

Mann, C.J. and Hoffman, L.R., 1984, Algal mounds in Storr's Lake, San Salvador, Bahamas, in Proceedings of the 2nd Symposium on the Geology of the Bahamas, ed. J.W. Teeter, p. 41-52, CCFL Bahamian Field Station.

Monty, C.L.V., 1972, Recent algal stromatolitic deposits, Andros Island, Bahamas, preliminary report, Geol. Rdsch. 61: P. 742-783.

Monty, C.L.V., 1976, The origin and development of cryptalgal fabrics, In Stromatolites, ed. M. R. Walter, p. 193-250, Elsevier, Amsterdam.

Monty, C.L.V. and Hardie, L.A., 1976, The geological significance of the freshwater blue-green algal calcareous marsh, in Stromatolites, Ed., M.R. Walter, p. 447-478, Elsevier, Amsterdam.

Pentecost, A. 1982, A quantitative study of calcareous and Tintenstriche algae from the Malham district, Northern England, Br. phycol. J. 17: p. 443-456.

Pentecost, A. & Riding, R., 1986, Calcification processes in cyanobacteria, in Biomineralization processes in Lower Plants and Animals, Eds. B.S.C. Leadbeater & Riding, p. 73-90, Clarendon Press, Oxford.

Shinn, E.A., Lloyd, R.M., and Ginsburg, R.N., 1969, Anatomy of a modern carbonate tidal flat, Andros Island, Bahamas, J. Sed. Petrol., 39: p. 1202-1228.

Thunmark, S., 1926, Bidrag till kannendomen om recenta Kalktuffer, Geol. För Stock. Förh., 48: p. 541-583.

DISCOVERY OF *PHORMIDIUM* STROMATOLITES IN GRAHAMS HARBOUR, SAN SALVADOR, BAHAMAS

Allan Pentecost
Division of Biosphere Sciences
King's College
Kensington Campus
Camden Hill, London W8 7AH

ABSTRACT

Stromatolitic buttons of the cyanobacterium *Phormidium hendersonii* have been found in Grahams Harbour, San Salvador. The buttons were attached to subtidal beachrock and measured 15-40 mm in diameter. Sagittal sections revealed fine daily laminae of entrapped carbonate sand with an estimated daily growth rate of 0.18 mm; lateral growth exceeded vertical growth by a factor of 4. The buttons had a low carbonate content (12 vol %) and were apparently short-lived.

INTRODUCTION

During July, 1987, orange discoid buttons were collected from subtidal beachrock opposite the CCFL Field Station in Grahams Harbour. Microscopic examination revealed *Phormidium hendersonii* Howe, a widely-distributed stromatolite-forming cyanobacterium of shallow Caribbean and subtropical Atlantic waters. The buttons were almost circular in plan, of diameter 15-40 mm, and maximum thickness 8-11 mm, with a diameter:thickness ratio of 3.7:1 indicating more rapid lateral than vertical growth. Young specimens were, however, more hemispherical in shape. The gelatinous stromatolites were found in 0-100 cm water at low tide and some were exposed briefly to the air, but were not observed to dry out. They covered about 5% of the beachrock below the intertidal zone of *Rivularia bullata* and may well occur in deeper water.

Sagittal sections revealed white laminae consisting of entrapped carbonate sand (Fig. 1) with 34-93 laminae in 10 mm (mean 56). Within these laminae the *Phormidium* trichomes were prostrate, but they were vertically arranged in the broader hyaline laminae which alternated with the white laminae. The form-birefringence of this *Phormidium* is high and the arrangement of trichomes in the layers is exceptionally clear in the polarizing microscope. *Phormidium hendersonii* is a motile, filamentous cyanobacterium with fine trichomes 1.0-1.7 μm wide (mean 1.26 μm, s.d. 0.3 μm, n=20). The cells are cylindrical and 2.7-11 μm in length (mean 6.06 μm, s.d. 2.7 μm) and match precisely the dimensions given for the species by Golubic & Focke (1978). However, samples taken from smaller, paler cushions were narrower with a mean diameter of 0.8 μm. A t-test indicated a highly significant difference ($p<0.001$) between the two samples suggesting more than one species might be involved although internally, the buttons appeared identical in form.

Golubic & Focke (1978) found that the laminae of this *Phormidium* are formed daily. According to their hypothesis, carbonate accretion in the white laminae occurs at night when the filaments assume a prostrate form. During the day, the filaments assume an approximately vertical position and grow as a result of photosynthesis thus 'diluting out' the rain of carbonate sediment which is continuously trapped onto the sticky sheaths surrounding the trichomes. This diel rhythm is clear in the Grahams Harbour

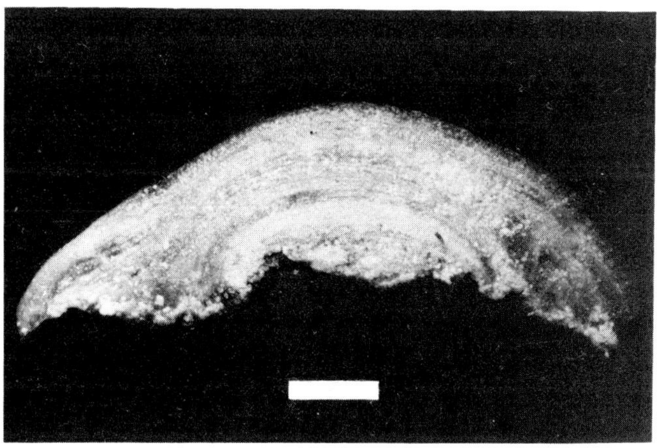

Fig. 1. Sagittal section across a Phormidium *button from Grahams Harbour showing the calcareous and hyaline daily laminae. Bar 5 mm.*

buttons although the lamina density is rather higher than the values previously reported. This higher density implies lower growth rates of around 180 μm per day, but the estimated age of mature buttons, averaging 46 days is comparable to the estimates made by Golubic & Focke (1978). The growth of the colonies is intriguing. Material collected in July, 1987, possessed very few living trichomes preventing identification, whilst samples collected in May appeared to be in a state of vigorous growth. Golubic & Focke (1978) noted that growth of colonies collected from Andros Island occurred between April and June whilst material from the Florida Keys grew from November to January. This apparently erratic seasonality has not been explained.

The trapped carbonate grains ranged from 10-100 (400) μm in diameter, but the buttons have never been reported lithified. With such rapid growth, the photosynthetic precipitation of carbonate within the colonies should be observable although it is not apparent from casual observation. The average calcium carbonate and organic matter content of the water-saturated colonies was 27 and 0.9wt% respectively. Small amounts of carbonate are also trapped and bound by *Rivularia bullata* and *Dichothrix seriata* further up shore, but daily laminae are absent in this material.

ACKNOWLEDGEMENTS

Dr. D. T. Gerace is thanked for providing the May material.

REFERENCES CITED

Golubic, S. & Focke, J. W., 1978, *Phormidium hendersonii* Howe: identity and significance of a modern stromatolite building microorganism J. Sed. Petrol. 48: p. 751-764.

PRELIMINARY INTERPRETATION OF A SEISMIC SECTION FROM THE NORTH END OF SAN SALVADOR, BAHAMAS

Rene Rodriguez, Frank R. Ettensohn, and Peter T. Goodmann
Department of Geological Sciences
University of Kentucky
Lexington, Kentucky 40506

ABSTRACT

A nearly north-south, 100-meter experimental seismic line was run on the north end of San Salvador using an analog seismic-enhancement seismograph. Over a depth of 130m from the surface, eight apparently good reflectors and a few less definite ones were interpreted. Most of these surfaces can be correlated with exposure horizons or major facies changes observed from the Supko core, less that 2.0km to the west.

The apparent geometry of these surfaces suggests beach-ridge complexes. If this interpreation is correct, most surfaces represent maximum beach-ridge development or offlap, followed by a period of sea-level drop, exposure and erosion, and succeded then by a period of sea-level rise and transgression or onlap. The lower six surfaces show a progressive seaward migration of beach-ridge complexes through time except for an apparent transgressive interval represented by the 45-57m surface. Overall these six surfaces suggest a net lowering of sea level and regression. The uppermost of these surfaces, at a depth of 27 to 41m is extremely prominent and apparently represents a major period of sea-level fall, regression and exposure.

Two overlying surfaces suggest beach-ridge migration of approximately 40-60m shoreward of those underlying the prominent surface at 2-41m, suggesting that major sea-level rise and transgression followed the formation of this surface.

Subsequently, however, sea-level rise and transgression apparently resumed, for a reef complex developed just seaward of the last beach-ridge complex and migrated shoreward to a position atop the beach ridge where it is exposed (Mosquito Marsh reef) along our section. What we have described above are net sea-level rise and fall episodes; each may be represented by several interpreted surfaces. Our data are so preliminary that possible time constraints must be viewed with caution. Nonetheless, possible correlations with the magnetostratigraphic dating from the Supko core, suggest that the 27 to 41m surface reflects a major period of late Pliocene sea-level decline. Thus, most of the overlying sediments, as well as five included reflecting surfaces, apparently represent Pleistocene or Holocene events.

INTRODUCTION

Establishing a stratigraphic framework in the Quaternary sediments of San Salvador has been rather difficult because of the plethora of subtidal, intertidal and supratidal facies, the similarity of any respective carbonate facies in the geologic record despite age, and the effects of multiple erosional events during late Tertiary and Quaternary sea-level fluctuations. Marker beds are essentially absent because of the facies mosaics and erosion, whereas the relatively short time spans involved and abrupt marine-to-non-marine facies transitions prevent the use of fossils for correlation. Moreover, the relative youth of the sediments, the geochemical mobility of carbonate and organic matter in the present climatic and geologic setting, as well as expense, obviate the large-scale use of absolute dating techniques. Although correlation using unconformities has been attempted, in situations where even ephemeral exposure may produce seemingly major unconformities rapidly, such correlations must remain uncertain.

In view of the problems in trying to establish a stratigraphic framework for the island by traditional means, we decided to experiment with the possibility of using seismic stratigraphy. Our short seismic line shows the type of resolution to be expected as well as some of the necessary constraints and some of the possible ways of interpreting the results.

METHODS

We chose the location of the seismic line

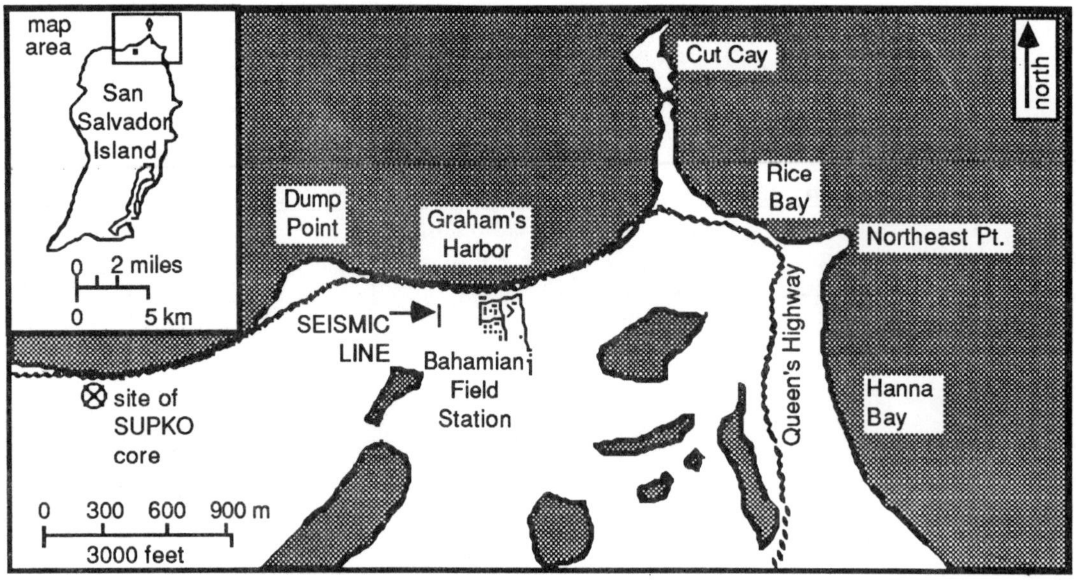

Fig. 1. Location map showing the position of the seismic line relative to the site of the Supko core.

(Fig. 1) to satisfy three criteria: 1) it was close to a core hole (Supko, 1970, 1977) where correlation with known lithologies and surfaces might be possible (Fig. 2); 2) it was in a near-shore area where beach-ridge facies were likely to predominate; and 3) it crossed the outcrop of a unique lithofacies, in this case, a Pleistocene patch reef (Mosquito Marsh reef; see Vierma et al., 1984).

The experimental seismic reflection line was 100m long and oriented N5°W; it is located approximately 250m west of the CCFL Field Station (Fig. 1). Reflections were recorded analogically on an OYO-NIMBUS 1220 seismic signal enhancement seismograph at a rate of 10,000 points per second. Single vertical geophones with a 14 Hz natural frequency were spaced at 10m intervals along the line. Five shots, at geophones 1,4,6,8, and 11, were set off separately in order to stack data using common depth points. The reflections were recorded analogically on light-sensitive Lynagraph paper as shown in Fig. 3. The seismograms obtained were analyzed for reflection arrivals from each geophone for each shot and by plotting the square of the arrival times (t^2 in msec **2). The best straight-line fit of the resulting data points indicates the different reflectors and their slope, as well as the RMS velocity for layers above the reflector. Fig. 4 shows an example of the analysis. The reflection times selected in this way were used to transform the two-way vertical reflections (TWVR) below each geophone by correcting for normal move out and migration. Initial statics corrections were made using refraction velocities from the overlying layers. The TWVR were used to produce the seismic section shown in Fig. 5. The resulting seismic section was tested by comparison with nearby exposures and with a nearby core in order to check for correspondence between reflectors and major geological discontinuities.

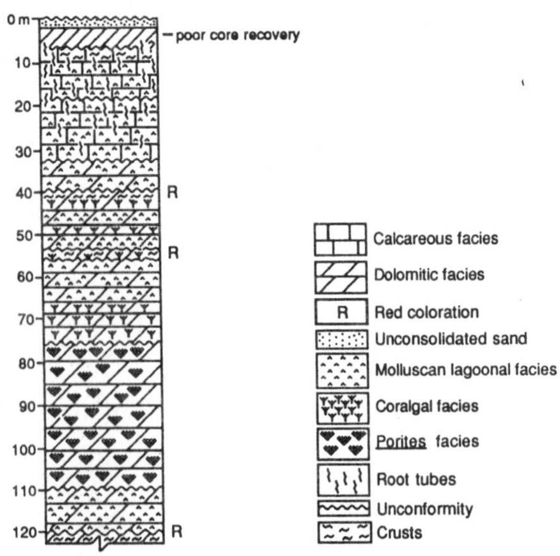

Fig. 2. Schematic stratigraphic column from the Supko core showing the vertical distribution of major facies, sedimentary features and unconformities.

BASES FOR INTERPRETATION

Based on correlations with the Supko core

BAHAMAS RETRACED ORIGINAL RECORDS

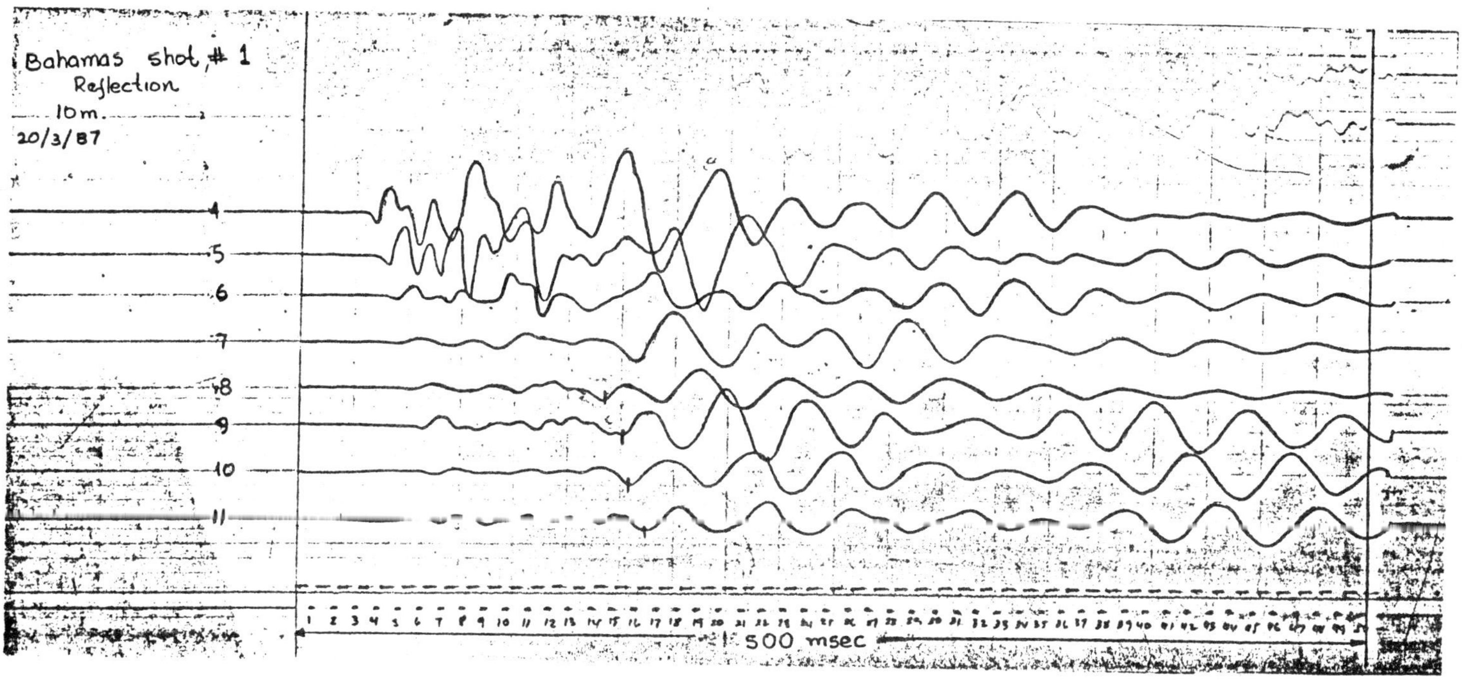

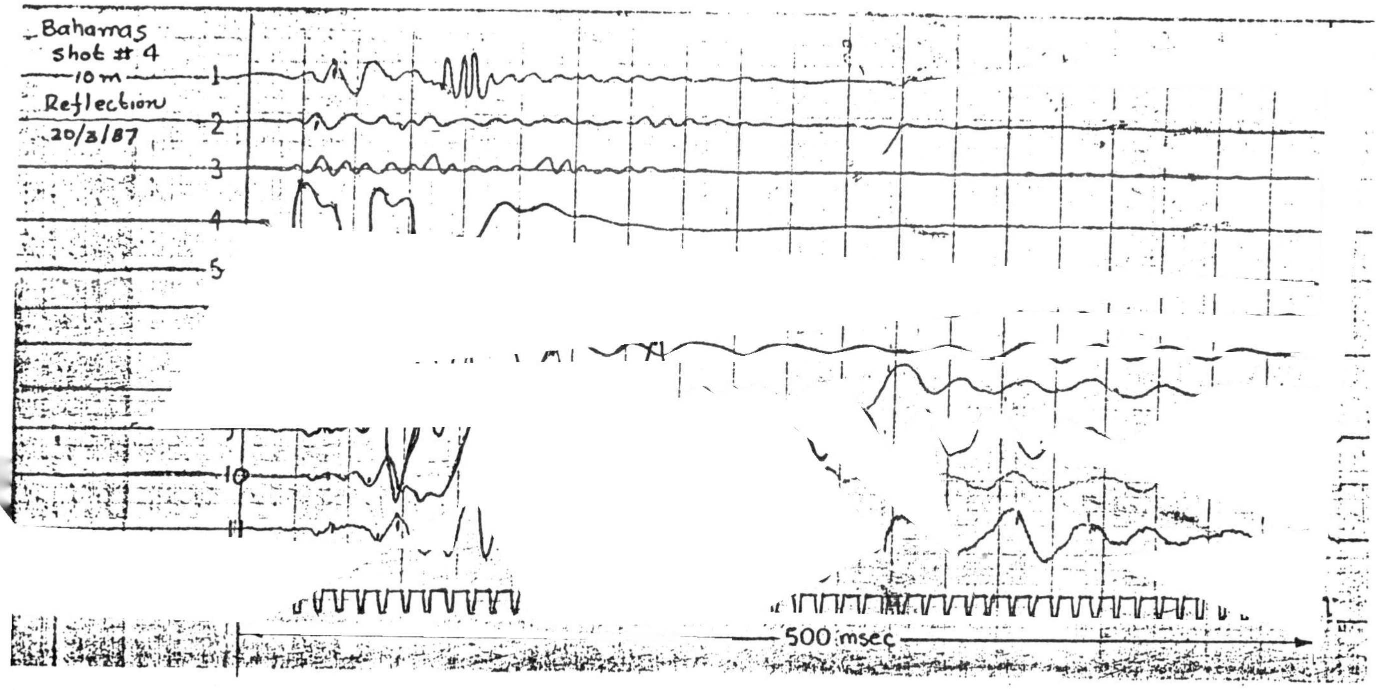

Fig. 3. Examples of two retraced analog seismographs recorded on light-sensitive Lynagraph paper. Individual peaks and troughs were picked and x^2 vs t^2 analysis performed to detect all possible reflectors.

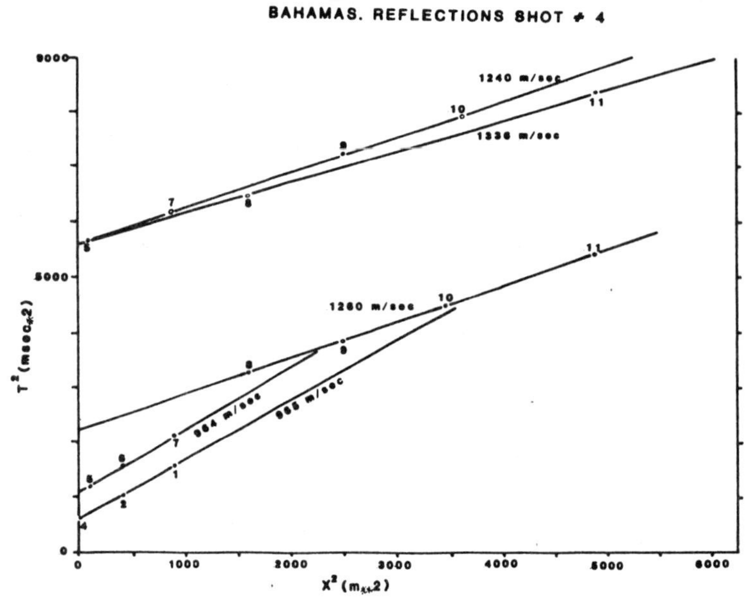

Fig. 4. Example of the straight-line fits resulting from shot #4, showing reflectors, their slopes, and RMS velocities.

(Supko, 1970) (Fig. 2), most of the reflection surfaces shown in Fig. 5 represent discontinuities or unconformities. The surfaces apparently contrast with adjacent lithologies because of thin, but dense, accumulations of micritic, pedogenic carbonates (see Brown, 1984, Hale and Ettensohn, 1984). Each such surface apparently represents regression of seas and exposure, followed at a later time by transgression. In a few other instances, however, the reflections are not so much related to indurated erosion surfaces as they are to contrasting lithosomes. The best example of this is the contrast between the Pleistocene patch reef and the encompassing

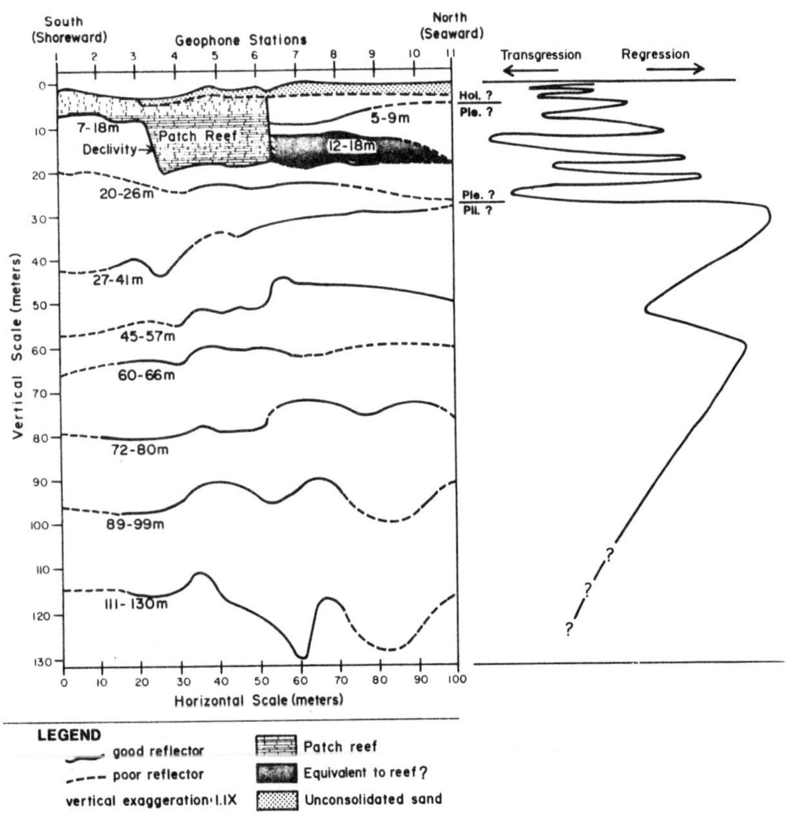

Fig. 5. Seismic section showing major reflectors to a depth of 130m. Curve on the right is interpreted and shows the probable sequence of transgressions and regressions based on coastal onlap and offlap indicated by surface profiles. Parts of the curve below the probable Pliocene-Pleistocene boundary are highly generalized and do not show smaller transgressions or regressions associated with individual surfaces. Numbers below each surface reflect the depth in meters below the surface for each reflector. Hol.=Holocene; Ple.=Pleistocene; and Pli.=Pliocene.

sediments from various beach-ridge complexes. Fortunately, the reef was exposed so that its origin was easily determined (Fig. 5).

Aside from correlating surfaces in the section with surfaces in the nearby core (Fig. 2), interpretations of net sea-level rise and fall were based on the geometry, or shape and slope of the individual surfaces and on the assumption that shoreward-or seaward-dipping parts of successive surfaces reflect parts of a beach-ridge profile (Fig. 6) which shifted either shoreward or sea ward with time. What we are effectively examining in the relative positions of seaward-versus-shoreward parts of the surfaces, are patterns of coastal onlap and offlap, or transgression and regression respectively. Although each surface apparently represents a regressive transgressive event, a net seaward shift of successive surfaces is assumed to indicate a larger period of overall regression or sea-level fall; an apparent shoreward shift of successive surfaces, in contrast is taken to represent a period of net transgression and sea-level rise. The interpreted pattern of transgression and regression suggested by the apparent migration of the surfaces is shown on the right of Fig. 5. Of course, in light of late Tertiary and Quaternary glaciation, we have assumed that transgressions and regressions are products of glacio-eustatic sea-level rises and falls respectively, but subsidence and uplift could have produced similar results.

SURFACE INTERPRETATION

The deepest reflector observed on our seismic line was an irregular surface occurring at 111-130m. This is probably the same surface noted by Supko (1970) at about 121m (397 ft). Supko indicated that the surface was an unconformity characterized by aphanitic, reddish crusts. The dense, aphanitic crusts apparently indicate pedogenic calcretes or subaerial exposure crusts formed during a time of exposure. Up to 19m of relief is present on this surface in our section, and at least three isolated ridges are apparent (Fig. 5) suggesting extensive erosion of accreted beach ridges or accretion of beach ridges during a series of sea-level drops.

The surface at 89-99m apparently is not represented in the Supko core unless it is equivalent to the unconformity at 110m (Fig. 2). The same depth interval in the Supko core is represented by a shallow subtidal *Porites* facies present throughout an interval over 30m thick. The predominance of this facies over that vertcal distance must indicate that at this time the area of the Supko core was seaward of any beach-ridge complex. In contrast, the surface at 72-80m suggests a seaward shift of the beach-ridge facies and is represented by a possible hiatus and an abrupt change from shallow subtidal to coralgal lagoonal facies in the Supko core. The hiatus, and the change to a more restricted, shallow-water facies may reflect the same regressive event indicated by the apprent seaward shift of beach-ridge facies in the section.

The surface at 60-66m is a very poorly defined reflector. The shoreward dip and low relief on the surface suggest a back-beach-ridge setting and that regression has moved the beach farther seaward beyond the seismic section. No equivalent surface is present in the Supko core, but the interval in the core is represented by a change from a coralgal lagoonal facies to an even more restricted, and possibly shallower, mollusc-algae-dominated lagoonal facies.

The geometry of the overlying 45-57m surface indicates that the surface was the product of a period of transgression, because a seaward-dipping, shoreface-foreshore surface appears to be present (Fig. 5). The same interval in the Supko core is represented by a change to a slightly deeper water facies (molluscan to coralgal), which is truncated by an apparent unconformity with reddish subaerial exposure crusts (Fig. 2). The sequence of events reflected in the core is essentially the same as indicated by the succession of surfaces in the seismic line at this

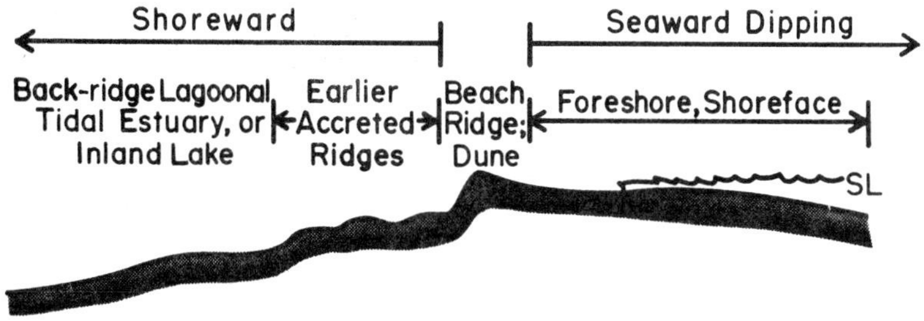

Fig. 6. Generalized beach-ridge profile based on the surface at 45-57m in Figure 5 used to interpret onlap versus offlap relations for successive surfaces. SL=sea-level.

point.

The reflecting surface at 27-41m is one of the most prominent along the line, and its shoreward dip suggests the back side of the beach-ridge complex. If this is the case, then the beach ridge and more marine shoreface environments were much farther seaward compared to the underlying surface, indicating that the 27-41m surface was the product of regression or major sea-level fall. The same vertical interval in the Supko core is very complex. The upper 8m (26-34m) of this core interval exhibits extensive soil features, breccias, and solution pipes and represents a major period of exposure. The middle 6m (34-40m) consists of restrictive lagoonal, intertidal and supratidal facies unconformably overlying a thin, reddish crust zone. Although the core indicates that this regression was somewhat more complicated that the single reflecting surface would suggest, both lines of evidence indicate that this was a major episode of sea-level fall.

The shape and profile position of two overlying seaward-dipping surfaces (20-26m and 7-18m) suggest that subsequent beach ridges migrated 40 to 60m shoreward of their probable position on the 27-41m surface, indicating net sea-level rise and transgression. The lower 20-26m surface is probably represented by an unconformity and soil zone at 18m (60ft.) in the Supko core (Fig. 2). The upper surface (7-18m) exhibits a prominent seaward facing declivity near the southern end of the seismic line (Fig. 5) and seems to reflect seaward progradation of a beach ridge, and hence sea-level fall compared to the underlying surface. However, the declivity and the relatively level seaward part of the surface (at 18m) suggests wave erosion and formation of a wave-cut bench during a succeeding period of sea-level rise and transgression. The modification of this surface by later transgression and erosion makes it difficult to relate this surface to one in the Supko core; the unconformity at 7m (22ft.) in the core (Fig. 2) is a possible equivalent.

The occurrence of sea-level rise at this time is also supported by the presence of a patch reef in front of the declivity (Fig. 5). Although the reef is exposed at the southern end of the seismic line (Fig. 5), its presence seaward of the declivity is indicated by geometry and seismic velocities identical to those for exposed parts of the reef. The apparent declivity may have been part of an older beach ridge modified by wave erosion into a promontory or point, inasmuch as patch reefs will preferentially develop just seaward of such points because wave energy is concentrated here. Nonetheless, as sea-level rise continued, the reef apparently migrated shoreward to a position atop the beach ridge where it is exposed (Mosquito Marsh reef, Vierma et al., 1984) along our section.

The remaining two surfaces (12-18m and 5-9m) are present only seaward of the patch reef and are rather poorly developed. If these are unconformities, the fact that they apparently do not extend into the reef suggests that they developed after reef formation. The 12-18m surface probably developed atop a package of subtidal sediment that was deposited simultaneously with reef formation. Because both surfaces are lower than the top of the reef, reef formation must have been succeeded by a period of sea-level fall of at least 9m, the vertical distance to the lower of the two surfaces. The two surfaces then may represnt periods of sea-level fall and subaerial exposure during Pleistocene sea-level fluctuations. The upper 5-9m surface is at approximately the same elevation as is the unconformity in the "Dump", Barkers Point and Rocky Point. If this is the same unconformity, the 2-6m of limestone above the surface may be the same as what has been called the "Grahams Harbour Formation" (Titus, 1980, 1981, 1983, 1984) or the Rice Bay Formation (Carew and Mylroie, 1985).

One to three meters of unconsolidated carbonate sands veneer these rocks along most of the seismic line and at the Supko core site.

AGE

No reasonable way exists to develop time constraints for the rocks along the seismic line from our data alone. However, possible correlations with the magnetostratigraphic dating from the Supko core (McNeill et al., 1988) suggest that the Pliocene-Pleistocene boundary interval can be approximated. Magnetostratigraphic dating of the Supko core indicates that this boundary occurs at a depth of approximately 25m. If any kind of elevational equivalence is present between the Supko core and our seismic line, then the Pliocene-Pleistocene boundary interval should occur near 25m in depth. More specifically, in our section we believe that the boundary occurs at approximately 27m in depth or along the 27-41m surface, because according to Harland et al. (1982, Fig. 5.1), the Pliocene ended with a major period of coastal offlap or regression, and the Pleistocene began with a major episode of coastal onlap or transgression, which is precisely what the surface at 27-41m on our seismic line indicates. Hence, we believe that the partial coincidence of elevations, as well as the fact that this surface apparently corresponds to a major period of sea-level fall or regression, approximately locates the Pliocene-Pleistocene boundary on our line. Moreover, in the presence of many transgressive-regressive fluctuations above the Pliocene-Pleistocene boundary and in the general transgression or onlapping apparent in the curve below the boundary, our transgressive-regressive curve (Fig. 5) is very similar to the sea-level curves or Harland et al. (1982, Fig. 5.1). If this is the case, then all rocks lower than the 27-41m surface are probably Pliocene in age. McNeill et al. (1988) only reported on rocks to a depth of 90m in the Supko core, but by this depth they were already well into the Lower Pliocene rocks.

As for other ages, Vierma et al. (1984) attributed the patch reef along our line to a Sangamonian high stand of sea level based on amino-acid racemization analysis. Furthermore, if the surface at 5-9m is the same as the unconformity at many of the nearby points (Dump, Barkers and Rocky points), then it may well represent the Pleistocene-Holocene boundary as suggested by Titus (1984) and Carew and Mylroie (1985).

CONCLUSIONS

The present study suggests that it is possible to determine the presence and configuration of reflecting surfaces in late Tertiary and Quaternary carbonates on San Salvador using very simple seismic reflection techniques. The presence of nearby exposure or cores, however, is essential in interpreting such a seismic line. If it is assumed that the configuration of the surfaces approximates some part of the depositional profile of a beach-ridge complex, then it is possible to delimit episodes of coastal onlap and offlap, or transgression and regression respectively, by observing the seaward or shoreward migration of various parts of the surfaces, even on short seismic lines like ours.

Where some surface control is present, these techniques and assumptions may be useful in explicating some of the stratigraphic complexities on San Salvador.

ACKNOWLEDGEMENTS

We wish to thank Don and Kathy Gerace who helped us acquire the explosives and the personnel necessary to set off our shots. The Department of Geological Sciences, University of Kentucky, supported part of this research. Beth M. Ettensohn and Margie Palmer helped with manuscript preparation.

REFERENCES CITED

Brown, T.W., 1984, Formation and development of caliche profiles in eolian deposits: San Salvador, the Bahamas, in Teeter, J.W. (ed.), Proceedings of the Second Symposium on the Geology of the Bahamas, CCFL Bahamian Field Station, p. 245-264.

Carew, J.L., and Mylroie, J.E., 1985, The Pleistocene and Holocene Stratigraphy of San Salvador Island, Bahamas, with reference to marine and terrestrial lithofacies at French Bay, in Curran, H.A. (ed.), Pleistocene and Holocene carbonate environments on San Salvador Island, Bahamas-Guidebook for Geological Society of America, Orlando Annual Meeting Field Trip: Ft. Lauderdale, Florida, CCFL Bahamian Field Station, p. 11-61.

Hale, A.P., and Ettensohn, F.R., 1984, Micromorphological features observed in pedogenic carbonates (caliche) on San Salvador Island, Bahamas, in Teeter, J.W., (ed.), Proceedings of the Second Symposium on the Geology of the Bahamas, CCFL Bahamian Field Station, p. 265-277.

Harland, W.B., Cox, A.V., Llewellyn, P.G., Pickton, C.A.G., Smith, A.G., and Walters, R., 1982, A geologic time scale: Cambridge, Cambridge University Press, 131 p.

NcNiell, D.F., Ginsburg, R.N., Chang, S.B.R., and Kirkschvink, J.L., 1988, Magnetostratigraphic dating of shallow-water carbonates from San Salvador, Bahamas: Geology, v. 16, p. 8-12.

Supko, P.R., 1970, Depositional and diagenetic features in subsurface Bahamian rocks (Ph.D. dissertation): Coral Gables, The University of Miami, 168 p.

Supko, P.R., 1977, Subsurface dolomites, San

Salvador, Bahamas: Journal of Sedimentary Petrology, v. 47, p. 1063-1077.

Titus, R., 1980, Emergent facies patterns on San Salvador Island, Bahamas, in Gerace, D.T. (ed.) Field guide to the geology of San Salvador (First Edition): CCFL Bahamian Field Station, p. 92-105.

Titus, R., 1981, Emergent facies patterns on San Salvador Island, Bahamas, in Gerace, D.T. (ed.) Field guide to the geology of San Salvador (Second Edition): CCFL Bahamian Field Station, p. 92-105.

Titus, R., 1983, Emergent facies patterns on San Salvador Island, Bahamas, in Gerace, D.T. (ed.) Field guide to the geology of San Salvador (Third Edition): CCFL Bahamian Field Station, p. 97-116.

Titus, R. 1984, Physical Stratigraphy of San Salvador Island, Bahamas, in Teeter, J.W. (ed.) Proceedings of the Second Symposium on the Geology of the Bahamas: CCFL Bahamian Field Station, p. 209-228.

Vierma, L., Kwolek, J., Heidt, D.A., Hattin D.E., Hasenmueller, W.A., and Feldman, H.R., 1985, Stratigraphic analysis of a newly discovered Pleistocene reef, San Salvador Island, Bahamas: The Compass, v. 62, p. 16-30.

CONTROLS ON THE RATE AND DISTRIBUTION OF CARBONATE BEDROCK DISSOLUTION IN THE BAHAMAS

Peter L. Smart and Fiona F. Whitaker
Department of Geography
University of Bristol
Bristol BS8 1SS
England

ABSTRACT

Three 'water-controlled' reactions are of importance for bedrock dissolution in the Bahamas: generation of carbonic acid in the soil, *in situ* oxidation of organic matter and mixing of water of different composition. All share certain common controls, which depend primarily on climate, and specifically on the balance between actual evapotranspiration and rainfall which directly controls groundwater flux. However this balance is also critically dependent on relief, because at high sea-levels the saturated zone is not decoupled from the effects of evaporation. Under arid conditions caliche crusts may develop, further reducing net groundwater recharge, but also causing it's concentration at breaks in the crust. Where mineral soils are not present a similar increase in concentrated recharge with time may result from the strong positive feedbacks between micro-topography, organic soil accumulation and soil CO_2, involved in the formation of solution pockets (banana holes) and shafts. In the saturated zone mixing at the water table and transition zone generates further dissolution, which is enhanced by oxidation of organic matter introduced in fissures. Major fracture systems and pre-existing cavernous horizons exert a strong control on the thickness of the fresh water lens and the degree of mixing at the transition zone. There is thus a change with time in the distribution of dissolution from a shallow, relatively homogeneous distribution to the deeper, heterogeneous pattern characteristic of karst terrains.

INTRODUCTION

Given the importance of the Bahamas as a natural laboratory for the study of carbonate sedimentation and diagenesis, it is surprising that relatively little work has been undertaken on the distribution and magnitude of bedrock dissolution. Such studies are needed both to understand the evolution of landforms in the area, and to provide models of the extent and distribution of secondary porosity within the platform (Craig 1988).

In this paper we present a brief review of some of the factors controlling bedrock dissolution based on results of field work on Andros Island and San Salvador. Emphasis is placed on 'water-controlled' dissolution reactions (James and Choquette 1984), which in the Bahamas include dissolution by carbonic acid derived from the soil atmosphere, dissolution due to mixing of waters of differing chemical composition, and dissolution due to *in situ* microbial oxidation of organic matter. Only brief attention is focused on 'mineral-controlled' reactions which are important in recent carbonate sediments, and involve the conversion of metastable high-magnesium calcite and aragonite to low-magnesium calcite (Walter 1985). Although this process gives rise to a change in the size distribution of porosity, creating larger vuggy pores with a heterogeneous distribution in the unsaturated zone (Harrison 1975), recent studies on Schooner Cays in the Bahamas (Budd 1988) suggest that it is a conservative process, with little loss of carbonate (overall efficiency c. 90%).

For both stable and unstable carbonate minerals the rate of bedrock dissolution is predominantly controlled by two sets of factors: 1. those controlling the inorganic and organic processes generating undersaturation with respect to calcite - the chemical potential - and 2. those controlling the supply and removal of reactants and products - the groundwater flow regime. Information on the former may be readily obtained by sampling and analysis of surface and groundwaters, but it is much more difficult to convert these concentrations to rates of dissolution because of the difficulties of measuring groundwater flux.

1. CLIMATE - PRECIPITATION

At present the major climatic gradient in the Bahamas is between the wetter, cooler north-western islands and the drier, warmer south-eastern islands. The contrast in temperatures (mean annual temperature 24-27°C) is however much smaller than that in rainfall (mean annual rainfall 1550-690mm). It is therefore the latter which is most important in controlling both spatial differences in present bedrock dissolution, and it's temporal variation with paleoclimate. The balance between annual rainfall and annual actual evaporation (currently about 1150mm), has an important control on soil type and on the magnitude and form of the freshwater lens. These both affect dissolution and are discussed further below. However, effective precipitation is the major direct control on the rate of carbonate dissolution because of it's effect on the net groundwater flux. This is demonstrated by the strong positive relationship between net denudation rates and the mean annual runoff for karst areas, as shown in Fig. 1 (derived from the literature by Smith and Atkinson 1976). This predominant effect is also clear from the theoretical model developed by White (1984).

Supporting data for the Bahamas are not available, but Pierson and Shinn (1985) report preservation of high-magnesium, calcite and aragonite cements in sediments greater than 300 ka in age at Hogsty Reef in the dryer southern Bahamas. At the much wetter Joulters Cay, to the north of Andros Island, more rapid leaching occurs and stabilization would be complete within 10 to 20ka (Halley and Harris 1979). Note however that the latter sediments have not been exposed to freshwater and Budd (1988) reports rates of stabilization between two and three times lower in the saline transition zone than in fresh water. Furthermore, mixed mineralogy sediments such as these transform more slowly than wholly aragonitic sediments (Halley and Harris 1979).

2. ORGANIC SOILS - DEPTH AND DISTRIBUTION

The presence of a soil cover has a significant effect upon the rate of denudation, this being higher for soil covered than for bare terrain (Fig. 1). True mineral soils are rare in the Bahamas at present except in areas improved for agriculture. Most soils comprise a thin litter layer underlain by an Ah horizon of decomposing humus, often with many roots. This may continue down into a rubbly C hoizon with corroded limestone blocks, or have a sharp basal transition to the limestone. Soil depth is very variable with the thicker soils accumulating in lows on the bedrock surface, such as topographic basins, pockets and joints. In the case of the larger basins, laterally continuous soils may develop, and often a discontinuous red clay-loam mineral horizon is observed above the bedrock surface.

Carbonic acid, formed by dissolution of carbon dioxide generated by microorganisms and root respiration in the soil, provides the major chemical potential driving dissolution in most karst areas. The PCO_2 of soil gas is controlled both by the rate of CO_2 production (discussed below), and the rate of removal by diffusion to the soil surface. Soil PCO_2 therefore increases with depth and with the bulk density of the soil matrix. This is shown for organic soils from North Andros Island in Fig. 2. The humified organic soils have higher CO_2 concentrations than those comprising only highly porous undecomposed litter, but for both types PCO_2 increases with depth.

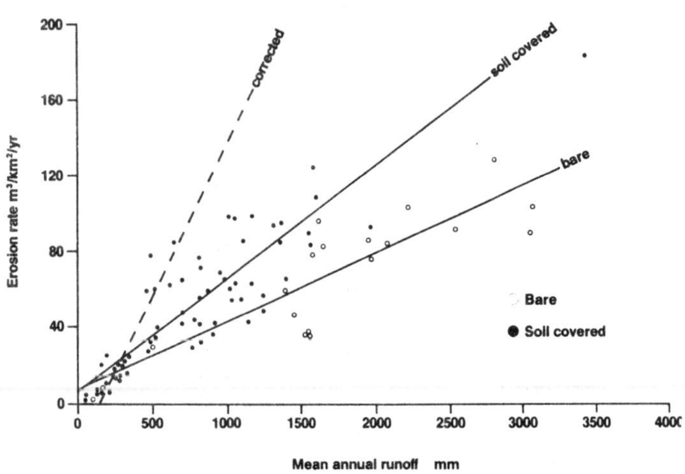

Fig. 1. Denudation rates for soil covered and bare limestone terrains (Smith and Atkinson 1976). Note the change to the observed relationship if solution rate is corrected for the presence of non-carbonate rocks in the catchment, the relationship which would apply in the Bahamas.

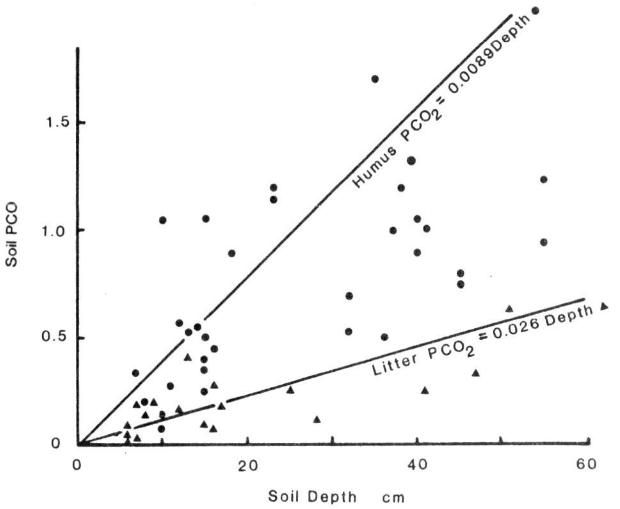

Fig. 2. Relationship between soil depth and PCO_2 measured during the wet season (July) on North Andros using a Miotke soil CO_2 probe with Gastec gas analysis tubes. Quoted regression lines are constrained to pass through the origin, and give R^2 values of 0.84 (humus) and 0.76 (litter).

In the Bahamas decomposition of organic matter by soil microorganisms is limited not by temperature but by soil moisture. During dry periods the litter layer and thinner organic soils become desiccated, and only the deeper soils retain sufficient moisture for decomposition to continue. An elevated PCO_2 is therefore maintained for longer in deeper sites. Furthermore, erosion of the desiccated soils by rainsplash and runoff along the bedrock surface occurs, emphasizing the contrast in soil thickness between 'accumulating' and 'shedding' sites. Observations on North Andros indicate that much of the thin organic mat is lost from the bedrock surface after fires, and only that in deeper moister pockets survives. These sites also accumulate ash and mineral particles washed down from the bare bedrock, providing a possible source for the mineral material observed at the base of these profiles.

Thus the direct effect of soil depth on PCO_2, the persistance of decomposition at deeper, moister sites, the tendency for organic material to erode from micro-topographic highs and accumulate in lows, and for vegetation to preferentially colonize these accumulating sites, gives a series of positive feedback mechanisms, developing and amplifying topographic lows in the bedrock surface (Fig. 3A). This gives rise to the development of 'banana holes' (Fig. 3B), soil filled pockets. These are initially small with a diameter of 0.2-1.0m and depth of 0.2-0.5m, but with continuing bedrock dissolution, increase in size to 1.0-2.0m diameter and 0.5-2.0 m depth. With development, capture of adjacent shallower pockets may occur, increasing the catchment of the main pockets where soil becomes concentrated. Penetration of tree roots into cracks and fissures at the base of the pocket encourages dissolution

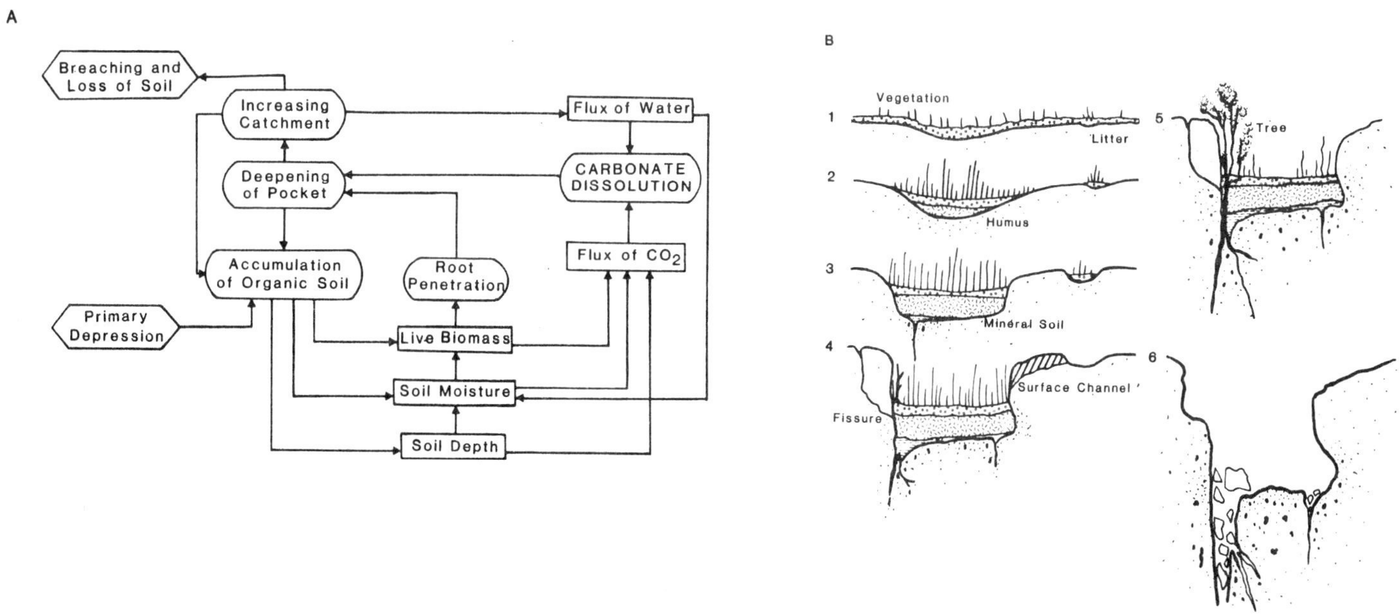

Fig. 3. (A) Process response diagram illustrating strong positive feedback mechanisms in the development of solutional pockets (banana holes). (B) Schematic diagram showing evolution of a solution pocket (banana hole). Density of stipple is proportional to vuggy and mouldic secondary porosity in the limestone. Note the important role of penetration of tree roots towards the water table in draining the pocket, eventually leading to breaching and soil loss.

of a drain, which discharges aggressive soil water from the pocket. This concentrated input rapidly dissolves fissures in the unsaturated zone, so that soil will eventually be lost downwards, and the development of the pocket by sub-soil dissolution will cease.

3. MINERAL SOILS - EVIDENCE FROM PALEOSOLS

There is considerable evidence that mineral soils derived from the accumulation of atmospheric dust (Glaccum and Prospero 1980) were widespread in the Bahamas in the past (Foos 1987, McCartney and Boardman 1987, Carew and Mylroie 1989). Although the preservation of these soils is often poor, they may be up to 1m deep, and unlike present organic soils were laterally continuous, as demonstrated by the paleosol at Crab Cay, San Salvador (Brown 1984). The paleosols are predominantly of the terra rossa type, indicating leaching of silica and accumulation of iron and aluminum rich residues. A second type of paleosol has also been reported from Singer Bar Point, San Salvador (Foos 1987, McCartney and Boardman 1987). In this the presence of the mineral boehmite is indicative of even more intense leaching under a more humid climate, giving the development of a latosol.

The high bulk density of these soils would give a much higher soil PCO_2 than the present organic soils; the relation of Brook et al (1983) predicting a mean growing season soil PCO_2 of 1.78 % compared to the mean of 0.74 ± 0.37 % for the soils sampled on North Andros. Thus the combination of more intense leaching, high soil PCO_2 and a continuous soil cover would be expected to generate both more uniform and more intense sub-soil dissolution than occurs at present.

Preservation of the sub-humid terra rossa soils is often due to subsequent cementation by caliche (Foos 1987), indicating a change in the soil-water balance to much drier semi-arid or arid conditions, with evapotranspiration exceeding rainfall. Accumulation of pedogenic calcium carbonate results in sealing of the bedrock surface by development of laminated micrite crusts. Thus, even during periods with a positive water balance, recharge is restricted and the groundwater flow is substantially reduced (Brown 1984). In the Yucatan Ward (1973) reported preservation of high-magnesium calcite in Pleistocene aeolliantes 25-30ka old, which are overlain by a caliche crust. In contrast, on Holocene deposits where a crust is absent, stabilization will occur in 20ka.

Development of a protective caliche crust under arid conditions will thus have a marked effect on bedrock dissolution rates, augmenting the direct reduction due to a decreased groundwater flux.

4. HYDROLOGICAL ROUTING IN THE UNSATURATED ZONE

The primary intergranular permeability of the relatively coarse grained aeolianites found on many Bahamian islands is very high, although in finer grained deposits it is much lower (Enos and Sawatsky 1981). Thus on initial exposure, recharge is via diffuse routes with the development of vuggy and mouldic dissolutional porosity, which may be locally concentrated (Harrison 1975). Under these conditions bedrock dissolution will be limited to a shallow irregular zone immediately below the developing soil. There will be little or no dissolution at depth as recharge waters rapidly reach equilibrium, and there is no renewal of their dissolutional potential. The dominance of diffuse recharge and low runoff will allow mineral soils to accumulate providing there is a sufficient supply of atmospheric dust, as has occurred on Bermuda (Ruhe et al 1961).

With time, recharge will switch to more concentrated routes, with cementation and case-hardening of exposed carbonates shedding water laterally into pockets and fissures (Fig. 3B). Even where mineral soils are present the development of an impermeable caliche crust may result in flow concentration. The role of tree roots may be significant in both these cases. In the latter, disruption and brecciation of the micrite crust by penetration of roots provides concentrated outlets for locally perched soil waters, with the development of rhizomorphs into the underlying porous carbonates (Brown 1984). In the former, direct bedrock dissolution and the mechanical action of roots growing downward toward the saturated zone may give rise to discrete continuous channels on root decay, which bypass the diffuse-flow part of the unsaturated zone. Such concentrated recharge routes are characteristic of mature karst terrains, and are best seen where there is a thick unsaturated zone, as in the Sandy Point area of San Salvador. Here Mylroie (1988) has mapped the resulting shaft complexes, and the dominant role of shallow tributary runnels and shallow sub-surface fissures is clearly apparent.

Concentrated recharge routes, once established, will tend to be self-perpetuating and will give a dramatic change in the distribution of bedrock dissolution. Rapid recharge will allow chemically aggressive water to penetrate the unsaturated zone, resulting in a very heterogeneous development of secondary porosity to a considerable depth. Rapid recharge may also result in mixing corrosion where chemical contrasts exist between the recharge waters and long residence evolved groundwaters of the saturated zone.

Finally transport of particulate organic matter into the upper part of the saturated zone via fissure flow may occur. Here aerobic decomposition generates additional chemical potential for dissolution. A plot of the calcium concentration and PCO_2 of present day fresh groundwaters from North and South Andros, shows a parallel increase in PCO_2 and calcium concentration as water passes from the surface to the saturated zone (Fig. 4). This suggests open system dissolution with a PCO_2 similar to that measured in the organic soils. In practice however recharge waters are actually saturated, and two of the fresh water samples have a PCO_2 substantially higher than the maximum measured in the organic soils. The clear conclusion is therefore that carbon dioxide is generated by microbial decomposition of organic matter within the fresh water lens.

5. DEPTH AND DISTRIBUTION OF THE TRANSITION ZONE

In carbonate islands, such as the Bahamas, the contrast in density between fresh and saline waters results in the development of a lens-shaped body of fresh water underlain by saline water. The interface between the two waters is known as the transition zone. Residence times within the fresh water lens (Vacher and Bengtsson 1989) are long compared with times to reach chemical equilibrium, and bedrock dissolution is therefore limited unless chemical potential is renewed. However in the transition zone considerable potential for dissolution may be generated by mixing of waters of different salinity, demonstrated theoretically by Plummer (1975).

Mixing in the transition zone is now widely recognised as a potent process for carbonate dissolution. Back and others (1979) have suggested that the major coastal embayments and underwater cavern systems of the Yucatan Peninsula were formed by this process. Mylroie (1988) has advanced similar arguments to explain the distribution of subaerial caves on San Salvador, which are developed along the flanks of exposed dunes, and stressed the great rapidity of the process. On South Andros Island pervasive dissolution of bedrock in the transition zone has been observed in the walls of blue holes (Smart

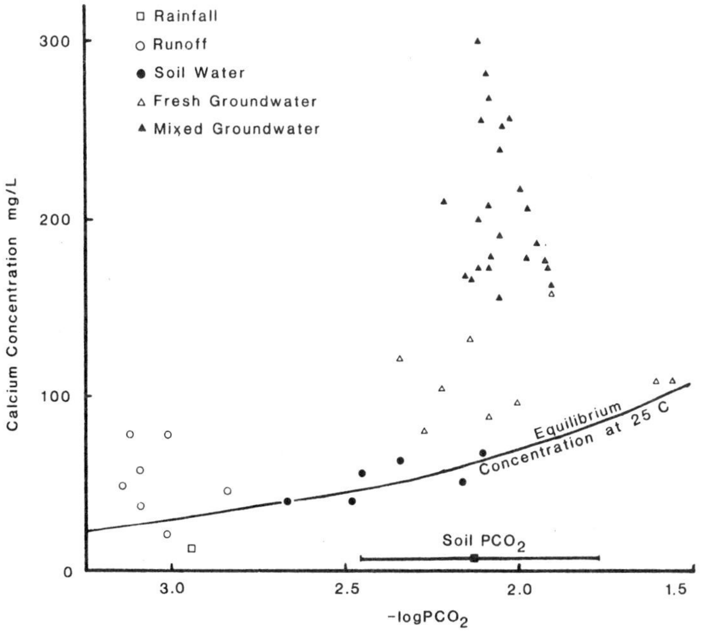

Fig. 4. Relationship between PCO_2 and calcium concentration for samples of rain water, runoff, soil water, and fresh and transition zone groundwaters. Points above the equilibrium line have lost CO_2 by evaporation (runoff) or degassing (groundwaters). Note that mixed transition zone waters have a PCO_2 comparable to fresh groundwaters, although a decrease with increasing percentage seawater would be expected for conservative linear mixing.

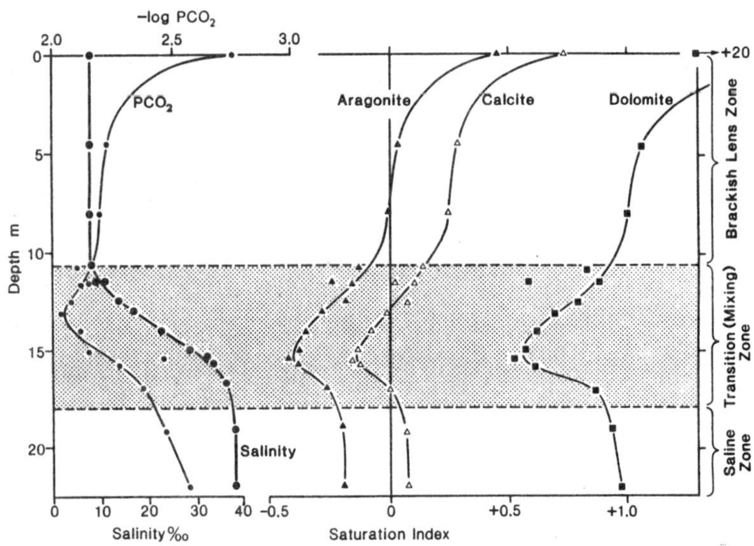

Fig. 5. Vertical variations in salinity, PCO_2, and aragonite, calcite and dolomite saturation indices in Evelyn Green's blue hole South Andros Island (Smart et al 1988). Note chemical undersaturation with respect to both aragonite and calcite, and increasing PCO_2 due to organic decomposition processes in the transition zone.

et al 1988), in association with waters that are undersaturated with respect to both calcite and aragonite (Fig. 5). In these cavernous sites organic particles penetrate the fresh water lens, but due to the density gradient remain suspended in the transition zone. Carbonate dissolution due to inorganic mixing is thus augmented by microbial generation of carbon dioxide. This is also demonstrated by the maintenance of PCO_2 at a level comparable to that in the fresh water lens (Fig. 4), contrary to the decreasing trend expected for linear conservative mixing.

Where a true fresh water lens is not developed, chemical contrasts in the groundwater body are more limited and thus the potential for mixing dissolution is reduced. Two factors are important in this respect; the balance between rainfall and potential evapotranspiration, and the relief. At present the relief of many Bahamian islands is relatively low, and large areas of open water in continuity with the groundwater permit direct evaporation at the potential rate. However lower sea-levels, as occurred during a substantial part of the Pleistocene, allow the development of a significant unsaturated zone, isolating the groundwater body from direct evaporation. Thus the dry southern islands such as Great Inagua, less than 1 % of which is currently underlain by fresh water (Cant and Weech 1986), would have had well developed fresh water lenses during periods of lower sea-level.

Under static groundwater conditions the contact between fresh and saline groundwaters would be sharp, with mixing generated only by diffusion. In practice there is considerable dispersion at the transition zone, generated by fluctuations in the position of the interface due to semi-diurnal tides (Mather and Buckley 1973) and variations in recharge which are accommodated by expansion and contraction of the lens (Kohout and Kline 1978). These factors enhance groundwater mixing and lead to the development of a zone of transition, the thickness of which may exceed that of the fresh water lens. The reduced density of the mixed waters also leads to the development of a buoyant circulation system in the transition zone, with the entrainment of saline water from below (Cooper 1959). In general there is therefore more active circulation and a thickening of the transition zone towards the coast. A similar effect is observed beneath the creeks which bisect the well developed fresh water lens on North Andros, and discharge large volumes of brackish water. However a small fall in sea-level (in the order of only 2m) would isolate these outlets, and cause the development of a much thicker, more extensive fresh water lens, with reduced mixing of groundwaters and a thinner transition zone.

6. CAVERNOUS POROSITY

It is apparent that bedrock dissolution in the saturated zone is concentrated both at the water-table, and especially in the transition zone. Mylroie (1988) has argued that rates of cave development in the latter may be very rapid. Thus cavernous porosity developed during earlier sea-level still-stands may be present within the bedrock. Because of the extremely high permeability of such zones, they may act to restrict the development of subsequent fresh water lens systems. Thus on Grand Bahama the

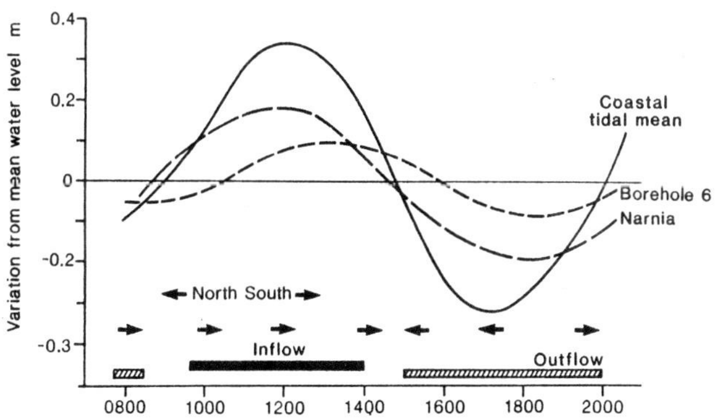

Fig. 6. Comparison of water-level fluctuations over a tidal cycle on South Andros Island, for monitoring stations at the east coast, at Narnia (an inland fracture blue hole) and an adjacent borehole. Arrows indicate the direction of flow observed in the brackish lens at Narnia, shaded bars show periods of inflow and outflow of saline water at an oceanic blue hole where the fracture runs offshore.

present position of the transition zone appears to be constrained by the open passages of Lucayan Caverns, which were developed during earlier high sea-level stands. Indeed Cant and Weech (1986) argue that the depth of the fresh water lens throughout the Bahamas is limited by the thickness of the Pleistocene Lucayn Limestone Formation, which is less cavernous than the older limestones beneath.

On South Andros and Grand Bahama cavernous porosity is preferentially developed along major fractures paralleling the bank margin (Palmer 1986). During every tidal cycle considerable volumes of saline water are exchanged between the fracture and the open ocean (Fig. 6). The caverns thus transmit tidal head differences with little attenuation, as shown by the comparable amplitude of water-level fluctuations in blue holes and at the coast. However attenuation of the tidal cycle occurs in the adjacent aquifer penetrated by borehole Number 6. This will cause pumping of saline and brackish waters into the adjacent fresh water diffuse flow aquifer. There is thus considerable potential for bedrock dissolution by groundwater mixing in such systems.

CONCLUSIONS

A major theme in the previous discussion has been the evolution from the homogeneous diffuse intergranular flow found in freshly exposed carbonate rocks, to concentrated recharge via fissures and cavernous flow with time. The resulting concentration of dissolution increases heterogeneity, and also allows the locus of dissolution to move from the surface to greater depth. However climate, and in particular the balance between evapotranspiration and precipitation, is a dominant control on the distribution and rates of bedrock dissolution in the Bahamas. This climatic control is both direct, via the control of groundwater flux, and indirect, via soil type and the extent of the fresh water lens. The latter is also influenced by relief, which is largely dependent on sea-level. Therefore before we can adequately understand the intrinsic changes in the distribution of dissolution (and therefore porosity) with time, it is important to identify the effects of climatic change. The presence of

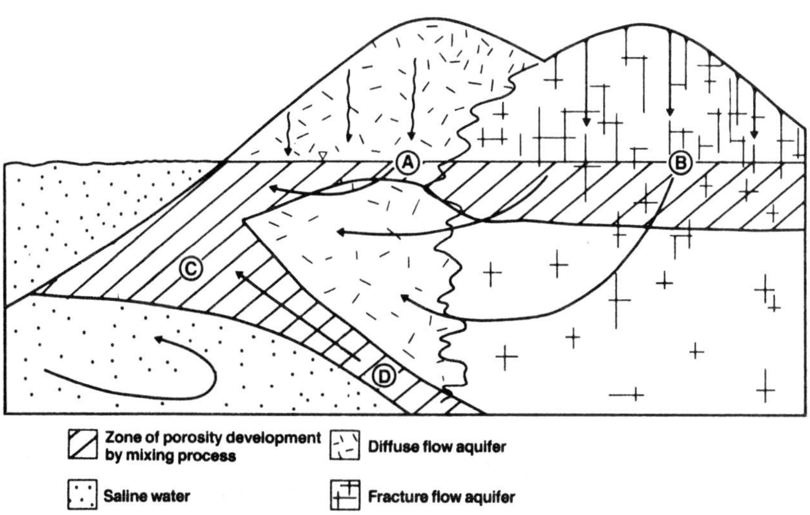

Fig. 7. Schematic diagram illustrating the zones where intense mixing may be expected for intergranular and fracture flow aquifers. At the water table there is limited potential for dissolution by the mixing of diffuse recharge waters and the fresh water lens (A), while fissure flow gives greater chemical contrasts, enhancing dissolutional potential (B). Tidal pumping and buoyant circulation thicken the transition zone near to the coast (C), compared to inland (D) where mixing is reduced.

paleosols ranging from sub-humid to arid in character provides a clear indication of the potential range palaeoclimatic conditions. What is now needed is a radiometrically dated record of these changes, such as may be obtained by the examination of growth periods and the isotopic composition of speleothems from the Bahamas.

ACKNOWLEDGEMENTS

This paper has benefitted from extensive discussion with Richard Cant, John Mylroie, Jim Carew, Neil Sealey, Rob Palmer and J.D. Crummock, and was made possible by financial support from Shell Exploration and Production Ltd, Amoco Ltd, and the National Environmental Research Council (Grant GR3/6694). Logistic support from Virgin Airways, Shell Bahamas, Steve Hobbs, Commander Ian Anderson Royal Navy, and Sue and Jim Phillips is gratefully acknowledged. Finally we thank the Government of the Bahamas for permission to undertake this study, and particularly the Department of Lands and Surveys for their help.

REFERENCES CITED

Back W., Hanshaw B.B., Pyle T.E., Plummer L.N. and Weide A.E., 1979, Geochemical significance of groundwater discharge and carbonate solution to the formation of Caleta Xel Ha, Quinta Roo, Mexico: Water Resources Research 15, 1521-1535.

Brook G.A, Folkoff M.E. and Box E.O., 1983, A world model of soil carbon dioxide: Earth Surface Processes and Landforms 8, 79-88.

Brown T.W., 1984, Formation and development of caliche profiles in aeolian deposits, in Teeter J.W. (Ed) Proceedings of the Second Symposium on the Geology of the Bahamas, College Centre of the Finger Lakes, San Salvador, Bahamas, June 1984, 265-278.

Budd D.A., 1988, Aragonite to calcite transformation during fresh water diagenesis of carbonates: insights from porewater chemistry: Geological Society of America Bulletin 100, 1260-1270.

Cant R.V. and Weech P.S., 1986, A review of the factors affecting the development of Ghyben-Hertzberg lenses in the Bahamas: Journal of Hydrology 84, 333-343.

Carew J.L. and Mylroie J.E., 1989, Preliminary report on the geology of eastern South Andros Island, Bahamas, in Mylroie J.E. Proceedings of the Fourth Symposium on the Geology of the Bahamas, College Centre of the Finger Lakes, San Salvador, Bahamas, June 1988.

Cooper H.H., 1959, A hypothesis concerning the dynamic balance of freshwater and saltwater in a coastal aquifer, Journal Geophysical Research 64, 461-467.

Craig D.H., 1988, Caves and other features of Permain karst in San Andres Dolomite,Yates Field Resevoir, West Texas, in James N.P. and Choquette P.W. Palaeokarst, Springer-Verlag, 342-363.

Enos P. and Sawatsky L.H., 1981, Pore networks in Holocene carbonate sediments: Journal of Sedimentary Petrology 51, 961-985.

Foos A., 1987, Paleoclimatic interpretation of paleosols on San Salvador Island, Bahamas, in Curran H.A. (Ed.) Proceedings of the Third Symposium on the Geology of the Bahamas, 67-72.

Glaccum R.A. and Prospero J.M., 1980, Saharan aerosols over the tropical North Atlantic - mineralogy: Marine Geology 37, 295-321.

Halley R.B. and Harris P.M., 1979, Freshwater cementation of a 100 year oolite: Journal of Sedimentary Petrology 49, 969-988.

Harrison R.S., 1975, Porosity in Pleistocene grainstones from Barbados: some preliminary observations: Bulletin of Canadian Petroleum Geology 23, 383-392.

James N.P. and Choquette P.W., 1974, The meteoric diagenetic environment: Geoscience Canada 11, 161-194.

Kohout F.A. and Kline H., 1978, Effect of pulse recharge on the zone of diffusion in the Biscayne aquifer, in Gunay G. and Karanjac J., Proceedings of the Symposium on Karst Hydrogeology, D.S.I., Ankara, Turkey , 151-169.

McCartney R.F. and Boardman M.R., 1987, Bahamian Paleosols: Implications for stratigraphic correlation by R.F., in Curren H.A. (Ed.) Proceedings of the Third Symposium on the Geology of the Bahamas, 99-108.

Mather J.D. and Buckley D.K., 1973, Tidal fluctuations and groundwater conditions in the Bahamian archipelago, Estratto da: Atti del 2^{0} Convegno Internationale Sulle Acque Sotterranee.

Mylroie J.E. (Ed), 1988, Field Guide to the Karst Geology of San Salvador Island, Bahamas, College Centre of the Finger Lakes, San Salvador, Bahamas, 83-96.

Palmer R.J., 1986, Hydrology and speleogenesis beneath Andros Island: Proceedings of the British Cave Research Association 13, 7-12.

Pierson B.J. and Shinn E.A., 1985, Cement distribution and carbonate mineral stabilization in Pleistocene limestones of Hogsty Reef, Bahamas, Society of Economic Paleontologists and Mineralogists Special Publication 38, 153-168.

Plummer L.N., 1975, Mixing of seawater with calcium carbonate groundwater: Geology Society America Memoirs 142, 219-236.

Ruhe R.V., Cady J.G. and Gomez R.S., 1961, Paleosols of Bermuda: Geological Society of America Bulletin, 72. 1121-1142.

Smart P.L., Dawans J.M. and Whitaker F.F., 1988, Carbonate dissolution in a modern mixing zone, South Andros, Bahamas: Nature 335, 811-813.

Smith D.I. and Atkinson T.C., 1976, Process, landforms and climate in limestone regions, in Derbyshire E. (Ed), Geomorphology and Climate, John Wiley, 367-409.

Vacher H.L. and Bengtsson T.O., 1989, Residence time of fresh groundwater in small lenses of Bermudian- and Bahamian-type islands: effect of hydraulic conductivity, in Mylroie J.E. Proceedings of the Fourth Symposium on the Geology of the Bahamas, College Centre of the Finger Lakes, San Salvador, Bahamas, June 1988.

Walter L.M., 1986, Relative efficiency of carbonate dissolution and precipitation during diagenesis: a progress report on the role of solution chemistry: Society of Economic Paleontologists and Mineralogists Special Publication 38, 1-12.

Ward W.C., 1973, Influence of climate on early diagenesis of carbonate aeolianites: Geology 1, 171-174.

White W.B., 1984, Rate processes: chemical kinetics and karst landform development, in LaFleur R.G. (Ed), Groundwater as a Geomorphic Agent, Allen and Unwin, 227-248.

PLEISTOCENE STRATIGRAPHY AND GEOCHRONOLOGY SOUTHWESTERN SAN SALVADOR ISLAND, BAHAMAS

Robert E. Stowers, II, and John E. Mylroie
Department of Geology and Geography
Mississippi State University
Mississippi State, Mississippi 39762

James L. Carew
Department of Geology
College of Charleston
Charleston, South Carolina 29424

ABSTRACT

A suite of Pleistocene marine and terrestrial limestone deposits crops out at several localities at the southern end of San Salvador. The uppermost unit at all outcrop localities in this area consists of oolitic eolian calcarenites belonging to the Grotto Beach Formation. At Watling's Quarry this eolianite has been dated at approximately 100,000 years, based on amino acid racemization of *Cerion* sp. found within the dunal deposits. At several localities these oolitic eolianites can be seen to grade downward into marine units which include fossil reefs that yield U/Th ages of at least 125,000 years (oxygen isotope substage 5e). At the Grotto Beach Sea Cliffs in the western portion of the field area, these reef deposits in growth position can be seen to overlie an erosionally truncated bioclastic eolian calcarenite that has been assigned to the Owl's Hole Formation. To the south, at the Owl's Hole type section, the oolitic dune deposits overlie a calcrete paleosol which caps a bioclastic eolianite unit (Owl's Hole Fm.). Traditional paleomagnetic analysis of that paleosol yielded a reversed remnant paleomagnetic signature, indicating an age of 740,000+ years (Matuyama Reversed Epoch). A stratigraphic sequence identical to that at Owl's Hole exists at the Watling's Quarry outcrop. However, the paleosol that caps the bioclastic unit at this locality yielded a normal paleomagnetic polarity. These data suggested that an additional stratigraphic unit might exist between the Owl's Hole and Grotto Beach formations.

A detailed and quantitative petrographic analysis was undertaken to attempt to distinguish these units from one another. The data show striking similarities in the abundance, texture and type of constituents that comprise the bioclastic eolianites at the Grotto Beach Sea Cliffs, Owl's Hole and Watling's Quarry sections; and also among the oolitic eolianites at those three sections. Based on these petrographic similarities, it is interpreted that no new stratigraphic unit exists at this end of the island. This interpretation is supported by cryogenic paleomagnetic analyses which indicate that no paleomagnetic reversals are present at the Owl's Hole, Watling's Quarry or Grotto Beach sea cliffs sections. The cryogenic paleomagnetic results thus suggest that the Owl's Hole Formation is younger than previously reported; however, this formation is still believed to contain the oldest rocks exposed in the Bahamas.

INTRODUCTION

San Salvador Island is an isolated part of the Bahamas archipelago. The subaerially exposed rocks of San Salvador are composed of three main types: 1) terrestrial limestones, consisting primarily of eolian calcarenites; 2) marine limestones, usually shoal, beach, back-beach and fossilized reef deposits; and 3) calcrete paleosols, mainly fossilized residual soils or caliche. Carew and Mylroie (1985) developed a physical stratigraphy of the island based on field observations, Uranium/Thorium dating of fossil corals and speleothems, ^{14}C dating of whole rock samples, amino acid racemization dating of *Cerion* sp., petrographic analyses of thin sections, and stratigraphic relationships of the various limestone units and interlayered paleosol horizons (Fig. 1). The purpose of this study was to determine whether the Pleistocene portion of that stratigraphy needed modification.

The study area, Sandy Point, is located at the southwestern end of San Salvador Island,

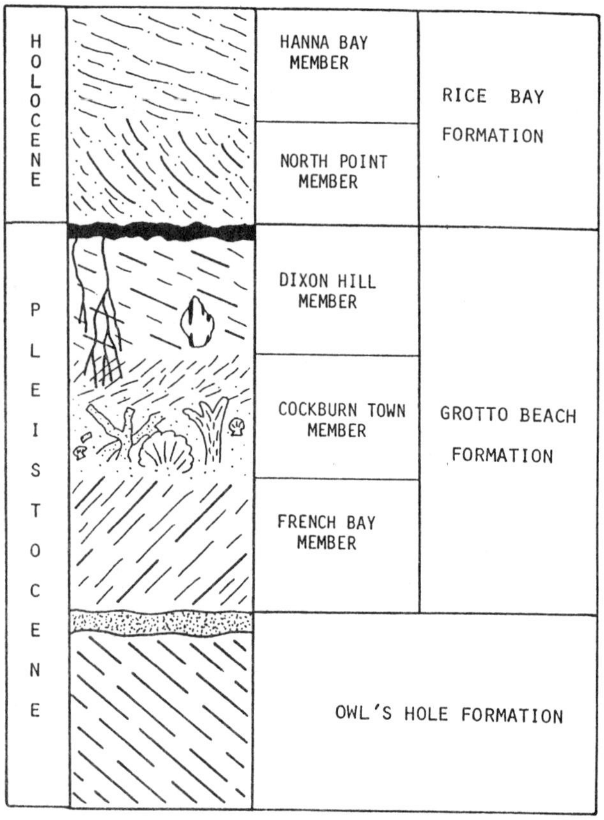

Fig. 1. Physical Stratigraphy of San Salvador. (from Carew and Mylroie, 1985).

where the Pleistocene rock units in question crop out. The rocks exposed in the central field area consist of a series of Quaternary eolianites and paleosols that form a northeast-southwest trending topographic high. The perimeter of the field area is characterized by subtidal, beach and reef deposits that interfinger with the eolianites and paleosols of the central area.

Initial paleomagnetic analyses yielded a reversed polarity for the lower paleosol at Owl's Hole, implying that this paleosol and the underlying bioclastic rock unit (Owl's Hole Formation) at this locality was at least 740,000 years old (Matuyama Reversed Epoch) (Carew and Mylroie, 1985). This reversed signature for the Owl's Hole paleosol was detected in five samples taken at three different times. As such, those authors interpreted the Owl's Hole Formation to be the oldest exposed rock unit in the Bahamas. Based on gross lithologic similarities and an identical stratigraphic sequence at the Owl's Hole and Watling's Quarry sections, Carew and Mylroie (1985) assigned the bioclastic unit below the lower paleosol at Watling's Quarry to the Owl's Hole Formation. Carew and Mylroie (1985) suggested that informal status be assigned to this formation because the base of the unit is not exposed at any outcrop. However, a paleomagnetic analysis of the lower paleosol at Watling's Quarry, done in 1986, revealed a normal polarity. This suggested that if the paleomagnetic data were correct, then an additional stratigraphic unit that was younger than the Owl's Hole Formation, but older than the oolitic Grotto Beach Formation (which has yielded ages of 150,000 to 85,000 years), existed at the base of the section in Watling's Quarry (Fig. 2).

METHODS

A detailed and quantitative petrographic analysis was performed on 64 thin sections. Fifty-six of these samples were collected at 1m intervals from nine measured sections. The other eight samples were taken from surface rock exposures. Figure 3 is a map of Sandy Point showing the localities of all samples collected; figure 4 contains a legend for the petrographic illustrations included in this report. Only three of the measured sections, the Owl's Hole (OH 0-9), Watling's Quarry southwest (WQSW 1-8) and the Grotto Beach sea cliffs (GBC 0-5) sections, will be discussed in this paper as they were the only outcrop localities observed to contain deposits of both the Owl's Hole and Grotto Beach formations. The other six measured sections and the eight surface rock samples were found to consist entirely of deposits belonging to the Grotto Beach Formation.

A total of six samples were collected for sensitive cryogenic paleomagnetic analysis. Four of them were from the Owl's Hole Formation; including the basal bioclastic eolianite and the overlying calcrete paleosol at Owl's Hole, and bioclastic eolianite samples, one each from the base of the Grotto Beach sea cliffs and Watling's Quarry sections. The remaining two samples were from the Grotto Beach Formation, including the upper oolitic eolianite at Watling's Quarry and Owl's Hole.

RESULTS

Owl's Hole Formation

The Owl's Hole Formation crops out at three different localities within the study area. At two of the localities, Owl's Hole (type section) and Watling's Quarry (see Fig. 3), this formation consists of a bioclastic eolianite capped by a calcrete paleosol. At the third locality, the Grotto Beach sea cliffs, the Owl's Hole Formation is

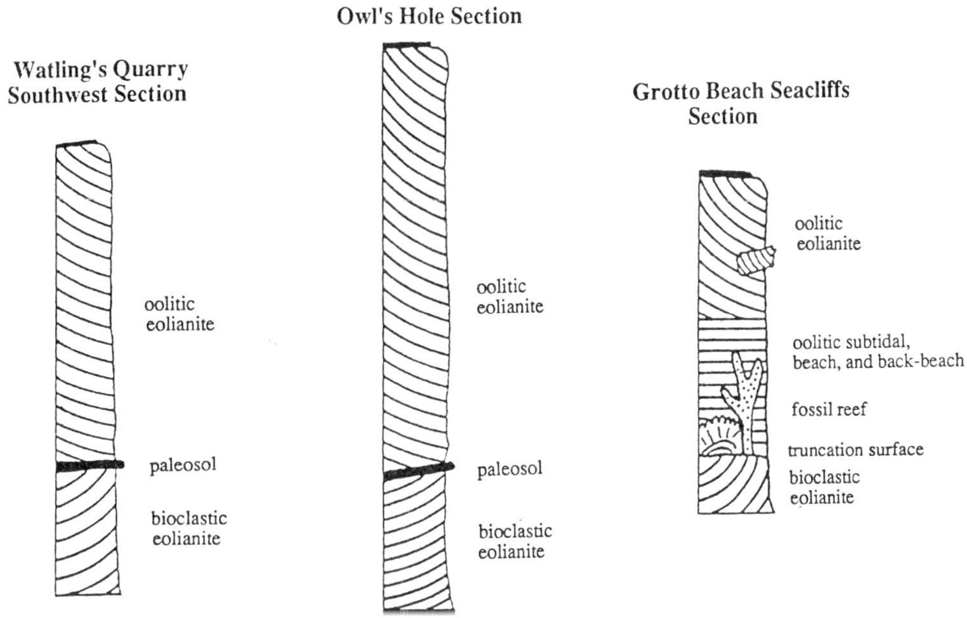

Fig. 2. Diagrammatic sketch of three key outcrops in the study area. Initial (non-cryogenic) paleo-magnetic analyses showed a reversed polarity for the lower paleosol at Owl's Hole and a normal polarity for the lower paleosol at Watling's Quarry. This suggested that the bioclastic units at Owl's Hole and Watling's Quarry were not correlative.

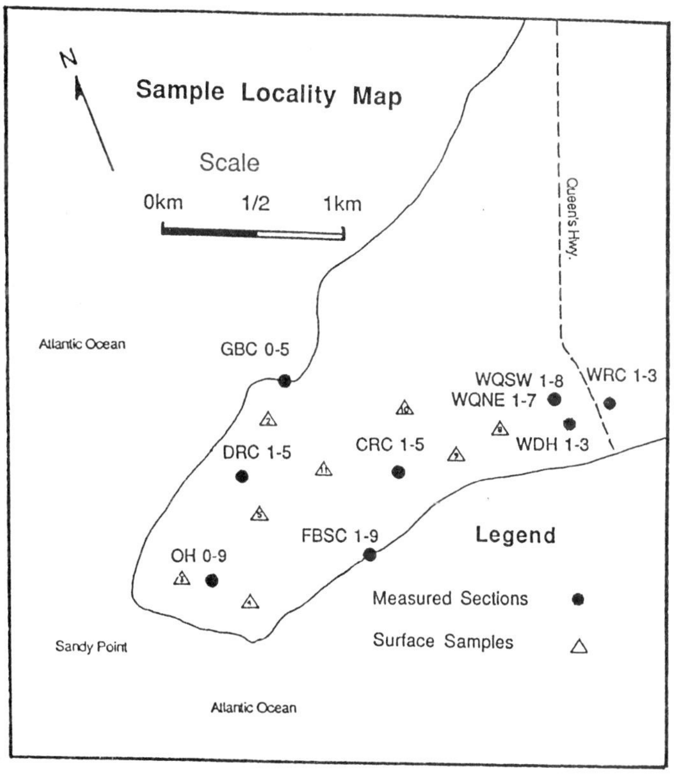

*Fig. 3. Sandy Point sample locality map. Capital letters designate the following sampled outcrops: GBC=Grotto Beach cliffs; DRC=Dripping Rock cliffs; OH=Owl's Hole; FBSC=French Bay sea cliffs; CRC=*Cerion *roadcut; WDH=Watling's dry hole; WQSW=Watling's Quarry southwest; WQNE= Watling's Quarry northeast; WRC=Watling's roadcut. Triangles numbered 2 through 5 represent samples SSJA-2 to SSJA-5; 8 through 11 represent samples SSMA-8 to SSMA-11.*

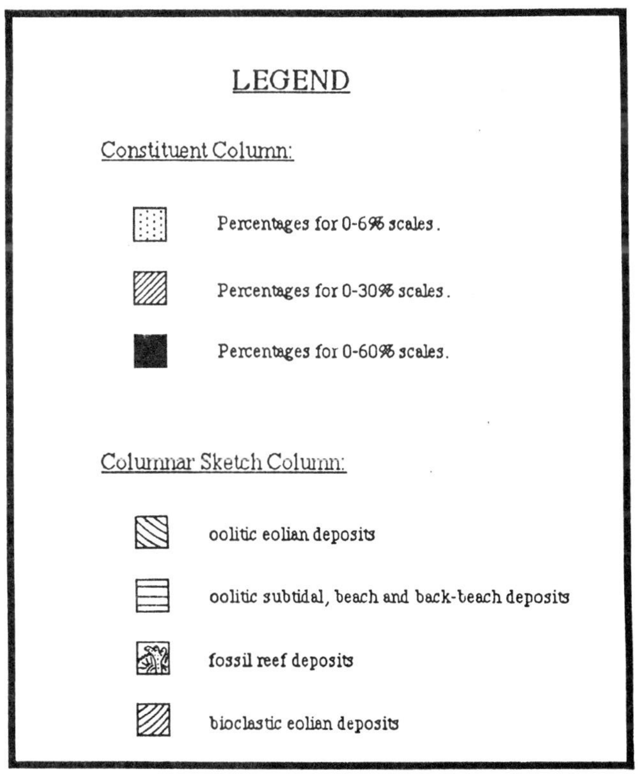

Fig. 4. Legend for petrographic illustrations and measured sections.

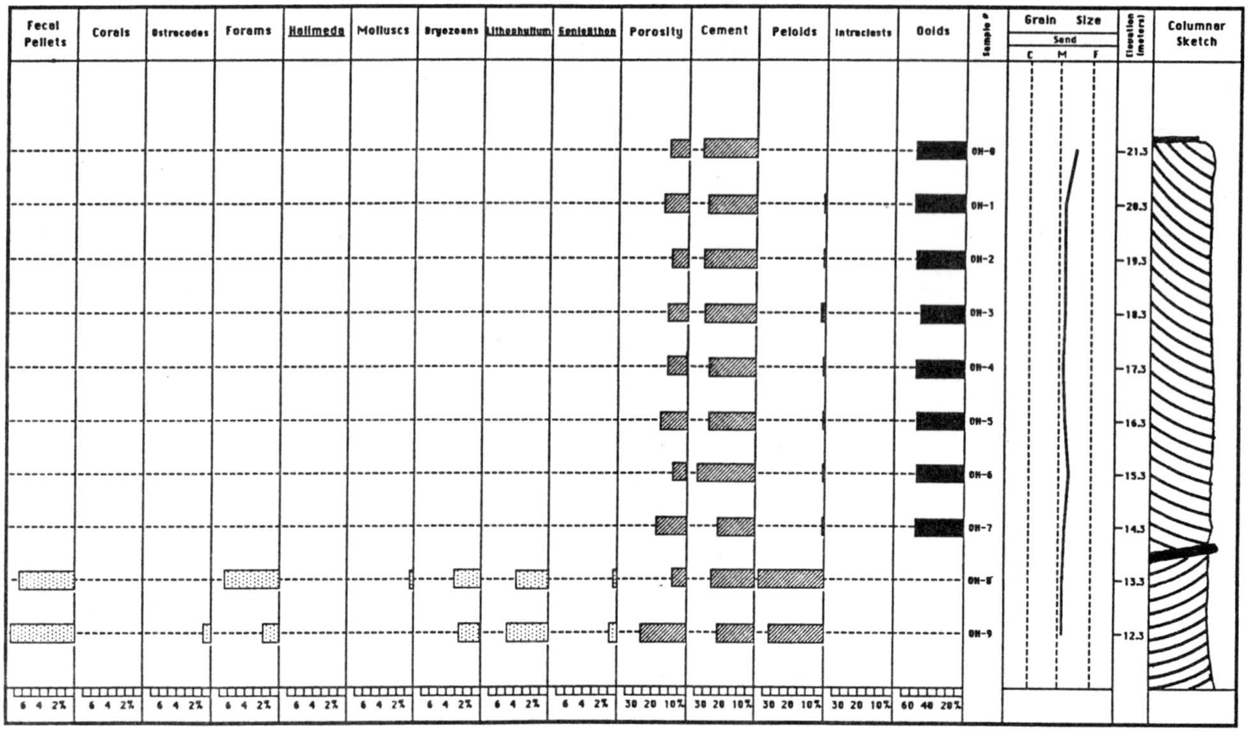

Fig. 5. Petrographic illustration of the Owl's Hole Section.

composed of an eolian calcarenite that has been erosionally truncated, and has no capping paleosol. However, the presence of alveolar texture in the upper meter of this unit is indicative of previous subaerial exposure. Carew and Mylroie (1985) suggested that the paleosol may have been removed by erosion that truncated the unit during the subsequent transgression that resulted in the formation of a 125,000 years old (oxygen isotope substage 5e; Shackleton and Opdyke, 1973) *in situ* fossil reef that is observed to immediately overlie the truncated eolian surface (hardground). It is obvious that the bioclastic eolianite (Owl's Hole Fm.) at the Grotto Beach sea cliffs has been inundated at least twice; during oxygen isotope stage 5e (125,000 years), and at present.

It is important to note that the Owl's Hole Formation consists predominantly of bioclastic allochems. At the type section at Owl's Hole the majority of the allochems consist of fine to medium sand-size skeletal fragments and pellets (Fig. 5). The high percentage of peloids observed in the thin sections from this outcrop of the Owl's Hole Formation is due to the high degree of micritization that this unit has experienced.

Rocks of the Owl's Hole Formation at Watling's Quarry and the Grotto Beach sea cliffs (Figs. 6 and 7 respectively) contain slightly different allochems than were observed at Owl's Hole. These two outcrops of the Owl's Hole Formation contain minor amounts of ooids (2.7%) intermixed with bioclastic grains. In addition, a significant percentage of intraclasts were observed in the Owl's Hole Formation at the Grotto Beach sea cliffs section. The average grain size at these two localities is in the medium sand range.

Four samples collected from the Owl's Hole Formation revealed a normal magnetic polarity by cryogenic paleomagnetic analysis. This suggests that the previously reported age of the Owl's Hole Formation (740,000+ years) (Carew and Mylroie, 1985) is in question.

Grotto Beach Formation

The Grotto Beach Formation overlies the Owl's Hole Formation and is divided into three members (Carew and Mylroie, 1985). From oldest to youngest they are the French Bay, Cockburn Town, and Dixon Hill members (see Fig. 1). The French Bay and Cockburn Town members crop out in the study area.

At the Grotto Beach sea cliffs, the stratigraphic sequence consists of an *in situ* fossilized patch reef. That reef grew on a truncation surface (hardground) that developed on an underlying bioclastic eolianite assigned to the Owl's Hole Formation. Uranium/Thorium age determination of the fossil reef coral yielded a date of 125,000 years (oxygen isotope substage 5e). This

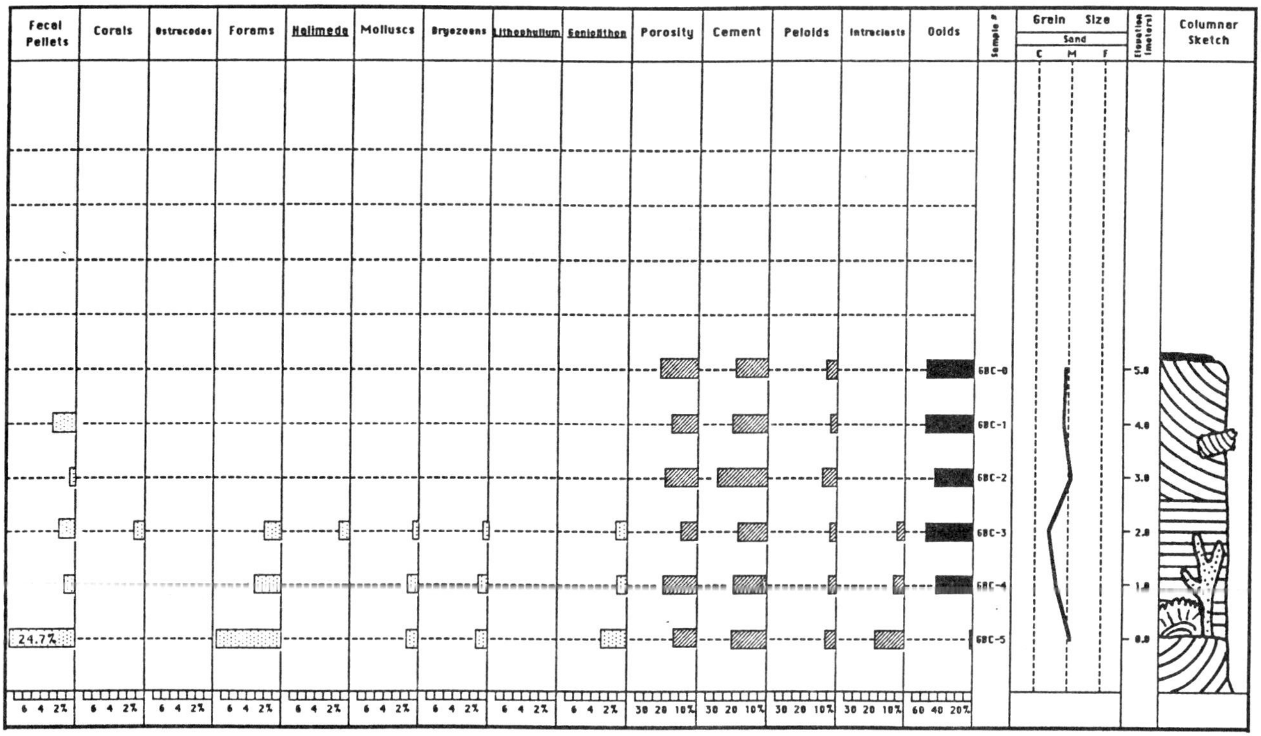

Fig. 6. Petrographic illustration of the Watling's Quarry Southwest Section.

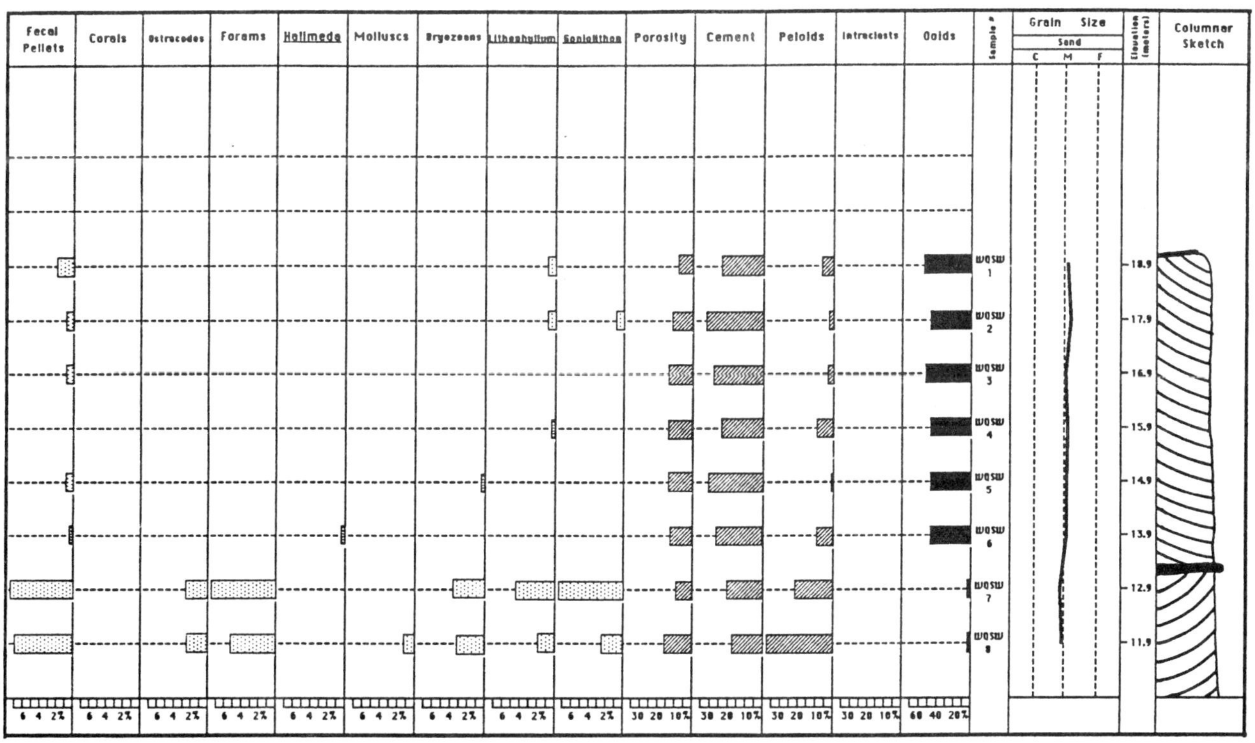

Fig. 7. Petrographic illustration of the Grotto Beach Seacliffs Section.

sequence was interpreted by Carew and Mylroie (1985) as evidence of the Sangamon transgression and still-stand of sea level. The fossil reef is surrounded by and infilled with subtidal, beach and back-beach calcarenites. This is interpreted as a regressive sequence that was deposited by strandline progradation at a sea level still-stand or subsequent marine regression associated with the end of oxygen isotope sub-stage 5e (sub-stage 5d). The rocks assigned to the Grotto Beach Formation at the Watling's Quarry and Owl's Hole sections are composed entirely of dunal eolian calcarenite facies.

All outcrops of the Grotto Beach Formation in the study area are capped by a paleosol. Carew and Mylroie (1985) considered this paleosol to be the Pleistocene-Holocene boundary.

The dominant allochems observed at all Grotto Beach Formation outcrops in the study area were ooids (Figs. 5, 6 and 7). Interestingly, in addition to the ooids a small percentage of bioclastic grains and fecal pellets were observed in the basal portion of several Grotto Beach Formation outcrops including the Grotto Beach sea cliffs and Watling's Quarry southwest. In addition, minor amounts of intraclasts were observed near the base of the Grotto Beach Formation at the Grotto Beach sea cliffs. Both samples from the Grotto Beach Formation that were subjected to cryogenic paleomagnetic analysis revealed a normal paleomagnetic signature.

DISCUSSION

Owl's Hole Formation

The Owl's Hole Formation at all three outcrop localities is interpreted to be eolian based on the presence of steeply dipping (30^0) large amplitude foreset beds and very good sorting. At the type section for this formation, the presence of rhizomorphs and fossilized *Cerion* sp. also lend support to this interpretation.

Considering the paucity of ooids observed in thin sections of the Owl's Hole Formation from the Watling's Quarry and Grotto Beach sea cliffs sections, any one of the following interpretations may apply.

1) The outcrops represent two bioclastic eolian units that were deposited during different sea level events. In this interpretation, the bioclastic units at Watling's Quarry and Grotto Beach sea cliffs represent units that are younger than the bioclastic units at Owl's Hole, but older than the Grotto Beach Formation.

2) The outcrops represent two different units that were deposited at different times during the same sea level event (e.g., transgressive vs. regressive).

3) The outcrops represent one unit that was deposited during the same sea level event, but each had a slightly different source. Sandy Point is surrounded on three sides by water and each of the three outcrops is located on a different side of Sandy Point easily permitting subtle differences in the sediment sources.

Based on the current data, and what was previously believed to be a 600,000 year lacuna, the most likely interpretation is that the Owl's Hole Formation is younger than originally reported by Carew and Mylroie (1985), but still probably represent the oldest rocks exposed in the Bahamas.

The new paleomagnetic data in conjunction with the petrographic results from this study also suggest that the best interpretation would be that the bioclastic eolianites at Watling's Quarry and Grotto Beach sea cliffs sections belong to the Owl's Hole Formation.

Grotto Beach Formation

As was mentioned earlier, the Grotto Beach Formation is predominantly oolitic, particularly in its eolian facies. When this is compared to the dominance of bioclastic material in the Owl's Hole Formation, two possible interpretations are implied:

1) The conditions that existed during the time of deposition of the Grotto Beach Formation were different than those that existed during deposition of the Owl's Hole Formation (i.e. dominantly bioclastic sediment production vs widespread ooid production.)

2) The depositional environments of the areas adjacent to these three localities were different during deposition of the Owl's Hole Formation than they were during deposition of the Grotto Beach Formation.

Of these, the former seems the much more likely.

Geochronology of the Sandy Point Area

The depositional history of San Salvador Island has been discussed by Titus (1980, 1981, and 1983), Carew (1983), Carew and Mylroie (1985, 1987), and Sims (1987). The discussions are based mainly on stratigraphic relationships, petrographic characteristics and absolute ages reported by Carew and Mylroie (1985, 1987) from Uranium-/Thorium analyses of fossil corals and speleothems, amino acid racemization of *Cerion* sp. and ^{14}C analyses of whole rock samples obtained from localities throughout the entire island.

This study is only concerned with the Quaternary deposits of the Sandy Point area. Therefore, the depositional history that follows deals with interpretations made only from rocks in the study area.

It was originally indicated by Carew and Mylroie (1985) that the oldest rocks exposed on San Salvador Island, and possibly the entire Bahamas, were those of the Owl's Hole Formation. These rocks were interpreted to be at least 740,000 years old (Matuyama Reversed Epoch) based on repeated paleomagnetic analyses that yielded a reversed paleomagnetic polarity in the capping paleosol at the type section for this formation (Owl's Hole; Fig. 1). Carew and Mylroie (1985) hypothesized that the Owl's Hole rocks, mainly bioclastic eolianites, represented a 740,000+ year old nucleation point or "ancient" San Salvador to which the modern San Salvador Island accreted, both vertically and laterally. This process was first proposed by Carew (1983), and was later termed catenary development by Garrett and Gould (1984).

However, the cryogenic paleomagnetic analyses reported in this study on the Owl's Hole rocks indicate that no reversed paleomagnetic signature is present, and therefore, the previously reported age of these rocks (740,000+ years) is questionable. The most likely interpretation is that the Owl's Hole rocks are younger than previously reported, but older than the Grotto Beach Formation and are therefore still considered the oldest rocks exposed in the Bahamas (Stowers, 1988). Even though the absolute age of the Owl's Hole Formation is uncertain, as both laboratories stand by their conflicting paleomagnetic results, the sequence of events responsible for the development of the Sandy Point area, as proposed by Carew and Mylroie (1985), remains the same. Only the absolute age of events is reinterpreted in this paper. The proposed sequence of Pleistocene events is as follows:

1) Sometime prior to 150,000 years ago "ancient" San Salvador Island consisted of bioclastic eolian ridges that were deposited in the southwest portion of present or modern San Salvador Island. This period of deposition was followed by climatic conditions conducive to the dissolution of limestone and to paleosol development. These rocks belong to the Owl's Hole Formation.

2) During the initial oxygen isotope stage 5e transgression modern San Salvador began to accrete, both vertically and laterally, to the "ancient" San Salvador. These rocks consist of oolitic subtidal, beach, back-beach and dune deposits and have been assigned to the French Bay Member of the Grotto Beach Formation.

3) During the subsequent still-stand of sea level (approximately +6m above psl), abundant carbonate sediment production and reef growth occurred. U/Th analysis of fossil corals yield ages of approximately 125,000 years (oxygen isotope substage 5e). These rocks comprise the lower section of the Cockburn Town Member (Grotto Beach Formation).

4) During that still-stand, or as sea level fell, the fossil reefs were entombed by subtidal, beach and back-beach deposits, and thus the coastline prograded outward.

5) The final episode in the southwest portion of the island's Pleistocene depositional history is characterized by widespread paleosol development that occurred at the end of Pleistocene time (approximately 15,000 years ago).

SUMMARY AND CONCLUSIONS

This study provides a detailed and quantitative petrographic analysis of the two Pleistocene stratigraphic units exposed in the Sandy Point area of San Salvador Island. These units are the Owl's Hole and Grotto Beach formations. Based on previous paleomagnetic analyses, it was thought that a new stratigraphic unit possibly existed between the older Owl's Hole Formation and younger Grotto Beach Formation. Based on the petrographic similarities between the rocks that could possibly have been assigned to a previously undescribed unit and the Owl's Hole Formation, it is suggested here that no additional stratigraphic unit exists between these two formations. Results of cryogenic paleomagnetic analyses reported here support this interpretation. These results also reveal that the 740,000+ year age for the Owl's Hole Formation is uncertain. The most likely interpretation is that the Owl's Hole Formation is younger than previously reported.

ACKNOWLEDGEMENTS

We are indebted to the Bahamian Field Station and its staff on San Salvador Island, Bahamas, for logistical and financial support for this research. Special thanks to Dr. Donald Gerace and Kathy Gerace for their support of and interest in this research.

Paleomagnetic analyses were run by Charles Barton and cryogenic paleomagnetic analyses were run by Don McNeill of the University of Miami; their efforts are deeply appreciated. Daryl Clark provided valuable field assistance.

REFERENCES CITED

Carew, J.L., 1983, Geochronology of San Salvador, Bahamas: in Gerace, D.T., ed., Field Guide to the Geology of San Salvador (3rd edition), Ft. Lauderdale, FL, College Center of the Finger Lakes Bahamian Field Station, p. 160-172.

Carew, J.L., and Mylroie, J.E., 1985, The Pleistocene and Holocene Stratigraphy of San Salvador Island, Bahamas, with Reference to Marine and Terrestrial Lithofacies at French Bay: in Curran, H.A., ed., Pleistocene and Holocene Carbonate Environments on San Salvador Island Bahamas, Geological Society of America Field Trip Guidebook, p. 11-61.

Carew, J.L., and Mylroie, J.E., 1987, A Refined Geochronology for San Salvador Island, Bahamas in Curran, H.A., ed., Proceedings of the Third Symposium on the Geology of the Bahamas: Ft. Lauderdale, Florida, College Center of the Finger Lakes Bahamian Field Station, p. 35-44.

Garrett, P., and Gould, S.J., 1984, Geology of New Providence Island, Bahamas: Geological Society of America Bulletin, v. 95, p. 209-220.

Gerace, D.T., ed., 1983a, Field Guide to the Geology of San Salvador: Ft. Lauderdale, Florida, College Center of the Finger Lakes Bahamian Field Station, 172 p.

Gerace, D. T., ed., 1983b, Proceedings of the First Symposium on the Geology of the Bahamas: Ft. Lauderdale, Florida, College Center of the Finger Lakes Bahamian Field Station, 62p.

Shackleton, N.J., and Opdyke, N.D., 1973, Oxygen Isotope and Paleomagnetic Stratigraphy of Equatorial Pacific Core V28-238: Oxygen Isotope Temperatures and Ice Volumes on a 10^5 year and 10^6 year scale: Quaternary Research, v. 3, p. 39-55.

Stowers, R.E., 1988, Stratigraphy and Geochronology of Pleistocene Carbonates, Sandy Point area, southern San Salvador Island, Bahamas: unpbl. Master's Thesis, Mississippi State University, 103p.

Teeter, J., ed., 1984, Proceedings of the Second Symposium on the Geology of the Bahamas: Ft. Lauderdale, Florida, College Center of the Finger Lakes Bahamian Field Station, 296 p.

Titus, R., 1980, Emergent Facies Patterns on San Salvador Island, Bahamas: in Gerace, D.T., ed., Field Guide to the Geology of San Salvador (First Edition), Ft. Lauderdale, Florida, College Center of the Finger Lakes Bahamian Field Station, p. 92-105.

Titus, R., 1981, Emergent Facies Patterns on San Salvador Island Bahamas: in Gerace, D.T., ed., Field Guide to the Geology of San Salvador (2nd edition), Ft. Lauderdale, Florida, College Center of the Finger Lakes Bahamian Field Station, p. 92-105

Titus, R., 1983, Emergent Facies Patterns on San Salvador Island Bahamas: in Gerace, D.T., ed., Field Guide to the Geology of San Salvador (3rd edition), Ft. Lauderdale, Florida, College Center of the Finger Lakes Bahamian Field Station, p. 97-116.

Titus, R., 1984, Physical Stratigraphy of San Salvador Island, Bahamas: in Teeter, J.W., ed., Proceedings of the Second Symposium on the Geology of the Bahamas, Ft. Lauderdale, Florida, College Center of the Finger Lakes Bahamian Field Station, p. 209-228.

REFINEMENT AND TIMING OF SALINITY FLUCTUATIONS IN WATLING'S BLUE HOLE, SAN SALVADOR ISLAND, BAHAMAS

James W. Teeter
Department of Geology
The University of Akron
Akron, Ohio 44325

ABSTRACT

Previous study of the Holocene of Watling's Blue Hole has revealed the presence of four ostracode assemblages: 1. *Cyprideis americana* Zone; 2. Pitted *C. americana* Zone; 3. *Hemicyprideis setipunctata* Zone; and 4. *Perissocytheridea bicelliforma* Zone in ascending order. Each assemblage reflects a prevailing salinity based on the observed salinity range of each zone's dominant species. Other lakes on San Salvador reveal similar, parallel salinity trends through the Holocene.

There are problems in the above application of taxonomic uniformitarianism. Three of the four common lacustrine ostracode species have broad overlapping salinity tolerances. Also, in the Holocene of Salt Pond the dominant ostracode is the brackish species *P. bicelliforma*, although there is good reason to believe that this lake has been hypersaline during much of its history.

An alternative method of paleosalinity interpretation is minor element chemistry of ostracode shells. Because of its wide salinity tolerance (10.8-98.5 ppt), *C. americana* was chosen. In this species, magnesium exhibits an inverse relationship with salinity. Overall, the paleosalinities based on assemblages and magnesium concentrations are similar for Watling's Blue Hole with two differences. In the latter method, each zone exhibits a range in salinity and a salinity minimum occurs at the boundary between each of the four zones. Similar salinity minima have also been observed in Salt Pond. Salinity fluctuations may be in response to variation in sea level or climate.

Accelerator mass spectrometry radiocarbon dates on molluscs at the boundaries between the four zones are 2680±80, 1900±80, and 1360±90 RCYBP. The earliest salinity minimum (2680±80 RCYBP) coincides with a lower stand of sea level. At this time sea level on San Salvador stood between 2.7 and 3.2 m below present. The cause of the salinity minima at 1900±80 and 1360±90 RCYBP is uncertain.

The time frame established by the youngest date and the lack of any younger obvious sedimentologic or paleontologic evidence indicating increased erosion rates during colonial times suggests that pre-colonial soils were neither thicker nor more extensive than present.

INTRODUCTION

The saline lakes of San Salvador Island have supported a low diversity collection of ostracode species throughout the Holocene. These species occur in distinct assemblages representing fresh water, marine and lacustrine environments. The lacustrine assemblage is by far the most common and occurs primarily in brackish and hypersaline conditions.

Crotty and Teeter (1984) recognized four ostracode zones, based mainly on changing abundances of lacustrine species, throughout the Holocene of Watling's Blue Hole. The four zones in ascending order are: 1. *Cyprideis americana* Zone; 2. Pitted *C. americana* Zone; 3. *Hemicyprideis setipunctata* Zone; 4. *Perissocytheridea bicelliforma* Zone. The *C. americana* Zone was interpreted to have a prevailing salinity of approximately 25 ppt whereas the Pitted *C. americana* experienced decreased salinity, probably in the low to mid teens. The ostracode assemblage of the *H. setipunctata* Zone reflects the highest prevailing Holocene salinity, probably between 25 and 35 ppt. The youngest interval, the *P. bicelliforma* Zone, experienced decreased salinity, possibly in the high teens. In their interpretation, Crotty and Teeter (1984) diagrammed the salinity as being relatively constant within each zone. However, in view of observations on salinity of the saline lakes of San Salvador during the past decade it is likely that fluctuation has occurred over prolonged intervals during the Holocene.

Little Lake (Sanger and Teeter, 1982) and Reckley Hill Pond (Luginbill, 1983) also exhibit successive Holocene ostracode assemblages which reflect salinity changes parallel to those interpreted from Watling's Blue Hole.

Salinity interpretations are most easily accomplished when stenohaline fresh water and marine assemblages are present. The much more frequently encountered lacustrine assemblage consists of four species. *Perissocytheridea bicelliforma* is apparently restricted to brackish water whereas the remaining three species, *Cyprideis americana*, *Hemicyprideis setipunctata* and *Dolerocypria inopinata* range from brackish to hypersaline conditions. Although each of the latter three species has a salinity range where it dominates, their overlapping tolerances make it difficult to interpret paleosalinities using taxonomic uniformitarianism.

A related problem was demonstrated in Salt Pond by Teeter et al (1987). The Holocene record here is dominated by the brackish water ostracode *P. bicelliforma* although the current salinity of approximately 300 ppt. and the presence of gypsum beds suggest hypersaline conditions. Clearly, an independent method of paleosalinity determination would help to resolve the above problems.

PALEOSALINITY REFINEMENT USING MAGNESIUM-SALINITY RELATIONSHIP IN *Cyprideis americana*

Because of its broad salinity tolerance (10.8 - 98.5 ppt.) *C. americana* was selected to determine whether any relationship existed between the magnesium content of its low magnesian calcite shell and salinity. Teeter (1988) reported on the inverse relationship between magnesium content, determined by electron microprobing, and the salinity of the water in which individuals of this species were living.

Five adult ostracode valves were selected from each of 23 intervals throughout a 1.98 m core from Watling's Blue Hole. Polished thin sections of these specimens were prepared for WDX electron microprobing. Instrument conditions were as follows: 15 KV; $.5\times10^{-5}$ma; beam diameter of approximately 10 microns. Each specimen was probed at 3 points midway through the shell thickness. The results of the analyses are recorded in Figure 1. Figure 2A, a spline curve constructed throughout the mean values of Figure 1, clearly illustrates fluctuating paleosalinity during the Holocene of Watling's Blue Hole. Figure 2B represents paleosalinity interpreted from ostracode assemblages. For direct comparison this latter figure has been modified from Crotty and Teeter (1984) to the same paleosalinity scale as Figure 2A.

At first glance the pronounced fluctuations in the spline curve make it appear distinctly different from the curve based on assemblage data. The fluctuations undoubtedly represent a more realistic situation. If the paleosalinities, determined on weight per cent magnesium oxide, are averaged for the intervals within each of the four zones and compared with those determined from ostracode assemblages the results are quite similar (Table 1). Thus paleosalinity determined from magnesium concentrations in *Cyprideis americana* has independently confirmed the previous interpretation (Crotty and Teeter, 1984) based on ostracode assemblages.

The remarkable feature about the salinity fluctuations (Figure 2A) is the occurrence of distinct salinity minima at the three zone boundaries established by ostracode assemblages. Two similar paleosalinity minima were observed by Teeter et al (1987) in Salt Pond. The zone between these two minima was correlated with the *Hemicyprideis setipunctata* zone of Crotty and Teeter (1984), the upper *Xestoleberis* zone of Sanger and Teeter (1982) and the normal marine zone of Luginbill (1983). The presence of the upper two paleosalinity minima in Watling's Blue Hole supports this correlation.

In Little Lake the lower two Holocene zones (lower *Xestoleberis* and *Cyprideis "ovata"* zones) are correlated with the *Cyprideis americana* and Pitted *C. americana* zones respectively in Watling's Blue Hole. In one of the cores from Little Lake the boundary between the two lower zones is marked by a fresh water ostracode assemblage, indicating a salinity minimum (Sanger and Teeter, 1982). This correlates nicely with the oldest salinity minimum in Watling's Blue Hole.

Varying salinity events may have been caused by change in climate, in sea level or interaction between the two (Crotty and Teeter, 1984). Increased precipitation or lowered temperature causing decreased evaporation would produce a lower salinity event. Lowering sea level acts to reduce salinity. As sea level drops, so does the water level in the saline lakes. With shrinking

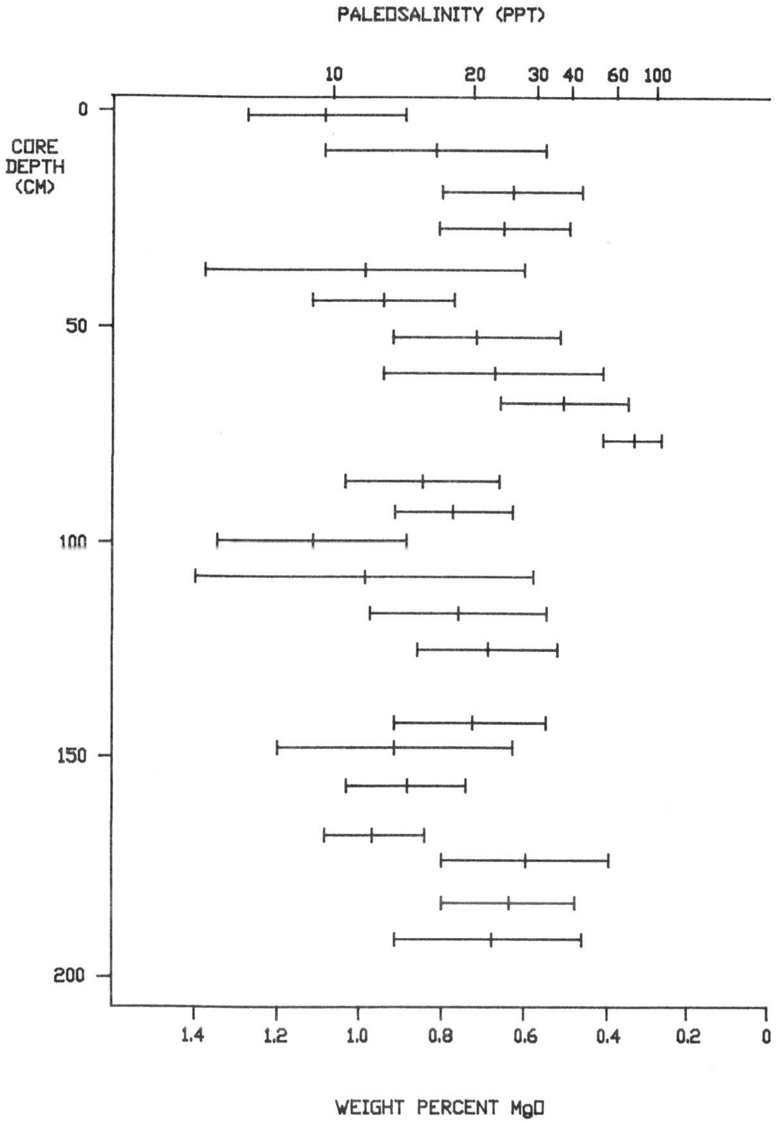

Fig. 1. Distribution of weight percent MgO in C. americana *and correlative paleosalinity, core 2, (Crotty and Teeter, 1984) Watling's Blue Hole. Mean and standard deviation indicated.*

volumes of saline water in the lake basins, precipitation and fresh ground water seepage exert an increasing control on salinity. If sea level should fall below the floor of a lake basin and if that basin were underlain by relatively impermeable materials a fresh water lake would result.

TIMING OF SALINITY-MINIMUM EVENTS

Accelerator mass spectrometry radiocarbon dates (RCYBP) on molluscs at the boundaries between the four ostracode zones are 2680±80, 1900±80 and 1360±90 RCYBP (Figure 2). The youngest two dates were obtained on the pelecypod *Polymesoda maritima* and the oldest date on a collection of Rissoacean gastropods. A live specimen of *Polymesoda maritima* was also dated in order to determine the influence of incorporation of "dead" carbon from the limestone bedrock. This factor was applied to the carbon dates from the core to produce the corrected values above.

DePratter and Howard (1981) presented evidence for a sea level lowstand between 3000 and 2400 years B.P. on the Georgia and adjacent South Carolina coasts. They state that sea level fell to at least 3 to 4 m below present mean sea level. Martin, Suguio and Flexor (1986) record a lowstand of sea level at approximately 2700 years B.P. from the Salvador region of coastal Brazil. Although they do not indicate the precise amount of sea level lowering their figure suggests approximately 1 m.

Therefore, the salinity minimum at 2680±80 RCYBP appears to represent a sea level lowstand. In core 2 from Watling's Blue Hole this horizon

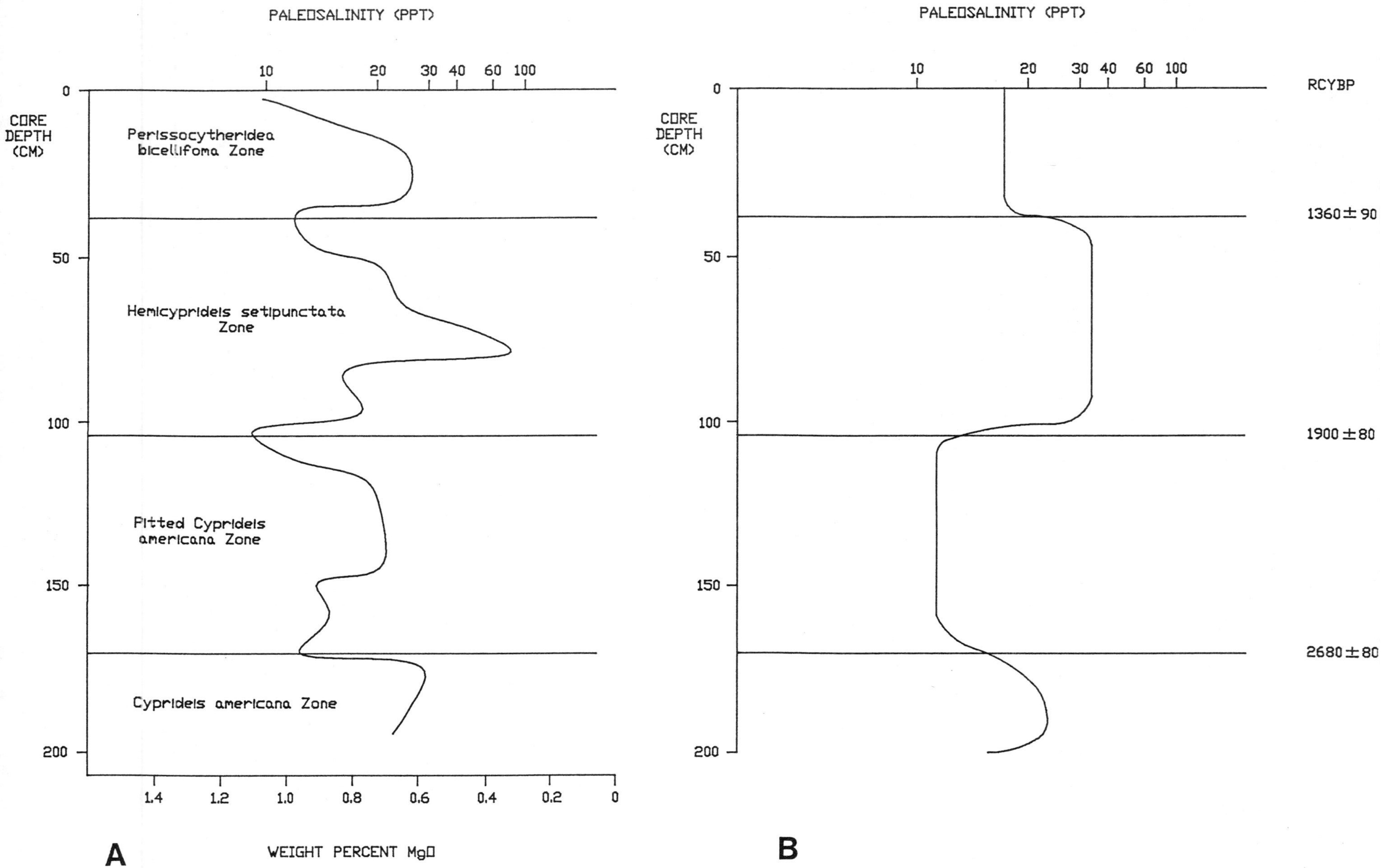

Fig. 2. Paleosalinity interpretations of the Holocene of Watling's Blue Hole. Horizontal lines at approximately 168, 104 and 38 cm represent the boundaries between the four ostracode zones and are dated at 2680±80, 1900±80 and 1360±90 RCYBP respectively. ***A****. Spline curve based on means of weight percent MgO in* C. americana. ***B****. Curve based on ostracode assemblages (modified from Crotty and Teeter, 1984).*

ZONE	PALEOSALINITY (ppt) FROM Mg CONTENT	FROM OSTRACODE ASSEMBLAGES
Perissocytheridea bicelliforma	18	high teens
Hemicyprideis setipunctata	28	25-35
Pitted *Cyprideis americana*	16	low to mid teens
Cyprideis americana	27	approximately 25

Table 1. Comparison of paleosalinities based on magnesium content of *Cyprideis americana* and ostracode assemblages

lies approximately 3.2 m below M.S.L. The brackish water ostracode assemblage here indicates that this elevation lay at or slightly below sea level at this time. In core 9 from Little Lake (Sanger and Teeter, 1982) the correlative horizon lies approximately 2.7 m below lake level which presently lies at or very close to M.S.L. The fresh water ostracode assemblage at this horizon indicates that this part of the lake lay above sea level at that time. Thus, at approximately 2700 years B.P. sea level on San Salvador lay between approximately 2.7 and 3.2 m below present sea level.

The cause of the salinity minima at 1900±80 and 1360±90 RCYBP is uncertain. If these were caused by lowered sea level the absence of any fresh water ostracode assemblages at correlative horizons in other lakes indicates that sea level lowering was less than 2.5 m and 1.9 m respectively.

ADVENT OF SLASH AND BURN AGRICULTURE

The colonization of the Bahamas by British loyalists in the late 1700's brought the introduction of clear cutting and slash and burn agriculture. It has been assumed that such practice would have greatly increased the rate of erosion, flushing soils into the lakes and surrounding marine environments.

Bowman and Teeter (1982) suggested that in Little Lake the advent of the Upper *Quinqueloculina bosciana* zone marked the beginning of slash and burn agriculture. However correlation with Watling's Blue Hole indicates that the Upper *Q. bosciana* zone began prior to 1360 RCYBP. In their study of Salt Pond, Teeter et al (1987) selected a horizon below the youngest salinity minimum as the possible initiation of slash and burn agriculture. This selection was based on increased insoluble inorganic residue and aluminum content.

Clearly both of the above events predate British colonization of the Bahamas by several hundred years. In the absence of any obvious younger sedimentologic or paleontologic evidence of increased erosion rates it is questionable whether the soils exposed by initial clear cutting were any more thick or extensive than present soils.

CONCLUSIONS

Paleosalinities determined by ostracode assemblage data and the application of the magnesium-salinity relationship in *Cyprideis americana* for each of the four ostracode zones in Watling's Blue Hole are closely similar. Fluctuation in paleosalinity, determined from magnesium concentration, is revealed in each of the four zones. Distinct salinity minima occur at each of the boundaries between the four zones. Accelerator mass spectrometry radiocarbon dates from each of the three boundaries are 2680±80, 1900±80 and 1360±90 RCYBP. The oldest boundary marks a lowstand of sea level between 2.7 m and 3.2 m below present. Cause of the younger two salinity minima is uncertain.

Previous speculation on the effect of clear cutting and slash and burn agriculture on erosion rates has been shown to be incorrect. In the absence of any clear sedimentologic or paleontologic evidence of increased erosion rates caused by colonization it is questionable whether soils were any thicker or more extensive than now.

ACKNOWLEDGEMENTS

I am indebted to Mr. Thomas Quick for his assistance with the electron microscope and especially with computer graphics. The University of Akron Faculty Research Grant 944 paid for the accelerator mass spectrometry radiocarbon dates. Dr. Annabelle Foos and Dr. Gary Kunze kindly reviewed this paper and offered helpful suggestions.

REFERENCES CITED

Bowman, P.A. and Teeter, J.W., 1982, The distribution of living and fossil Foraminifera and their use in the interpretation of the post Pleistocene history of Little Lake, San Salvador Island, Bahamas: College Center of the Finger Lakes, Bahamian Field Station Occasional Papers, No. 2, 23p.

Crotty, K.J., and Teeter, J.W., 1984, Post-Pleistocene salinity variations in a blue hole, San Salvador Island, Bahamas, as interpreted from the ostracode fauna: in Teeter, J.W., ed., Proceedings of the Second Symposium on the Geology of the Bahamas, College Center of the Finger Lakes, Bahamian Field Station, San Salvador, Bahamas, p. 3-16.

DePratter, C.B., and Howard, J.D., 1981, Evidence for a sea level lowstand between 4500 and 2400 years B.P. on the southeast coast of the United States: Journal of Sedimentary Petrology, Vol. 51, p. 1287-1295.

Luginbill, C.P., 1983, Ecology of living Ostracoda from selected lakes and post-Pleistocene history of Reckley Hill Pond, San Salvador Island, Bahamas: Ohio Academy of Sciences Annual Meeting, v. 83, p. 27.

Martin L., Suguio, K., and Flexor, J.M., 1986, Shell middens as a source of additional information in Holocene shoreline and sea-level reconstruction: examples from the coast of Brazil in Van de Plassche, O., ed., Sea level research: a manual for the collection and evaluation of data: Geo Books, Norwich, U.K., p. 503-521.

Sanger, D.B., and Teeter, J.W., 1982, The distribution of living and fossil Ostracoda and their use in the interpretation of the post-Pleistocene history of Little Lake, San Salvador Bahamas: College Center of the Finger Lakes, Bahamian Field Station Occasional Papers, No. 1, 26 p.

Teeter, J.W., 1988, Salinity-controlled Mg content in the ostracode *Cyprideis americana* and its application to Holocene lake history, San Salvador Island, Bahamas. Abstracts with programs, North Central Section, Geological Society of America, Vol. 20, No. 5, p. 391.

Teeter, J.W., Beyke, R.J., Bray, Jr., T.F., Brocculeri, T.F., Bruno, P.W., Dremann, J.J., and Kendall, R.L., 1987, Holocene depositional history of Salt Pond, San Salvador, Bahamas: in Curran, A., ed., Proceedings of the Third Symposium on the Geology of the Bahamas, College Center of the Finger Lakes, Bahamian Field Station, San Salvador, Bahamas, p. 145-150.

EFFECT OF HYDRAULIC CONDUCTIVITY ON THE RESIDENCE TIME OF METEORIC GROUND WATER IN BERMUDIAN- AND BAHAMIAN-TYPE ISLANDS

H. L. Vacher
Department of Geology
University of South Florida
Tampa, Florida 33620

T. O. Bengtsson
Southwest Water Management District
2379 Broad Street
Brooksville, Florida 33512

ABSTRACT

In isolated islands consisting of young limestones, it is common for units with higher hydraulic conductivity (K) to underlie units with lower K. The different hydrostratigraphic units are commonly separated by paleosols or emergence surfaces, and the increasing K with age reflects the more extensive development of secondary permeability in the older units. The contrast in K is commonly one or more orders of magnitude.

In the larger islands of the Bahamas and in Big Pine Key, Florida, fresh-water lenses are large enough that the fresh-water/sea-water interface reaches to the buried high-K units; the lenses in these "Bahamian-like islands" are truncated, with strong geologic control of the base of the lens. In Bermuda, the succession of different-K units is more lateral, reflecting the lateral relative-age pattern resulting from the shoreward accretion of successive interglacial beach-dune complexes during the Pleistocene buildup of that island; lenses in "Bermudian-type islands" are strongly asymmetric, with greatest thicknesses toward the lower-K shoreline. In small islands, such as Holocene oolitic cays of the Bahamas, lenses tend to be symmetric because the islands tend to be composed of homogeneous, single-K sediment. An exception is the asymmetry of atoll-island lenses that are skewed in accordance with a lagoon-to-reef increase in grain size.

The average residence time of ground water in island lenses can be calculated from Dupuit-Ghyben-Herzberg analysis given the areal geometry of the island, distribution and magnitude of K, and distribution and magnitude of recharge. For strip islands with uniform recharge, the calculation is by integration of equations for the water-table elevation and depth to interface (which gives the volume of the lens). Then the average residence time is the volume of fresh water divided by total recharge. For homogeneous-K islands, average residence times are in the range of 5 to 20 years, if K is in the range of 10-100 m/day, porosity is 20-40 percent, recharge is a few tens of cm/yr, and the strip island is 1000 m wide; the result is directly proportional to the width of the island. In Bermudian-type islands, the effect of a strip which is half the width of the island and an order of magnitude more permeable is to reduce the average residence time by about 50 percent from that of the lens in the homogeneous-K island composed only of the low-K unit. In Bahamian-type islands, the effect of the buried high-K unit is not as large. For example, if the contact were at about 30 or 50 percent of the depth of the lens that would be present if only the lower-K unit were present, then the average residence time would be reduced by about 35 and 20 percent, respectively, given an order-of-magnitude contrast of hydraulic conductivity between the superposed layers.

INTRODUCTION

Fresh ground water in small oceanic islands typically occurs in the form of a Ghyben-Herzberg lens (Fig. 1). The typical picture of a Ghyben-Herzberg lens is like that of Figure 1A, where the lens is symmetric and completely contained in a single rock unit. Different geometries occur in Bermuda (Fig. 1B) and the larger islands of the Bahamas (Fig. 1C). These different geometries reflect differences in the spatial variation of hydraulic conductivity in these islands.

The shape and size of an island lens can be calculated given the areal geometry of the island, the distribution and magnitude of recharge, and the distribution and magnitude of hydraulic con-

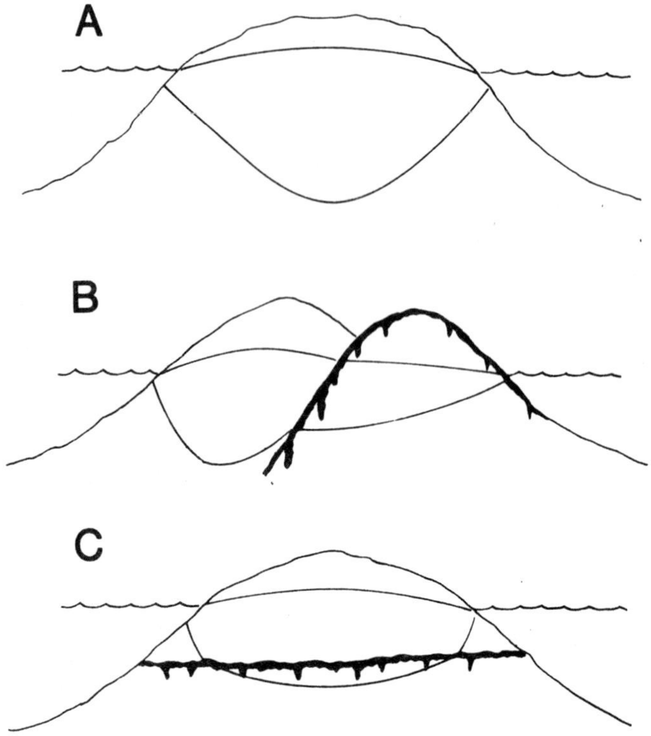

Fig. 1. Three types of island lenses considered in this paper: In A, the island is composed of homogeneous sediment. In B, the island consists of a younger, less permeable unit that onlaps a more permeable limestone. In C, the older, more permeable limestone completely underlies the less permeable unit.

ductivity (Fetter, 1972; Vacher, 1988). Equations for infinite-strip islands with the kinds of hydraulic-conductivity variations shown in Figure 1 are derived by Vacher (1988). The purpose of this paper is to extend the analysis to assess the residence time of fresh ground water in these types of islands.

BACKGROUND

Lenses in General

The fresh ground water is derived from recharge on the island and flows shoreward to discharge at the shoreline. Because of the shoreward flow, the elevation of the water table decreases toward the shoreline in accordance with Darcy's Law. Because of the difference in densities between the fresh ground water and underlying seawater, the column of fresh ground water extends to some depth below sea level. If there is no flow in the salt-water region (i.e., salt-water head is zero), the depth to the fresh-water/salt-water interface is related to the elevation of the water table by the well-known Ghyben-Herzberg Principle. This stipulates that the ratio of the interface depth to the water-table elevation at opposite ends of an equipotential (Hubbert, 1940; Vacher, 1988) is equal to the density-difference ratio (the density of the fresh water divided by the difference in densities between the fresh water and seawater). This ratio is usually about 40:1. With horizontal ground-water flow, equipotentials are vertical, so the Ghyben-Herzberg Principle applies along a vertical line. Then, the shape of the interface is simply a 40-fold-exaggerated mirror image of the water table. The water table and interface, therefore, define a lens-shaped geometry, the Ghyben-Herzberg lens.

The residence time of fresh ground water in the lens refers to the length of time for water to flow from its point of entry at the water table to its discharge at the shoreline. As illustrated in Figure 2, this amount of time depends on where the water enters the lens, if recharge is distributed across the island. For example, the residence time of water entering at A is larger than that of water entering at B, and it is largest for water that enters in the vicinity of the

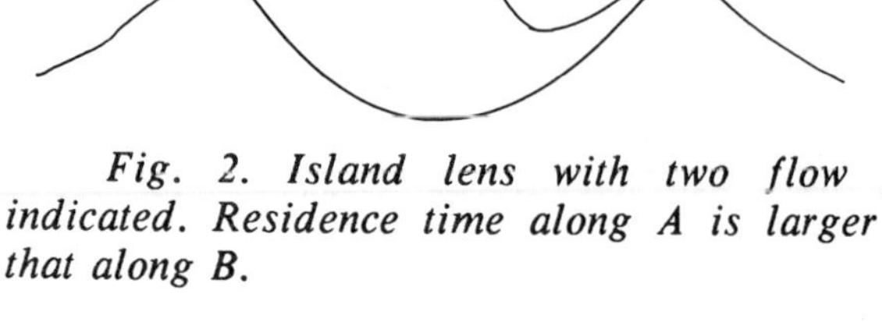

Fig. 2. Island lens with two flow paths indicated. Residence time along A is larger than that along B.

divide. Despite this variability, the overall average is the volume of water in the lens divided by the total flux through it (Vacher et al., in review). For an infinite-strip island,

$$\tau = nV/RL \qquad (1)$$

where is the average residence time, n is porosity, V is the volume of the lens (per unit length parallel to the shoreline), R is recharge, and L is width of the cross section.

The diagrams of Figures 1 and 2 are oversimplifications, because they portray the fresh-water/salt-water contact as a sharp interface. Actually a gradational transition or mixing zone is present (Fig. 3). The presence of such a transition zone implies that there is a discharge of brackish ground water to the shoreline. From the conservation of salt, this implies a circulation of seawater (Cooper, 1959). The inland flow of seawater, in turn, implies that salt-water heads beneath the lens are negative, that is, less than sea level. For such cases of non-zero salt-water heads, the relationship between the depth to the interface and the elevation of the water table is given by the Hubbert equation (Hubbert, 1940; Vacher, 1988). This equation can be considered a two-term, or expanded version of the Ghyben-Herzberg Principle. The negative salt-water heads mean that the ratio of interface depth to water-table elevation is somewhat larger than 40; therefore, the depth to the interface (the midline of the transition zone, in this case) is deeper than that found by not considering the negative salt-water heads. Residence times of fresh ground water would be somewhat larger, because V is larger in equation 1.

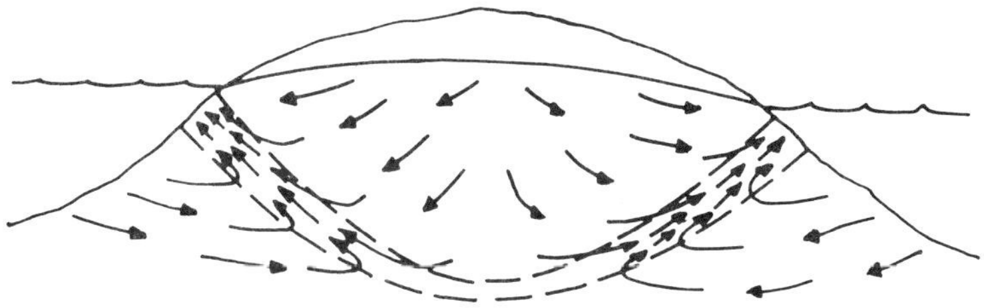

Fig. 3. Diagrammatic flow system in an island lens and associated transition zone.

Method of Analysis

First-cut analyses of Ghyben-Herzberg lenses assume that (1) salt-water head is zero and (2) equipotentials of the fresh-water flow are vertical. The first assumption means that the presence of the transition zone is ignored and that sea level all around the island is at the same elevation; this allows the one-term Ghyben-Herzberg Principle to be used. The second assumption, that flow is horizontal, means that the Ghyben-Herzberg Principle can be applied to equipotentials, rather than curved ones. The justification for this assumption is analyzed by Vacher (1988), and follows from the fact that these lenses are very thin relative to their width -- much more so than the cartoons that portray these ground-water bodies would suggest. The horizontality of the flow system allows for Dupuit analysis, a standard analytical procedure that greatly simplifies application of Darcy's Law to flow problems.

Combination of Dupuit analysis with the one-term Ghyben-Herzberg Principle leads to Dupuit-Ghyben-Herzberg analysis (Bear, 1972; Fetter, 1972; Vacher, 1988), which is used here to evaluate the volume of lenses in islands with the kinds of hydraulic-conductivity variations as occur in Bermuda and the larger Bahamian islands (Fig. 1). From the volume, estimates of residence time follow immediately from equation 1. The results underestimate the average residence time of fresh ground water, because Dupuit-Ghyben-Herzberg analysis ignores the presence of the mixing zone. To take the mixing zone into account would require variable-density solute-transport modeling, which is a much more difficult procedure. It is suspected that the results from this more sophisticated approach, for fresh-water residence times, would not be significantly larger than those calculated here.

It is assumed throughout this paper that the islands are infinite strip, meaning that the bounding shorelines are parallel and the islands are very long relative to their width. It is also assumed that recharge is constant, so that the lenses are steady-state. Because the intent of the paper is to focus on the effects of variations in hydraulic conductivity, recharge is taken also to be areally uniform.

THE BASIC LENS: HOMOGENEOUS HYDRAULIC CONDUCTIVITY

Configuration of the Lens

Mathematically and conceptually, the simplest strip-island lens is one in an island with homogeneous and isotropic hydraulic conductivity (K) and areally uniform recharge (R). The eleva-

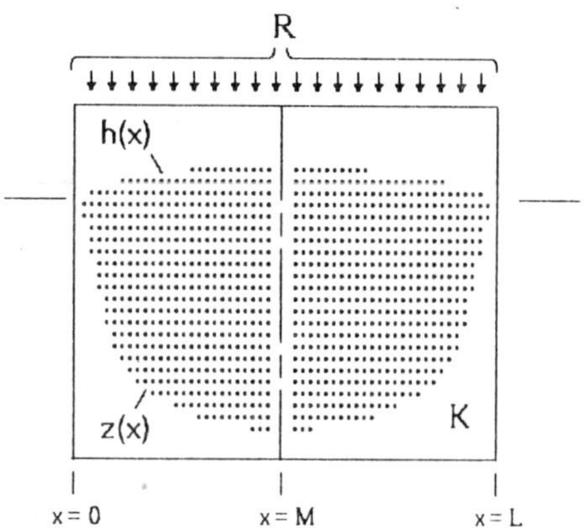

R = recharge
K = hydraulic conductivity
L = island width
M = coordinate of divide

x = distance from divide
ρ_f = fresh-water density
ρ_s = seawater density
$\alpha = \rho_f / (\rho_s - \rho_f) \cong 40$

Then:

(1) Elevation of water table above sea level: $h^2 = \frac{R}{K(\alpha + 1)} (Lx - x^2)$

(2) Depth of interface below sea level: $z^2 = \frac{R\alpha^2}{K(\alpha + 1)} (Lx - x^2)$

(3) Thickness of fresh-water column: $H^2 = \frac{R(\alpha + 1)}{K} (Lx - x^2)$

(4) Divide: $M = L/2$

Fig. 4. Parameters and equations for water table, interface and fresh-water thickness as function of distance from shoreline in homogeneous, strip-island lens from Dupuit-Ghyben-Herzberg analysis. Derivations are in Vacher (1988).

tion of the water table, the depth to the interface, and the thickness of the fresh-water column are easily calculated from Dupuit-Ghyben-Herzberg expressions included in Figure 4. As indicated by these equations, the highest elevation of the water table, the largest depth of the interface and thickest column of fresh ground water occur along the central axis of the island. The lens is symmetric about this axis.

Some features of these homogeneous-island solutions are worth noting. First, the thickness/width ratio of the lens is a function of the ratio of recharge to hydraulic conductivity. That is, the thickest column of fresh water (H_m) is given by

$$H_m{}^2 = R(\alpha + 1)L^2/4K \quad (2),$$

where the terms are as defined in Figure 4. Therefore, the thickness/width ratio of the lens is

$$H_m/L = (\alpha + 1)^{1/2}(R/K)^{1/2}/2 \quad (3).$$

Because R/K is commonly in the range of 10^{-4} to 10^{-6}, this equation implies that island lenses are commonly very thin compared to their width -- on the order of 1:30 to 1:100 (Budd, 1984; Vacher, 1988). Typical pictures of island lenses, where they are drawn as fat, squat icebergs, are really cartoons; real lenses, in fact, are long, thin slivers. As an example, consider an island 1 km across, with K=10 m/day, and R=0.3 m/yr (which represent a fine sand and a Bermudian recharge, respectively). According to equation 2, the thickest column of fresh water is 28.3 m; from equation 3, the thickness/width ratio is about 3 percent. The conclusion to be drawn is that fresh-water, phreatic diagenesis is clearly limited to only a relatively thin zone below sea level in the exposed limestones -- in small islands, at least.

The second point illustrated by equation 3 is the effect of hydraulic conductivity on the thickness of the fresh-water column. Hydraulic conductivity ranges over several orders of magnitude. Islands composed of limestones with relatively large hydraulic conductivity will have relatively thin lenses. For example, if K=90 m/day (corresponding to upper Pleistocene calcarenites of Bermuda), rather than the 10 m/day assumed in our hypothetical example, then the thickness and thickness/width ratio would both be reduced to a third, 9.4 m and about 1 percent, respectively.

The third point is the linear dependence of the thickness on the width of the island (equation 2). If our hypothetical island were 10 km rather than 1 km wide (with K=10 m/day), the thickness of the lens then would be 10 times larger, or 283 m (and the thickness/width ratio would still be 2.8 percent). In real islands, however, this greater depth of the interface in larger islands generally means that the interface reaches down to some older unit that has a different hydraulic conductivity. If this occurs, then hydraulic conductivity is not homogeneous within the fresh-water phreatic zone, and the equations of Figure 4 do not apply. This is exactly the case documented by Cant and Weech (1986) for the larger islands of the Bahamas. These islands are sufficiently large that the interface intersects deeper units that have a significantly larger hydraulic conductivity (e.g., Fig. 1C).

Residence Time

The volume of the strip-island lens with homogeneous K and uniform R is found by integrating Hdx from x=0 to x=L, where H(x) is from Figure 4. Combination of this volume with equation 1 leads to an expression for the average residence time of fresh ground water in strip-island lenses

$$\tau = (n \pi L/8) (\alpha + 1)^{1/2}/(R K) \quad (4).$$

For the hypothetical example where L=1000 m, K=10 m/d, and R=0.3 m/yr, the average residence time works out as 15.2 yrs, if the porosity were 0.2.

Figure 5 shows the dependence of residence time on recharge and hydraulic conductivity. The standard island is the example used above, where the maximum interface depth is 28.3 m and the average residence time is 15.2 yrs. From equation 4, it is seen that the residence time would be half as large if either the recharge or hydraulic conductivity were increased by a factor of 4. The reason for this is obvious in the case of hydraulic conductivity: if K were increased by a factor of 4, the lens would be everywhere thinned by a factor of 2 (Fig. 4), so V/RL also would be decreased by a factor of 2. The dependence on recharge is not as immediately apparent. If the recharge were increased by a factor of 4, the lens would be twice as large, but, because the total flux, RL, would also be four times as large, V/RL would be half as large. Accordingly, the residence time varies with the -1/2 power with both hydraulic conductivity and recharge, as is indicated by equation 4 and shown in Figure 5.

Because residence time varies linearly with respect to both porosity and island size, the residence times in Figure 5 are easily converted to give the residence times in islands with different L and n than those used in Figure 5 (1000 m and 0.2, respectively). For example, given the same K and R as in the hypothetical example where the residence time is 15.2 yrs, the residence time would be 30.4 yrs, if the island were 2000 m wide, or the porosity were 0.4.

SPATIAL VARIATION IN HYDRAULIC CONDUCTIVITY: BERMUDIAN- AND BAHAMIAN-TYPE ISLANDS

Increase of Hydraulic Conductivity with Age

Smith et al. (1976, Fig. 6.3) give a useful summary plot showing the geological variability of hydraulic conductivity of limestones (Fig. 6). The hydraulic conductivities shown in the figure were calculated by Smith et al. (1976) by assuming the pore system to be a bundle of straight, parallel tubes. As shown in the figure, the hydraulic conductivities of limestones (not counting the "marble") range from less than 10^{-3} m/day in "massive limestones" to more than 10^{6} m/day in cave systems -- more than 9 orders of magnitude. The important point for the present

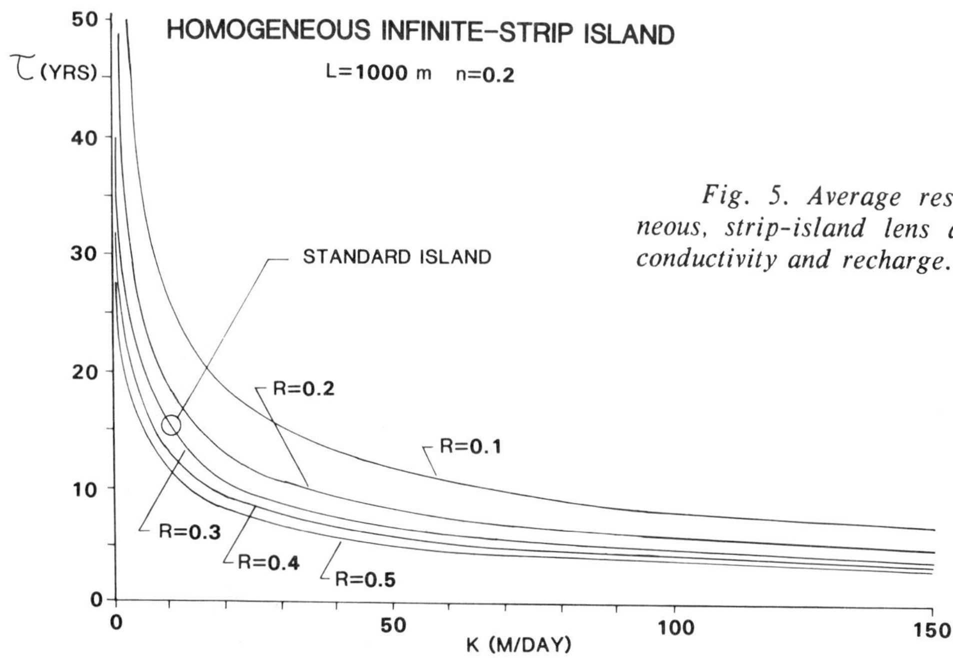

Fig. 5. Average residence time in homogeneous, strip-island lens as function of hydraulic conductivity and recharge.

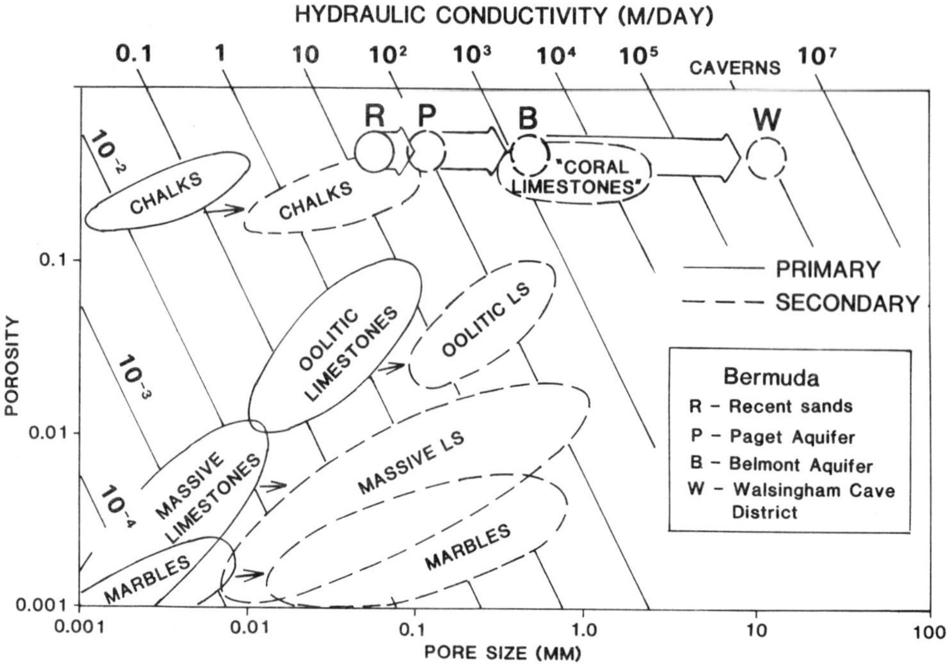

Fig. 6. Hydraulic conductivity of limestones as function of development of secondary porosity. The base figure is from Smith et al. (1976). The added data from Bermuda indicate the increase in hydraulic conductivity as a function of age due to meteoric diagenesis and karstification.

discussion is the orders-of-magnitude difference in hydraulic conductivity as secondary porosity is developed. According to the numbers of Smith et al. (1976), the hydraulic conductivity of oolitic limestones with primary porosity is in the range of 10^{-1} to 10^{1} m/day, and, with the development of secondary porosity ("fissured oolites," according to Smith et al., 1976), the hydraulic conductivity of the oolitic limestones is 1 to 2 orders of magnitude higher, 10^{1} to 10^{2} m/day. Young "coral limestones" are higher, 10^{3} to 10^{4} m/day, according to Smith et al. (1976), and truly cavernous limestones are some 2 orders of magnitude higher still.

The numbers calculated by Smith et al. (1976) are similar to those found from hydrogeologic studies in Bermuda. Representative Bermudian hydraulic conductivities are included in Figure 6. These numbers are obtained by a variety of methods that involve large-scale, in-situ stress tests, such as response to pumping and dampening of ocean tides (Vacher, 1978). As such, they are large-scale, bulk parameters, representing an average over a significant volume of rock. They are not comparable to the lower permeabilities determined on the scale of perm plugs, for example, which give only the permeability between the more important, secondary pores.

Typical Bermuda hydraulic conductivities are on the order of 10^{1} to 10^{3} m/day for the Pleistocene eolianites that occur in the region of the fresh-water resource. A fundamental fact is that the lower values occur in younger stratigraphic units. For example, the two principal hydrostratigraphic units are the Paget and Belmont aquifers (Vacher, 1978; Rowe, 1984). The Paget unit was deposited during the Sangamon Interglacial, 130 to 80 thousand years ago. The hydraulic conductivity of most of this unit is about 80 to 100 m/day. Younger portions of the Paget aquifer correspond to the Southampton Formation (Land et al., 1967; Vacher and Harmon, 1987) which was deposited during post-Substage 5e sea-level events, and in these areas the Paget aquifer appears to have a hydraulic conductivity of about 30 m/day. In contrast to the Paget, the Belmont aquifer has a hydraulic conductivity of about 1000 m/day; this heterogeneous unit includes the Belmont and Town Hill Formations (Vacher and Harmon, 1987), which were deposited in earlier, mid-Pleistocene interglacials (probably during Isotope Stages 7 to 11). At the low extreme of the range of hydraulic conductivity in Bermuda, the recent beach and dune sediment has a hydraulic conductivity of about 20 m/day, judging from its grain-size characteristics and absence of secondary porosity. At the high extreme of the range, there is the Walsingham cave district, which occurs in the oldest limestone formation exposed on the island. Water-table fluctuations in the caves (measured by Tom Iliffe, pers. comm., 1982) suggest a hydraulic conductivity comparable to the 10^6 and 10^7 m/day that Smith et al. (1976) estimated for "caverns" in their summary figure (Fig. 6).

All these Bermudian limestones consist of the same material (bioclastic calcarenite) and were deposited in the same environment (coastal dunes and associated beaches). The differences in hydraulic conductivity reflect only differences in age -- i.e., the extent of alteration by meteoric diagenesis and karstification. This pattern of increasing hydraulic conductivity with increasing age reflects the redistribution of porosity that occurs as the limestones are subjected to circulation of meteoric ground water through island lenses and associated mixing zones. This alteration in fresh-water lenses and mixing zones occurred discontinuously, as the limestones were exposed to phreatic processes only during the peaks of sea-level highstands; therefore, the correlation of larger hydraulic conductivity with age can also be viewed as a correlation of hydraulic conductivity with number of phreatic soakings. In any case, the transformation is from intergranular to cavernous porosity, and it is accompanied by a reduction in the ratio of surface area to pore volume within the pore system (although bulk porosity may actually be reduced; Land et al., 1967). The result is the observed trend: older limestones are sequentially more permeable, by orders of magnitude as shown in Figure 6.

Bermudian- and Bahamian-type Islands

Given the pattern of increasing hydraulic conductivity with increasing geologic age for limestones altered by coastal and island ground waters in the zone of eogenetic diagenesis (Choquette and Pray, 1970), we can define 3 patterns of hydraulic-conductivity variation in small limestone islands such as occur in Bermuda and the Bahamas:

1. Small islands composed of a single hydrostratigraphic unit. Small, in this case, is a relative term, meaning that the areal dimensions are sufficiently small that the interface does not intersect another hydrostratigraphic unit. Obvious examples include Bahamian oolitic cays such as Joulters and the Schooners studied by Halley and Harris (1979) and Budd (1984), respectively, and the many erosional outliers off the 5 main islands comprising Bermuda.

2. Larger islands in which there is a lateral succession of stratigraphic units. Larger islands of Bermuda are examples (Vacher, 1987). These islands are composed of a lateral succession of Pleistocene beach-dune complexes deposited during successive interglacials. Beach-dune complexes of more recent interglacials onlap the seaward margin of earlier-deposited, more diagenetically altered limestones. Older limestones border inshore water bodies (sounds and reaches off the North Lagoon), and younger limestones border the external shoreline. The overall pattern of cross-island sections, then, is one of lateral accretion of successively less-altered limestones, from large hydraulic conductivities along interior shorelines to smaller hydraulic conductivities along external shorelines. These islands with a lateral variation in hydraulic conductivity will be called "Bermudian-type islands."

3. Larger islands in which there has been vertical buildup rather than lateral accretion. In these islands, older, more permeable units completely underlie the less altered units that occur at the water table. According to the analysis by Cant and Weech (1986), larger islands of the northern Bahamas, (e.g., Grand Bahama and Abaco) are examples. The main hydrostratigraphic contact is the base of the Pleistocene Lucayan Formation. Cant and Weech (1986) also note the presence of discontinuity or subaerial-exposure surfaces within the Pleistocene section that correspond to abrupt downward increases in hydraulic conductivity. Such islands with low-K layers above high-K layers will be called "Bahamian-type islands."

SHAPES OF LENSES IN BERMUDIAN- AND BAHAMIAN-TYPE ISLANDS

1. Small Cays

Small islands composed of a single uniform lithology (e.g., Joulters, Schooners Cays) are homogeneous islands. They contain the simple, symmetric lenses like that shown in Figure 1a. Equations of Figure 4 apply.

2. Bermudian-Type Islands

If the upper saturated zone consists of laterally adjoining sectors of different hydraulic conductivity, the fresh-water lens is asymmetric. Equations for the water table, depth to interface,

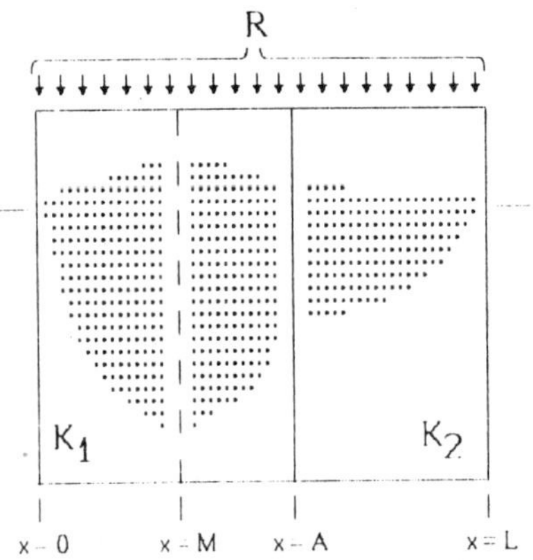

K_1, K_2 hydraulic conductivities A = coordinate of geologic contact

Then:

(1) Divide: $M = (1/2)\dfrac{K_1(L^2 - A^2) + K_2A^2}{K_1(L - A) + K_2A}$

(2) Water table

(a) $0 \leq x \leq A$ $\quad h^2 = \dfrac{R}{K_1(\alpha + 1)}(2Mx - x^2)$

(b) $A \leq x \leq L$ $\quad h^2 = \dfrac{R}{K_1(\alpha + 1)}(2MA - A^2) + \dfrac{R}{K_2(\alpha + 1)}[2M(x - A) - (x^2 - A^2)]$

(3) Interface: $z = \alpha h$

(4) Fresh water thickness: $H = (\alpha + 1)h$

Fig. 7. Dupuit-Ghyben-Herzberg equations for configuration of lens in a strip island with Bermudian-type variation in hydraulic conductivity. Parameters same as in Figure 4. Derivations are in Vacher (1988).

and thickness for the two-sector case (Vacher, 1988) are given in Figure 7.

Cross sections calculated from these equations are shown in Figure 8. The set of cross sections shows two ratios of hydraulic conductivities (K2/K1 at 10 and 100) and three positions of the geologic contact (at 0.25, 0.5, and 0.75 of the island width). The scales showing island width, depth to interface, and elevation of the water table in meters are based on an island where L=1000 m, R=0.3 m/yr, and K1=10 m/day. In addition to these scales, a dimensionless scale for the depth to the interface (z/zc) is also shown. This scale indicates the ratio of the depth of the interface (z) in the two-K lens to the depth of the interface at the midline (x=L/2) of the lens that would occur if the island were composed only of K1 (zc). The dimensionless cross sections (h/hc and z/zc vs x/L) apply to any lens with the same ratio of A/L and K2/K1 (i.e., they do not depend on L, R or K1; Vacher, 1988).

From Figure 8, it is obvious that the amount of asymmetry of the lens in a Bermudian-type island depends on the contrast in hydraulic conductivity between the different-K segments and the location of the contact. The thickest column of fresh ground water is in the lower-K unit, which acts as a relative dam to the shoreward flow of fresh ground water; the high-K unit acts as a relative drain. Accordingly, the groundwater divide occurs relatively close to the shoreline bordered by the lower-K rocks, and the majority of recharge exits at the higher-K shoreline.

The Bermuda example is shown in Figure 9.

3. Bahamian-Type Islands

If the upper saturated zone consists of horizontal layers of contrasting hydraulic conductivity, the effect of the refraction of the interface as it passes from one layer into the other will be a thinning of the lens from the geometry that would occur if there were no geologic contact. The thinning is reflected in a lower water table, as seen from the Ghyben-Herzberg Principle. Equations for the two-layer case (Vacher, 1988) are shown in Figure 10. Examples calculated from these equations are shown in Figure 11. The examples are analogous to those of Figure 8 in the K2/K1 ratios and the relative position of the geologic contact.

As in Figure 8, the scales in Figure 11 giving the depth to the interface in meters are based on an island with L=1000 m, R=0.3 m/yr, and K1=10 m/day. Also as in Figure 8, the dimensionless scales in Figure 11 reflect the comparison of interface depth in the two-layer case to the corresponding midline value in the comparison island with the same L and R, and a single K=K1. In the two-layer case, the depth below sea level of the geologic contact (b) is made dimensionless relative to zc; the three examples are where the geologic contact lies at 25, 50, and 75 percent of the depth of the center-line interface in the one-K comparison island. Then, with these conventions, the dimensionless cross sections apply to any Bahamian-type two-layer lenses with the same K2/K1 and b/zc, regardless of recharge, size of the island, and numerical value of the hydraulic conductivities (Vacher, 1988).

As shown in Figure 11, the amount of thin-

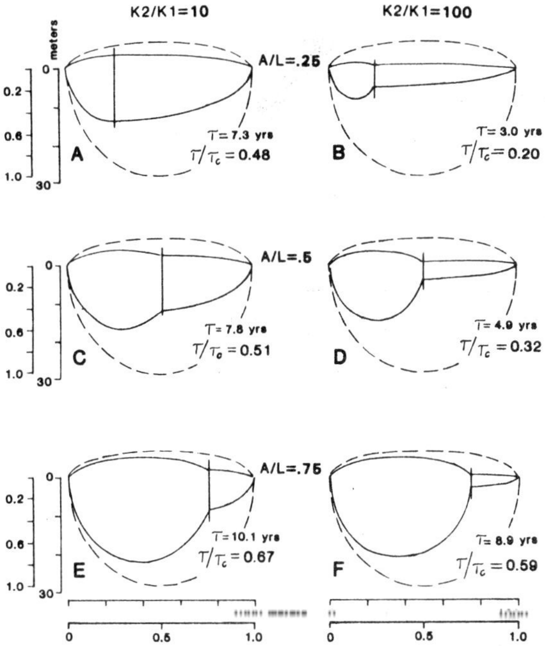

Fig. 8. Shape and average residence times in selected Bermudian-type lenses. Dashed lines indicate the comparison lens, which would be present if the island were composed only of the K1 unit.

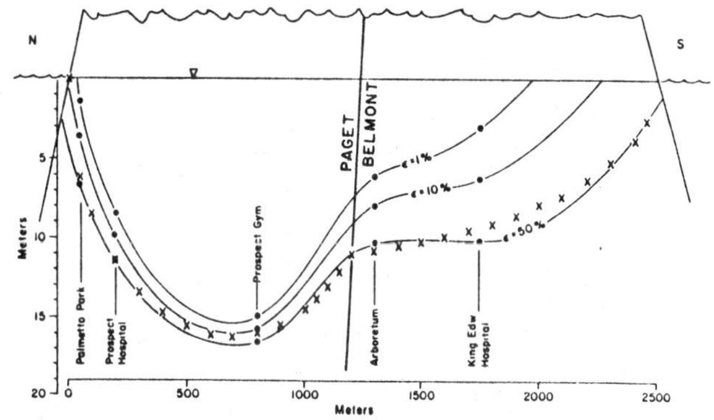

Fig. 9. Configuration of Devonshire lens Bermuda (Vacher, 1978). The 50%-seawater line indicates the position of the interface, which is at a depth 40 times the elevation of the water table (Rowe, 1984). The x's indicate the calculated position of the interface using Dupuit-Ghyben-Herzberg analysis with Bermudian parameters for recharge and hydraulic conductivities (Vacher, 1974, 1978).

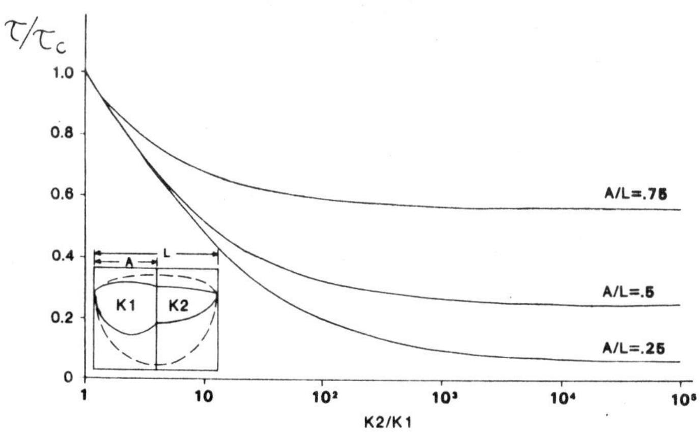

Fig. 10. Dupuit-Ghyben-Herzberg equations for configuration of lens in a strip island with Bermudian-type variation in hydraulic conductivity. Parameters same as in Figure 4. Derivations are in Vacher (1988).

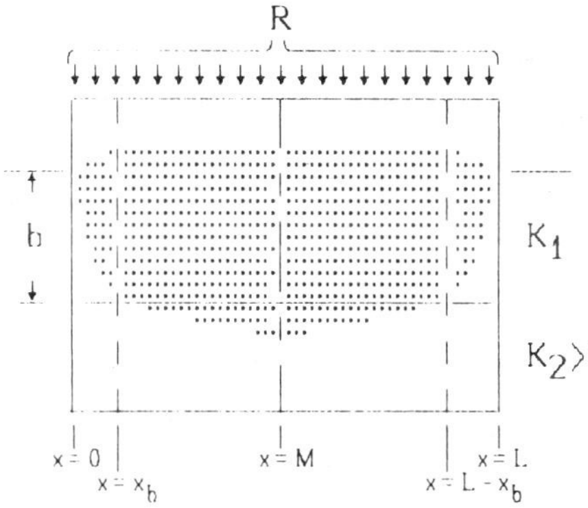

K_1, K_2 hydraulic conductivities

b = depth of geologic contact below sea level

x_b = coordinate of intersection of interface and geologic contact

h_b = elevation of water table at $x = x_b$

Then:

(1) Divide: $M = L/2$

(2) Intersection coordinate from: $2Mx_b - x_b^2 = K_1(\alpha + 1)b^2/RA^2$

(3) Water table:

(a) $0 < x < x_b$ $\quad h^2 = \frac{R}{K_1(\alpha + 1)}(Lx - x^2)$

(b) $x_b < x < M$

$$(K_1 + \alpha K_2)(h^2 - h_b^2) + 2(K_2 - K_1)b(h - h_b) = R[L(x - x_b) - (x^2 - x_b^2)],$$

where $h_b = (R/K_1)(Lx_b - x_b^2)/(\alpha + 1)$

Fig. 11. Shape and average residence times in selected Bahamian-type lenses.

ning due to the deeper high-K unit is related to the contrast in hydraulic conductivities and the depth of the geologic contact. In the extreme (K2/K1 infinite), the lens would be truncated at the geologic contact, so fresh ground water would occur only in the uppermost layer.

Such cases of strong geologic control by deeper, large-K units are described in the summary of Bahamian lenses by Cant and Weech (1986). For example:

> *"The Pleistocene formation, named the Lucayan Limestone..., is the main freshwater aquifer of the Bahamas.... No older formations are known to be able to prevent freshwater from mixing with saline water. Because of this, the base of the Lucayan Limestone represents the maximum thickness to which the freshwater lens can develop....*
>
> *"...It is the shallow depth of the Lucayan Limestone in the Little Bahama Bank that effectively limits the freshwater lens thickness in Grand Bahama and Abaco islands. These islands are larger than Eleuthera and have greater rainfall, but the freshwater lenses are thinner than on Eleuthera; this is due simply to the fact that the Lucayan Limestone extends down to about 40 m in Eleuthera as compared to 21 m below sea level in the Little Bahama Bank."*

Cant and Weech (1986, p. 341) also describe the similar effects of changes in hydraulic conductivity within the Lucayan:

> *Throughout the Lucyan Limestone there are discontinuity surfaces that are interpreted as horizons of subaerial exposure.... Rapid changes in rock character can occur across a discontinuity surface, and in particular at the major one that occurs at a depth ranging from 8 to 10 m in the Central Bahamas. This particular change demarcates a significant change in rock permeability downwards, and it acts as a common barrier to the development of freshwater lenses in the smaller islands. This is a well known constraint on the island of New Providence...*

From the cross sections in Figure 11 for K2/K1=100, it is easily seen how the base of the Lucayan Limestone can exert the kind of geologic control that Cant and Weech (1986) document. With two orders of magnitude contrast in hydraulic conductivity, the critical control on the thickness of the lens would be the depth to the contact.

Comments

The classification of Bermudian-type and Bahamian-type hydraulic-conductivity variations and lens geometries is intended to characterize endmembers only, and it is somewhat arbitrary. From the descriptions by Cant and Weech (1986), Bermudian-type hydraulic-conductivity variations and lenses are present locally in the Bahamas, and there are also lens geometries that do not fit either pattern. For example, Cant and Weech (1986, p. 342) note the following Bermudian-type relation between the lower-K Holocene sands and the higher-K Lucayan:

> *"In addition to the Lucayan Limestone, freshwater can also occur in Holocene sands. Very large lenses are known to occur in parts of Eleuthera, Abaco, Cat Island, Exuma, and in many other islands where such sand bodies occur in coastal areas. Quite often freshwater will occur in a sand where it would not occur in a rock formation, and this is because permeabilities are so low in the sand.... On many of the long thin islands of the central Bahamas, freshwater only occurs where there are sands that can effectively reduce seepage losses to sea."*

Figure 8B shows the kind of geometry that seems to be implied by the descriptions of Cant and Weech (1986) for islands where a package of low-K Holocene sand abuts against the highly permeable limestone bedrock that comprises the main part of the island.

Cant and Weech (1986) also indicate that in the southern, drier Bahamas, there is sufficiently low recharge that the thin lenses are effectively "broken up" by hypersaline ponds which, from their descriptions, act as evaporation sinks. The resulting lens configuration then would be the result of lateral variations in recharge (with negative R at the sinks), rather than a control by the distribution of hydraulic conductivity. The descriptions from Cant and Weech (1986) that we quote above to illustrate lenses with a Bahamian-type variation in hydraulic conductivity are said by Cant and Weech (1986) to be characteristic in the northern Bahamas which receive more than some 115 cm/yr of rainfall.

It should be noted also that a Bahamian-type variation in hydraulic conductivity is present in Bermuda as well as in the Bahamas. Deep drilling beneath the Bermudian lenses reveals cavernous limestone near the contact with the underlying volcanic seamount. From the limited data available, tidal fluctuations in head are significantly larger in these deeper units than are the fluctuations of the water table, meaning there is a significantly larger hydraulic conductivity of these deeper limestones. These deeper limestones, however, are well below the depth reached by the interface in the Paget and Belmont aquifers, so these deeper layers play no role in the configuration of the lens. This would not be the case at a lower stand of sea level, however, when the exposed Bermuda would have the dimensions similar to a present-day Bahamian island.

RESIDENCE TIME IN BERMUDIAN- AND BAHAMIAN-TYPE ISLANDS

The residence time for each of the patterns of hydraulic conductivity variation can be found from the analytical solutions for the across-island variation of the thickness of the lens. The volume of the lens is found from numerical integration of Hdx using the equations in Figures 7 and 10.

Bermudian-Type Islands

Calculated results for the six cases of Figure 8 are shown in that figure. For example, the average residence time (τ) in a two-sector lens with K1=10 m/day, K2=100 m/day, R=0.3 m/yr, n=0.2, and L=1000 m is 7.8 yrs, if the geologic contact is at A=500 m (Fig. 8C). The residence time (τ_c) of the comparison homogeneous island (K=10 m/day, R=0.3 m/yr, L=1000 m, and n=0.2) is 15.2 yrs, from equation 4. Therefore the dimensionless residence time (τ/τ_c) for this two-layer island is 0.51. This dimensionless residence time holds for any two-layer island with the same K2/K1 and A/L, independent of size, recharge, porosity and hydraulic conductivity. That is, the effect of the high-K sector in a Bermudian-type island is a reduction of the residence time by 49 percent, if the hydraulic conductivity of the higher-K sector were an order of magnitude larger and the sector occupied 50 percent of the cross-island width of the lens.

Dimensionless residence time is shown as a function of K2/K1 and A/L in Figure 12. Obviously, the amount of reduction of residence time is related to the contrast in hydraulic conductivity and the percentage of the island that is made up of the higher-K rock.

Bahamian-Type Islands

Calculated average residence times for the six Bahamian-type islands of Figure 11 are shown in that figure. As in Figure 8, the results are given both in terms of years for the particular cases calculated and dimensionless residence times formed by comparison with the homogeneous lens with K=K1. These dimensionless residence times are the same in all lenses with the same K2/K1 and b/zc, where the latter is the ratio of the depth to the geologic contact to the depth of the center-line interface (x=L/2) in the comparison island. For example (Fig. 11 C), the effect of the buried high-K layer would be to reduce the residence time by 19 percent if there were an order-of-magnitude contrast in hydraulic conductivity, and the geologic contact were to occur half way to the deepest interface that would be present if the deeper layer were not present.

Dimensionless residence time for Bahamian-type islands is shown as a function of K2/K1 and b/zc in Figure 13. As in the Bermudian-type island, the amount of reduction of residence time varies directly with the permeability contrast and inversely with the percentage of the comparison-island lens that is occupied by the lower-K unit. If the depth to the geologic contact is relatively large, so that the deeper unit "chops off" only the lower part of the homogeneous-island lens, then the reduction in residence time, obviously, is relatively small.

Comments

Other examples than those given here (Figs. 8 and 11) can be easily calculated from the properties of dimensionless residence time and the graphs of dimensionless residence time vs ratios of hydraulic conductivity (Figs. 12 and 13). One need not get into the equations and numerical integration; one only needs to recall that

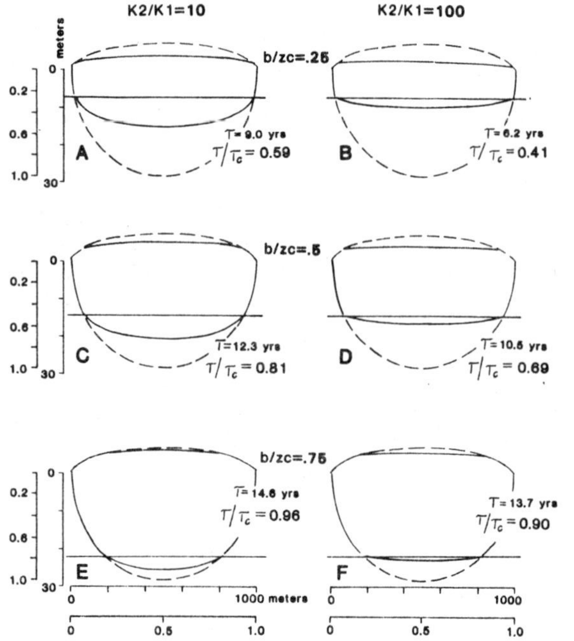

Fig. 12. Dimensionless average residence time in a two-sector, strip-island lens with a Bermudian-type hydraulic-conductivity variation.

average residence time is 15.2 yrs in the standard island of Fig. 5 (a homogeneous island with K=10 m/day, R=0.3 m/yr, n=0.2, and L=1000 m). For Bahamian-type islands, it is useful to recall that the deepest interface is 28.3 m for this same island. To illustrate the technique, consider the following problem set of two "unknowns."

a. Suppose the island is 3000 m wide and consists of permeable Pleistocene limestones onlapped by a package of Holocene sand. Suppose the Holocene unit is 1000 m wide, and pump tests and/or tide studies indicate K=2000 m/day in the Pleistocene and K=40 m/day in the Holocene. Suppose that it is also known that recharge is 0.2 m/yr and the porosity of both units is 0.25.

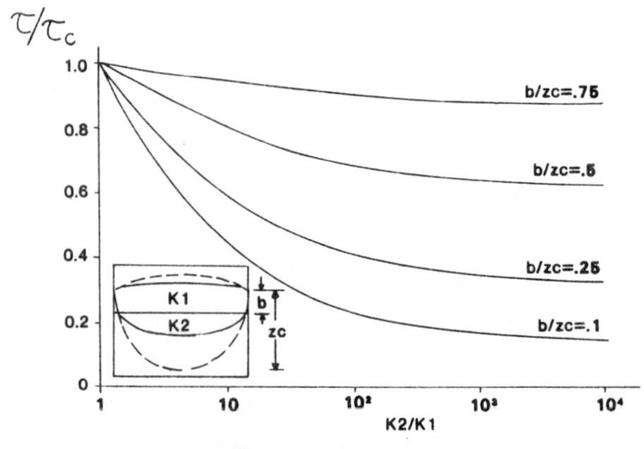

Fig. 13. Dimensionless average residence time in a two layer, strip-island lens with a Bahamian-type hydraulic-conductivity variation.

Then, remembering the effect of the variables in equation 4, the residence time of the comparison island consisting only of Holocene sand can be found from the ratios of the controlling variables:

$$\tau_c/15.2 = (3000/1000)(0.25/0.2)(0.2/0.3)^{-1/2}(40/10)^{-1/2}$$

Then τ_c=32.9 yrs. Next, the critical ratios for our problem island (K2/K1=50 and A/L=.33) are used with the graphs of Figure 12 to estimate the dimensionless residence time for the problem island (approximately 0.3, by interpolation). This result, with τ_c, indicates that the average residence time in the problem lens is about 10 yrs.

b. Suppose the island is 8000 m across and consists of Pleistocene rocks (K=500 m/day) overlying older limestones (K=80,000 m/day). Suppose the contact is horizontal at 15 m below sea level, that R=0.4 m/yr, and n of both units is 0.3. Then, by the same technique as in the preceding example, the residence time of the comparison island (consisting only of Pleistocene rocks) is 22.3 yrs. In order to read the dimensionless residence time for this island from Figure 13, one needs the permeability ratio (160) and b/zc. The latter is found by remembering the effect of the controlling variables in equation 1; that is,

$$zc/28.3=(0.4/0.3)^{1/2}(500/10)^{-1/2}(8000/1000) ,$$

Then zc=37 m, and b/zc=0.41. With these ratios, the dimensionless residence time for the problem island is interpolated on Figure 13 (approximately .6). The average residence time in the problem island is 8.5 years.

DISCUSSION

1. The residence time discussed in this paper is the overall average. Considering the individual parcels of water that pass through the lens, there is a wide range of residence times depending on the travel lengths of the various flow paths (Fig. 2). Residence time as a function of entry point can be found either by flow-net analysis or by a mathematical technique to find the stream function in the lens (Vacher et al., in review). Such work indicates that some 95% of the water that passes through the lens resides in the lens for less than about 6 times the overall

average residence time. Therefore, given the results of the preceding problem as an example, it is probably safe to say that practically all the fresh ground water passes through that island in less than 75 yrs, while most of it passes through in less than 10 yrs.

2. Although we have used the term "Bahamian-type lens" and it seems appropriate from the descriptions by Cant and Weech (1986), the Bahamian-type variation in hydraulic conductivity, where less-permeable limestones overlie more-permeable limestones, is certainly not limited to the Bahamas. In fact, it is widespread, and perhaps a characteristic of islands composed of relatively young limestones. The presence of buried highly permeable units in Enewetok Atoll is well known (Wheatcraft and Buddemeier, 1981), and they play an important role in the propagation of the tidal signal there. However, these units are below the lens, because the emergent atoll island is very small. The same is true in other atolls, such as Pingelap (Ayers and Vacher, 1986) and Kwajalein (Hunt and Peterson, 1980). In Pingelap, the shape of the lens is controlled partly by a lateral variation in hydraulic conductivity which is a reflection of grain-size differences between reef-bordering and lagoon-bordering sediments of the Holocene aquifer (and partly by the presence of very low-permeability reef plate near the reef). The atoll-island lens, therefore, is asymmetric as in Bermuda; however, it is less "bulbous" than in Bermuda, because the change in hydraulic conductivity is gradual rather than abrupt. But, in all these islands -- the three atolls and Bermuda as well -- a Bahamian-type downward increase in hydraulic conductivity is present or suspected from the tidal data. Similarly, in Big Pine Key (Stewart, 1988), a Bahamian-type change in hydraulic conductivity is present where the Miami Oolite overlies the Key Largo Limestone, and that contact controls the thickness of the lens.

To the extent that the configuration of the lens in limestone islands can be generalized, therefore, it appears that the size of the island is the principal control on the shape of the lens. Higher-permeability units tend to occur at depth below the sediments and rocks that occur at the water table. The cross-island geometry of the lens, then, depends on whether the size of the island (and, to a lesser extent, the recharge) is large enough to cause the interface to encounter these higher-permeability rocks. The larger Bahamian islands and Big Pine Key are large enough, so there is the kind of geologic control documented by Cant and Weech (1986) in the Bahamas. In smaller islands, where the lens's shallow interface is unlikely to encounter buried higher-permeability rocks, the cross-island variation in lens geometry depends on whether there is a lateral variation in hydraulic conductivity. In Bermuda, such a variation is present in the larger islands because of the lateral accretion pattern of the sedimentary buildup of that island. In atoll islands, a lateral variation in hydraulic conductivity is present because of the spread of a wide range of grain sizes in the storm-deposited reef debris. In both cases, the resulting lenses are asymmetric -- with thickest fresh-water columns near the young-rock shoreline in Bermuda and near the lagoon in atoll islands. In small islands that are composed of a single-age sediment or rock consisting of uniform grain sizes, the simple symmetric lens occurs. The small Bahamian cays of Joulters and the Schooners are examples.

3. The equations for a homogeneous island imply that the thickness of the lens is a few percent of the width of the island. This rule of thumb is limited to the small, young-limestone islands where homogeneous hydraulic conductivities are present. In the larger young-limestone islands, higher hydraulic conductivities are apt to be encountered by the interface, and the lenses are even thinner. This is true regardless of whether Bermudian-type or Bahamian-type variations in hydraulic conductivity are present; in both cases, the larger hydraulic conductivities lead to a smaller overall R/K ratio than the 10^{-6} that is consistent with the thickness/width ratio of about 1 percent. Given the widespread occurrence of high-permeability units that occur in the subsurface of present-day, isolated carbonate islands such as the Bahamas, Bermuda, and the atolls, it is difficult to see how Ghyben-Herzberg lenses, and associated diagenesis, can extend to below a few tens of meters below sea level, unless significant lower-permeability, presumably clastic units, are also present.

CONCLUSIONS

The average residence time of fresh ground water in an island lens depends on the volume of ground water in the lens. This volume is fundamentally related to the hydraulic conductivity of the sediments and rocks containing the lens. The effect of Bermudian-type high-permeability sectors and Bahamian-type high-permeability buried layers is to thin the lens and thereby reduce the residence time. These average residence times can be calculated from Dupuit-Ghyben-Herzberg analysis given knowledge of the recharge, location of geologic contacts, and magnitude of the hydraulic conductivities. If the island is an infinite strip and recharge is uniform, the shape of the lens and the average residence time of fresh ground water in it can be calculated from the equations and graphs of this paper.

REFERENCES CITED

Ayers, J.F., and Vacher, H.L., 1986, Hydrogeology of an atoll island: A conceptual model from detailed study of a Micronesian example: Ground Water, v. 24, p. 185-198.

Bear, J., 1972, Dynamics of fluids in porous media: New York, American Elsevier, 764 p.

Budd, D.A., 1984, Freshwater diagenesis of Holocene ooid sands, Schooner Cays, Bahamas: Ph.D. Dissertation, University of Texas at Austin, 492 p.

Cant, R.V., and Weech, P.S., 1986, A review of the factors affecting the development of Ghyben-Hertzberg lenses in the Bahamas: Journal of Hydrology, v. 84, p. 333-343.

Choquette, P.W., and Pray, L.C., 1970, Geological nomenclature and classification of porosity in sedimentary carbonates: Bulletin American Association of Petroleum Geologists, v. 54, p. 207-250.

Cooper, H.H., Jr., 1964, A hypothesis concerning the dynamic balance of fresh water and salt water in a coastal aquifer, in Cooper, H.H., Jr., Kohout, F.A., Henry, H.R., and Glover, R.E., Seawater in coastal aquifers: U. S. Geological Survey Water Supply Paper 1613-C, p. C1-C12.

Enos. P., and Sawatsky, L.H., 1981, Pore networks in Holocene carbonate sediments: Journal of Sedimentary Petrology, v. 51, p. 961-986.

Fetter, C.W., Jr., 1972, Position of the saline water interface beneath oceanic islands: Water Resources Research, v. 8, p. 1307-1314.

Halley, R.B., and Harris, P.M., 1979, Freshwater cementation of a 1000 year-old oolite: Journal of Sedimentary Petrology, v. 49, p. 969-988.

Hubbert, M.K., 1940, The theory of ground-water motion: Journal of Geology, v. 48, p. 785-944.

Hunt, C.D., and Peterson, F.L., 1980, Ground-water resources of Kwajalein Islands, Marshall Islands: University of Hawaii, Water Resources Research Center Technical Report no. 126, 91 p.

Land, L.S., Mackenzie, F.T., and Gould, S.J., 1967, The Pleistocene history of Bermuda: Geological Society of America Bulletin, v. 78, p. 993-1006.

Rowe, M., 1984, The freshwater "Central Lens" of Bermuda: Journal of Hydrology, v. 73, p. 165-176.

Smith, D.I., Atkinson, T.C., and Drew, D.P., 1976, The Hydrology of limestone terrains, in Ford, T.D., and Cullingford, C.H.D. (eds), The Science of Speleology: London, Academic Press, p. 179-212.

Stewart, M., 1988, Electromagnetic mapping of fresh-water lenses on small oceanic islands: Ground Water, v. 26, p. 187-191.

Todd, D.K., 1980, Groundwater Hydrology: New York, John Wiley & Sons, 525 p.

Vacher, H.L., 1974, Ground-water hydrology of Bermuda: Public Works Department, Bermuda Government, Hamilton, 87 p.

Vacher, H.L., 1978, Hydrogeology of Bermuda--Significance of an across-the-island variation in permeability: Journal of Hydrology, v. 39, p. 207-226.

Vacher, H.L., 1987, Nature of the lithostratigraphic record in Bermuda: Geological Society of America Annual Meeting, Abstracts with Programs, p. 874.

Vacher, H.L., 1988, Dupuit-Ghyben-Herzberg analysis of strip-island lenses: Geological Society of America Bulletin, v. 100, p. 580-

591.

Vacher, H.L., and Harmon, R.S., 1987, Penrose Conference field guide to the geology of Bermuda, 58 p.

Vacher, H.L., Bengtsson, T.O., and Plummer, L.N., in review, Hydrology of meteoric diagenesis -- Ground-water residence times and aragonite-calcite stabilization rates in island fresh-water lenses.

Wheatcraft, S.W., and Buddemeier, R.W., 1981, Atoll island hydrology: Ground Water, v. 19, p. 311-320.

FIELD GUIDE TO THE SUE POINT FOSSIL CORAL REEF SAN SALVADOR ISLAND, BAHAMAS

Brian White
Department of Geology
Smith College
Northampton, Massachusetts 01063

INTRODUCTION

Pleistocene coral reefs and associated facies are exposed in the vicinity of Sue Point on the west coast of San Salvador Island between Bonefish Bay and Victoria Hill Settlement. The Sue Point reefs provide excellent fossil examples of *Montastrea-Diploria-Porites* patch reefs. Such patch reefs are common at the present time in the shallow waters of the shelf around San Salvador Island and at numerous other localities in the Caribbean and warm water areas of the Atlantic Ocean.

As part of an ongoing study to determine the age of fossil corals by uranium-thorium dating (see Curran et al., this volume), three fossil corals from the main part of the more northerly patch reef were dated and gave the following results (Jim Chen, unpublished data, 1988): *Montastrea annularis* 121.5±1.9ky; *M. annularis*: 120.7±1.4ky; *Diploria strigosa*: 121.0±1.6ky. The Sue Point fossil reef has the same age as the Cockburn Town fossil reef on San Salvador Island and the Devil's Point reef on Great Inagua (Curran et al., this volume) and flourished during the Sangamon high stand of sea level.

In his survey of the geology of San Salvador, Adams (1983) mentioned the presence of small scattered heads of *Diploria* and *Montastrea* at Sue Point. Carew and Mylroie (1987) report a uranium-thorium date of 135±8ky for a sample of *Montastrea annularis* from Sue Point.

As part of the Keck Geology Consortium project to study fossil coral reefs, the Sue Point fossil reef area was mapped in June 1987 by Doug Cattel (Colorado College), Storr Nelson (Whitman College), Molly Stark (Smith College), and the author. Topographic base maps were surveyed on a scale of 1:200 using a combined electronic theodolite and laser distance measuring instrument. Map datum was the bench marks located near the dock in Cockburn Town. These bench marks have been tied to accurately measured mean sea level (Adams, 1983; Curran and White, 1985). Rock samples collected in June 1987 were studied by Cattel (1988), Nelson (1988), and Stark (1988a, b), as part of undergraduate independent research projects, to determine the diagenetic history of the Sue Point reef. An area of outcrop at the northeast end of the fossil reef that was covered by beach sand in 1987 was revealed in June 1988 and this was mapped by Kathy White (Smith College) and the author.

DESCRIPTION OF FIELD STOPS

Introduction

The Sue Point fossil patch reefs and associated facies are located along the west shore of San Salvador south of Victoria Hill Settlement (Fig. 1). The shoreline in this vicinity follows a zig-zag pattern and two linear fossil patch reefs are exposed. The geology of these reefs is shown in two separate maps, the northern one on Map 1 and the southern one on Map 2. Although the entire area was mapped in the same detail, the

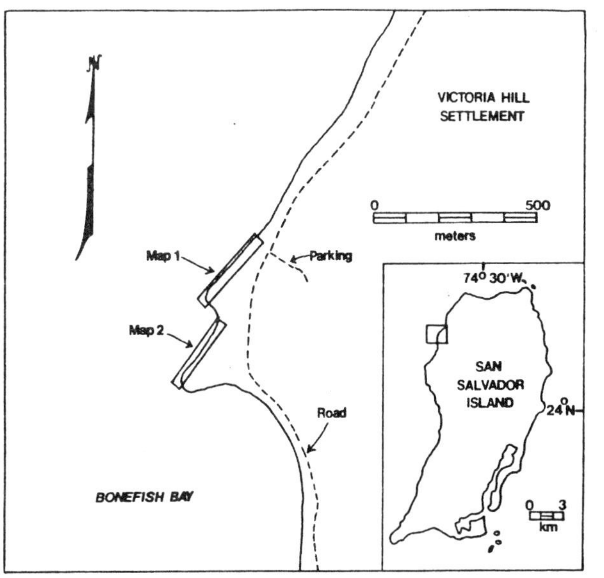

Fig. 1. Location of the fossil coral reefs at Sue Point.

MAP 1

GEOLOGY OF SUE POINT PLEISTOCENE CORAL REEF, SAN SALVADOR, BAHAMAS

NORTHERN PART

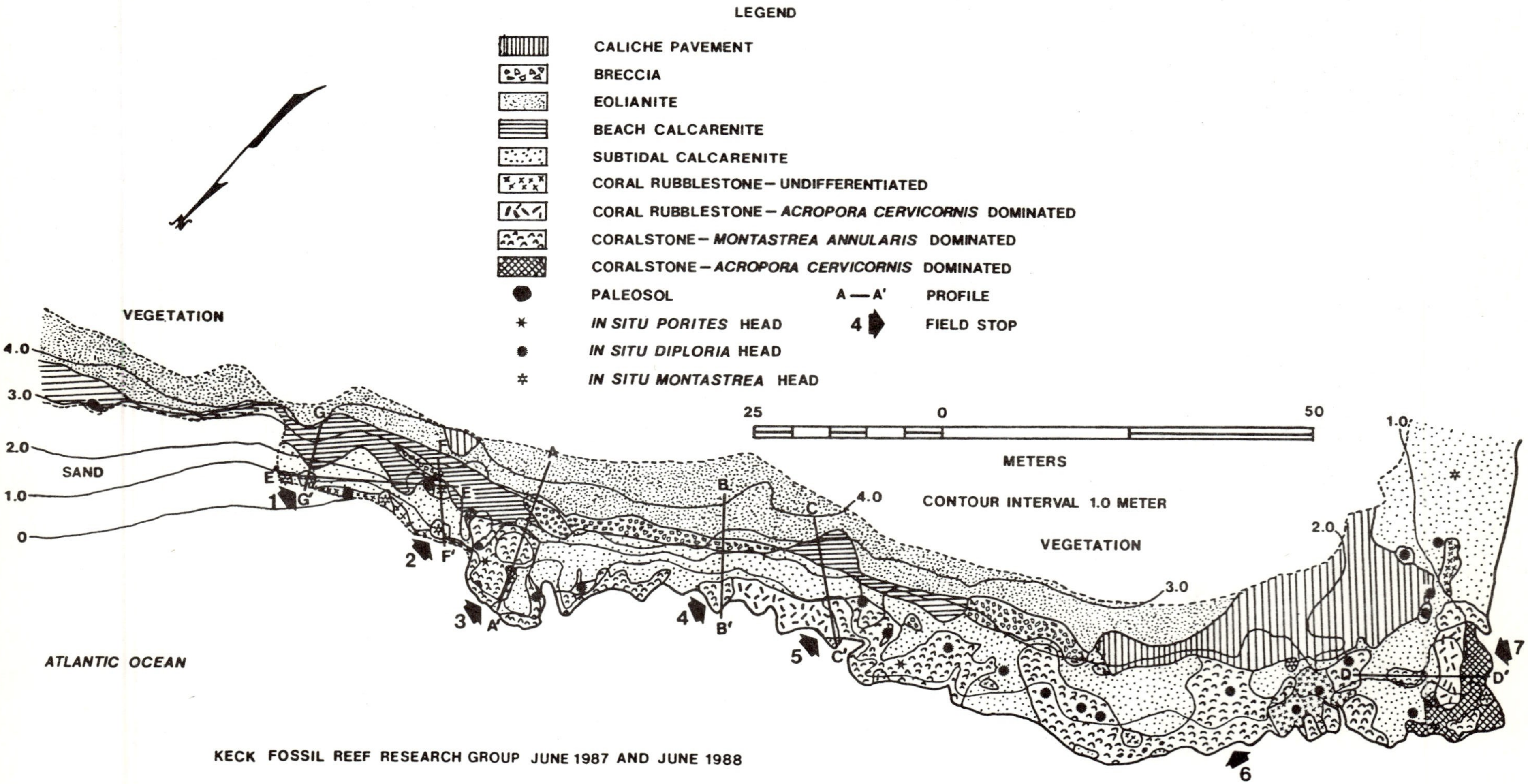

KECK FOSSIL REEF RESEARCH GROUP JUNE 1987 AND JUNE 1988

MAP 2

GEOLOGY OF SUE POINT PLEISTOCENE CORAL REEF, SAN SALVADOR, BAHAMAS

SOUTHERN PART

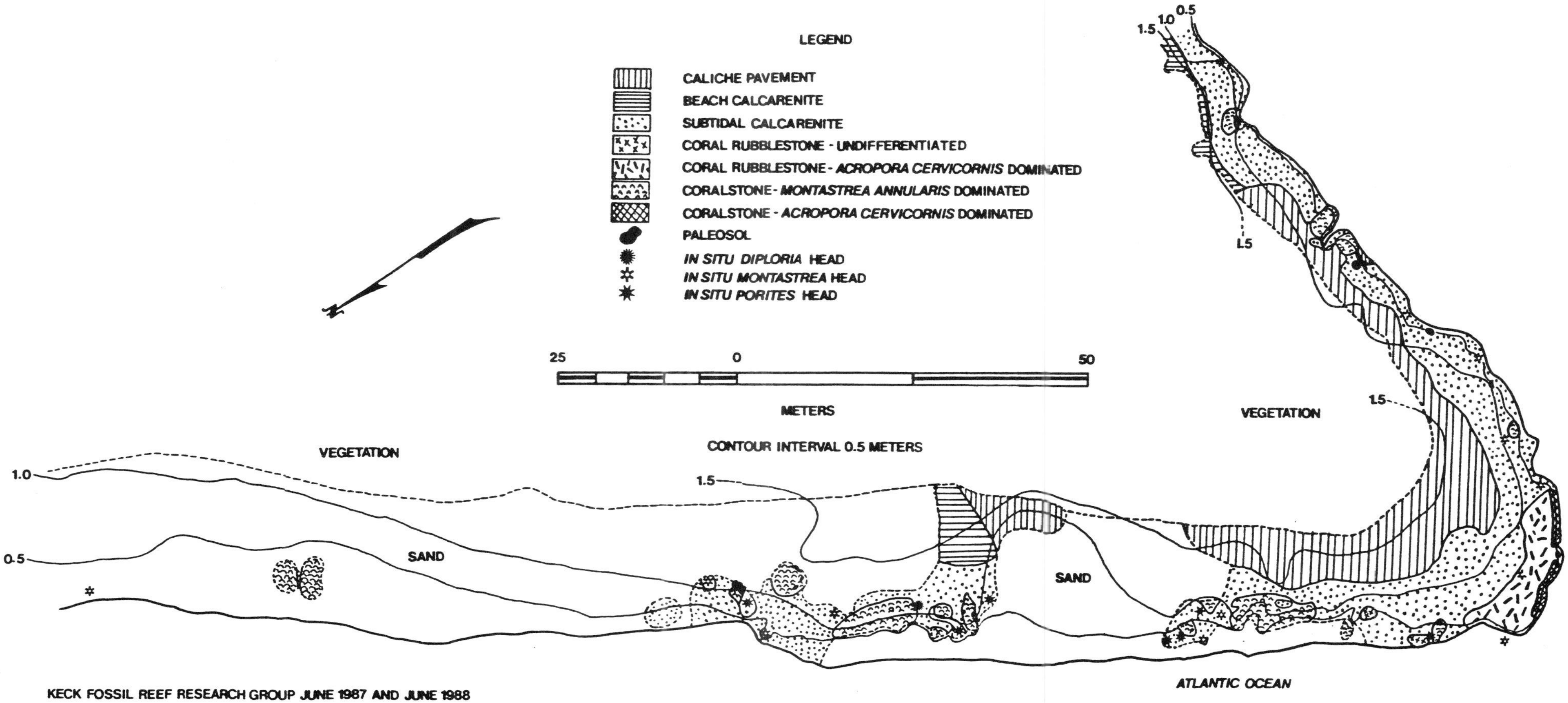

Fig. 2a. View of the northern fossil coral reef and associated facies south from Stop 3. The arrow marks the wave-cut notch.

Fig. 2b. View of Stop 1 showing in situ *fossil corals (1), subtidal calcarenites (2), and beach calcarenites (3).*

Fig. 2c. Paleosol on in situ Montastrea annularis. *Stop 1. Scale = 15cm.*

northern part was chosen for the measurement of profiles and the collection of rock samples because it is better and more continuously exposed (Fig. 2a). This field guide is for the northern part of the area where examples of all the elements of the fossil reef and associated facies are to be found. Some aspects of the southern part are discussed below and a geologic map of the area (Map 2) is included in this guide for those who wish to extend their field trip.

Much geologic research remains to be done on the Sue Point fossil reef, and it would be appreciated if no rock hammers or spray paint are used on the rocks. Your help in preserving these fine examples of fossil patch reefs will be valued by geologists and biologists who visit them in the future.

Some rock exposures described in this field guide extend into the present day intertidal zone, and the optimum time to visit the fossil reef is during the two hours around low tide. The upper, gray part of the rocky intertidal zone commonly is very rough with sharp edges and jagged points that are a result of bioerosion. The lower, yellow part of the rocky intertidal zone commonly is very slippery because of a thin cover of algae. Great care should be taken when walking in the intertidal zone, and sturdy footwear is recommended.

Visitors to the reef should park their vehicles along the small road located on the east side of the main road near the crest of the hill south of Victoria Hill Settlement (see Fig. 1). A short trail starts on the west side of the main road opposite the above mentioned small road and visitors should follow this westward down to the first outcrops at the extreme northeast end of the fossil reef. The locations of measured profiles and field stops are shown on Map 1.

In the area of Stops 1 and 2 and profiles E-E', F-F', and G-G', the amount of rock exposed varies greatly depending on the most recent events in the erosion/deposition cycle of modern beach sand. In June 1987 when the initial mapping of the fossil reef was done, most of this area was buried by sand. In early June 1988 recent extensive beach erosion had revealed considerable rock outcrop, which was surveyed and added to the 1987 map. A few days later the outcrop was being buried progressively by beach sand accretion.

Fig. 3. Profile E-E' and legend for all profiles.

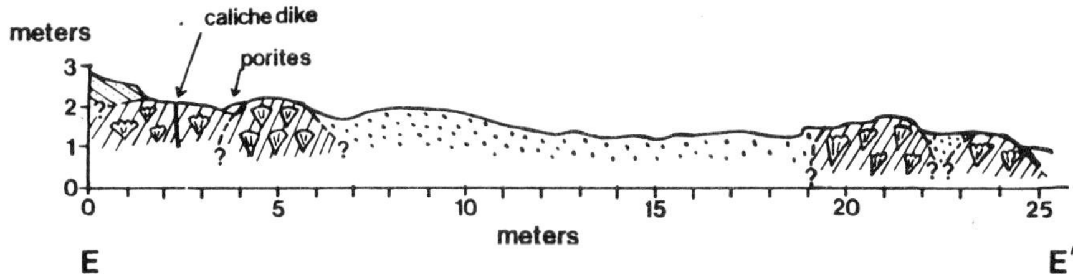

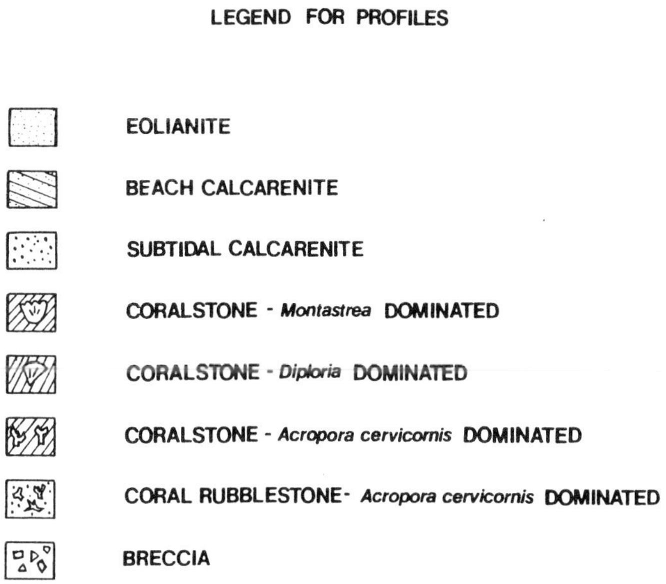

These exposures do illustrate important aspects of the geology of the Sue Point fossil reef and associated facies, as described in this guide, but they may not be visible always to visiting geologists.

Stop 1

The sequence in the vicinity of Stop 1 is illustrated by the NE end of profile E-E' (Fig. 3), by profile G-G' (Fig. 4), and by Fig. 2b. The lowest exposed rocks are coral rubblestones and coralstone with heads of *in situ Montastrea annularis* and some *Diploria strigosa*. Small patches of reddish-brown paleosol occur on the coralstone heads (Fig. 2c). Coarse, shelly subtidal calcarenites surround and overlie the fossil corals. The lowest of these subtidal rocks contain specimens of the trace fossil *Ophiomorpha* sp., which represent the fossilized burrows of callianassid shrimps (Curran, 1984). Above the burrowed calcarenites the rocks contain well-developed trough cross beds. A similar sequence was described from Great Inagua Island (White and Curran, 1987), and subtidal calcarenites with *Ophiomorpha* sp. burrows and trough cross beds also occur in association with *in situ* fossil corals at the Cockburn Town fossil coral reef (Curran and White, 1985). The burrowed calcarenites indicate relatively immobile subtidal sands that permitted the formation and preservation of the callianassid burrows, whereas the trough cross beds indicate mobile sand transported and deposited by longshore currents.

The subtidal calcarenites are overlain with an overstepping contact by calcarenites deposited in the beach zone (Fig. 5a). These beach calcarenites are not as coarse as the subtidal calcarenites, and they contain far fewer fossil shells. They are characterized by gently dipping

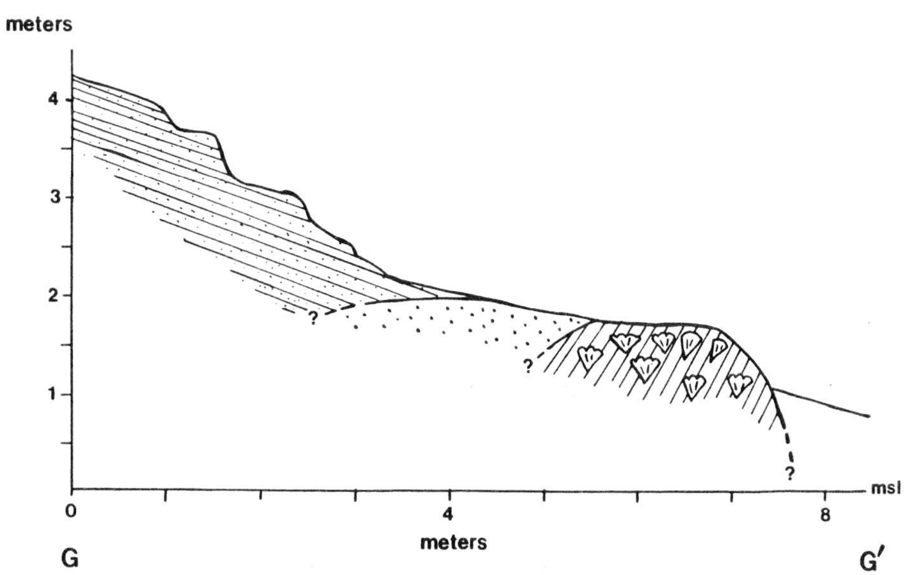

Fig. 4. Profile G-G'.

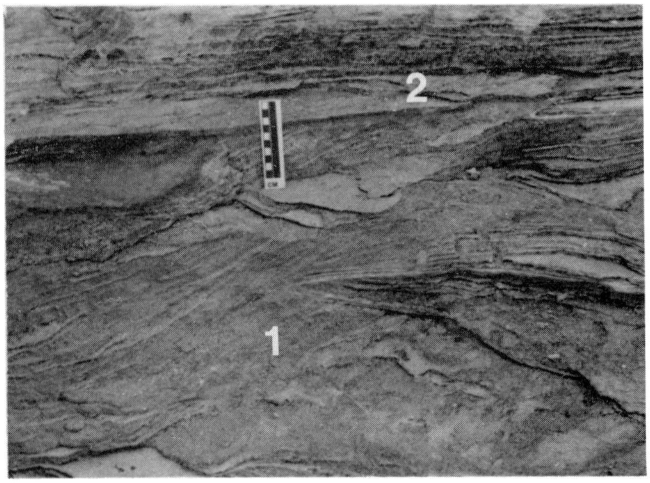

Fig. 5a. Subtidal calcarenites (1) overlain by overstepping beach calcarenites (2). Stop 1. Scale = 10cm.

Fig. 5b. Laminated beach calcarenites. The arrow marks keystone vugs. Stop 1.

Fig 5c. In situ Montastrea annularis. *Stop 2. Scale = 10cm.*

laminations typically produced by wave swash on the seaward-dipping foreshore (Inden and Moore, 1983). These laminations dip at low angles in a northwesterly direction, showing that the landward direction was to the southeast and implying the presence of an emergent ancient San Salvador. The direction of dip of the beach bedding is almost exactly perpendicular to the long axis of the linear fossil patch reef, suggesting that the reefs grew approximately parallel to the ancient coastline. Many of the beach calcarenites contain very porous layers with keystone vugs (Fig. 5b) that are indicative of the beach zone (Dunham, 1970). These features are well exposed in a small cliff that extends along the outcrop in this vicinity and which normally is exposed above the encroaching modern beach sands. The highest rocks exposed in the sequence are eolianites. These are fine-grained, well-laminated calcarenites with cross beds that have a wide range of dip directions and some convex-upward doming of laminations.

The rocks at Stop 1 illustrate a shallowing-upward sequence from shallow subtidal marine coral reefs and subtidal sands through beach sands to a capping of wind-blown dune deposits. Such regressive sequences are common in equivalent Sangamon interglacial deposits (Curran and White, 1985; White et al., 1984; White and Curran, 1987), and they result from falling sea level caused by the onset of Wisconsinan glaciation.

Stop 2

The SW end of profile E-E' (Fig. 3) and profile F-F' (Fig. 6) show the geologic relations in the area of Stop 2. The sequence is similar to that seen at Stop 1 with some differences that will be described here. The most landward *in situ* fossil corals are *Montastrea annularis* (Fig. 5c) and *Diploria strigosa* (Fig. 7a) with a small patch of *Porites porites* between the two more prominent species. These fossil corals differ from others in this vicinity in being directly overlain by beach calcarenites (Fig. 7b), reflecting a high point of the variable relief that typifies many patch reefs. Just landward of these corals, a patch of breccia drapes the front of a small cliff formed in the beach calcarenites (Fig. 7c). Exposures of this breccia occur at approximately the same elevation (2 to 3m above mean sea level) south along the next 100m of the shore. The possible significance of this breccia is discussed in the description of

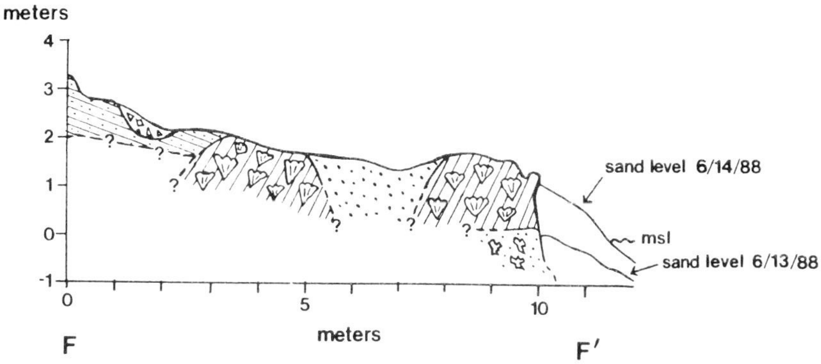

Fig. 6. Profile F-F'.

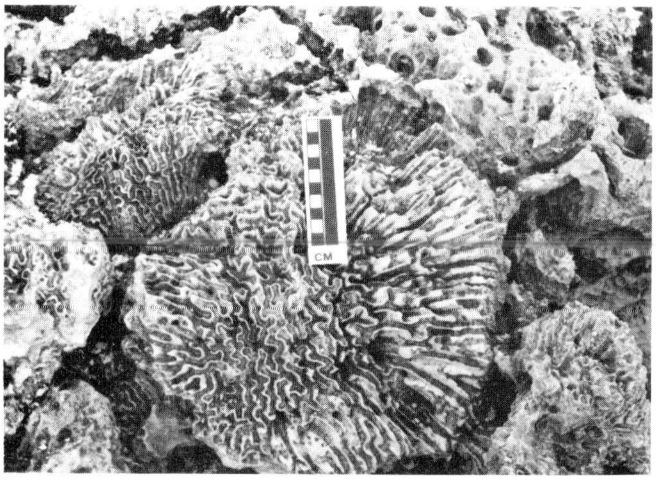

Fig. 7a. In situ Diploria strigosa. *Stop 2. Scale = 10cm.*

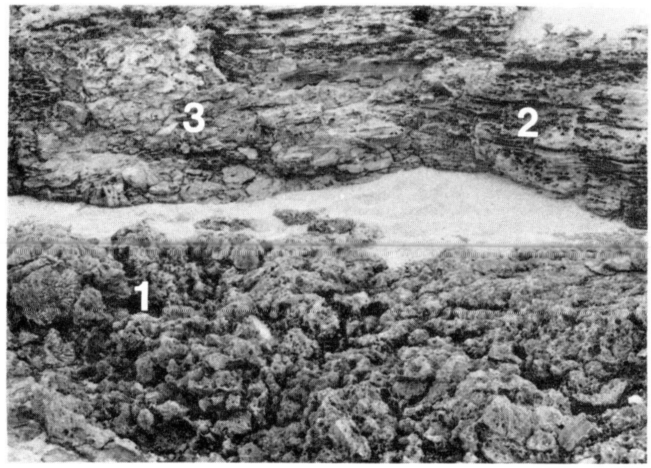

Fig. 7c. View of Stop 2 showing in situ *fossil corals (1), beach calcarenites (2), and breccia (3).*

Fig. 7b. In situ Montastrea annularis *overlain directly by beach calcarenites. Stop 2. Scale = 10cm.*

Stop 4. The seaward part of the outcrop near stops 1 and 2 is subject to rapid changes in the amount of modern beach sand present. This is illustrated by profile F-F' (Fig. 6) which was measured on 13 June 1988. The following morning fifteen hours and two high tides later, a 104cm thicker layer of sand had been added to the seaward end of the profile.

Stop 3

The rocks are exposed more reliably at Stop 3 where the shallowing-upward sequence from subtidal to subaerial lithologies can be seen, as illustrated in profile A-A' (Fig. 8). Prominent *in situ* fossil corals are exposed next to the ocean in this area. The dominant coral species is *Montastrea annularis*, but some searching will reveal several others, including *M. cavernosa* (Fig. 9a), *Diploria strigosa, Porites porites, P. furcata, P. astreoides, Agaricia* sp., and *Manicina aveolata.* Patches of reddish-brown paleosol occur directly on top of the fossil corals and provide graphic evidence of the change from marine to non-marine environments. Some patches of the paleosol contain blackened pebbles of limestone (Fig. 9b). Such dark pebbles commonly are found in paleosols, and their blackening has been ascribed to fire (Shinn et al., 1984) or to terrestrial exposure in the vicinity of hypersaline lakes (Ward et al., 1970). The origin of black micrite

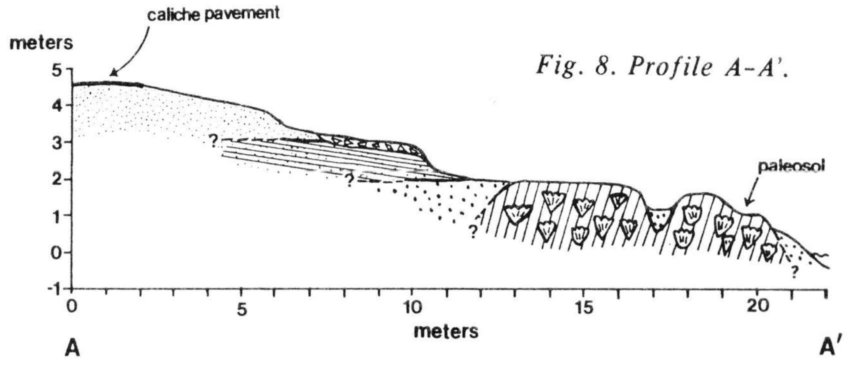

Fig. 8. Profile A-A'.

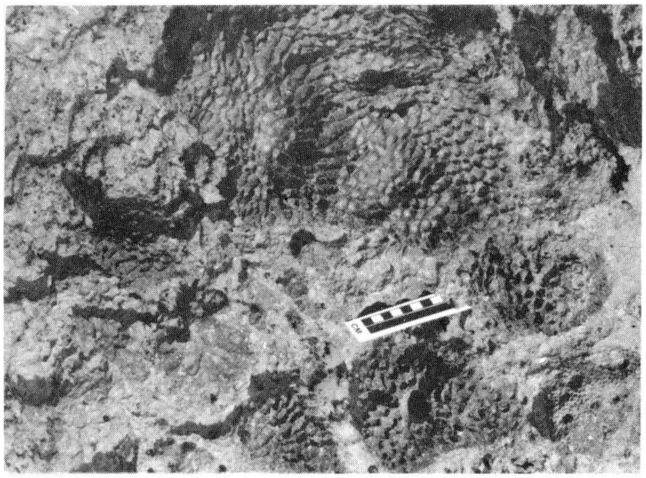

Fig, 9a. In place Montastrea cavernosa. *Stop 3. Scale = 10cm.*

Fig. 9c. Wave-cut notch and breccia. Stop 4.

Fig. 9b. Black micrite clasts in paleosol. Stop 3. Scale = 10cm.

pebbles on San Salvador and their stratigraphic significance was discussed at length by Bain (1984). He concluded that pre-existing limestones, commonly grainstones, underwent diagenetic micritization in the presence of fungi and algae in a subaerial environment. Carbonization of the contained organic matter produces the black color. Reworked fragments of these blackened micrities are incorporated into a variety of subsequent deposits, including paleosols as seen at Sue Point. Their presence is a clear indication of subaerial exposure, and they mark a disconformity or unconformity.

Stop 4

The shallowing-upward sequence is exposed continuously along the outcrop to Stop 4 in the vicinity of profile B-B' (Fig. 10). The breccia first described above under Stop 2 extends from Stop 3 to Stop 4, and along here it is associated with a wave-cut notch (Figs. 2a and 9c). At several localities this notch has a caliche layer draped over it, demonstrating that it was formed during a pre-modern sea level high stand. The breccia may have formed during this same period of higher sea level and may represent patches of talus debris associated with the erosion of small sea cliffs. The breccia clasts are angular and some have the characteristic microscopic texture of eolianites described by White and Curran (1988). This suggests that the wind deposited sands that formed during the post-Sangamon

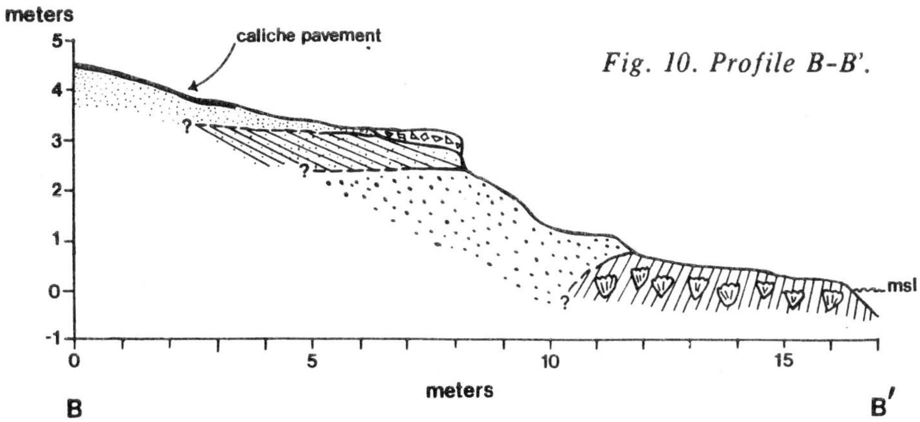

Fig. 10. Profile B-B'.

regression were lithified and later subjected to erosion during a high stand of sea level between the Sangamon and the present. Other workers have presented evidence from San Salvador for such a sea level high stand. Titus (1984) describes former sea cliffs, wave-cut notches, and lake deposits and attributes their formation to a latest Pleistocene sea level high stand of +2.5m relative to present sea level. He describes a wave-cut bench with a low sea cliff on the landward side that extends from Victoria Hill Settlement, just northeast of the Sue Point fossil reef, to Line Hole Settlement. Holocene rocks overlie the wave-cut bench. Using the age data of Carew (1983), Titus proposes that the wave-cut bench was cut into rocks that are approximately 75,000 years old during a sea level high stand 35,000 to 40,000 years ago. Based on their analysis of the formation of Lighthouse Cave and speleothems from within it, Mylroie and Carew (1988) propose two post-Sangamon sea level high stands. In the interval 37,000 to 49,000 years B.P. sea level temporarily reached approximately its present level, probably not high enough to cut the notch at Sue Point. An earlier high stand of +1 to +6m above present sea level occurred 70,000 to 85,000 years B.P. and this postulated sea level high stand seems the best candidate for the formation of the wave-cut notch at the Sue Point fossil reef locality.

Stop 5

The sequence in this area is shown by profile C-C' (Fig. 11). Exposed here is a rubblestone composed of broken "branches" of *Acropora cervicornis* (Fig. 12a). Some of the *A. cervicornis* may be in growth position or close to it, although it is difficult to be certain in this particular instance. There is a good cross section of a *Montastrea annularis* head in growth position overlying coral rubblestone.

Beyond Stop 5 the main *Montastrea-Diploria-Porites* patch reef extends for approximately 75m to the SW. In places the reef is cut by small gullies that are floored by shelly calcarenites (Fig. 12b) and that are reminiscent of similar features found in many modern patch reefs. Several small patches of paleosol occur along this stretch of the fossil reef. Landward there

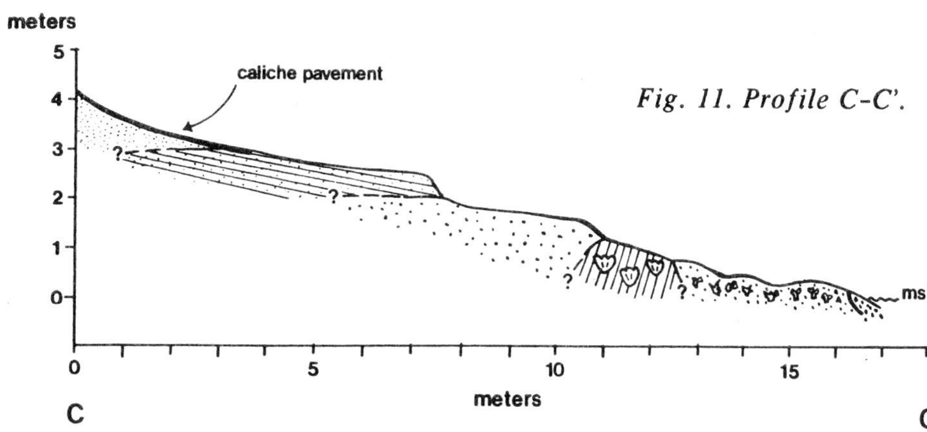

Fig. 11. Profile C-C'.

Fig. 12a. Coral rubblestone containing clasts of Acropora cervicornis. *Stop 5.*

Fig. 12b. Shelly calcarenites. Stop 5.

Fig. 12c. Caliche draped over breccia. Stop 5. Scale = 10cm.

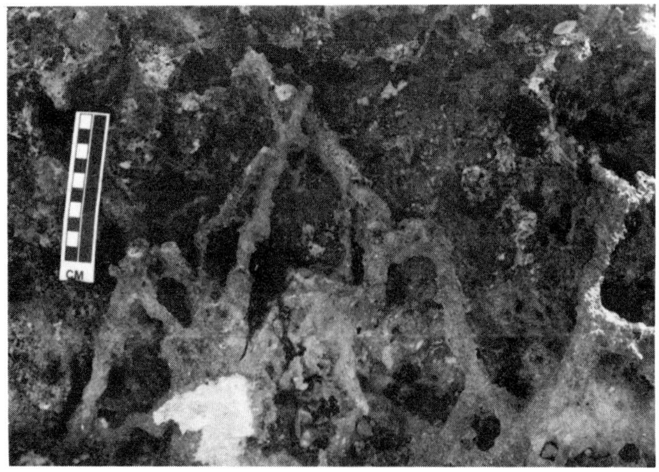

Fig. 13a. Rhizomorphs formed by the precipitation of micrite around the roots of terrestrial plants. Stop 5. Scale = 10cm.

Fig. 13b. Large in place heads of Montastrea annularis. *Stop 6.*

Fig. 13c. Acropora cervicornis *preserved in growth position. Stop 7. Scale = 10cm.*

are good examples of breccia (Fig. 12c), which commonly contain rhizomorphs (Fig. 13a) formed by the precipitation of micrite around the roots of terrestrial vegetation. Rhizomorphs also can be found directly on fossil corals, again demonstrating graphically the effects of falling sea level on the environments and associated biota. Similar plant trace fossils are abundant in many of the Quaternary carbonate rocks of the Bahamas and have been known for a century since they were described by Northrop (1890) and given the name rhizomorphs.

Stop 6

At this stop there are large heads of *Montastrea annularis* several meters in diameter preserved in their original growth position (Fig. 13b).

Stop 7

Near Stop 7 the coastline makes a sharp swing to the SE, and here *Acropora cervicornis* becomes the dominant coral species of the fossil reef, as illustrated in profile D-D' (Fig. 14). *A. cervicornis* in growth position is exposed prominently along the SE-facing cliff edge (Fig. 13c), where it may be examined most easily at low tide. *A. cervicornis* fragments are abundant in coral rubblestones exposed adjacent to the *in situ* corals.

Examples of all the important features of the Sue Point fossil coral reefs can be seen somewhere between Stops 1 and 7. Exposures of a second fossil patch reef extend along the coast south of Stop 7 (see Map 2). A similar lateral distribution of facies is evident as the one described above for the more northerly patch reef. From NE to SW there is a trend from scattered *in situ* heads of *Montastrea annularis* amidst subtidal calcarenites, to a more continuous *Montastrea-Diploria-Porites* fossil patch reef, ending with *Acropora cervicornis* in rubblestones and preserved in growth position in the SE-facing sea cliff. The zig-zag pattern of the modern coastline formed by the outcrops of the Sue Point fossil coral reefs is paralleled by living patch reefs. It is possible that the configuration of the Sue Point fossil reefs reflects a similar geography along a Sangamon shoreline.

The fossil *Acropora cervicornis* may represent the somewhat deeper water end of linear patch reefs. In several places along the fossil reef, heads of *Montastrea annularis* can be seen to have grown on top of *Acropora cervicornis* rubblestone, indicating that *A. cervicornis* was already present somewhere in the area. This may represent a succession of coral species as water depth decreased over the reef, either by upward growth of the reef or by absolute fall in sea level, or some combination of the two. In comparison with the Sangamon fossil reefs at Devil's Point, Great Inagua, and at Cockburn Town, San Salvador (Curran and White, 1985; Curran et al., this volume), fossil *Acropora palmata* is notably absent from the Sue Point fossil reefs. This implies that the Sue Point reefs were not located at the higher energy edge of a shallow shelf nor did they grow up to mean sea level. By comparison with Sangamon sea level elevations determined from the Cockburn Town fossil reef (Curran and White, 1985), the tops of *Montastrea annularis* heads reached to within 2 to 4m of mean sea level. Thus they were similar to living *Montastrea annularis* in the patch reefs of Fernandez Bay and elsewhere on the shelf around San Salvador.

The field trip can be continued by walking SE along the coast until reaching the main road, or by retracing one's path to the starting point.

ACKNOWLEDGEMENTS

I thank Don and Kathy Gerace and the staff of the Bahamian Field Station for full logistical support of field work on San Salvador. This field guide was made possible by the able field work of

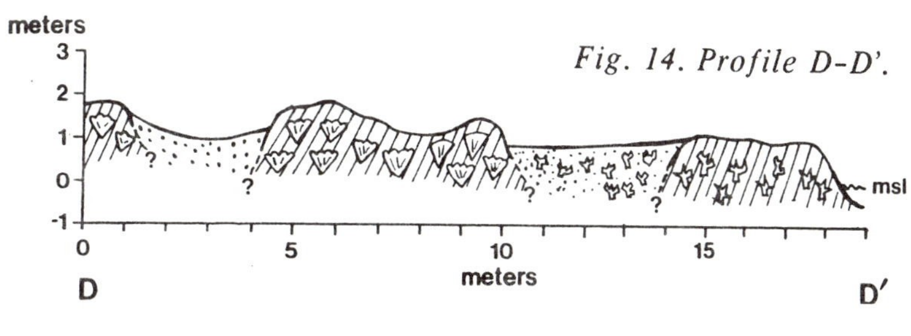

Fig. 14. Profile D-D'.

the following geology students: Doug Cattel (Colorado College), Storr Nelson (Whitman College), Molly Stark (Smith College), and Kathy White (Smith College). I thank C. M. Soja for carefully reading the manuscript and offering many suggestions for its improvement and David White for drafting the illustrations and Kathy Bartus for word processing. Field work was supported by a grant to the Keck Geology Consortium from the W. M. Keck Foundation, to whom I extend my thanks. Acknowledgement is made to the donors of the Petroleum Research Fund, administered by the American Chemical Society, for support of my research on the Pleistocene rocks of San Salvador.

REFERENCES CITED

Adams, R.W., 1983, General guide to the geological features of San Salvador, in Gerace, D.T., ed. Field guide to the geology of San Salvador: San Salvador, Bahamas (third edition), CCFL Bahamian Field Station, p. 1-66.

Bain, R.J., 1984, Diagenetic origin for San Salvador micrite pebbles, in Teeter, J.W., ed. Proceedings of the Second Symposium on the Geology of the Bahamas: San Salvador, Bahamas, CCFL Bahamian Field Station, p. 229-244.

Carew, J.L., 1983, Geochronology of San Salvador, in Gerace, D.T., ed. Field guide to the geology of San Salvador, Bahamas (third edition), CCFL Bahamian Field Station, p. 160-172.

Carew, J.L., and Mylroie, J.E., 1987, A refined geochronology for San Salvador Island, Bahamas, in Curran, H. A., ed. Proceedings of the Third Symposium on the Geology of the Bahamas: San Salvador, Bahamas, CCFL Bahamian Field Station, p. 35-44.

Cattel, D.A., 1988, A petrographic study of the diagenesis of the Sue Point fossil reef, San Salvador, Bahamas: First Keck Research Symposium in Geology, Beloit, Wisconsin, p. 26-28.

Curran, H.A., 1984, Ichnology of Pleistocene carbonates on San Salvador, Bahamas: Journal of Paleontology, v. 58, p. 146-159.

Curran, H.A., and White, B., 1985, The Cockburn Town fossil coral reef, in Curran, H.A., ed., Pleistocene and Holocene environments on San Salvador Island, Bahamas - Guidebook for Geological Society of America, Orlando annual meeting field trip: Ft. Lauderdale, Florida, CCFL Bahamian Field Station, p. 95-120.

Curran, H.A., White, B., Chen, J.H., and Wasserburg, G.J., 1989, Comparative analysis and geochronology for the development and decline of two Pleistocene coral reefs, San Salvador and Great Inagua islands, Bahamas, in Mylroie, J. E., ed. Proceedings of the Fourth Symposium on the Geology of the Bahamas: San Salvador, Bahamas, Bahamian Field Station.

Dunham, R.J., 1970, Keystone vugs in carbonate beach deposits (abstract): American Association of Petroleum Geologists Bulletin, v. 54, p. 845.

Inden, R.F., and Moore, C.H., Beach environment, in Scholle, P.A., Bebout, D.G., and Moore, C.H., Carbonate depositional environments: American Association of Petroleum Geologists, Memoir 33, p. 212-265.

Mylroie, J.E., and Carew, J.L., 1988, Solution conduits as indicators of Late Quaternary sea level position: Quaternary Science Reviews, v. 7, p. 55-64.

Nelson, S.L., 1988, The Sue Point fossil reef complex: A diagenetic history: First Keck Research Symposium in Geology, Beloit, Wisconsin, p. 43.

Northrop, J.I., 1890, Notes on the geology of the Bahamas: New York Academy of Sciences Transactions, v. 10, p. 4-22.

Shinn, E.A., Robbin, D.M., and Lindz, B.H., 1984, Fire and limestone: Origin of black pebbles (abstract): American Association of Petroleum Geologists Bulletin, v. 68, p. 527.

Stark, E., 1988a, A comparison of the diagenesis of two Bahamian fossil coral reefs: First Keck Research Symposium in Geology, Beloit, Wisconsin, p. 46-48.

Stark, E., 1988b, A comparison of the diagenesis of two Bahamian fossil coral reefs: Senior Honors thesis, Department of Geology, Smith College, 80 p.

Titus, R., 1984, Physical stratigraphy of San Salvador Island, Bahamas, in Teeter, J.W., ed. Proceedings of the Second Symposium on the Geology of the Bahamas: San Salvador, Bahamas, CCFL Bahamian Field Station, p. 209-228.

Ward, W.C., Folk, R.L., and Wilson, J.L., 1970, Blackening of eolianite and caliche adjacent to saline lakes, Isla Mujeres, Quintana Roo, Mexico: Journal of Sedimentary Petrology, v. 40, p. 548-555.

White, B., and Curran, H.A., 1987, Coral reef to eolianite transition in the Pleistocene rocks of Great Inagua Island, Bahamas, in Curran, H.A., ed. Proceedings of the Third Symposium on the Geology of the Bahamas: San Salvador, Bahamas, CCFL Bahamian Field Station, p. 165-179.

White, B., and Curran, H.A., 1988, Mesoscale physical sedimentary structures and trace fossils in Holocene carbonate eolianites from San Salvador Island, Bahamas: Sedimentary Geology, v. 55, p. 163-184.

White, B., Kurkjy, K.A., and Curran, H.A., 1984, A shallowing-upward sequence in a Pleistocene coral reef and associated facies, San Salvador, Bahamas, in Teeter, J. W., ed. Proceedings of the Second Symposium on the Geology of the Bahamas: San Salvador, Bahamas, CCFL Bahamian Field Station, p. 53-70.

FEASIBILITY OF USE OF VNIR REFLECTANCE SPECTROSCOPY FOR ANALYSIS OF WATER AND ORGANIC CONTENT IN SKELETONS OF SCLERACTINIAN CORALS

Victor P. Zabielski
Curriculum in Marine Sciences
University of North Carolina
Chapel Hill, North Carolina 27514

Susan J. Gaffey
Department of Geology
Rensselaer Polytechnic Institute
Troy, New York 12181

ABSTRACT

Spectral and petrographic data show that water and organics in coral skeletons are intimately associated in both modern and Pleistocene samples. The water itself appears to be entrapped body fluids of the coral rather than sea water.

Relative band areas provide a more accurate indicator of total water and organic content than the relative band intensities used in previous studies. While the area of the 1.4-um absorption band provides a useful measure of the water content of samples, the 1.9-um band appears to provide a more sensitive indicator of their organic content. Spectra of skeletons of colonies of *Diploria strigosa* which were live when collected show there is significant intraskeletal variation in both water and organic content. Thecae commonly contain more water and organics than septa, but water contents of different portions of septa and thecae also show meaningful variations. Water and organics are lost rapidly after death of the coral. Amounts of organics and water in septa and thecae become more uniform throughout the skeleton, and differences between septa and thecae decrease.

Although organics are still present in dead and fossil skeletons, trends in areas of 1.4-um and 1.9-um water bands in their spectra show that both the absolute abundance of organics and abundance of organics relative to water has decreased. Changes in organic content are also indicated by the progressive decrease in intensity of the 0.67 um chlorophyll band in spectra of dead and Pleistocene samples.

INTRODUCTION

The chemical and mineralogical composition of skeletal carbonates is more complex and varied than the usual designations of "aragonite" and "high" or "low magnesian calcite" would suggest. Previous work has shown that water (liquid and bound H_2O and OH^-) is an important and ubiquitous component of all carbonate skeletal material (Gaffey, 1983, 1984a,b, 1985, 1988). Water in carbonate skeletons affects their solubility and stability, and can serve as a medium for diagenetic reactions (see Gaffey, 1988 for summary and references).

Organic compounds can also have a marked effect on diagenesis of carbonates (e.g. Gautier, 1985; Mitterer and Cunningham, 1985). Although it has been known for many years that carbonate skeletons contain organic material (see Weiner et al., 1983, for review) there is no tool routinely in use by carbonate petrologists for determining organic content, or for determining the relationship between water and organics in skeletal samples, and data on organic contents of carbonate skeletons is limited. Reflectance spectroscopy provides a rapid, non-destructive tool for characterization of water and organic content *in situ* (Gaffey, 1983, 1984a,b, 1985, 1988; Holden, 1988). Like X-ray diffraction, vibrational spectroscopy is non-selective, detecting all spectrally active components in a sample whether or not they are anticipated. Unlike X-ray diffraction, however, reflectance spectroscopy can detect liquid or amorphous phases, as well as those which have an ordered crystal structure.

The purpose of this study is to test the applicability of reflectance spectroscopy to studies of the effects water and organics on early diagenetic alteration of carbonate sediments. A more detailed discussion of the spectral properties of water and organics can be found in Gaffey (1988 and this volume).

SAMPLES AND METHODS

All coral samples used in this study were collected during two visits to San Salvador, Bahamas, the first in January, 1988, and the second in June, 1988. Living corals were bleached in dilute clorox. Samples were identified at the generic, and, when possible, the specific level. Localities at which samples were collected and condition i.e., modern living, modern dead (whether still in place on the reef, in reef rubble, or in the intertidal zone on beaches), or fossil (from Pleistocene reefs on San Salvador) were recorded. Localities for and classification of samples used in this study are given in Table 1.

Thin sections of a radial section of coral heads, or longitudinal section of branches were prepared for each specimen. Acetate peels were also prepared for some samples. Septa of DSRBML-75, DSRBMD-92, and D-SPPL-84 were examined with SEM. Mineralogy of dead and fossil skeletons which might contain cements or sediment infill, or which might have been altered to calcite was determined using X-ray diffraction. Mineralogy of these samples is also given in Table 1.

Because grinding destroys some fluid inclusions, spectra were obtained from polished and bleached slabs. An 8-15mm thick slab was cut from the center of those specimens chosen for analysis. Each slab was polished to remove saw marks and bleached in 30% H_2O_2 (buffered to a neutral pH with NaOH) for 5 to 7 days to remove any organics freshly exposed by slabbing. The slabs were rinsed in tap water and dried in an oven at 37°C for at least one day before spectra were obtained. Samples were never subjected to temperatures above ~40°C as high temperatures can alter or destroy organics or hydrated mineral phases. Bidirectional VNIR (0.3 or 0.6 to 2.7 um) reflectance spectra were obtained with the spectrophotometer at RELAB at Brown University, described by Pieters (1983). Spectral resolution of all spectra is 5nm, and the area covered by each spectrum was about 2mm in diameter. Specimens were examined with hand lens and binocular microscope to ensure that all portions of the coral from which spectra were to be obtained were free of cements or sediment infill.

One species of scleractinian coral, *Diploria strigosa*, was studied in detail because good specimens from all three categories (living, modern dead, and fossil) were available. Also the structure and size of the corallites made it possible to study different structures, i.e. septa and thecae, within the skeleton to determine the amount of variability within one colony, enabling changes in water and organic content caused by diagenetic alteration to be distinguished from compositional variations inherent in the original

Table 1. Coral samples from which spectra were obtained.

Sample Number	Classification	Locality	Mineralogy
DSRBML-75	*Diploria strigosa*	Rice Bay, live specimen	aragonite
DSRBMD-92	" "	Rice Bay beach, marine vadose	thecae-aragonite septa-aragonite, minor HMC
DSQAPL-80	" "	Quarry A, Pleistocene	thecae-aragonite septa-aragonite, minor LMC
ACSTMD-86	*Acropora cervicornis*	Sandy Point, marine vadose	aragonite
DSBPML-77	*Diploria strigosa*	Bamboo Point, live specimen	aragonite
DLBPMD-81	*Diploria labyrinthiformis*	Bamboo Point, marine vadose	thecae-aragonite septa-aragonite + HMC
ACSPPL-82	*Acropora cervicornis*	Sue Point, Pleistocene	aragonite + LMC
D-SPPL-84	*Diploria sp.*	Sue Point Pleistocene	aragonite + LMC

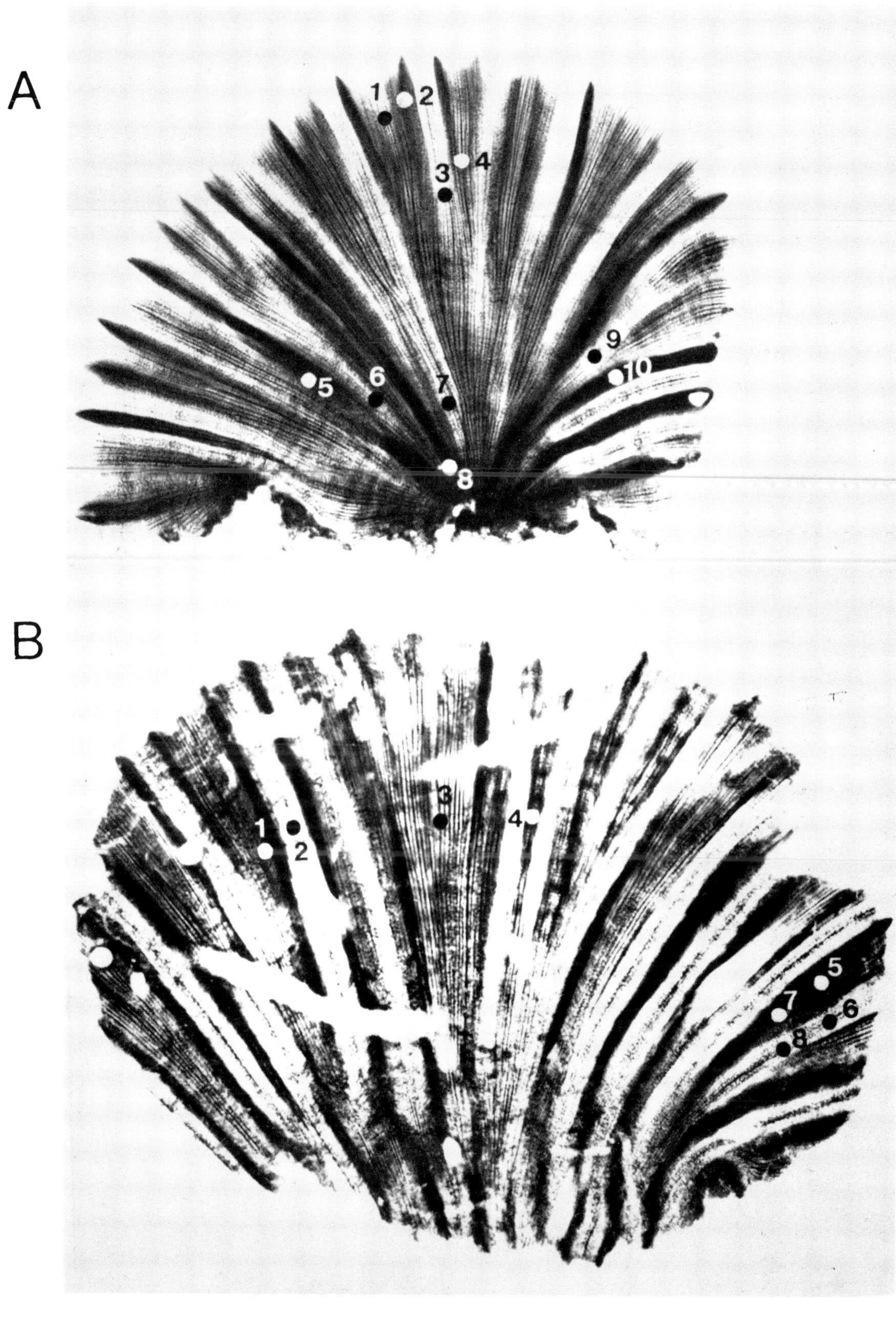

Fig. 1. X-ray radiographs of (A) skeleton of living coral collected in Rice Bay (DSRBML-75) and (B) dead coral skeleton collected on the beach, also at Rice Bay (DSRBMD-92). Dots indicate points from which spectra obtained.

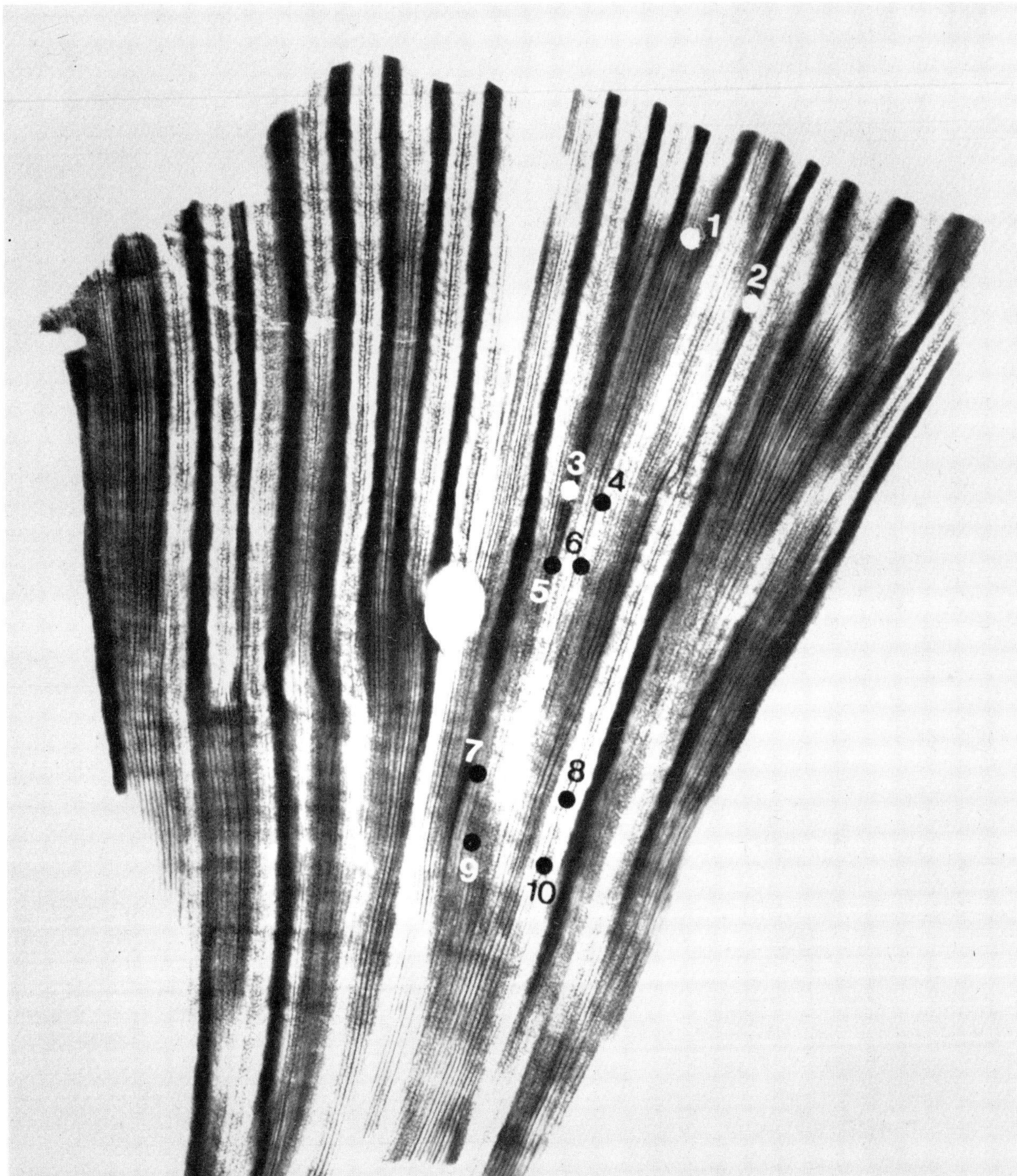

Fig. 2. X-ray radiograph of Pleistocene coral (DSQAPL-80) collected from Quarry A showing portions of skeleton from which spectra were obtained.

skeleton. X-ray radiographs were obtained of the three samples which were studied in greatest detail (DSRBML-75, DSRBMD-92, and DSQAPL-80) and points from which spectra were obtained are shown on the X-ray radiographs of these specimens (Figs. 1 and 2). Spectra were obtained of ten points, five on the septa, and five on the thecae for samples DSRBML-75 and DSQAPL-80. Eight points, four each on the septa and thecae, were measured on DSRBMD-92. Points from which spectra were obtained were distributed uniformly between the basal interior and the outer edges of the coral skeletons. Spectra of samples of *Diploria labyrinthiformis*, and *Acropora cervicornis* were obtained as well.

Spectra of Iceland spar, liquid H_2O, and of a bleached coral slab are shown in Figure 3. The liquid water spectrum was provided by Steve Pratt at RELAB.

Spectra of organics were also obtained and are shown in Figure 4. One of the advantages of VNIR reflectance spectroscopy is that measurements may be made of unstable phases *in situ*, without subjecting them to extraction processes which may destroy or alter some of the components of interest (Holden, 1988). Differences in spectral properties of organics obtained from extracted organics and from organics *in situ* can be seen in Figure 4. The first spectrum is that of organics extracted from a sample of *Diploria*. Approximately 20 grams of skeleton were treated with 1M HCl to remove the $CaCO_3$, following the procedures of Kimber and Griffin (1987). The residual organics were removed from the HCl by filtration and gently rinsed with deionized water. Material retained on the filter was collected and air dried prior to analysis.

Spectra were also obtained from DSBPML-77, which was not bleached after slabbing. The second spectrum shown in Figure 4 was obtained from a portion of a coral polyp which remained in the skeleton after preliminary bleaching with clorox. Spectra were also obtained of portions of this skeleton which had been bored by endolithic algae, as well as portions which had not. A spectrum was also obtained of an unbleached powder from this skeleton. Subsequently any coral referred to as "bleached" was treated in H_2O_2 after slabbing.

ANALYSIS OF SPECTRAL DATA

Intensity of an absorption band increases with the increase in concentration of the ab-

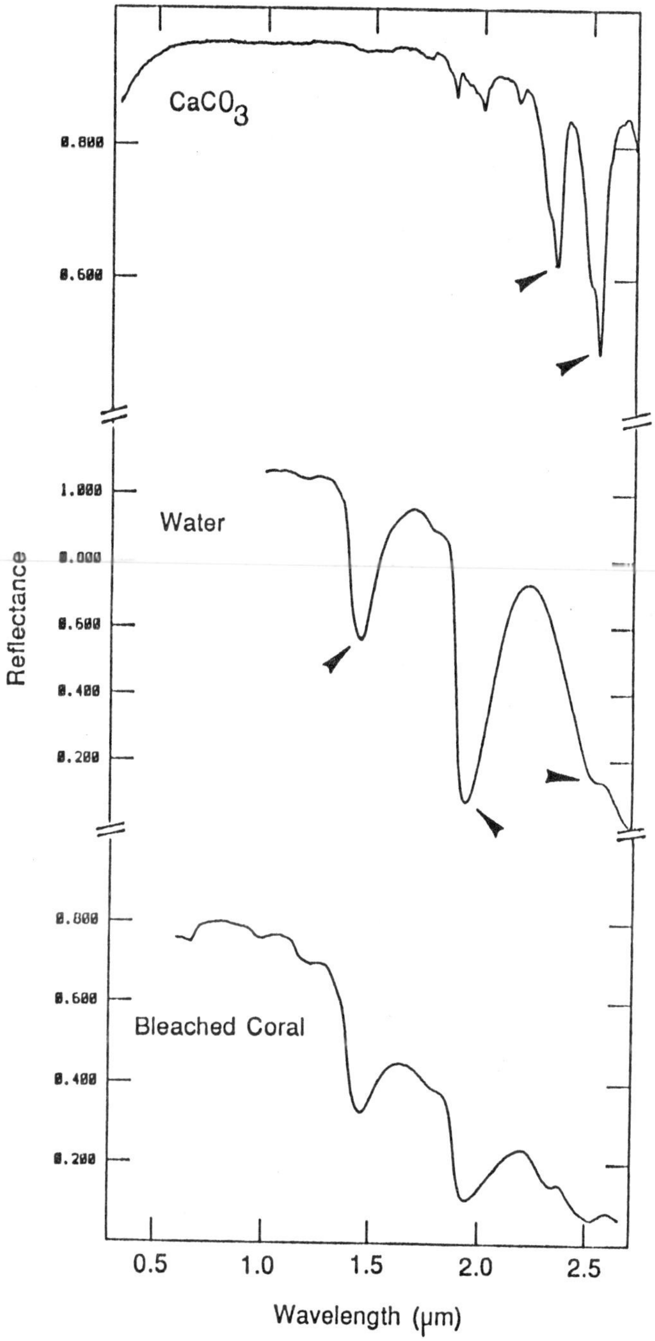

Fig. 3. Spectra illustrating absorption features due to vibrations of the carbonate ion and the H_2O molecule. The first spectrum is that of Iceland spar, the second spectrum is that of liquid H_2O. Arrows indicate the strong absorption features which indicate the presence of both these phases in the spectrum of the bleached coral also shown in this figure.

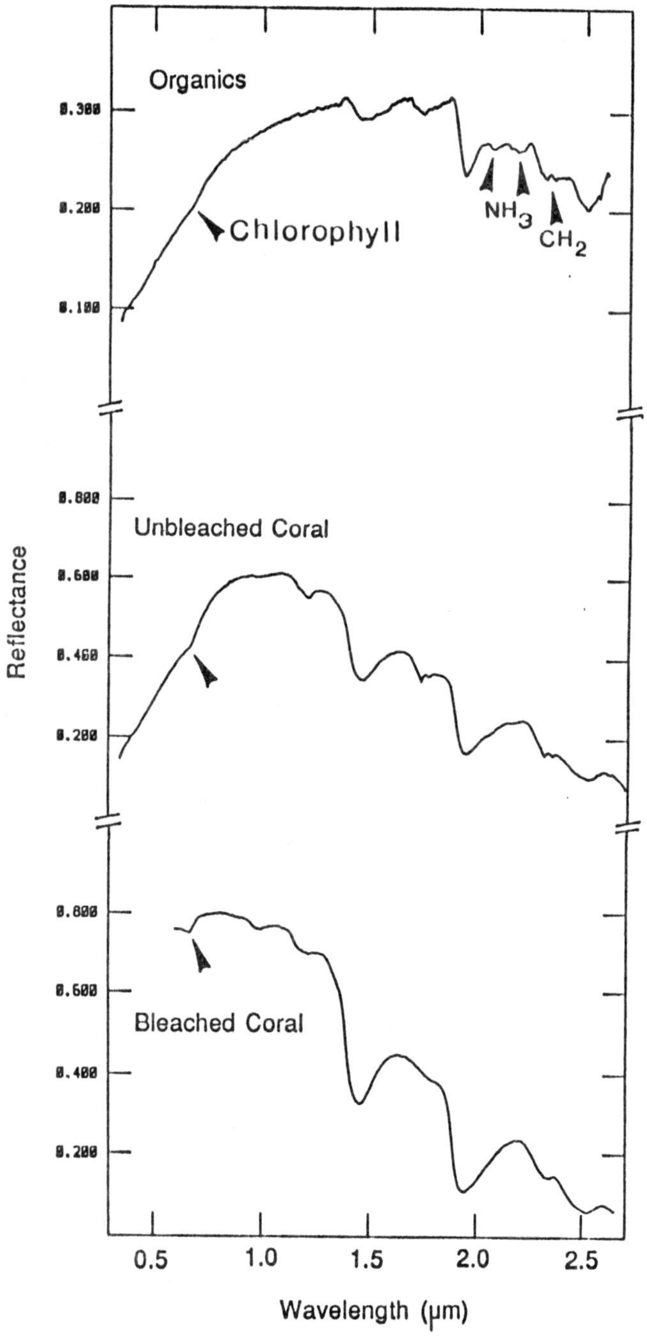

Fig. 4. Spectra illustrating absorption features due to organics. The first spectrum is that of organics extracted from a bleached Diploria strigosa *skeleton using HCl. The second spectrum is that of a portion of a coral polyp remaining after preliminary bleaching with clorox and exposed upon slabbing of the skeleton. The second spectrum contains a feature near 0.67 um due to organic pigments in symbiotic zooxanthellae. It also contains sharp narrow features near 1.2, 1.7, and 2.3 due to overtones and combination tones of the C-H stretching fundamental and broader bands near 2.1 and 2.2 um due to overtones and combination tones of the N-H stretching fundamental (see Gaffey, this volume for further discussion). The third spectrum is that of a coral skeleton which, despite prolonged bleaching in* H_2O_2 *still retains absorption features due to organic compounds near 0.67, 2.2 and 2.3 um.*

sorbing species (Herzberg, 1945; Burns, 1970). However, as previous work by Gaffey (1988) has shown that relative water band intensity alone does not accurately reflect the total water content of samples when water is present in more than one phase, it was decided that band areas would be used. Although calibration work still remains to be done which will allow water and organic contents to be quantitatively determined from band areas, these data show relative differences in water content within and between samples.

Two methods of band area calculation were employed in this study. The first used a straight line continuum, extended from the 0.6 to 1.3-um portion of spectra, as shown in Figure 5. Gaffey (1986) found the assumption of a straight-line continuum provided a useful model for determining positions, widths, and intensities of carbonate absorption bands in carbonate spectra. Log intensities were used because the Beer-Lambert law more accurately reflects behavior of absorption features than other models which have been employed (Clark and Roush, 1984; Gaffey,

CONTINUUM AREA METHOD

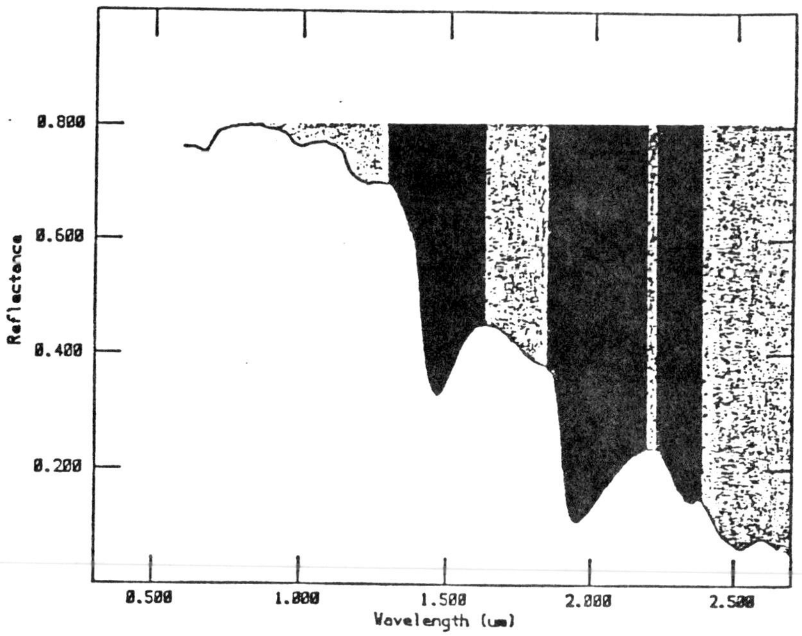

Fig. 5. Illustration the continuum-area method of calculating band areas.

1986). The edges of the absorption bands in the spectra in this study were determined by visual estimate on an expanded printout of the log of each spectrum. Several different criteria for defining band edges, e.g. a reflectance maximum, an inflection point on the curve, or a particular wavelength, may be used. However, the same criterion must be used for all analyses to minimize errors in calculation. The area below the continuum between the two wavelengths designated as band edges, colored black in Figure 5, represents the continuum area of the band.

The second method used was the point-line method. The area below a straight line connecting the two points on the spectrum which marked the edge of the band, was calculated (Fig. 6).

POINT-LINE AREA METHOD

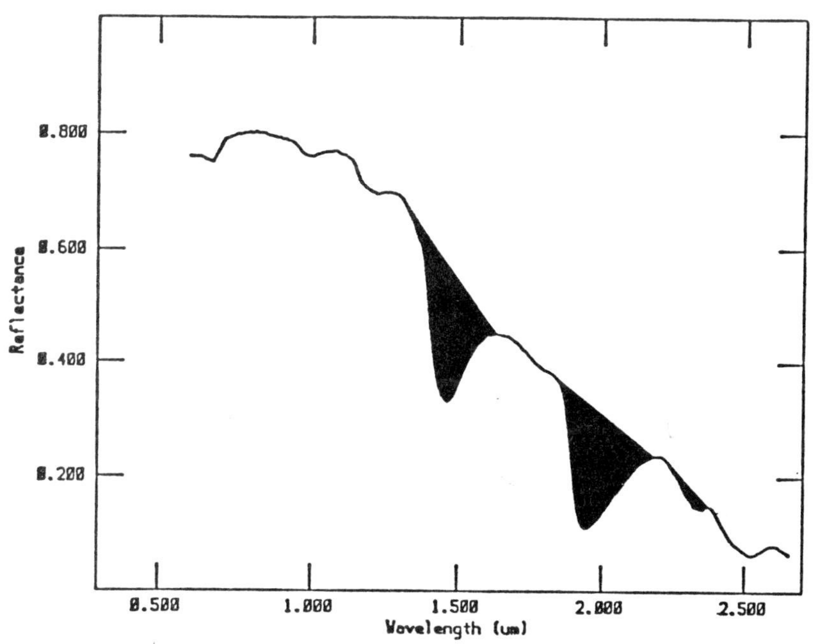

Fig. 6. Illustration of the point-line method for calculating band areas.

Comparison of data obtained by these two different methods showed that the continuum area method (Fig. 5) failed to show any consistent trends. The continuum-area method was therefore discarded and all band areas shown in this paper were calculated using the point-line method.

Areas were calculated for three bands in each spectrum: the 1.4- and 1.9-um water bands and the 2.3-um carbonate band. Band intensities and areas are a function not only of the concentration of the absorbing species, but of the particle size and packing or porosity of the sample as well. However, the intensity of an absorption feature relative to intensities of other features in the same spectrum remain constant, regardless of particle size (Gaffey, 1986). To correct for differences in grain size or porosity between samples, the area of each H_2O band in a spectrum was divided by the area of one of the strong carbonate absorption features in the same spectrum. Gaffey (1988) previously used the intensity of the 2.5-um carbonate band to correct for particle size effects. However, as can be seen from Fig. 3, liquid H_2O also has a strong 2.5-um feature. For this reason, areas of the 1.4- and 1.9-um water bands were ratioed to the area of the 2.3-um carbonate band to correct for particle size effects. As will be seen in the discussion of spectral features of organics below, and in Gaffey (this volume) use of the 2.3-um band is also less than ideal, as features due to CH_2 in organics occur in the 2.3-um region. However, CH_2 is a much weaker absorber than H_2O, so the apparent intensification of the 2.3-um carbonate feature is much less than that of the 2.5-um carbonate band.

Data on band areas for the three *Diploria* samples which are the focus of this study are listed in Table 2, with standard deviations and coefficients of variation. Data for individual spectra are illustrated in Figure 7, while Figure 8 shows the averages of all data presented in the histograms.

Ratio spectra, in which one spectrum is divided into another, yield a spectrum of differences in types and abundances of contained phases between two samples. Ratio spectra for the three *Diploria* samples are shown in Figure 9. The short-wavelength portion of three spectra (Fig. 10) show features due to organic pigments.

RESULTS AND DISCUSSION

Coral skeletons contain both intra- and

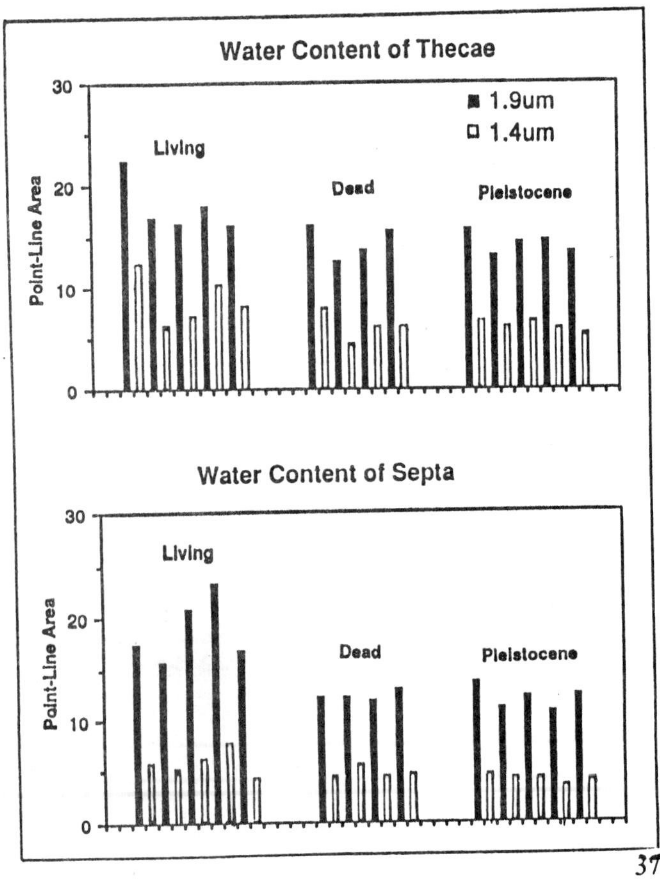

Fig. 7. Band areas calculated for the 1.4- and 1.9- um band in spectra obtained from living (DSRBML-75), dead (DSRBMD-92) and Pleistocene (DSQAPL-80) samples of Diploria strigosa. *Points from which spectra were obtained illustrated in Figures 2 and 3. Localities listed in Table 1.*

Table 2. Relative band areas calculated for each spectrum. These are ratios and therefore dimensionless numbers.

SAMPLE	Pt. No.	Description	1.4 um band	1.9 um band
DSRBML-75	1	septa	5.73	17.43
	2	theca	12.34	22.33
	3	septa	5.32	15.79
	4	theca	6.23	16.80
	5	theca	7.17	16.28
	6	septa	6.22	20.80
	7	septa	7.77	23.18
	8	theca	10.28	17.90
	9	septa	4.29	16.89
	10	theca	8.20	16.14
DSRBMD-92	1	theca	7.98	16.10
	2	septa	4.45	12.19
	3	septa	5.71	12.19
	4	theca	4.52	12.54
	5	theca	6.20	13.77
	6	septa	4.41	11.84
	7	theca	6.04	15.52
	8	septa	4.66	12.93
DSQAPL-80	1	septa	4.74	13.77
	2	theca	6.68	15.75
	3	theca	6.18	13.21
	4	septa	4.32	11.12
	5	theca	6.73	14.50
	6	septa	4.38	12.10
	7	theca	5.90	14.67
	8	septa	3.57	10.78
	9	theca	5.37	13.60
	10	septa	4.19	12.38

	Mean	Standard deviation	Coefficient of variation
DSRBML-75			
1.4-um band, septa	5.87	1.15	0.195
1.9-um band, septa	18.82	7.57	0.402
1.4-um band, thecae	8.84	4.87	0.550
1.9-um band, thecae	17.89	5.30	0.296
DSRBMD-92			
1.4-um band, septa	4.81	0.28	0.059
1.9-um band, septa	12.29	0.16	0.013
1.4-um band, thecae	6.18	1.51	0.243
1.9-um band, thecae	14.48	1.99	0.138
DSQAPL-80			
1.4-um band, septa	4.24	0.15	0.034
1.9-um band, septa	12.03	1.11	0.092
1.4-um band, thecae	6.17	0.26	0.042
1.9-um band, thecae	14.35	0.79	0.055

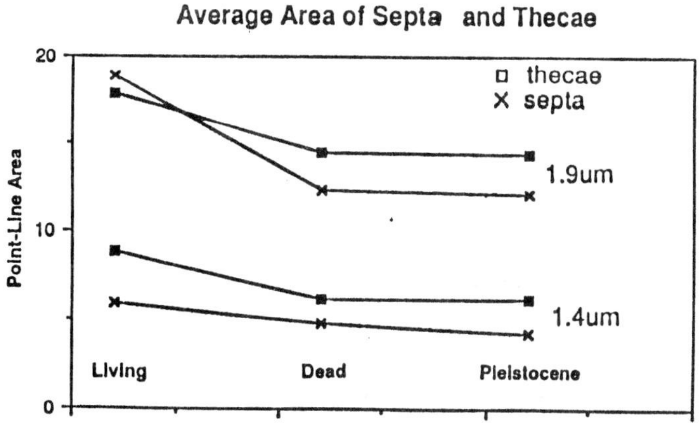

Fig. 8. Average band areas calculated for 1.4- and 1.9-um bands in spectra of septa and thecae of each specimen.

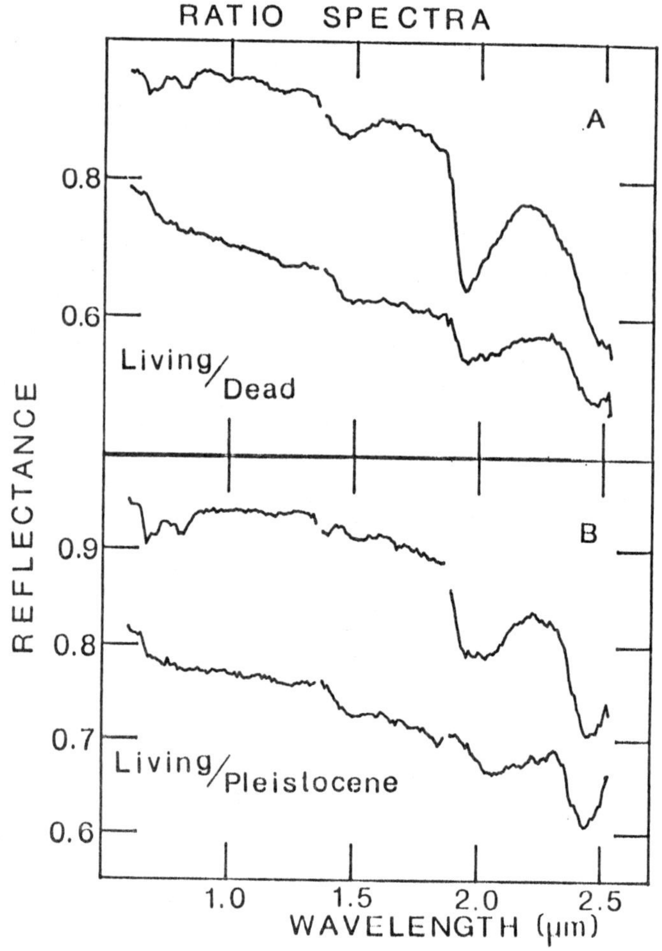

Fig. 9. Ratios of spectra obtained of skeletons of live (DSRBML-75) and dead (DSRBMD-92) (A) and live and Pleistocene (DSQAPL-80) (B) coral skeletons. Each spectrum is obtained by dividing a reflectance spectrum obtained from one skeleton into that obtained from another. Compare to spectra shown in Figures 1 and 4. Spectra show that the skeleton of the dead and Pleistocene samples have lost liquid H_2O as well as the organics which produce the CH_2 features near 1.7 and 2.3 um, the NH_3 features near 2.1 and 2.2 um, and organic pigments which produce features near 0.67 um. However, amount and form of organics lost vary within as well as between skeletons from the Recent and Pleistocene.

interskeletal voids described by Gvirtzman and Friedman (1977). Interskeletal voids, those portions of the corallite once occupied by the coral polyp, contain organic material (Gvirtzman and Friedman, 1977). The centers of sclerodermites, the bundles of radiating aragonite crystals which form the skeleton, contain small (<1 um) equant carbonate crystals surrounded by soluble organic material (James, 1974). In addition, the aragonite crystals making up the skeleton contain minute intracrystalline voids which contain organics (Green et al., 1980) and/or liquid water as fluid inclusions (Bruni and Wenk, 1985; Gaffey, 1983, 1984a,b; 1985, 1988). The organic matter in coral skeletons is composed of proteins and polysaccharides (Johnston, 1980; Wilbur and Simkiss,

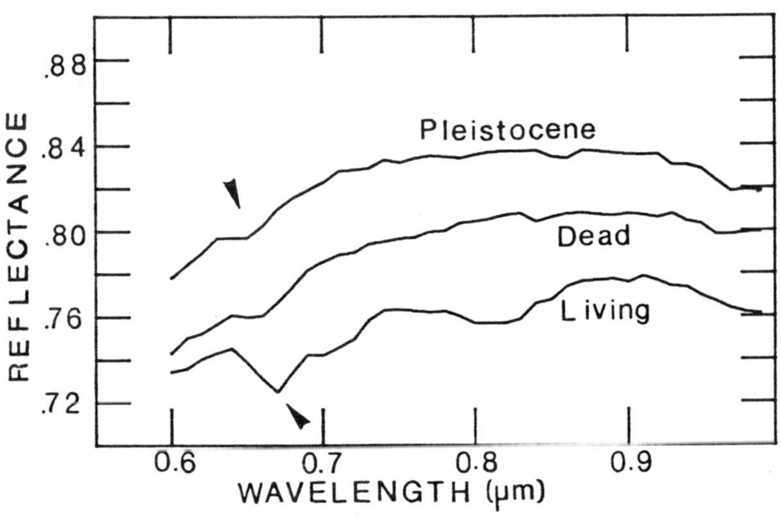

Fig. 10. Short-wavelength portion of spectra of living, dead and Pleistocene Diploria *skeletons showing progressive weakening of the 0.67- um chlorophyll absorption and loss of 0.7-um H_2O band.*

1979). These organic compounds may include some or all of the following: the organic matrix used in forming the skeleton, soft tissues of the coral polyp, and compounds produced by the endolithic algae. All of these components can contribute features to VNIR spectra of corals.

Spectral properties of carbonates and liquid H_2O are shown in Figure 3. The first spectrum in Figure 3 is that of Iceland spar. Spectra of all anhydrous carbonate minerals show these same absorption features, but precise positions of these features vary with mineralogy (Gaffey, 1986, 1987). Liquid H_2O (Fig. 3) has three strong absorptions in the VNIR region, near 1.4, 1.9 and 2.5 um (see Gaffey, 1988, for discussion and references to the chemical literature). Simple addition or multiplication of the $CaCO_3$ spectrum and the H_2O spectrum will approximate that of the bleached coral. The number, positions, and widths of the 1.4- and 1.9-um bands in the coral spectrum show that all the water is present as liquid H_2O.

Spectra of coral tissue and of organics isolated from the coral skeleton are shown in Figure 4. Spectra of organic compounds in the VNIR region contain features due to NH_3^+, NH_2, OH^-, and CH_2 (see Gaffey, this volume, for discussion and references to the chemical literature). Overtones and combination tones of the stretching vibrations of CH_2 in amino acids or polysaccharides produce the sharp narrow features near 2.35, 1.7 and 1.2 um. Overtone and combination tones of the stretching modes of NH_3^+ in proteins produce the broader, more diffuse features near 2.1 to 2.2 um. OH^- in organic compounds generally produces broad features near 1.45 and 2.1 um. In addition, absorptions occur at visible wavelengths (0.3 - 0.7 um) in coral spectra. These are due to organic pigments in symbiotic zooxanthellae, and endolithic algae.

Petrographic and Mineralogical Data

All coral skeletons studied appear to be composed primarily of aragonite, any high magnesian calcite (HMC) being present in marine cements and the bulk of low magnesian calcite (LMC) occurring in fresh water vadose cements similar to those described from the Cockburn Town reef by White et al. (1984) and from the Mosquito Marsh reef described by Vierma et al. (1984).

Examination with petrographic microscope shows DSRBML-75 contains minor amounts of fibrous aragonite cement. DSRBMD-92 contains both fibrous aragonite and micritic HMC cements. Pleistocene samples from Quarry A show some marine cements (fibrous aragonite and micritic calcite, presumably now LMC) and marine sediment infill, primarily composed of ooids. Septa and dissepiment have undergone considerable dissolution.

Examination of septa of DSRBML-75, DSRBMD-92, and D-SPPL-84 showed a reduction in amount of organics with time. Large areas of the septa of the living skeleton (DSRBML-75) are covered with organic films and filaments, which in many areas are penetrated by acicular crystals of marine aragonite cements. Areas of septa of the dead coral (DSRBMD-92) covered by organics are much smaller than those in the living sample, and organics appear to be partially calcified by subspheric nannograins, as described by Loreau (1982). The surfaces of septa from the Pleistocene sample are almost entirely free of organics. A few small, scattered rhombs can be seen replacing aragonite in the Pleistocene sample, as well as some platy crystals ~1.0 um in diameter which may be hematite. Additional work is needed to completely characterize the mineralogy of these samples.

Loss of Water and Organics

The histograms in Figure 7 and data in Table 2 show considerable variation in band areas for both the 1.4- and 1.9-um features in spectra of different portions of the skeleton of the live coral. In general there is a decrease both in individual and average band areas for both the 1.4- and 1.9- um bands in spectra of the dead and Pleistocene samples relative to those in spectra of the living sample. However, areas of the two bands, and band areas for septa and thecae do not decrease by the same amount or at the same rate. The decrease in area is greater for the 1.9-um band than for the 1.4-um band. The trend for both the 1.4- and 1.9-um band areas for the thecae are similar, but for the septa, the decrease in area of the 1.9-um band is more pronounced than that of the 1.4-um band. The area of the 1.9-um band is greater in spectra of the septa than of the theca in the live coral, but the reverse is true for the dead and Pleistocene samples.

Spectra of other samples of *Diploria* studied showed these same trends in band areas. Spectra of *Acropora* showed less variability in band areas between different portions of the skeleton of the same colony, probably because their skeletons have a more porous open structure throughout, and because the smaller corallite size made it impossible to isolate different structures within the skeleton for analysis.

In general, the standard deviations and coefficients of variation also show a decrease for data obtained from spectra of dead and Pleistocene samples relative to these values for spectra of the living sample, the 1.9-um band in the spectra of septa of DSRBMD- 92 being the exception.

Because there are a number of NH_3^+ and organic OH^- absorptions in the 2.0-2.2-um region which overlap with the 1.9-um water band, the areas determined for the 1.9-um band in part reflect the organic content of the skeletons. The 1.4-um band shows less overlap of absorption features due to different phases, and appears to reflect the H_2O content of the skeletons more accurately. Comparison of the data for living samples with that for dead and Pleistocene samples (Figs. 7 and 8) support this conclusion. Greater organic content in a sample is reflected by an anomalous increase in area of the 1.9-um band without a similar increase in the area of the 1.4-um band. Differences in trends in areas for the 1.9-um band in spectra of septa and thecae, and differences in the magnitude of changes in areas of the 1.4- and 1.9-um bands from one sample to another indicate that more than one phase is contributing to these spectral features. If only one phase, e.g. liquid H_2O were present, the 1.4- and 1.9-um bands would show the same trends in band area. Effect of organic content on band areas is most clearly illustrated in spectra of the septa of the living coral.

The marked decrease in areas of both the 1.4- and 1.9-um bands in spectra of the septa probably represents loss of organics which were part of the tissue of the coral polyp. The thecae would be expected to contain fewer organics, perhaps only those employed in formation of the coral skeleton, and those contributed by endoliths (Additional work will be needed to further characterize the organic contents of septa and thecae). The dense structure of the thecae, however, would prevent rapid loss of organics seen in the septa of the dead and Pleistocene samples. Water and organics trapped within intracrystalline voids and within the centers of sclerodermites remain in place until the voids are breached by dissolution or physical destruction of the skeleton.

The persistence of organics in corals of Pleistocene age is also shown by the chlorophyll absorption feature near 0.67 um in all the spectra in Figures 4 and 10, as well as in most of the spectra obtained in this study. This chlorophyll is probably the result of assimilation of products from the symbiotic zooxanthellae (Von Holt and Von Holt, 1968; Muscatine and Cernichiari, 1969) or from the alga *Ostreobium* which is the predominant endolith infesting coral skeletons (Lukas, 1973).

There is great variation in the amounts and types of phases being lost, both at different points within the same skeleton, and at different stages during its diagenetic history. Ratio spectra in Figure 9 illustrate some of these differences in organic and water content between skeletons. As mentioned above, when one spectrum is divided into another, the resulting curve is a spectrum of differences between the two samples, i.e. of the phases which one contains that the other does not. Comparison of these ratio spectra with the spectra of organics and of liquid H_2O shows that not only is H_2O being lost, but other phases containing OH^-, NH_3^+, and CH_2 as well. The first spectrum in Figure 9A shows that water is the principle phase lost in this portion of the septa of the dead coral. The 0.67-um region of the spectrum shows that chlorophyll has also been lost. The second spectrum in Figure 9A shows that at another location on the septa, proteins as well as water have been lost. The ratio spectra in Figure 9B show that water and organics (proteins and possibly polysaccharides) have all been lost in both portions of the septa from which these spectra were obtained, but in differing amounts.

The high degree of variability in water and organic content between different portions of the living and dead coral colonies, and the variation in amount and types of phases being lost, indicates that early alteration is taking place within microenvironments which are not in equilibrium with each other or with the ambient environment, a model proposed by Pingitore (1982). As diagenesis proceeds in the subaerial environment all voids in the skeleton are enlarged by dissolution until they are interconnected (Gvirtzman and Friedman, 1977). At this stage all portions of the skeleton tend toward equilibrium reflected in the low coefficients of variability for band areas, and in the similarity in band areas for septa and thecae in the Pleistocene sample.

The SEM and spectral data also indicate

that prolonged bleaching does not remove all of the organics in skeletons, a fact which has important repercussions for other types of chemical analyses.

Dissolution and Cementation

Water and organics in coral skeletons undoubtedly play an important role in their diagenesis. The centers of the sclerodermites and the seed crystals which they contain are preferentially dissolved, even when there are no visible conduits to the outer surface of the coral (Gvirtzman and Friedman, 1977; James, 1974; Pingitore, 1976). Gvirtzman and Friedman (1977) also noted that centers of the sclerodermites retain water longer than any other portion of the skeleton. Constantz (1986b) noted that aragonite crystals in coral skeletons are more susceptible to dissolution than adjacent aragonitic marine cements. It is our contention that the inclusions of organics and liquid water are in part responsible for the decreased stability of skeletal aragonites, and, as suggested by Gvirtzman and Friedman (1977) for the apparently anomalous dissolution of sclerodermites from the center outwards. Water is present to serve as a medium for diagenetic reaction, and breakdown of organics may increase the CO_2 content of solutions facilitating dissolution of skeletal aragonites.

Organics can serve as a substrate for precipitation of carbonate (e.g. Loreau, 1982; Mitterer and Cunningham, 1985). Calcification of organic films appears to have occurred in the skeleton of the dead colony. Decay of organics can also change pore water chemistry. For example NH_3^+ may be released, increasing the pH of pore waters and promoting precipitation of carbonate. While the occurrence of aragonite cements penetrating organic films does not provide conclusive evidence, it does suggest a relationship between the two.

It is important to note that the liquid H_2O and organic compounds involved in these reactions are an intrinsic part of the skeleton incorporated during growth. During formation of the skeleton, the sites of calcification are isolated from surrounding sea water (Wilbur and Simkiss, 1979; Crenshaw, 1982), and inclusions in the crystals of the skeleton probably contain the mineralizing fluids of the animal. Johnston (1980) notes that some of the organics in coral skeleton are intimately bound with and completely dispersed within the skeletal aragonite. Organics are probably trapped within the crystal during growth as suggested by Barnes (1970) and Constantz (1986a). This is not surprising as organic matrices are used to facilitate formation of the skeleton and water provides a medium for transport of Ca^{2+} and HCO_3^- to the site of deposition (Crenshaw, 1982).

CONCLUSIONS

Although both water and organics can potentially have a marked affect on the diagenesis of carbonate skeletons, there is at present no tool routinely in use by sedimentary petrologists for their detection and identification, or for determination of their abundances. Visible and near-infrared (VNIR) (0.3 - 2.7 um) reflectance spectroscopy can provide a rapid, inexpensive, non-destructive tool for characterizing water and organic contents of skeletons. This technique has the added advantage of allowing organics to be studied *in situ*, without deleterious effects of the standard techniques employed for their extraction (Holden, 1988).

In spectra of coral skeletons relative area of the 1.9-um absorption band calculated by the point-line method appears to reflect total H_2O and organic content, while the relative area of the 1.4-um band appears to more accurately reflect the H_2O content alone. At present, only relative changes in water and organic contents of skeletons can be studied in this way, and calibration curves must be developed so that water and organic contents of skeletons can be quantitatively characterized from spectral data. In addition, other techniques such as NMR, ESR, Raman spectroscopy, and coulometry will help to better characterize the water and organic phases contained in these skeletons. In addition, because organics are lost so rapidly after death of the colony, skeletons found in place on reefs and in reef rubble should be studied as well, to give a more complete picture of early diagenetic changes.

Water and organics are primary components incorporated during growth, and are intimately associated in skeletons of scleractinian corals. Some of the water and organics occur in interskeletal voids which are in communication with the environment surrounding the coral head. Organics in these portions of the skeleton are lost rapidly after death of the coral. Other organics, and the water with which they are associated, are isolated within intracrystalline and intraskeletal voids. These are retained for a longer period within the skeleton. Water and organics in intracrystalline voids are retained until the crystals are dissolved or physically destroyed.

ACKNOWLEDGMENTS

This work was begun as a senior research thesis at Rensselaer Polytechnic Institute under the Undergraduate Research Program. We are indebted to Dr. Donald T. Gerace and the staff at the College Center of the Finger Lakes Bahamian Field Station for their logistical support. We also thank Carle Pieters and Steven Pratt at RELAB at Brown University for the use of her spectrometer and their assistance in obtaining these spectral data. Ramona Lotti, Curator of the Deep-Sea Sample Repository at Lamont-Doherty Geological Observatory kindly provided the X-radiographs. Funding for this work was provided by the RPI Undergraduate Research Program and by NSF Grant No. EAR-8721094.

REFERENCES CITED

Barnes, D.J., 1970, Coral skeletons: an explanation of their growth and structure: Science, v. 170, p. 1305-1308.

Burns, R.G., 1970, Mineralogical Applications of Crystal Field Theory: Cambridge, England, Cambridge University Press, 224 p.

Bruni, S.F., and Wenk, H.-R., 1985, Replacement of aragonite by calcite in sediments from the San Cassiano Formation (Italy): Journal of Sedimentary Petrology, v. 55, p. 159-170.

Clark, R.N., and Roush, T.L., 1984, Reflectance spectroscopy: Quantitative analysis techniques for remote sensing applications: Journal of Geophysical Research, v. 89, no. B7, p. 6329-6340.

Crenshaw, M.A., 1982, Mechanisms of normal biological mineralization of calcium carbonates, in Nancollas, G.H., ed., Biological Mineralization and Demineralization, New York, Springer-Verlag, p. 243-257.

Constantz, B. R., 1986a, Coral skeleton construction: A physiochemically dominated process: Palaios, v.1, p. 152-157.

Constantz, B.R., 1986b, The primary surface area of corals and variations in their susceptibility to diagenesis, in Schroeder, J.H., and Purser, B.H., eds., Reef Diagenesis, Berlin, Springer-Verlag, p. 53-76.

Gaffey, S.J., 1983, Spectral reflectance of carbonate rocks and minerals [abs.]: American Association of Petroleun Geologists Bulletin, v. 67, p. 465.

Gaffey, S.J., 1984a, Fluid inclusions in skeletal carbonates and their loss during diagenesis [abs.]: 1984 Society of Economic Paleontologists and Mineralogists Annual Midyear Meeting Abstracts, p. 33.

Gaffey, S.J., 1984b, Spectral reflectance of carbonate minerals and rocks in the visible and near infrared (0.35 to 2.55 um) and its applications in carbonate petrology [Ph.D. thesis]: Honolulu, Hawaii, University of Hawaii, 236 p.

Gaffey, S.J., 1985, Reflectance spectroscopy in the visible and near-infrared (0.35-2.55um): Applications in carbonate petrology: Geology, v.13 p. 270-273.

Gaffey, S.J., 1986, Spectral reflectance of carbonate minerals in the visible and near infrared (0.35-2.55 um): Calcite, aragonite, and dolomite: American Mineralogist, v. 71, p. 151-162.

Gaffey, S.J., 1987, Spectral reflectance of carbonate minerals in the visible and near infrared (0.35-2.55 um): Anhydrous carbonate minerals: Journal of Geophysical Research, v. 92, No. B2, p. 1429-1440.

Gaffey, S.J., 1988, Water in skeletal carbonates: Journal of Sedimentary Petrology, v. 58, p. 397-414.

Gautier, D.L., 1985, Interpretation of early diagenesis in ancient marine sediments, in Gautier, D.L., Kharaka, Y.K., and Surdam, R.C., eds., Relationship of organic matter and mineral diagenesis: Society of Economic Paleontologists and Mineralogists Short Course Notes Number 117.

Green, H.W., Lipps, J.H., and Showers, W.J., 1980,

Test ultrastructure of fusulinid Foraminifera: Nature, v. 283, p. 853-855.

Gvirtzman, G., and Friedman, G.M., 1977, Sequence of progressive diagenesis in coral reefs, in Frost, S.H., Weiss, M.P., and Saunders, J.B., eds., Reefs and related carbonates - Ecology and sedimentology, Studies in geology Number 4: Tulsa, Oklahoma American Association of Petroleum Geologists, p. 357-380.

Herzberg, G., 1945, Molecular spectra and molecular structure, II, Infrared and Raman spectra of polyatomic molecules: New York, Van Nostrand Reinhold, 632 p.

Holden, P.N., 1988, Porphyrin absorption bands isolated from whole rock reflectance spectra [M.S. thesis]: Troy, New York, Rensselaer Polytechnic Institute, 76 p.

James, N.P., 1974, Diagenesis of scleractinian corals in the subaerial vadose environment: Journal of Paleontology, v. 48, p. 785-799.

Johnston, I.S., 1980, The ultrastructure of skeletogenesis in hermatypic corals: International Review of Cytology, v. 67, p. 171-214.

Kimber, R.W.L., and Griffin, C.V., 1987, Further evidence of the complexity of the racemization process in fossil shells with implications for amino acid racemization dating: Geochimica et Cosmochimica Acta, v.51, p. 839-846.

Loreau, J.-P., 1982, Sediments aragonitiques et leur genese: Memoires du Museum National D'Histoire naturelle, Serie C, v. 47, 312 p.

Lukas, K.J., 1983, Taxonomy and ecology of the endolithic microflora of the corals with a review of the literature on endolithic microphytes [Ph.D. thesis]: University of Rhode Island, 159 p.

Mitterer, R.M., and Cunningham, R., 1985, The interactions of natural organic matter with grain surfaces: Implications for calcium carbonate precipitation, in Schneidermann, N., and Harris, P.M., eds., Carbonate cements: Society of Economic Paleontologists and Mineralogists Special Publication 36, p. 17-32.

Muscatine, L., and Cernichiari, 1969, Assimilation of photosynthetic products of zooxanthellae by a reef coral: Biological Bulletin, v. 137, p. 506-523.

Pieters, C.M., 1983, Strength of mineral absorption features in the transmitted component of near-infrared reflected light: First results from RELAB: Journal of Geophysical Research, v. 88, No. B11, p. 9534-9544.

Pingitore, N.E., Jr., 1976, Vadose and phreatic diagenesis: Processes products, and their recognition in corals: Journal of Sedimentary Petrology, v. 46, p. 985-1006.

Pingitore, N.E., Jr., 1982, The role of diffusion during carbonate diagenesis: Journal of Sedimentary Petrology, v. 52, p. 27-39.

Vierma, L., Kwolek, J.M., Heidt, D.A., Hattin, D.E., Hasenmueller, W.A., and Feldman, H.R., 1984, Stratigraphic analysis of a newly discovered Pleistocene reef, San Salvador Island, Bahamas: The Compass, v. 62, p. 16-30.

Von Holt, C., and Von Holt, M., 1968, Transfer of photosynthesis products from zooxanthellae to coelenterate hosts: Comparative Biochemistry and Physiology, v. 24, p. 73-81.

Weiner, S., Traub, W., and Lowenstam, H.A., 1983, Organic matrix in calcified exoskeletons, p. 205-224, in Biomineralization and Biological Metal Accumulation, D. Reidel Publishing Company, Boston, 533 p.

White, B., Kurkjy, K.A., and Curran, H.A., 1984, A shallowing-upward sequence in a Pleistocene coral reef and associated facies, San Salvador, Bahamas, in Teeter, J.W., ed., Proceedings of the second symposium the geology of the Bahamas: San Salvador, Bahamas, College Center of the Finger Lakes Bahamian Field Station, p. 53-70.

Wilbur, K.M., and Simkiss, K., 1979, Carbonate turnover and deposition by metazoa, in Trudinger, P.A., and Swaine, D.J., eds., Biogeochemical cycling of mineral-forming elements: New York, Elsevier, p. 69-106.